Mechanical Separation Processes in the Food Industry

Unit Operations and Processing Equipment in the Food Industry

Mechanical Separation Processes in the Food Industry

Edited by

Seid Mahdi Jafari

Department of Food Materials and Process Design Engineering, Gorgan University of Agricultural Sciences and Natural Resources, Gorgan, Iran

Asli Can Karaca

Department of Food Engineering, Faculty of Chemical and Metallurgical Engineering, Istanbul Technical University, Maslak, Istanbul, Turkey

ELSEVIER

WP
WOODHEAD PUBLISHING
An imprint of Elsevier
elsevier.com/books-and-journals

Woodhead Publishing is an imprint of Elsevier
50 Hampshire Street, 5th Floor, Cambridge, MA 02139, United States
125 London Wall, London EC2Y 5AS, United Kingdom

ISBN: 978-0-128-19523-9

For Information on all Woodhead Publishing publications visit our website at https://www.elsevier.com/books-and-journals

Publisher: Mara Conner
Acquisitions Editor: Vaishali Singh
Editorial Project Manager: Bhavya Rattan
Production Project Manager: Sruthi Satheesh
Cover Designer: Vicky Pearson Esser

Working together to grow libraries in developing countries
www.elsevier.com • www.bookaid.org

Typeset by Aptara, New Delhi, India

Contents

Section II Application of mechanical separation operations in the food industry

Contributors

Samiye Adal, Department of Food Engineering, Faculty of Engineering, Afyon Kocatepe University, Afyonkarahisar, Türkiye

Nisa Akkuzu, Department of Food Engineering, Faculty of Chemical and Metallurgical Engineering, Istanbul Technical University, Maslak, Istanbul, Türkiye

Özgül Altay, Department of Food Engineering, Faculty of Engineering, Ege University, İzmir, Türkiye

Vinkel Kumar Arora, Department of Food Engineering, National Institute of Food Technology Entrepreneurship and Management-Kundli, Sonipat, Haryana, India

Ulaş Baysan, Department of Food Engineering, Faculty of Engineering, Ege University, İzmir, Türkiye; Science and Technology Application and Research Center, Dokuz Eylül University, İzmir, Türkiye

Nikita S. Bhatkar, Department of Food Engineering and Technology, Institute of Chemical Technology Mumbai, Bhubaneswar, India

Esra Capanoglu, Department of Food Engineering, Faculty of Chemical and Metallurgical Engineering, Istanbul Technical University, Maslak, Istanbul, Türkiye

Humeyra Cavdar Dincturk, Department of Food Engineering, Faculty of Chemical and Metallurgical Engineering, Istanbul Technical University, Maslak, Istanbul, Türkiye

Vimal Challana, Department of Food Engineering and Technology, Institute of Chemical Technology Mumbai, Bhubaneswar, India

Christian O. Díaz-Ovalle, Tecnológico Nacional de México/I.T. Roque, Departamento de Ingenierías, Celaya, Guanajuato, México

M. Eigenfeld, BioTechMed Graz, Mozartgasse, Graz, Austria

Ebru Firatligil, Department of Food Engineering, Faculty of Chemical & Metallurgical Engineering, Istanbul Technical University, Maslak, Istanbul, Türkiye

Rukiye Gundogan, Department of Food Engineering, Faculty of Chemical and Metallurgical Engineering, Istanbul Technical University, Maslak, Istanbul, Türkiye

Seid M. Jafari, Department of Food Materials and Process Design Engineering, Gorgan University of Agricultural Sciences and Natural Resources, Gorgan, Iran; Halal Research Center of IRI, Iran Food and Drug Administration, Ministry of Health and Medical Education, Tehran, Iran

Asli Can Karaca, Department of Food Engineering, Faculty of Chemical and Metallurgical Engineering, Istanbul Technical University, Maslak, Istanbul, Turkey

Figen Kaymak Ertekin, Department of Food Engineering, Faculty of Engineering, Ege University, İzmir, Türkiye

Celale Kirkin, Department of Food Engineering, Faculty of Chemical and Metallurgical Engineering, Istanbul Technical University, Maslak, Istanbul, Türkiye

Sanket Balasaheb Kokane, Department of Food Engineering, National Institute of Food Technology Entrepreneurship and Management-Kundli, Sonipat, Haryana, India

Bharati Kumari, Department of Food Engineering and Technology, Tezpur University, Assam, India

Özgün Köprüalan Aydın, Department of Food Engineering, Faculty of Engineering, Ege University, İzmir, Türkiye

Emrah Kırtıl, Department of Chemical Engineering, Gebze Technical University, Kocaeli, Turkey

Tanja Lužaić, Faculty of Technology Novi Sad, University of Novi Sad, Blvd. cara Lazara 1, Novi Sad, Serbia

Mondher Mzoughi, Department of Food Engineering, Faculty of Chemical & Metallurgical Engineering, Istanbul Technical University, Maslak, Istanbul, Türkiye

Geetanjali Pathania, Department of Food Science and Technology, Punjab Agricultural University, Ludhiana, Punjab, India

Ranko Romanić, Faculty of Technology Novi Sad, University of Novi Sad, Blvd. cara Lazara 1, Novi Sad, Serbia

S.P. Schwaminger, NanoLab Graz, Division of Medicinal Chemistry, Otto Loewi Reseaerch Center, Medical University of Graz, Neue Stiftingtalstraße, Graz, Austria; BioTechMed Graz, Mozartgasse, Graz, Austria

Özge Seçmeler, Department of Gastronomy and Culinary Arts, Faculty of Applied Sciences, Altınbaş University, İstanbul, Turkey

Shivanand S. Shirkole, Department of Food Engineering and Technology, Institute of Chemical Technology Mumbai, Bhubaneswar, India; Deprtament of Food Technology, School of Engineering and Technology, D. Y. Patil Agriculture and Technical University, Talsande, Kolhapur, India

Nandan Sit, Department of Food Engineering and Technology, Tezpur University, Assam, India

Özge Süfer, Food Engineering Department, Osmaniye Korkut Ata University, Osmaniye, Türkiye

Maria Turtoi, Life Sciences Department, Cross-Border Faculty, "Dunarea de Jos" University of Galati, Galati, Romania

Merve Yavuz-Düzgün, Department of Gastronomy and Culinary Arts, Faculty of Applied Sciences, Altınbaş University, İstanbul, Turkey

Gülşah Çalışkan Koç, Department of Food Processing, Esme Vocational School, Usak University, Usak, Türkiye

Yasemin Çelebi, Food Technology Program, Department of Food Processing, Esme Vocational School, Usak University, Usak, Türkiye

Sevil Çıkrıkçı-Erünsal, Department of Food Engineering, Konya Food and Agriculture University, Konya, Turkey

Işıl İlter Baysan, Department of Food Engineering, Faculty of Engineering, Ege University, İzmir, Türkiye; Department of Food Engineering, Faculty of Engineering, Manisa Celal Bayar University, Manisa, Türkiye

Introduction to mechanical separation processes in the food industry

1

Asli Can Karaca[a] *and Seid M. Jafari*[b,c]
[a] Department of Food Engineering, Faculty of Chemical and Metallurgical Engineering, Istanbul Technical University, Maslak, Istanbul, Turkey, [b] Department of Food Materials and Process Design Engineering, Gorgan University of Agricultural Sciences and Natural Resources, Gorgan, Iran, [c] Halal Research Center of IRI, Iran Food and Drug Administration, Ministry of Health and Medical Education, Tehran, Iran

1.1 Introduction

Mechanical separation processes are among the most important unit operations applied in the food industry (Pouliot et al., 2014). These processes involve the application of a physical force to the food product, resulting in separation of the product components. Mechanical separation processes include filtration, sedimentation, centrifugal separation, and sieving. Filtration is used to separate solids from liquids by passing the mixture through fine pores that are small enough to retain the solids but large enough to allow the liquid to pass through (Rushton et al., 2008). In the sedimentation process, a liquid and a solid, or two immiscible liquids with different densities are allowed to come to equilibrium by gravitational forces, resulting in their separation from each other (Sorsamäki & Nappa, 2015). The rate of sedimentation can be increased significantly by the application of centrifugal forces, which is also referred to as the centrifugal separation process. The sieving process is commonly used for classification of solid particles based on their size differences using a barrier. Various mechanical separation processes are applied in the food industry, such as in the manufacturing of dairy products, fruit juices, beverages, and refining of edible oils and sugar. This chapter aims to provide an overview of the basic concepts of the mechanical separation processes, the equipment necessary for mechanical separation unit operations, and various applications of mechanical separation processes in the food industry.

1.2 Different mechanical separation processes

As outlined in **Chapter 2**, flotation and sedimentation are the main operations used in the food industry for washing, cleaning, classification, clarification, and thickening purposes. Different flotation processes, such as natural, dissolved air, froth, electrolytic, and colloidal gas aphrons, as well as sedimentation processes, including solid-liquid and liquid-liquid separation, including thermal- or microwave-assisted,

Mechanical Separation Processes in the Food Industry: Unit Operations and Processing Equipment in the Food Industry.
DOI: https://doi.org/10.1016/B978-0-128-19523-9.00010-X

ultrasound-assisted, or chemical demulsification can be used in various applications, such as water or crude oil treatment. Floatation is a separation technique used for separating solid-liquid or liquid-liquid systems, specifically for particles with densities lower than that of the surrounding liquid (Kastali et al., 2021). Sedimentation is basically a unit operation that separates solids or liquid droplets from a fluid, which can be either gas or liquid, using the force of gravity.

As described in **Chapter 3**, centrifugation utilizes centrifugal force to accelerate the separation process, significantly enhancing the speed and effectiveness compared to the gravity-based methods. Centrifugal separation involves the mechanical separation of a fluid mixture having two or more immiscible phases with different densities by applying centrifugal force. Industrial centrifugation is classified into two groups: sedimenting and filtering (Luckiram, 2022). The basic parts of a centrifuge include motor, rotor, chamber, lid, latch and the sample to be centrifuged (Tantray et al., 2023). The centrifugation process has diverse applications in the food industry, including clarification of fruit juices and beverages, cream separation in dairy industry, oil and fat extraction, dewatering and concentration of food products, among others.

Chapter 4 outlines filtration and clarification processes. Filtration is a mechanical separation process that plays a vital role in the food industry, for ensuring the quality and safety of food products. It is used in a solid-liquid system to separate solids from liquids or gases (Tien, 2012). The suspension is forced to pass through filter media that allows the liquid to flow through it, while the suspended solids accumulate on the pores of filter media. The driving force needed for the flow of suspension may be generated due to gravity, vacuum, or pressure. Membrane technology has recently gained popularity due to its higher efficiency compared to other separation methods (Schroën & de Aguiar, 2023). Membrane filtration can be classified into mainly four types based on their pore size: microfiltration, ultrafiltration, nanofiltration, and reverse osmosis (Van der Bruggen, 2018). The membrane filtration method is commonly used for the filtration and clarification of beverages, which is a important stage for ensuring the quality, safety, and marketability of the products (Jain & De, 2019).

Chapter 5 describes sieving, a separation process that plays an important role in food and agricultural industries to ensure the quality of products. Sieving a fundamental separation method that allows the discrimination of various particles based on their sizes. Different sieving methods are gravitational, centrifugal, vibratory, and rotary sieving (Sanchez-Suarez et al., 2022). Commonly preferred sifters are vibratory, rotary, gyratory, and air-jet sifters for refining products. Entoleters, like sifters, are equipment used for separation but are selected for different purposes. They are among the equipment used to sterilize flour or grain group products, destroy insect eggs and larvae, and ensure food safety through high-speed impact technology (Bell, 2014). Each sieving technique and equipment provide unique advantages in industrial applications, ensuring the quality and safety of food products.

Chapter 6 outlines magnetic and metal detectors, which play a crucial role in ensuring food safety by detecting metallic contaminants in food products. Magnetic separators are commonly used for ferromagnetic waste, while eddy current metal detectors are suitable for nonferromagnetic materials (Cárcel-Carrasco et al., 2022). Magnetic detection hinges on the fundamental principles of magnetism, whereby ferrous particles are drawn towards magnetic fields. Utilizing the forces of magnetism and electromagnetic fields, these technologies efficiently identify and eliminate contaminants, protecting consumers from possible health hazards and thus play an essential role in preserving food safety standards.

As outlined in **Chapter 7**, cyclone separators recollect, totally or partially, the powder contained in a fluid stream by centrifugal force and gravity during flow-through. Cyclone separators recover powder or dust and are commonly used in the food industry for handling, storing, manipulating, operating, and modifying any solid, whether raw material, intermediate, or final product (Sorsamäki & Nappa, 2015). The efficient performance of the cyclone separators depends on the design elements including the geometrical shape and size and impacts the separation efficiency (Avci & Karagoz, 2003). Cyclone separators are commonly applied in the processing of food products in solid form, such as dairy powders, sugar, flavor powders, grains, and protein powders (Bhandari et al., 2013).

Chapter 8 outlines deaeration, which refers to the removal of oxygen or air from the surrounding environment. The deaeration process includes eliminating dissolved gases from the food matrix, headspace in the packaging material, and expelling air during the canning process. Vacuum deaeration, gas stripping, membrane techniques, ultrasonication, enzymatic deaeration, and centrifugal degassing are common methods of degassing or deaeration. Although the deaeration process is an intermediate step in food processing, it is critical for extending the shelf life and ensuring product quality. The process also ensures that the processing line is smooth and in continuous operation, which is important for maintaining the equipment.

1.3 Application of mechanical separation operations in the food industry

Numerous mechanical separation processes are applied in the food industry in production of dairy products, fruit juices, beverages, and refining of edible oils and sugar. For example, the sedimentation process is used for quality assessment, separation of components, and treatment of wastewater in the dairy industry (**Chapter 9**). On the other hand, sedimentation can naturally occur due to gravitational forces and can be observed in stored dairy products as a precipitate at the bottom of the container. Sedimentation in dairy products depends on several parameters, such as the method and temperature of the heat treatment, time and temperature of storage, type of milk, milk proteins, pH level, and calcium concentration (Anema, 2019). Sedimentation can generally be prevented by controlling the pH and concentration of ionic calcium, addition of stabilizers, and inactivating microorganisms and proteolytic enzymes (Rauh & Xiao, 2022). Another example of the application of mechanical separation operations in the dairy industry is separation of cream from milk by centrifugation (**Chapter 10**). The separation principle is based on the difference in density between skimmed milk and cream, and takes place in a milk centrifugal separator. The separation of cream from milk by centrifugation is essential for processing raw milk to obtain a large variety of dairy products with reduced and standardized fat content, as well as specific products with higher fat content, such as cream, butter, or anhydrous milk fat.

Filtration is an essential unit operation applied in the manufacturing of fruit juices and beverages. As outlined in **Chapter 11**, the filtration process ensures that the final product is not only clear and visually appealing, but also free from undesired particulate matter and microorganisms that can compromise its quality and shelf life. Various filtration techniques, ranging from mechanical methods to advanced membrane-based applications, are applied for enhancing the clarity, quality, and safety of fruit juices. These techniques efficiently remove solids, manage turbidity, and preserve essential nutrients, which are susceptible to degradation during processing (Sarbatly et al., 2023). In case of beverage processing, mechanical filtration methods can be used at different steps (**Chapter 12**). For example, a coarse filtering process may be used to clean the raw beverage. Then, by a fine filtration process, undesired solids can be completely removed. Mechanical filtration methods increase the clarity of beverages while also preserving the taste and aroma. The mechanical filtration process has a vital role in the beverage industry, and it is predicted that its importance will increase with the development of filtration processes with new equipment and systems.

Filtration is also an important separation process applied in refining of crude vegetable oils to produce high-quality edible oils (**Chapter 13**). The main goal of the refining process is to remove the undesired components, including free fatty acids, oxidation products, and photosensitizers that adversely affect oil quality, thereby reducing the shelf life and consumer acceptance (Wen et al., 2023). This critical process needs to be customized as the composition of crude oil is highly variable (Vaisali et al., 2015). The refining process includes degumming (pre-refining), neutralization (chemical and/or physical), bleaching, winterization (dewaxing), and deodorization. Certain oil refining stages, such as bleaching and winterization, include filtration as an essential part. Furthermore, filtration plays an important role in the sugar refining process as well. Achieving a consistently high level of clarity and low color in the juice production process is challenging. While conventional methods have served well to date, changing environmental regulations and the need for enhanced sugar quality at reduced costs demand adaptation. Membrane filtration, particularly microfiltration, ultrafiltration, and nanofiltration, emerges as a promising solution to address these challenges, as outlined in **Chapter 14**.

In conclusion, various mechanical separation processes, including filtration, centrifugal separation, sieving, and sedimentation are applied in the food industry in the manufacture of dairy products, fruit juices, beverages, the refining of edible oils and sugar. These processes are critical for ensuring the desired quality and safety of the end products. A key understanding of the equipment used in mechanical separation unit operations, along with a comprehensive knowledge of the composition and physicochemical characteristics of the products, is essential for developing new applications of these operations in the food industry.

References

Anema, S. G. (2019). Age gelation, sedimentation, and creaming in UHT milk: A review. *Comprehensive Reviews in Food Science and Food Safety, 18*(1), 140–166.

Avci, A., & Karagoz, I. (2003). Effects of flow and geometrical parameters on the collection efficiency in cyclone separators. *Journal of Aerosol Science, 34*(7), 937–955.

Bhandari, B., Bansal, N., Zhang, M., & Schuck, P. (Eds.). (2023). *Handbook of food powders: Processes and properties*. Elsevier.

Bell, C. H. (2014). Pest control of stored food products: Insects and mites. *Hygiene in food processing* (pp. 494–538). Woodhead Publishing.

Cárcel-Carrasco, J., Pascual-Guillamón, M., & Salas-Vicente, F. (2022). Composition of some metallic fragments found in food that are undetectable by magnetic or eddy currents equipment: A case study. *LWT, 153*, 112358.

Jain and De, & Alexandru Mihai Grumezescu and Alina Maria Holban (Eds.). (2019). Processing of beverages by membranes. *Processing and sustainability of beverages* (pp. 517–560). Woodhead Publishing.

Kastali, M., Mouhir, L., Saafadi, L., Yilmaz, L., & Souabi, S. (2021). Pretreatment of industrial wastewater by natural flotation: Application to pollution reduction from vegetable oil refinery wastewaters. *Environmental Science and Pollution Research, 28*, 34598–34610.

Luckiram, B., & Barry A. Perlmutter (Eds.). (2022). Centrifugation. *Integration and optimization of unit operations: Review of unit operations from R&D to production: Impacts of upstream and downstream process decisions.* Elsevier Inc.

Pouliot, Y., Conway, V., & Leclerc, P. L. (2014). Separation and concentration technologies in food processing. *Food processing: Principles and applications*, 33–60. https://doi.org/10.1002/9781118846315.ch3.

Rauh, V., & Xiao, Y. (2022). The shelf life of heat-treated dairy products. *International Dairy Journal, 125*, 105235.

Rushton, A., Ward, A. S., & Holdich, R. G. (2008). *Solid-liquid filtration and separation technology*. John Wiley & Sons.

Sanchez-Suarez, N., Orozco-Mendoza, G. L., Zartha-Sossa, J. W., Gafaro-Garcés, D. C., Melchor-Cahuana, L. G., & Gonzalez-Tovar, C. (2022). Trends in sieving and its applications in cereals. A literature review. *Frontiers in Sustainable Food Systems, 6*, 902147.

Sarbatly, R., Sariau, J., & Krishnaiah, D. (2023). Recent developments of membrane technology in the clarification and concentration of fruit juices. *Food Engineering Reviews, 15*(3), 420–437.

Schroën, K., & de Aguiar, I. B. (2023). The under-recognized fundamental effects of component interaction and deformability for membrane processes for food. *Separation and Purification Technology, 332*, 125844.

Sorsamäki, L., & Nappa, M. (2015). *Design and selection of separation processes. VTT Technical Research Centre of Finland Ltd.*

Tantray, J. A., Nissa, N. U., Wani, R. F. C., & Shafi, S. M. (2023). Basic Life Science Methods: A Laboratory Manual for Students and Researchers. Elsevier. doi:https://doi.org/10.1016/C2021-0-02283-5.

Tien, C. (2012). *Principles of filtration*. Elsevier.

Vaisali, C., Charanyaa, S., Belur, P. D., & Regupathi, I. (2015). Refining of edible oils: A critical appraisal of current and potential technologies. *International Journal of Food Science & Technology, 50*(1), 13–23.

Van der Bruggen, B. & Patricia Luis, (Ed.). (2018). Microfiltration, ultrafiltration, nanofiltration, reverse osmosis, and forward osmosis. *Fundamental modelling of membrane systems. Van der Bruggen, Patricia Luis* (pp. 25-70). Elsevier.

Wen, C., Shen, M., Liu, G., Liu, X., Liang, L., Li, Y., & Xu, X. (2023). Edible vegetable oils from oil crops: Preparation, refining, authenticity identification and application. *Process Biochemistry, 124*, 168–179.

Section I

Different mechanical separation processes

Flotation and sedimentation

2

Merve Yavuz-Düzgün[a], *Emrah Kırtıl*[b], *Sevil Çıkrıkçı-Erünsal*[c] *and Özge Seçmeler*[a]
[a]Department of Gastronomy and Culinary Arts, Faculty of Applied Sciences, Altınbaş University, İstanbul, Turkey, [b]Department of Chemical Engineering, Gebze Technical University, Kocaeli, Turkey, [c]Department of Food Engineering, Konya Food and Agriculture University, Konya, Turkey

2.1 Flotation

Flotation is a fundamental method for separating the three phases: solid, liquid, and gas. It is particularly applicable to particles (dispersed phase) with a density lower than the liquid (continuous phase) in which they are suspended (Kastali et al., 2021). In other words, it is a separation technique used for solid-liquid or liquid-liquid systems, specifically for particles with densities lower than the surrounding liquid.

The efficiency and speed of mechanical separation in natural flotation could be insufficient for the mass separation of smaller particles, which needs additional aided forces. The types of aided flotation systems: (a) dissolved air (pressure) flotation (DAF), (b) dispersed (induced) air flotation (IAF), (c) froth flotation (d) electrolytic flotation (EF), and (e) flotation by colloidal gas aphrons (CGA) (Rubio et al., 2002; Shammas & Bennett, 2010).

2.1.1 Natural flotation

Natural flotation can occur when the natural density difference is significant enough to allow for effective separation. On the other hand, aided flotation becomes applicable when external means are used to promote this separation process. As another type, in induced flotation, density of the particles is artificially decreased for floating (Degremont, 2007). There are several applications of natural flotation in various kinds of separation and cleaning processes. In this chapter, we will mainly focus on its utilization in decontamination and oil refinery wastewater treatments.

In food processing, cleaning of the raw materials is crucial to eliminate contaminants, ranging from harmless to hazardous substances. Key considerations during cleaning include maximizing efficiency, minimizing damage to the raw materials, and preventing recontamination. This is vital for consumer safety as well as for the protection of processing equipment. For example, removing sand, stones, or metallic objects from the grains before milling process is essential to prevent any damage of roller mills or contamination (Ortega-Rivas, 2012a). The contaminated food could then result in the food deterioration during storage. Physical hazards, such as mineral (sand, metal, stones, oil, etc.), plant (leaves, husks, etc.), animal (insects, feces, etc.), or

Mechanical Separation Processes in the Food Industry: Unit Operations and Processing Equipment in the Food Industry.
DOI: https://doi.org/10.1016/B978-0-12-819523-9.00006-8

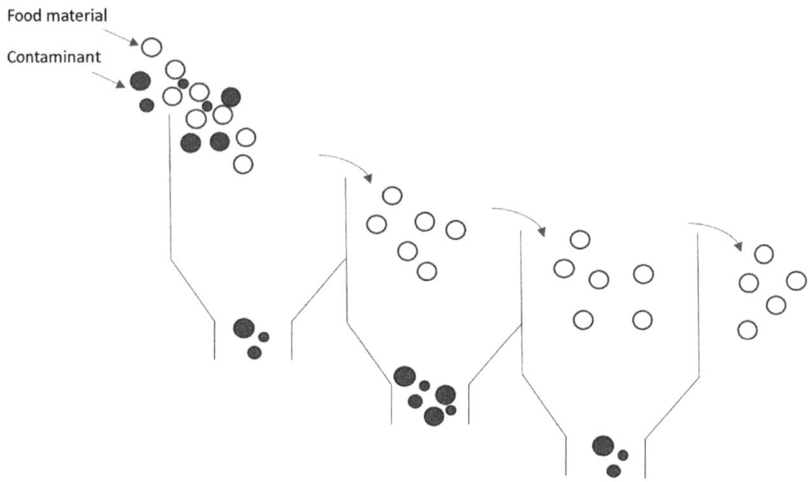

Figure 2.1 Simple representation of flotation washing.

microbial contaminants (bacteria, mold, yeast, etc.) and chemical hazards (fertilizers, pesticides, etc.) could be given among such kind of contaminants (Tzia et al., 2015).

Food manufacturing plants typically use two main methods for cleaning: dry and wet cleaning. While dry methods involve techniques; such as screening, abrasion, and magnetism, wet methods mainly include soaking, spraying, and flotation. Combination of these methods can also yield optimal results (Ortega-Rivas, 2012a). Herein, flotation washing as one of the wet techniques is based on the principle about buoyancy difference of raw food materials to separate desired components from undesired ones. For instance, when fruit is submerged in water, heavier particles like soil and metal sink, while the sound fruit pieces float. Passing fruit through water over a series of barriers effectively cleans them, along with other foods like peas and beans (Ortega-Rivas, 2012a). Fig. 2.1 represents a typical flotation washing. After the wet cleaning procedure, it is essential to remove the washing water at final stage. Centrifugation might be one technique for this purpose, but tissue might be damaged during the process. That's why vibration or rotary screens are mostly preferred for the dewatering step (Tzia et al., 2015).

Freshly harvested foodstuff may contain aflatoxin, or other kind of undesired compounds as contaminants during pre- or post-harvest activities, such as drying, storage, distribution, and processing if not managed properly. Fresh fruit and vegetables are one of the categories causing food illness due to pathogens, such as pathogenic *Escherichia coli, Salmonella* spp., etc. and require similar decontamination routes, such as washing, spraying, ultraviolet, or chemicals (Huang et al., 2018). Herein, advantages and limitations such as negative effect on food nutrition and quality, industrial speed, energy efficiency or regulations should be carefully considered to decide the applied techniques (Marshall et al., 2020).

As mentioned before, to make contaminated stuff safe for consumption, one method is to eliminate the contaminated parts of them such as the kernels or cobs (Ayalew et al., 2017). This can be conventionally done by sorting or flotation techniques. Sorting involves visual inspection of kernels for signs of fungal infection or biochemical changes. Size and density are also used to distinguish contaminated and uncontaminated kernels. Size sorting is typically done with sieves while winnowing and flotation rely on air or liquid to separate kernels based on weight. However, sorting and flotation result in two fractions: one with low aflatoxin levels and one with high levels. If the high aflatoxin fraction is not discarded properly, it can pose greater danger than the original material. Hand sorting, and density segregation efficiency depend on product type, quantity, and application method. Since hand sorting is labor-intensive, it may not be adopted by traders or storage facilities unless legally required or beneficial. However, it can be effective at the household level for small quantities processed into flour (Ayalew et al., 2017).

Mature macadamia is one of the types of nuts composed of approximately 75% oil, indicating its very high oil content. That's why any variation in the oil content of each kernel creates different eating qualities, so specific gravity can be presented as an index of kernel quality (Ross et al., 2002). At that point, flotation enables the classification of kernels according to their oil content. Ross et al. (2002) investigated the effect of flotation in water, ethanol, and salt solution on product quality and shelf life, immersing whole kernels into these mediums for 30 to 60 seconds, and then drying and storing under vacuum conditions. The findings pointed out that water had no effect on both quality and shelf life, whereas brine or ethanol solution caused a change in kernel flavor, although they didn't affect shelf life. As a result, flotation in water was highly recommended for the separation of kernels according to their oil content.

As well as nuts, cereals are one of the groups very susceptible to pre- and post-harvest contamination, especially with mycotoxins, which cause various negative health effects, including carcinogenicity, neurotoxicity, etc. (Matumba et al., 2015). Cereals are commonly consumed either as whole or processed grains, such as flour or bread. They are also used as animal feed, contributing to dairy, poultry, and meat production. The Food and Agriculture Organization (FAO) reports that a quarter of the world's cereal production is contaminated with mycotoxins. This contamination has led to a rise in mycotoxicosis cases in humans over recent decades, impacting the sustainability of agribusinesses (Kumar et al., 2022). Mycotoxins, including fumonisins, ochratoxin A, trichothecenes (type A and B), patulin, aflatoxin are produced by various fungal species and these toxins commonly contaminate cereal crops during farming, handling, transportation, or storage processes. Due to their heat-resistant nature, mycotoxins often persist through food processing, necessitating checks for contamination in processed products before consumption. Aflatoxins, in particular, are among the most prevalent mycotoxins worldwide, found in cereals and cereal products like whole grains, bread, cornflakes, breakfast cereals, and pasta (Kumar et al., 2022).

Maize is among these mycotoxin-contaminated cereals that should be studied to eliminate it. Matumba et al. (2015) examined the efficiency of different mycotoxin decontamination techniques which were flotation, dehulling, hand sorting and also their combined use in the shelled white maize. According to their regression analysis results, hand sorting was found to be the most efficient way of mycotoxin reduction as flotation gave the higher amount of remaining mycotoxin in the maze. It was reported that flotation or microwave heating are not utilized by the feed industry owing to various reasons, including high operational and investment cost, unavailable processing units for large scale, etc. (Colović et al., 2019). However, it should still be considered from other perspectives, including other kinds of foods.

In a recent study, various industrial-scale cleaning methods, including mechanical size separation, aspiration to remove dust and fine particles, gravity-based kernel separation, and optical sorting, were applied to detect defects in maize kernels (Pascale et al., 2020). Sampling during cleaning followed regulatory guidelines, and aflatoxin analysis was conducted using high-performance liquid chromatography and immunoaffinity column cleanup. The results showed significant reductions in aflatoxin levels, ranging from 65% to 84% when continuous cleaning lines were utilized. Combining the results from different cleaning steps resulted in an overall reduction in aflatoxin levels ranging from 55% to 94%. The rejected fractions contained notably high levels of aflatoxins, particularly in dust, and fractions were rejected during aspiration and optical sorting. In conclusion, Pascale et al. (2020) suggested that integrating both mechanical and optical sorting technologies into the cleaning process provided an effective solution for mitigating aflatoxin contamination in maize.

In overall, natural flotation becomes an alternative for different functions in the food industry, from washing to mycotoxin reduction. Its integration into other techniques can be proposed to increase the efficiency in the application areas.

2.1.2 Dissolved and dispersed air flotation

Over the last sixty years, DAF has gained widespread usage for removing suspended solids (70%–75%), oils (85%–95%), and biochemical oxygen demand (65%–85%) (Bennett & Peters, 1988). DAF includes bubble formation by a reduction in water pressure that is pre-saturated with gas (usually air) at pressures higher than the atmospheric level. The supersaturated water with the dissolved air is forced to pass through needle valves or special oriffices, and because of pressurization/depressurization, clouds of tiny bubbles at 10–100 μm in diameter are produced downstream of the constriction. Then, flotation is conducted by raising air bubbles to the surface with oil and suspended solids attached (Bennett & Peters, 1988; Rubio et al., 2002). The dissolved air flotation method, also known in the literature as dissolved gas flotation, is referred to in this chapter as dissolved air flotation (DAF).

DAF could be used both to clarify water and concentrate solid suspensions. In some cases, the solids may already be capable of being captured; in other cases, coagulation and flocculation may be needed (Sorsamäki & Nappa, 2015).

A typical DAF unit, shown in Fig. 2.2, comprises a contact cell or coagulation chamber, a flotation cell, a surface skimmer, a bottom skimmer, an effluent discharge baffle and chamber, and an air saturation system. DAF systems use chemical agents to flocculate (or aggregate) particles to be removed from water. Fine air bubbles are introduced to the water and attach to the flocculated particles, lifting them to the water surface (phase addition). Then, the floating particles are removed by a skimmer.

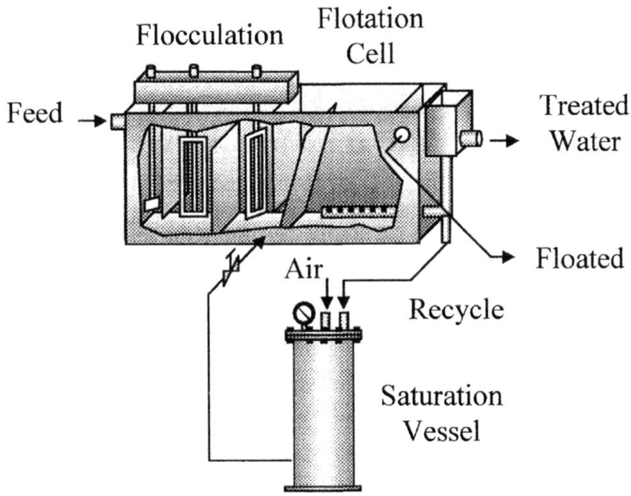

Figure 2.2 Dissolved air flotation unit.
(From Rubio et al. (2002)).

The DAF process consists of four stages: screening, equalization, chemical addition, and float handling (Ross et al., 2000). Design parameters include recycling pressure, saturation effficiency, recycle rate, tank detention, hydraulic loading rate, solids loading rate, and air/solids ratio.

In dispersed or induced air/gas flotation (IAF), a high-speed mechanical agitator/impeller, and an air injection system form bubbles mechanically. The technology uses the developed centrifugal force mainly for oil-water separation. The gas introduced at the top and the liquid thoroughly intermingled, forming many bubbles (700–1500 μm diameter) after passing through a disperser outside the impeller (Rubio et al., 2002).

The advantages of DAF over IAF are that it requires relatively inexpensive equipment, and an effective separation process is implemented. However, DAF also requires flocculation aids, which could contaminate byproducts, large tanks, and additional separation equipment (e.g., skimming). Being time-consuming and lower effficiency related to DAF are other disadvantages of it (Bowser, 2012). Ghasemi Naghdi and Schenk (2016) validated that DAF is suitable for separating microalgal oil and debris from a wet slurry of captured Chaetoceros mueller. They reported that DAF is less energy-intensive and a lower-cost process, making it advantageous to centrifugation for cell debris separation and oil recovery.

Pre-oxidation-enhanced flotation is also used to separate algal cells, dissolved organic carbon, polysaccharides, and proteins. Nguyen et al. (2013) showed that pre-oxidation by ozone in combination with IAF was suitable for separating algal cells from water. Ndikubwimana et al. (2016) mentioned that IAF was also less energy-consuming than other algae harvesting methods for biofuel production.

The main applications of DAF are wastewater pretreatment in mining, petroleum, mineral, and metallurgy industries, including residual reagents, powders, chemicals, metal ions, oils, organics, and the treatment of heavy metal-containing water and emulsiffied oil wastes (oil-in-water emulsions) (Rubio et al., 2002). DAF systems also provide product recovery and thickening of biological solids in industries ranging from food processing to pulp and paper to petrochemicals (Ross et al., 2000). Microalgae, drinking water, and dairy industry are the food industries in which air flotation is used. Some application areas of DAF are listed below:

Wastewater treatment/clariffication of reffinery wastewater, wastewater reclamation: Food processing wastewater, such as that from tuna, beef, poultry, seafood processing, and the dairy industry, is of considerable environmental concern. Potential added-value products can be recovered from these wastes. Casein is an example of these products, which can be harvested in large quantities from dairy plants. Ryder et al. (2018) compared the fundamental characteristics of casein derived by DAF from dairy plant wastewater with skim milk and skim milk powder. The casein profile of DAF has been found similar to that of skim milk powder, where it differs in its hydrophobic properties. Skim milk casein prepared by acid coagulation exhibited hydrophobic properties like those of DAF-casein but differed in rheological properties. The researchers suggested combining the DAF processing with the dairy industry cleaning-in-place procedures to produce DAF-casein.

Separation of solids and others in drinking water treatment plants: After coagulation and flocculation pre-treatment, low-density solid particles can be separated by bubble attachment to these floes (Newcombe & Dixon, 2006). Water supplies containing (a) natural color, (b) algae, (c) low turbidity (<10 nephelometric turbidity units [NTU]) and low total organic carbon (TOC), (d) moderate mineral turbidities (10-50 NTU) are candidates for DAF treatment compared to filtration and sedimentation due to ease in removal of low density flog and pathogen. DAF treatment provides an additional process for particle and pathogen removal with a less temperature dependence. In addition, biological materials (bacteria, fungi, yeasts, activated sludge, grape stalks) can be tested and floated efficiently to remove metal ions from dilute aqueous solutions (Kyzas & Matis, 2014). DAF could also be used to remove toxic or non-toxic organic solids and dissolved oils in water treatment and separate air and volatile organic compounds (VOCs) from spray-dried, or powdered foods like milk, juices, coffee, and butter (Bowser, 2012).

Sludge thickening and separation of biological flocs: Singh et al. (2007) reported that DAF could be used to separate colloidal gas aphrons (CGAs) from bulk liquids. CGAs consist of microbubbles encapsulated by surfactant multilayers (Singh et al., 2007). It was also stated that DAF can be used for the concentration and recovery of proteins, enzymes, and cellular products such as *Saccharomyces cerevisiae* or baker's yeast, and for enhancement of gas transfer in cell bioreactor and bioremediation processes (Hashim et al., 1999; Hashim et al., 1995b; Hashim et al., 1995a; Jauregi et al., 1997; Kyzas & Matis, 2014; Save & Pangarkar, 1995).

Removal of algae or microalgae or algal biomass harvesting: Flotation is one of the popular methods for low-density algae separation. The suspended solid material is the solid-liquid algae culture in the dewatering and drying algal cultures (Jacob-Lopes et al., 2020). Air flotation is an alternative method to other dewatering methods of algae, such as centrifugation, flocculation, filtration, gravity sedimentation, and electrophoresis. The algae's size ($<500\mu m$) and surface charge are the main efficiency parameters. Moreover, Ghasemi Naghdi and Schenk (2016) confirmed that pH and slurry densities are the parameters of the selectivity of the DAF processes.

Treatment of dairy effluents with organic coagulants: Dairy wastewater characterized by high biochemical oxygen demand (BOD), chemical oxygen demand (COD), high dissolved solids, suspended solids, oils, greases, and nutrients is dangerous for the environment. Before DAF treatment, the chemically enhanced primary treatment that uses coagulants to enhance the removal of pollutants was applied to clarify dairy effluents. Pereira et al. (2020) used organic, non-toxic, low-cost, combined polyacrylamide and Tanfloc coagulants with high molecular weight and water solubility for chemically enhanced primary treatment (CEPT) level and optimized the process of dissolved air flotation for the parameters of coagulant dosing and pH.

Stoke's Law is used as the theoretical basis for air flotation systems. Stoke's Law, whose equation is given below, is used to calculate the rate of rise of bubble drops, agglomerates, and bubble oil aggregation in gravity-controlled processes (Bennett & Peters, 1988).

$$v_t = \frac{gD_p^2(\rho_P - \rho)}{18\mu} \tag{2.1}$$

where υ_t (m/s) is the terminal velocity, g (9.80665 m/s^2) is the gravitational constant, D_p (m) is the diameter of the spherical particle, ρ_p (kg/m^3) is the density of the particle, ρ is the density of the liquid, and μ (kg/ms) is the viscosity of the fluid.

Bennett and Peters (1988) reported that a reduction in the density of the target particle (oil or solid) or agglomerate causes an increase in the rate of rise of bubble/solid or bubble/oil agglomerates in aerated systems over the rise in unaerated systems. This is accomplished by the attachment or encapsulation of an air bubble onto or into the target particles.

The design variables that affect the separation efficiency of air flotation are reported by Bennett and Peters (1988) as the air input rate and the volume of air entrained per unit volume, the amount of air formed by reducing the pressure, bubble size distribution and degree of dispersion, surface properties of the suspended target matter, hydraulic design of the flotation chamber, concentration and type of dissolved materials, chemicals added (surfactants, hydrocolloids etc.), the temperature of the aqueous medium and pH of the aqueous medium.

The air input rate and the volume of air entrained per unit volume of liquid under pressure depends on air solubility in water based on Henry's law:

$$p = Hc \tag{2.2}$$

where p is the partial pressure of air, c is the concentration of the dissolved air, and H is Henry's Law constant. It is reported that the solubility of air depends on temperature and the concentration of dissolved solids (Bennett & Peters, 1988).

The amount of air formed by reducing the pressure to 1 atm is calculated by the following equation:

$$S = S_a\left(\frac{fP}{14.7}\right) \tag{2.3}$$

where S is the air formed at 1 atm (mg/L), S_a is the air saturation at 1 atm (mg/L), P is the absolute pressure of the initial pressurization (psi), and f is the fractional system dissolving-efficiency factor.

DAF systems have been designed based on solids loading, hydraulic loading, and the ratio of air (or gas) to suspended solids (or oil) (G/S) where it is calculated as (Bennett & Peters, 1988).

$$G/S = \frac{RC_a f(P/101.3)}{S_0 Q - S_e R} \tag{2.4}$$

where R is the pressurized liquid flow rate (L/day), C_a is the air saturation concentration at 1 atm (mg/L), f is the saturation efficiency, P is the gauge pressure or saturation (kN/m^2), S_0 is the concentration of the suspended solids (or oil) in wastewater (mg/L), Q is the raw wastewater flow rate (L/day), S_e is the concentration of the suspended solids (or oil) in the pressurized liquid stream (mg/L).

Bubble size is the most critical dependent variable in air flotation systems (Bennett & Peters, 1988; Wang et al., 2022). The average bubble sizes in three different aeration systems are 170, 75, and 100 μm for dispersed, dissolved, and electrolytic flotation, respectively. The number and size of air bubbles depend on the aqueous medium's physical system and chemical content.

Surface properties of the suspended target matter are the other critical factors for air flotation systems. The surface properties of the oil droplet (hydrophilicity or hydrophobicity) depend on the degree of the oil droplet's wetness where the wetness depends on the size of the contact angle between the oil droplets and water bubbles affected by flocculant, pH, and salinity (Wang et al., 2022). pH of the aqueous medium is also an important parameter as it affects the interaction between oil droplets and water bubbles. At a higher pH value, oil droplets are more hydrophilic, which leads to a lower probability of binding to bubbles than those at a lower pH value (Wang et al., 2022). The oil removal efficiency was reported to be high in the pH range of 6–9. If the anaerobic digestion of DAF slurries to produce biomethane is concerned, an initial pH adjustment is needed for biotechnological reasons (Logan et al., 2021).

The hydraulic design of the flotation chamber, hence the turbulence size, is a critical parameter for flotation performance due to its effects on oil removal efficiency. The probability (P) of successful collision of bubbles and oil droplets in the flotation tank can be expressed by the following equation (Wang et al., 2022):

$$P = P_c . P_a . (1 - P_d) \tag{2.5}$$

Pc, Pa, and Pd are the collision, attachment, and detachment probability of bubbles-oil droplets, respectively.

As a result, dissolved air flotation is a highly effective process in applications ranging from water treatment to algae or natural color treatment. The design of the system requires consideration of various parameters such as air flow rate, input power, bubble size, retention time, hydraulic surface, and presence of surfactants.

2.1.3 Aqueous two-phase plotation

With the significant advancements in biotechnology, biological products are increasingly utilized across diverse sectors including food, pharmaceuticals, cosmetics, and textiles. As biotechnology applications progress, industries are shifting towards extracting and purifying protein products from biological sources. The emphasis is primarily on ensuring the purity of these biological products, underscoring the importance of downstream processing in biomolecule production (Sankaran et al., 2019). The recovery and purification of biomolecules involve a complex and costly process, often accounting for up to 70% of the total product cost. Proteins, being delicate materials, necessitate gentle processing conditions to prevent denaturation. Traditional methods such as organic solvents, batch adsorption, membrane chromatography, and crystallization are commonly used for separation but face challenges like scalability, high production costs, and a lack of suitable solvents. This isolation and purification process can contribute substantially, ranging from 30% to 80%, to the overall production expenses. Consequently, there is a growing interest in developing efficient and cost-effective downstream processes for biomolecule recovery and purification (Sankaran et al., 2019). In addition to biotechnological products, pharmaceutical low-molecular-weight compounds like penicillin (Bi et al., 2011), high-molecular-weight compounds (ex-enzymes) (Kee et al., 2021), or lipase recovery by its combined version with fermentation (Sankaran et al., 2018b) are among other promising applications.

As understood from the previous discussion, bioseparation area is still open to innovations and to increase efficiency with the help of new and combined techniques. Although liquid-liquid extraction (LLE) is one of the popular methods, high consumption of toxic solvents (chloroform, tetrachloromethane, etc.), low yield and purity may become some challenges in this field (Sankaran et al., 2019). At this point, the aqueous two-phase flotation (ATPF) technique as a special type of LLE can be seen as an alternative to overcome these challenges; however, large-scale applications with the integrated techniques should still be discovered in detail. Its advantages can be listed as an environmentally friendly method giving high purity and high recovery yield in the separation and recovery of biomolecules (Sankaran et al., 2019). On the other hand, it has some drawbacks such as slow separation process under gravity and recycling problems of phase forming chemicals. Some efforts have been made, including modification of chemical compositions, or changing conventional mass transfer mode, etc. (Sankaran et al., 2019).

ATPF, distinguished from ATPE, is recognized as an innovative method characterized by reduced usage of organic solvents and elevated concentration coefficients (Jiang et al., 2019). As mass transfer is based on diffusion in ATPE, the transportation

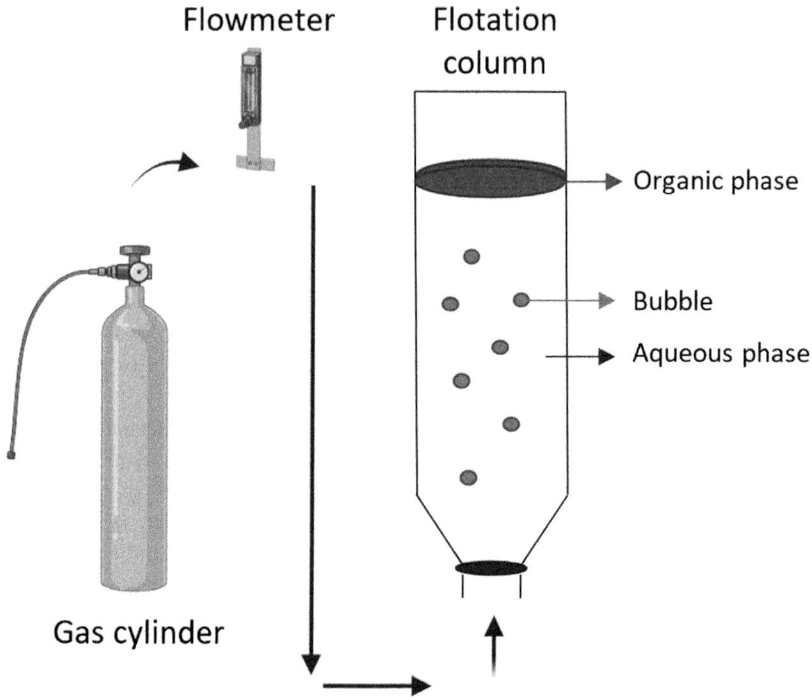

Figure 2.3 Representative diagram of ATPF apparatus.

is held on by flotation in ATPF. In other words, it is the integrated form of ATPE and flotation (Jakob et al., 2022). In detail, the ATPF system is a unique form of liquid-liquid extraction, integrating the principles of both aqueous two-phase systems (ATPS) and solvent sublation (SS). Comprised mainly of an aqueous solution, ATPF maintains the native structure of purified biomolecules (Sankaran et al., 2019). There are several studies conducted on ATPF for protein or herbal medicine enrichment/separation (Jiang et al., 2020), as well as for lipase (Sankaran et al., 2018c), flavonoids and bromelain isolation (Pakhale et al., 2013).

ATPF, a method combining ATPS and SS principles, involves generating two immiscible liquid phases by dissolving hydrophobic liquids above critical conditions. Target bioproducts are transferred from the aqueous-rich phase to a less dense solvent phase via bubble attachments (Fig. 2.3). SS enhances ion flotation without foam formation by adsorbing hydrophobic compounds onto rising gas bubbles, accumulating in an organic solution layer. ATPF utilizes similar mass transfer mechanisms, employing two non-miscible aqueous phases with a higher volume ratio of salt to polymer. As based on the same theory, three main ATPF types exist: hydrophilic organic solvent/salt, polymer/salt, and ionic liquid/salt. Mainly, biomolecules are extracted by introducing air bubbles into a flotation column, interacting with them as they ascend, and accumulating in the top organic phase. Mass transport occurs primarily through air bubbles and diffusion between aqueous phases. Additionally, water sheath drag aids in transferring solutes to the upper organic phase. As shown in Fig. 2.3, gas bubbles, produced by sparging gas through a porous sintered glass disk at the column bottom, play an important role. In addition to basic working principle, the ways to increase ATPF efficiency should be considered. Various influencing factors including phase-forming components, flow rate, gas bubble size, flotation time, and pH could be optimized for the enhancement of this efficiency (Sankaran et al., 2019).

As an innovative approach, the application of ATPF with sonication and electrolysis have also been demonstrated as recent efforts for cell disruption process. Electricity integrated ATPF was explored to extract betacyanins from peel and flesh (Leong et al., 2019). They obtained satisfying results obtaining efficiency up to 98%. Another attempt to integrate sonication process was done by Sankaran et al. (2018a) for the extraction of proteins from microalgae.

To increase purification capacity, there is another factor that should be considered for ATPF: process type as batch, semi-continous or continuous. The majority of research has been conducted at the laboratory scale, typically using volumes smaller than 300 mL and employing glass filters for gas input, lacking descriptions of scalable processes or integration into larger systems (Jakob et al., 2022). The largest reported scale for ATPF experiments to date is seen approximately 2.5 L, utilizing a non-scalable converted water dispenser as a flotation cylinder (Mathiazakan et al., 2016). On the other hand, to enable ATPF for industrial applications, optimization of processes and apparatus is imperative. It has been proposed by Jakob et al. (2022) that the potential benefit of alternative process strategies, such as stepwise or continuous exchange of the top phase could be obtained to enhance ATPF efficiency. The transfer rate observed in semi-continuous ATPF served as the basis for implementing continuous phase exchange using peristaltic pumps, referring the first use of continuous ATPF. In continuous ATPF, the first 30 mins were similar to batch and semicontinuous ATPF and after 30 mins, using peristaltic pumps took the role for continuous phase exchange. The findings pointed out that continuous ATPF exhibited higher efficiency compared to batch ATPF, offering significant advantages for industrial downstream processes in biotechnology. Continuous purification in upstream processes reduces the need for large apparatus equipment, such as storage tanks, facilitates the use of smaller ATPF cells, and enables direct further processing of the product-loaded top phase. Future process development for ATPF should focus on optimizing the apparatus to enable direct flow of continuous phase exchange, ensuring high concentration gradients and faster separation. Integration of suitable gas input and online measurements of product concentration, and phase stability will enable precise process control. The study offered a design for establishing an efficient industrial ATPF process (Jakob et al., 2022).

Proteins have a great interest for the industry owing to their nutritional value and unique functional properties (Jiang et al., 2021). Therefore, their separation from the complex sources becomes very attractive. Protein separation from whey (byproduct during cheese production from the milk) is one of them as an example. Since whey contains around 1% protein mostly composed of α-lactalbumin with molecular weight of 14.2 kDa and β-lactoglobulin (MW of 18.4 kDa), their separation techniques receive high attention (Jiang et al., 2021). Jiang et al. (2021) investigated the separation of

α-lactalbumin from the whey by ATPF method as a promising method for simplicity and low cost. They perfomed ATPF with the help of poly (ethylene glycol-ran-propylene glycol) monobutyl ether (UCON) and KH_2PO_4. The optimum operating parameters were found to be as 20 min flotation time, 30 mL/μLmin of flow velocity giving satisfied resμts and without causing structural change in the protein.

As a purpose to reuse food waste product which is salted egg white, Jiang et al. (2019) examined two phase flotation using polyethylene glycols (PEG 1000) and $(NH_4)_2SO_4$ to obtain ovalbumin from salted egg white. A response surface method (RSM) analysis was performed to see the effect of concentration, time and flow rate on ovalbumin purity and flotation efficiency. 28 mL of 28% $(NH_4)_2SO_4$ (w/w), 5 mL of 80% PEG 1000 (w/w), 30 mins flotation time and 35 mL/min of flow rate were found to be the optimal conditions reaching up to efficiency and purity of protein as around 82% and 92%, respectively (Jiang et al., 2019).

Ionic liquid–salt aqueous two-phase flotation (ILATPF) represents another innovative method for pretreating samples that is characterized by its novelty, environmentally friendly nature, lack of toxicity, and high sensitivity. Aqueous two-phase system (ATPS), as a good performance extraction method becomes very important in liquid-liquid extraction and in the use of room temperature ionic liquids (ILs) and water structuring salts in the separation and purification of biological products such as proteins, hormones, drugs from complex mixtures (Han et al., 2011). On the other hand, Solvent Sublation (SS) acts a non-foaming flotation technique or adsorptive bubble separation, involves hydrophobic compounds in water being adsorbed onto the surfaces of ascending gas bubbles. These compounds are then collected in a separate immiscible liquid layer, typically an organic solvent lighter than water, positioned atop the bulk aqueous phase in the column. This method offers several advantages, including simultaneous separation and concentration, high efficiency in separation, a high concentration coefficient, minimal organic solvent usage, a gentle separation process, and straightforward operation. It has found widespread application in dealing with surface-active and hydrophobic compounds, including active ingredients in natural products, metal ions, organic pollutants, and more (Han et al., 2011). Han et al. (2011) investigated the combined performance of new IL–salt ATPSs (ILATPSs) and SS which is ILATPF for the separation and detection of trace level of Chloramphenicol (an effective antibiotic but toxic for human consumption) from feed water, milk and honey. For this purpose, the samples were mixed with Chloramphenicol solution and then centrifuged and finally filtered before the use. After putting the samples to flotation cell with the use of additional chemical solutions, it was floated by bubbling nitrogen gas from the bottom of the cell and extracted at the surface. HPLC (high-performance liquid chromatography) was also performed for the determination of Chloramphenicol. According to the results, so many variables such as concentration, pH, gas flow rate, flotation time played a crucial role in the efficiency. Here, ILATPF was comprised of two mechanisms: IL–salt ILATPS formation and solvent sublation. In contrast to liquid-liquid extraction, solvent sublation, and ionic liquid aqueous two-phase extraction, ILATPF achieves both low cost and high sublation efficiency for chloramphenicol separation and concentration while minimizing ionic liquid wastage. This innovative method is not only simpler, and more environmentally sustainable but

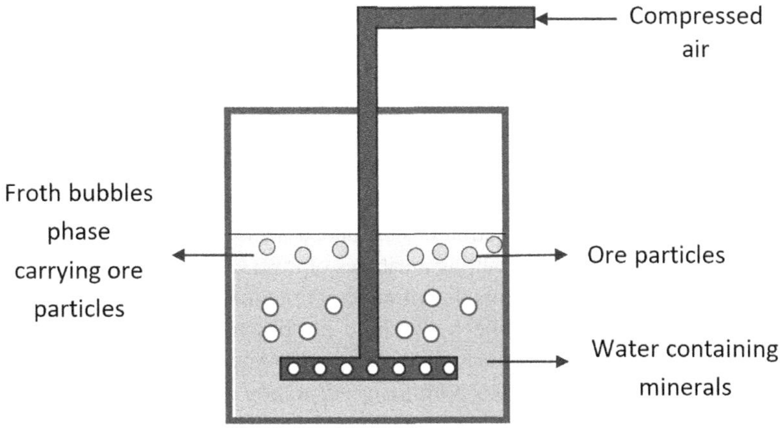

Figure 2.4 Froth flotation process.

also holds promise for concentrating and separating various small biomolecules (Han et al., 2011).

As seen from the examples, the performance of the flotation depends on so many factors such as type of material, flotation time, flow rate, concentration and combined technologies integrating different mechanisms. Therefore, it should be optimized for each process conditions and foodstuffs, accordingly. ATPF methodology emerges a viable alternative to tackle some obstacles; nevertheless, comprehensive exploration of large-scale applications utilizing integrated techniques still need to be studied further.

2.1.4 Froth flotation

Froth flotation is a method utilized in mineral processing and industrial sectors for separating hydrophobic materials from hydrophilic ones. Initially developed as a technique for mineral beneficiation, it has since found extensive applications in areas, such as wastewater treatment, recycled paper deinking, and environmental cleanup efforts (Gao et al., 2021)

At its essence, froth flotation capitalizes on the differences in surface properties of materials. This technique involves a series of intricate processes occurring within a flotation cell, where solids, aqueous medium, and air bubbles interact. The principle underlying froth flotation hinges on the attachment of hydrophobic particles to air bubbles, facilitating their flotation and subsequent recovery into the froth product, while hydrophilic particles remain in the liquid phase (De Witt, 1940; Penaloza et al., 2023). Fig. 2.4 depicts a basic flotation unit for recovery of minerals from water.

Surface energy emerges as a critical determinant in shaping the wettability behavior of particles, thereby delineating their hydrophobic or hydrophilic nature. The surface energy of solids, reminiscent of the surface tension in liquids, is divided into dispersive (non-polar) and specific (polar) components. This wettability characteristic, discerned through surface energy calculations and techniques, such as contact angle measurements or inverse gas chromatography, governs the feasibility of particle flotation (Jung et al., 2023).

Numerous factors influence the flotation response of materials, spanning particle size, shape, surface chemistry, pulp chemistry, and froth structure and stability. Adjustments to these parameters can significantly impact flotation selectivity and recovery. For instance, surface-reactive reagents act as frothers and collectors, crucial in generating and stabilizing air bubbles while enhancing the hydrophobicity of target solids. In recent years, froth flotation has expanded its scope to include plastic separation, leveraging the selective wetting attributes of diverse polymers. The mechanism underlying plastic flotation mirrors that of mineral flotation, with hydrophobic plastics being lifted by air bubbles while hydrophilic plastics remain in the liquid phase. Techniques such as pH adjustments and surface treatments have been explored to manipulate plastic wettability, thus optimizing separation efficiency.

Despite its efficacy, froth flotation encounters challenges in efficiently separating very fine particles, often smaller than 10 μm, owing to the low probability of bubble-particle collision. Recent strides in nanobubble flotation hold promise in augmenting the recovery of ultrafine particles, offering avenues to surmount these obstacles and enhance flotation efficiency further (Jung et al., 2023).

The principle of operation of froth flotation is as follows: The ore undergoes comminution, where it is ground into particles. Ideally, individual minerals are fully liberated within a size range of 2–500 μm in diameter. Froth flotation involves treating an aqueous slurry of the ground ore with a frothing agent, such as sodium ethyl xanthate. This collector, for instance, separates galena from sphalerite by attaching its polar part to ore particles and forming a hydrophobic layer with its non-polar hydrocarbon part. Efficient separation typically requires around 300 g/t of ore. The length of the xanthate's hydrocarbon chain affects hydrophobic efficiency and ore type selectivity; shorter chains, like in sodium ethyl xanthate, exhibit high selectivity to certain ores. Flotation cells aerate the slurry, causing hydrophobic particles to attach to air bubbles and rise to form a froth, which is then skimmed off to produce a mineral concentrate (Farrokhpay et al., 2021).

Unfloated minerals, termed flotation tailings, undergo further flotation stages, known as scavenging, to recover valuable particles missed initially. The final tailings from scavenging are typically disposed of as mine fill, or stored in tailings disposal facilities for long-term storage. Flotation is conducted in multiple stages to optimize mineral recovery and concentrate concentration while minimizing energy input (Gallios & Matis, 1992).

To elucidate the theoretical principle, extensive research has explored the behavior of frothers and free liquid films over the past decades. Findings suggest that pure liquids do not froth; surface-active agents are necessary to induce frothing. When added to the liquid, these agents decrease the solution's surface tension due to their heteropolar nature, leading to preferential adsorption at the air-water interface. This adsorption is governed by the Gibbs adsorption reaction.

$$\Gamma = -\frac{a}{RT}\frac{d\rho}{d\alpha} \tag{2.6}$$

where Γ is the surface excess concentration of the surface-active agent, ρ is the surface tension and α is the activity of solute.

In dilute solutions, common in flotation, the surface tension serves as an indicator of solute surface activity. Froth dynamics, however, are complex and cannot be explained by a single theory, as properties depend on numerous variables. Froths can vary from unstable to persistent, with some collapsing due to higher surface energy than the segregated gas and liquid. Unstable froths break down as liquid drains between bubbles, with weak surface forces unable to prevent coalescence entirely. Metastable froths, on the other hand, may form planar lamellas of uniform thickness, lasting indefinitely under stable conditions. In such froths, gas diffusion between bubbles, influenced by bubble size, can contribute to eventual breakdown (Bulatovic, 2007).

In metastable froths, the liquid drainage between bubbles can form planar lamellas with uniform thickness, which can endure indefinitely in the absence of external disturbances, such as vibration or evaporation. Within this type of froth, there occurs a gradual diffusion of gas from smaller bubbles into larger ones due to pressure differences. This diffusion rate is inversely related to the radius of the bubble and can ultimately lead to the breakdown of the froth structure.

Moreover, experimental research has introduced another perspective on froth stability. It suggests that the positive disjoining pressure plays a crucial role in hindering the thinning of lamellae. According to Gibbs theory, during the stretching of foam lamellae, a localized increase in surface tension arises in thin layers due to the depletion of surfactants by adsorption into the stretched portion, leading to a change in surface elasticity. The coefficient of static surface elasticity (E) is defined as the stress divided by strain per unit area. This theory primarily applies to static systems and thin lamellae, where there is an insufficient number of surface-active molecules within the film body to diffuse to the surface and reduce surface tension.

Under dynamic conditions, the surface tension within a froth film exhibits variations during extension and compression compared to its equilibrium value. This phenomenon, known as the Marangoni effect, operates on an expanding or contracting surface with an adsorbed layer, regardless of the underlying liquid's thickness. Marangoni effect explains the foam stability by highlighting the ability of a frothing solution to withstand variations in surface tension within a fraction of a second. Despite these insightful theories, measuring the magnitude of the Marangoni effect remains a challenge, and the theoretical treatment of this phenomenon is still incomplete (Bulatovic, 2007; Rajendran & Murty, 2023).

Flotation reagents play a crucial role in inducing hydrophobicity in floatable particles, and modifying their surfaces for effective separation. These reagents, along with their dosages, significantly influence the flotation rate constant. Collectors, typically surface-active organic compounds like oils, render mineral surfaces hydrophobic by displacing polar water molecules. They are categorized into ionizing (anionic and cationic) and nonionizing compounds, each serving specific flotation requirements.

Frothers, composed of polar and nonpolar groups, stabilize air bubbles in the flotation process. By adsorbing at the gas/liquid interface, they prevent bubble bursting and reduce interfacial tension, crucial for efficient particle-bubble attachment. Proper frother dosage is essential to prevent micelle formation, typically ranging from 5 ppm

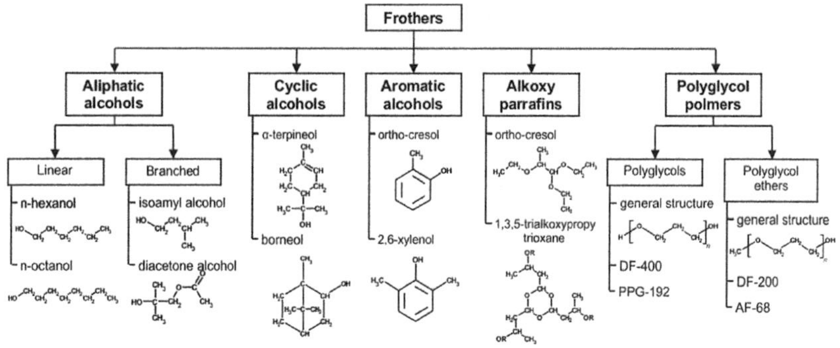

Figure 2.5 The categorization of flotation frothers, accompanied by pertinent exemplars, contingent upon their molecular compositions.
(From Pawliszak et al. (2024)).

to 10 ppm in flotation. Some collectors may also possess frother properties, obviating the need for additional frothers.

Regulators, including pH modifiers, activators, and depressants, enhance selectivity among minerals being separated. pH modifiers create optimal flotation conditions, stabilizing collectors and promoting selective mineral adsorption. Activators facilitate collector adsorption on valuable minerals, while depressants inhibit the flotation of unwanted minerals. Optimizing the reagent suite, comprising collectors, frothers, and modifiers, is paramount for effective ore processing. Bench-scale studies are conducted to fine-tune the reagent composition, ensuring optimal performance before implementation in plant operations (Bulatovic, 2007; Pattanaik & Venugopal, 2018).

Frothers used in flotation processes are classified based on their chemical structure and their function (see Fig. 2.5) Structurally, they are surfactants composed of a hydrophilic head and a hydrophobic tail. They can be categorized into anionic, cationic, and non-ionic frothers. Non-ionic frothers are predominantly used in industrial settings due to their versatility across a wide range of pH levels.

Non-ionic frothers are further divided into aliphatic, cyclic, and aromatic alcohols, as well as alkoxy hydrocarbons, and polyglycol-based polymers. Aliphatic alcohols, such as n-hexanol and methyl isobutyl carbinol (MIBC), are commonly employed in flotation due to their selectivity and sensitivity to pH. Cyclic alcohols, derived from natural sources like eucalyptus and pine oils, serve as secondary collectors in certain flotation applications. Phenols, despite their toxicity, find use in coal flotation due to their ability to produce less dense froths with larger bubbles, crucial for coal processing (Cho & Laskowski, 2002; Drzymala & Kowalczuk, 2018).

Alkoxy hydrocarbons, including compounds like 1,1,3-triethoxybutane, exhibit high selectivity and stability but suffer from low solubility in water. Polyglycol-based polymers, like polyglycols and polyglycol ethers, offer versatility by allowing modifications to their properties based on molecular weight. While higher molecular weight polymers create more stable froths, they are less selective. However, they have better solubility and can be tailored to specific flotation processes.

Each frother type has its advantages and limitations, making their selection crucial for optimizing flotation efficiency. Factors, such as pH, solubility, selectivity, and stability must be considered when choosing the appropriate frother for a given flotation application (Cho & Laskowski, 2002; Gupta et al., 2007a; Melo & Laskowski, 2006).

Functional classification categorizes frothers into two groups based on their properties and behavior in flotation processes. The first group comprises standard frothers, which primarily aid in dispersing bubbles and forming froth. These frothers contain hydroxyl groups and hydrophobic chains, allowing them to interact with the air-water interface while avoiding adsorption at solid-liquid interfaces. They lack collecting properties and are versatile across various minerals. Their properties can be adjusted by modifying their molecular structure, including the number and positioning of hydroxyl groups, as well as the length and branching of the hydrocarbon chain.

The second category includes frother-collectors, which possess both frothing and collecting capabilities. These molecules not only interact with the air-water interface but also adsorb at the solid-liquid interface, increasing the hydrophobicity of the solid. Frother-collectors are often ionic surfactants containing carboxylic, sulfur-, or nitrogen-based groups. Examples include long-chain amines and amine salts, which are used in quartz mineral flotation. Sulfates with long alkyl chains, such as sodium dodecyl sulfate and sodium petroleum sulfonate, are also utilized in specific mineral flotation applications (Drzymala & Kowalczuk, 2018; Pawliszak et al., 2024).

Additionally, long-alkyl chain fatty acids and derivatives, along with naturally occurring mixtures like pine oil or cresylic acid, exhibit collecting activity. These compounds are employed in various mineral flotation processes, including those involving apatite, calcite, or spodumene. Pine oil, for instance, is widely used in zinc and iron flotation due to its cost-effectiveness and flexibility in application. However, the compositional heterogeneity of natural frothers may pose challenges in process control, and overdosing can lead to excessive froth formation or reduced collection efficiency due to their high interfacial activity (Gupta et al., 2007b; Pawliszak et al., 2024).

The stability of froth mainly depends on the timely arrival of surface-active molecules at the interface It is crucial with surface elasticity becoming relevant within a range of 10^{-1} to 10^{-3} s. This timeframe often determines the optimal concentration for maximizing frothing. Additionally, the stability of froth is influenced by the viscosity or plasticity of the adsorbed layer to surface-active minerals. Researchers have identified three types of froth stability:

1. Transient foams form at concentrations where initial black-film formation occurs, characterized by critical thickness values.
2. Highly persistent froth is observed in solutions of higher concentrations, with rupture attributed to surface fluctuation under mechanical stimuli.
3. Froths of intermediate stability emerge at low frother concentrations, featuring a sharp transition from low to high stability; both thinning and black-spot formation contribute to these froths (Bulatovic, 2007; Cho & Laskowski, 2002; Gupta et al., 2007a).

Flotation finds widespread industrial applications in mineral and chemical engineering for separating and concentrating various substances, including minerals, coal, precipitates, inorganic waste, effluents, and even microorganisms and proteins. Annually, over two billion tons of ores and coal are treated worldwide through flotation, accounting for approximately 85% of mined ores. With the depletion of high-grade ore deposits, this figure is expected to rise in the future. The mechanization of mining methods has led to increased coal flotation, particularly to handle large quantities of fine coal particles. Furthermore, flotation technology has expanded to include deinking wastepaper for recycling, water treatment, and plastics separation. Flotation deinking alone contributes approximately 130 million tons of recovered paper annually to global paper production, representing around 50% of the annual papermaking capacity (Nguyen, 2013).

Wastewater purification using froth flotation is an effective method for removing suspended solids, oils, greases, and other contaminants from water. In this process, the wastewater is first treated with chemicals to promote the attachment of air bubbles to the suspended particles. These chemicals, known as frothers and collectors, help create a stable froth layer on the surface of the water. Air is then introduced into the wastewater, either through mechanical agitation or by releasing dissolved air under pressure. As the air bubbles rise to the surface, they carry the attached contaminants with them, forming a froth layer that can be easily skimmed off. The froth containing the contaminants is removed from the top of the flotation tank, leaving behind cleaner water. Froth flotation is particularly effective for treating wastewater in industries, such as mining, paper mills, food processing, and oil refineries, where the removal of fine particles and organic matter is crucial for environmental compliance (Nguyen, 2013).

The future of froth flotation is shaped by advancements in technology and modeling to enhance its efficiency and adaptability in mineral processing. Research in the last five years highlights several key developments:

Advanced modeling for control: There is a growing focus on developing more sophisticated models for froth flotation control, particularly using model predictive control (MPC) frameworks. These models aim to improve flotation performance by accurately representing the dynamics of the process, including both the pulp and froth phases. However, current models primarily focus on the pulp phase, with less attention on the froth phase, which is critical for operational efficiency (Quintanilla et al., 2021).

Incorporation of Machine Learning: Machine vision systems using convolutional neural networks (CNNs) are increasingly applied to froth image analysis, significantly enhancing the accuracy of froth classification in real time. This aids in monitoring and controlling flotation plants more effectively (Zarie et al., 2020).

Scale-up Challenges: As flotation equipment scales up to process more minerals at reduced capital costs, new challenges in tank design and operation have emerged. Reviews suggest a need for better scale-up rules that consider both kinetic and machine design aspects to enhance froth stability and transport in larger flotation cells (Mesa & Brito-Parada, 2019).

Greener Processes: Environmental considerations are becoming crucial in flotation processes. Studies are exploring more sustainable frother chemistries, such as biodegradable polymers, to reduce the environmental footprint while maintaining high separation efficiencies (Nuorivaara & Serna-Guerrero, 2020).

Froth flotation, as detailed in this chapter, emerges not just as a pivotal technique in mineral processing but also as a versatile technology applicable across various industrial realms, from wastewater treatment to plastic separation. This technique's enduring relevance and adaptability are rooted in its fundamental ability to distinguish and separate based on the surface properties of materials—be they hydrophobic or hydrophilic. The ongoing refinement of froth flotation through advances in chemical reagents, particle size manipulation, and surface energy modifications underscores its critical role in enhancing recovery rates and operational efficiencies in numerous sectors. However, the journey of froth flotation is not devoid of challenges, particularly in handling fine particles and extending its application to emerging materials. The forward path is illuminated by promising research into nanobubble technology and advanced surface treatment methods, aiming to optimize and expand the scope of this essential process. As we look to the future, it's clear that froth flotation will continue to play a key role in meeting the demands of efficient resource utilization and environmental sustainability, ensuring its place as a cornerstone of industrial separation technologies.

2.1.5 Electrolytic flotation

Electrolytic flotation (EF) represents an electrochemical adaptation of conventional flotation techniques, demonstrating heightened effectiveness in the separation of solids from liquids or differing liquid phases. This method stands out for its ability to remove even the minutest of particles through the generation of uniformly small hydrogen and oxygen bubbles at the electrode surfaces during water electrolysis. This critical feature elevates the electrodes to a pivotal role within the EF system, often described as its core (Chen et al., 2002).

The primary objective of EF is to facilitate the removal of suspended matter from water solutions, ensuring the resultant effluent meets environmental standards and poses no risk. Additionally, the separated suspended matter can be repurposed as feed, fuel, or for mineral recovery, or disposed of more effectively (Martínez-Huitle et al., 2018).

The primary distinction between EF and traditional flotation methods lies in the bubble generation mechanism. EF achieves bubble formation directly at the cathode and anode surfaces, unlike in dispersed (induced) air flotation (IAF), where bubbles are mechanically created through a high-speed mixer combined with air injection, relying on centrifugal force to produce bubbles ranging from 700 to 1500 μm in diameter. Conversely, in the dissolved air flotation (DAF) technique, bubble creation occurs through the depressurization of air-saturated liquid beyond atmospheric pressure, where the liquid, forced through narrow valves or nozzles, generates bubbles sized between 30 and 1000 μm in diameter (Rubio et al., 2002).

EF's standout features, distinguishing it from DAF or IAF, include its ability to generate extremely small bubbles, enhancing particle flotation. Moreover, EF offers advantages like total tank surface coverage, facilitated flocculation even in surfactant presence, easy operational control, and safe equipment management, without reliance on effluent ionic charge (Mook et al., 2014).

In electroflocculation (EF) reactors, typically, the configuration entails positioning an anode at the base of the reactor and a cathode around 10–50 mm away from the anode (see Fig. 2.6). Nonetheless, this setup does not facilitate direct interaction between bubbles generated at the base and the effluent flow. Consequently, bubbles dispersed on

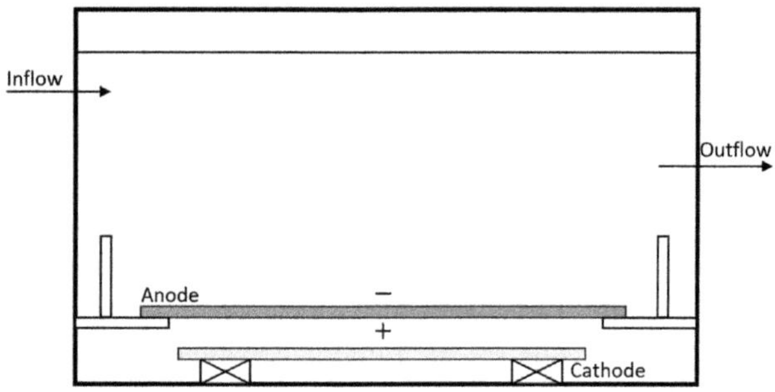

Figure 2.6 A conventional EF system.

the bottom electrode are not promptly distributed throughout the effluent. Additionally, there exists a potential risk of flock ruptures, which can compromise the efficiency of the electroflocculation process.

Efficiently removing oil, dyes, heavy metals, and small particles from various wastewater streams, EF finds applications in diverse sectors, such as industrial water treatment, mineral recovery, and biodiesel wastewater treatment (de Oliveira da Mota et al., 2015; Kyzas & Matis, 2016; Mook et al., 2014; Rubio et al., 2002). It is particularly effective against highly toxic contaminants. Combining electrocoagulation (EC) with EF in a unified system enhances effluent treatment. Unlike EF, EC utilizes sacrificial anodes like Al and Fe, generating coagulant agents that bind to contaminants, subsequently removed by EF (Mook et al., 2014).

The basic principles of EF closely resemble those of flotation, with the key distinction lying in the method of bubble generation, achieved through electrochemical processes during water electrolysis. In this process, the electrodes, serving as cathodes and anodes, play a crucial role. While stainless steel is commonly employed as a cathode (Alam & Shang, 2017; da Cruz et al., 2016), other materials like nickel (Janssen et al., 1984) and titanium (Qin et al., 2013) can also be utilized. Various materials serve as anodes, including graphite (Maksimov & Ostsemin, 2015) PbO_2 (Hosny, 1996), Pt (Sarkar et al., 2010), and dimensionally stable anodes (DSA). During electrolysis, electrochemical reactions occur, resulting in the production of oxygen and hydrogen at the anode and cathode surfaces, respectively (reactions 1 and 2).

$$2H_2O \rightleftharpoons O_2 + 4H^+ + 4e^- \quad (1)$$

$$2H_2O + 2e^- \rightleftharpoons H_2 + 2OH^- \quad (2)$$

Bubble generation through electrolysis involves three distinct phases: nucleation, growth, and detachment (Khosla et al., 1991; Venkatachalam, 1992). The initiation of bubble nucleation is directly impacted by the magnitude of the current density administered. At the onset of nucleation, dissolved gas accumulates on the electrode surface until it reaches a critical state of supersaturation, triggering a physical transformation (da Cruz et al., 2016). In instances of low current density, the dissolved gas disperses from the electrode surface and diffuses through the liquid medium via molecular diffusion and convection, preventing the formation of a gas phase. Similarly, at elevated current densities, the dissolved gas undergoes rapid and homogeneous reaction, impeding nucleation. Therefore, an appropriately high current density is imperative to facilitate bubble formation at predetermined nucleation sites on the electrode surface (Venkatachalam, 1992).

The rate of mass transfer of dissolved gas governs the duration of bubble formation on the electrode surface. This stage involves the interplay of two competing processes: firstly, the generation of dissolved gas molecules through chemical reactions taking place on the electrode's surface, and secondly, the diffusion of these molecules towards the bulk solution. The authors examined how changes in current density influenced the concentrations of dissolved gas on the electrode surface and the timing of nucleation (Tawfik & Diez, 2014).

As a whole, electrochemical flotation (EF) mirrors conventional flotation methodologies but diverges in its bubble generation process, driven by electrochemical reactions during water electrolysis, where electrodes serve as pivotal components. While stainless steel remains a common cathode material, alternatives like nickel and titanium are viable options, alongside an array of anode materials including graphite, PbO_2, Pt, and dimensionally stable anodes (DSA). Bubble formation through electrolysis unfolds across nucleation, growth, and detachment phases, with the critical onset of nucleation heavily influenced by the magnitude of current density. The duration of bubble formation hinges on the rate of dissolved gas mass transfer, dictated by the balance between gas generation and diffusion processes, shedding light on the intricate dynamics underlying EF systems (Martínez-Huitle et al., 2018).

Bubble size in electrolysis-driven flotation (EF) is intricately influenced by multiple factors, each playing a significant role in shaping the characteristics of the bubbles produced. One of the primary advantages of EF over conventional flotation methods lies in its ability to generate small, finely dispersed bubbles, typically ranging from 8 to 15 μm in diameter. This size range is notably smaller than the average diameter of bubbles formed in conventional flotation processes, which typically fall within the range of 0.9 to 2.0 mm. The smaller size of bubbles in EF is particularly advantageous as it enhances flotation efficiency by facilitating more effective particle-bubble collisions (Matis & Peleka, 2010).

The control over bubble size and distribution in EF is achieved through careful manipulation of various parameters. Firstly, the choice of electrode material plays a crucial role, with materials like stainless steel, nickel, and titanium being commonly utilized for cathodes, and materials like graphite, PbO_2, Pt, and dimensionally stable anodes (DSA) for anodes (Alam & Shang, 2017; Casqueira et al., 2006; da Cruz et al., 2016; Ghernaout et al., 2014; Sarkar et al., 2011). The surface morphology of the electrodes, their curvature, and surface treatment also contribute to bubble size control. Interestingly, the electrode diameter has a direct impact on bubble size, with larger diameter electrodes generally producing larger bubbles. This relationship between electrode size and bubble size is particularly evident when other parameters, such as current density and pH are held constant.

Speaking of current density, its effect on bubble size is multifaceted. While some studies have observed an increase in bubble diameter with higher current densities due to bubble coalescence, others have reported the opposite effect (Sarkar et al., 2010; Venkatachalam, 1992). Moreover, the surface characteristics of the electrodes also play a role, with smoother surfaces generally leading to larger bubbles at higher current densities, while rougher surfaces may exhibit the opposite trend. pH also exerts a significant influence on bubble size, with hydrogen bubbles tending to be larger in acidic environments compared to neutral or alkaline mediums. In contrast, oxygen bubbles typically exhibit larger sizes in neutral or alkaline mediums. This pH-dependent variation in bubble size underscores the importance of pH control in EF processes to achieve desired flotation outcomes. Additionally, bubble size is subject to variation based on factors, such as bubble position, reaction time, dissolved gas concentration profile, and electrode surface treatment. This variability contributes to the wide range of reported bubble sizes observed across different studies and experimental setups (Sarkar et al., 2010).

Efficient dispersion of small bubbles into the wastewater flow is crucial for optimizing EF efficiency. Rapid dispersion of bubbles facilitates the prevalence of tiny bubbles throughout the wastewater, thereby enhancing flotation efficiency by promoting effective particle-bubble interactions. Conversely, systems where bubble dispersion is delayed may experience increased coalescence and the formation of larger bubbles, thereby diminishing the availability of effective small bubbles and consequently reducing flotation efficiency (Chen et al., 2002). Therefore, the design and operation of EF systems should prioritize mechanisms that ensure rapid and effective dispersion of bubbles into the wastewater flow to maximize flotation efficiency and overall process performance.

In electrochemical treatment of contaminated waters, pollutants are separated from water by attaching to tiny bubbles of hydrogen and oxygen and floating to the surface. This process is one facet of the electro-fenton (EF) treatment, which also involves other complex mechanisms. For instance, in the presence of chloride ions, highly oxidizing species like chlorine gas, hypochlorous acid, and hypochlorite are formed (reactions 3 to 5). These agents oxidize organic pollutants into CO_2, H_2O, and inorganic acids. Additionally, atomic oxygen generated at the anode (reaction 1) acts as an oxidizing agent for organic compounds, carried from the anode by convection.

$$2Cl^- \rightarrow Cl_2 + 2e^- \tag{3}$$

$$Cl_2 + H_2O \rightarrow HOCl + Cl^- + H^+ \tag{4}$$

$$HOCl \rightarrow OCl^- + H^+ \tag{5}$$

During treatment, effluent passes between electrodes of opposite charges, creating an electric field that promotes the coagulation of particulate matter without chemical reagents, through negative and positive particle agglomerates (Matis & Peleka, 2010). Utilizing sacrificial electrodes, such as aluminum or iron, further allows in situ generation of coagulants. At the cathode, hydrogen gas production aids the flotation of flocculated particles. Thus, the treatment combines initial electrocoagulation with an EF step (Chen, 2004). For effective EF treatment, the fluid must exhibit certain characteristics like adequate conductivity to minimize electrolyte use and floatable suspended matter; otherwise, an additional coagulation step is required. Typically, EF follows a first-order kinetic model, which depicts the rate of flotation over time (Matis & Peleka, 2010).

The EF process offers several advantages over traditional flotation methods, including rapid and easy control of operating conditions, reliable and safe equipment, and independence from the ionic strength of the effluent. Additionally, it facilitates the recovery of valuable by-products, such as proteins, fats, oils, and metals from waste streams without additional chemical reagents (Matis & Peleka, 2010). The high transport efficiency, control over bubble size and distribution through current density adjustments, and the stability of flocs due to the absence of mechanical agitation are notable benefits. The process also supports high temperatures, minimal chemical use, and possible anodic disinfection depending on the electrolyte used. Moreover, EF requires simple, compact equipment with few accessories. However, it has limitations, including the need for DC power, conductivity in effluents, restricted tank heights (<1 m), potential electrode corrosion, and reduced effectiveness with high specific gravity materials (Mook et al., 2014; Tawfik & Diez, 2014).

The application of EF as wastewater treatment is described below:

Industrial wastewater treatment: Industrial wastewater poses significant challenges due to its diverse pollutant composition, including high COD, BOD, toxicity, and intense coloration. Electrolytic flotation (EF) offers a non-selective approach to enhance effluent degradability, facilitating subsequent biological treatment or complete contaminant mineralization. Various studies (Belkacem et al., 2008; García-Montoya et al., 2015; Mansour & Chalbi, 2006; Nahui et al., 2008; Zouboulis & Matis, 2012) have investigated EF applicability for different industrial wastewaters, yielding promising results. For instance, Liakos and Lazaridis (2014) examined melanoidin degradation using EF, achieving up to 90% removal in 120 min. with aluminum electrodes. Skender et al. (2010) focused on removing Tween 20 surfactant, attaining a maximum removal of 74.79% after 30 minutes using stainless steel plates. Additionally, Gonzales et al. (2013) explored EF for treating small particle contaminants from mining activities, demonstrating efficient bubble formation with higher current densities and bacterial concentrations.

Heavy Metals in Wastewater: Heavy metals, notorious for their toxic nature and high solubility, pose significant environmental risks, often stemming from industrial activities. Studies have extensively explored electrolytic flotation (EF) for treating heavy metal-contaminated wastewater, emphasizing its efficacy even at low concentrations. de Oliveira da Mota et al. (2015) demonstrated simultaneous removal of up to 97% of pollutants, including Pb, Ba, and Zn, under specific EF conditions. Moreover, the addition of flocculants has been investigated to enhance EF efficiency, particularly for aggressive contaminants like Cr^{3+}. Perfil'eva et al. (2016) observed a significant increase in Cr^{3+} particle size and 99% extraction with the addition of flocculants. However, the choice of flocculant type is crucial, as nonionic flocculants may hinder heavy metal adhesion to gas bubbles, affecting removal efficiency. Such studies underscore the importance of optimizing treatment conditions for effective heavy metal removal.

Organic and Inorganic Compound Removal: Phosphate ions in wastewater are linked to the growth of bacteria, fungi, and algae, posing risks to water potability. Ricordel et al. (2010)

demonstrated a mild reduction in phosphate, nitrate, hardness, and suspended solids through EF combined with electrocoagulation, reducing the need for biocides and enhancing groundwater safety. Nitrate, associated with health risks, necessitates effective treatment methods to mitigate methemoglobinemia and hypertension. Majlesi et al. (2016) highlighted electrocoagulation/electroflotation as a promising approach for nitrate removal, achieving up to 96% removal from solutions containing 100 mg/L. pH adjustments in the system enable control over co-product formation. Refractory pollutants like ethylenediamine tetraacetic acid (EDTA) and cyanide require specialized treatments to enhance biodegradability. Khelifa et al. (2013) found that electrolyte composition significantly influences EDTA degradation, with chloride medium promoting substantial degradation due to active chlorine species. Combining electro-chlorination with EF proved effective in degrading EDTA and freeing metal ions for subsequent EF treatment, particularly beneficial for effluents from metal finishing industries.

In conclusion, EF emerges not only as a potent tool in wastewater treatment but also as a versatile technology with applications across diverse sectors. Its capability to generate fine bubbles through electrolysis renders it invaluable in numerous industries beyond wastewater treatment. In mining, EF aids in the separation of small particle contaminants, while in mineral processing, it facilitates ore beneficiation and the recovery of valuable metals from industrial wastewater. Environmental remediation efforts benefit from EF's ability to treat contaminated soil and groundwater. Moreover, EF plays a crucial role in purifying drinking water, ensuring access to safe and potable water in urban and rural settings alike. In research laboratories, EF finds utility in tasks, such as bubble size measurement and analysis. As EF continues to evolve, advancements in electrochemistry, including the development of suitable electrode materials and enhanced EF cell design, pave the way for its broader industrial application. Therefore, beyond its effectiveness in wastewater treatment, EF stands as a multifaceted technology poised to address diverse challenges and contribute to sustainability across various sectors.

2.1.6 Flotation by colloidal gas aphrons

Aphrons, coined by Sebba (1971), are stable bubbles or droplets encompassed by a thin surfactant film. The term "aphron" derives from the Greek word for foam, reflecting its foamy nature. These structures, also known as biliquid foams (Sebba, 1971, 1979), consist of a spherical core, whether gas or liquid, enveloped by a surfactant shell. Within this shell, the hydrophobic ends of surfactants face inward toward the core, while the hydrophilic ends extend into a second phase, typically water, serving as an aphron stabilizer alongside a thickening agent. Moreover, surfactant molecules in the outermost boundary extend into a third outer phase, further stabilizing the structure. This outer phase prevents the coalescence of neighboring aphrons (Sebba, 1971, 1979).

Colloidal gas aphrons (CGAs), termed microfoam systems by Sebba (1987), are characterized by closely packed spherical bubbles ranging from 10 to 100 μm, classified as kugelschaums or 'ball foams' (Larmignat et al., 2008). These structures consist of a gaseous core surrounded by a thin multi-layered aqueous surfactant shell, as illustrated in Fig. 2.7 (Larmignat et al., 2008). CGAs are typically generated using techniques,

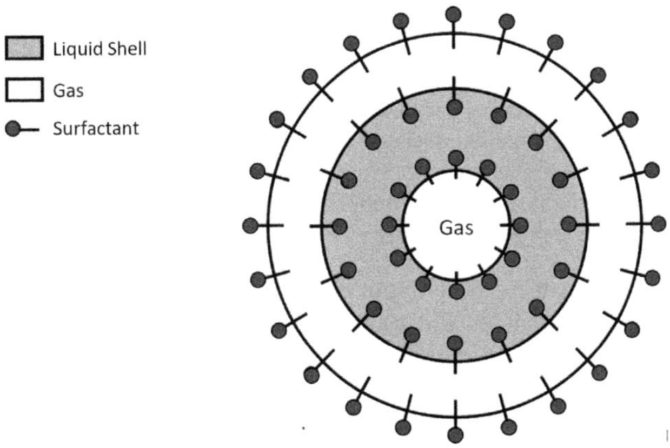

Figure 2.7 Illustration of the proposed model of colloidal gas aphrons.

such as stirring in a baffled vessel with a high-speed spinning disk impeller or through sonication and homogenization (Bjorndalen et al., 2014; Sebba, 1971; Xu et al., 2008). The properties of CGAs, including their large interfacial area, relatively high stability due to the multilayer surfactant film, and similar flow properties to pure water, make them suitable for various applications (Jauregi et al., 1997; Larmignat et al., 2008). With up to 65% air volume and a specific gravity of approximately 0.35, CGAs exhibit remarkable compressive strength, surviving pressures of at least 27.3 MPa (Bhatia et al., 2005; Hashim et al., 2012). These properties have led to diverse applications, such as metal and dye removal from aqueous solutions, protein recovery, flotation of cells and minerals, and use as carriers or mass transfer agents in various processes, including in-situ biodegradation and dust collection (Ciriello et al., 1982; Hashim et al., 2012; Jauregi et al., 1997). Characterization of CGAs primarily focuses on stability, gas holdup, and size distribution.

The properties of CGAs which are important for the effectiveness of the process are described below:

Stability: CGAs exhibit higher stability compared to conventional foams due to their multilayer structure. The momentum required to break the barrier of six surfactant-stabilized interfaces in CGAs is higher than that for conventional foams, resulting in enhanced stability. Stability is measured in terms of liquid drainage rate, often following first-order kinetics (Bhatia et al., 2005; Sebba, 1971, 1979). The half-life (t) is defined as the time required for half the original liquid volume to drain.

Buoyancy and separation: The buoyancy of encapsulated gas allows for easy separation of the aphron phase from the bulk liquid phase without requiring centrifuges. Stirring CGAs at a rate ensuring lateral movement greater than upward buoyancy avoids creaming when needed.

Adherence to particles and molecules: CGAs' encapsulating shell facilitates adherence of particles and molecules. Surface properties can be modified using different surfactants, resulting in outer surfaces with varying charges (anionic, cationic, or non-charged). This enables selective adsorption by altering the surfactant type.

Low viscosity: CGAs exhibit low viscosity similar to water, facilitating easy pumping regardless of surfactant type. This advantageous property remains as long as the gas hold-up stays below 65%. CGAs retain consistent flow properties during pumping, unlike conventional foams, which undergo changes due to the elastic nature of bubbles.

Small bubble size and large interfacial area: CGAs feature small bubble sizes, leading to a larger interfacial area per unit volume. This characteristic enhances their capacity for adsorption of molecules and fine particles. Bubble size varies based on surfactant concentration, ionic strength, and the presence of other molecules or particles.

These properties enable various applications and have been investigated extensively by researchers (Larmignat et al., 2008; Tseng et al., 2006). Additionally, CGAs have been utilized in industrial applications, such as in well-drilling fluids by MASI Technologies LLC, benefiting from their improved shear thinning properties (Brookey & House, 2002).

Conventional flotation processes involve sparging air into a solution containing the target compound (colligend), where the compound adheres to gas bubbles and is subsequently removed by flotation. Surfactants are often added to enhance removal efficiency, sometimes causing precipitation of the colligend followed by flotation, known as precipitate flotation. CGAs, a more advanced approach, have been introduced as an improvement over conventional methods, offering benefits. such as smaller bubble sizes, maximizing interfacial area, and enhanced stability (Noble et al., 1998; Sebba, 1971, 1979).

Subramaniam et al. (1990) pioneered the use of colloidal gas aphrons (CGAs) for clarifying industrial effluents, employing CGAs generated from anionic (Aerosol OTTM, SDBS [sodium dodecylbenzene sulfonate], LUX flakes) and cationic (BDHA [benzyldimethylhexadecylammonium]) surfactants. They investigated various operating variables, such as pH, temperature, air to solid ratio, and solid concentration, finding that CGAs produced from ionic surfactants were most effective at pH values close to the respective p*K*a values. Hashim et al. (1995a) utilized CGAs in a flotation column to recover yeast from aqueous solutions, employing a cationic surfactant solution of BHDC. They found that CGA flow rate was critical in controlling process efficiency, with high separation efficiency achieved at high CGA flow rates and relatively low residence time of the feed. Hashim et al. (1998) further investigated the clarification of baker's yeast suspension by CGAs prepared from benzyldimethyl hexa-decylammonium chloride (BHDC), optimizing parameters, such as CGA flow rate, air flow rate, and yeast concentration to achieve up to 95% separation efficiency. They also elucidated the mechanism of particle-bubble attachment as monolayer adsorption of yeast cells on the CGAs.

Basu and Malpani (2001) employed colloidal gas aphrons in a flotation column to remove methyl orange and methylene blue dyes from water. They investigated various operating variables, including surfactant type and concentration, CGA diameter and gas holdup, flow rate, residence time, pH, and salt concentration. Dye removal ranged from 95% to 98%, with the efficiency influenced by the surfactant charge. Alkaline conditions favored methyl orange removal when using hexadecyltrimethyl ammonium bromide (HTAB)-generated CGAs, while acidic conditions were more effective for methylene blue removal with SDBS-generated CGAs. Increasing CGA flow rate and gas holdup, and decreasing CGA diameter and residence time, improved dye removal efficiency.

The utilization of colloidal gas aphrons (CGAs) across various applications demonstrates their effectiveness in industrial processes. Subramaniam et al. (1990) pioneered CGA usage for clarifying effluents, highlighting the importance of ionic surfactants and operational variables. Hashim et al. (1995a) and Hashim and Gupta (1998) showcased CGAs' potential in yeast recovery and pulp separation, respectively, emphasizing flow rate optimization. Basu and Malpani (2001) demonstrated CGAs' efficacy in dye removal, revealing the influence of surfactant charge and process parameters. Mansur et al. (2006) further underscored CGAs' versatility by achieving significant removal of fine particles, illustrating the impact of surfactant type and concentration. These findings collectively highlight CGAs' promising role in enhancing industrial processes through efficient particle separation and wastewater treatment.

Roy et al. (1992) reported successful removal of synthetic dyes from wastewater using CGAs, achieving removal rates exceeding 95%. Chaphalkar et al. (1994) compared CGAs with conventional flotation techniques for pentachlorophenol removal, highlighting minimized solute redistribution due to CGAs' gas dispersion, which reduces processing time and enhances efficiency.

Caballero et al. (1989) compared CGAs with conventional solvent-sublation processes for methyl orange removal, noting significantly reduced processing time with CGAs while maintaining high removal yields.

CGAs have been utilized for clarifying palm oil mill effluents and flotation of cellulose fiber from paper mill wastewater, achieving successful removal rates of solids and microbial cells (Hashim & Sen Gupta, 1998; Subramaniam et al., 1990). The efficiency and speed of recovery processes were enhanced compared to conventional flotation techniques.

In conclusion, CGAs, with their unique structure of multilayer surfactant films and small, stable bubbles, provide enhanced interfacial area and stability, which are critical in processes, such as mineral and particle separation, and water purification. The remarkable stability of CGAs, as highlighted in the text, is a direct result of their ability to withstand significant mechanical stresses, which conventional foams cannot. This enhanced stability is not only crucial in maintaining the integrity of the bubbles but also in ensuring consistent and reliable outcomes in separation processes. The small bubble size further augments the efficacy of CGAs by increasing the surface area available for interaction with target molecules or particles, thereby maximizing the efficiency of separation and adsorption processes. Furthermore, the versatility of CGAs is evident in their varied applications, from mineral processing to dye removal and water purification. The studies cited demonstrate significant improvements in efficiency and processing times when using CGAs. For instance, the rapid and efficient removal of heavy metals and synthetic dyes from aqueous solutions underscores CGAs' capability to handle a range of substances with high recovery rates, often surpassing traditional methods. In environmental applications, CGAs show immense potential in addressing pollution and recovery of valuable materials from industrial effluents. Their ability to operate under varied conditions—adjustable through changes in surfactant type, concentration, and operational parameters—allows for tailored solutions to specific environmental challenges. This adaptability is crucial in applications like the recovery of yeast cells from solutions or fine particles from industrial wastewater, where conventional methods may fall short. The exploration and utilization of colloidal gas aphrons represent a promising frontier in both industrial and environmental engineering. By harnessing the unique properties of CGAs, industries can achieve more sustainable practices, showcasing an innovative approach to solving traditional problems of separation and purification processes. The continued investigation and adaptation of CGA technology will undoubtedly enhance its applicability and effectiveness, cementing its role as a pivotal technology in modern industrial applications.

2.2 Sedimentation

2.2.1 Solid-liquid separation

Sedimentation is basically a unit operation that provides separation by utilizing the force of gravity. Some researchers use the term of sedimentation instead of settling (Earle, 1983; Heldman & Singh, 1981) while others define sedimentation as a settling process that provides liquid-solid separation (Saravacos & Kostaropoulos, 2002; Sorsamäki & Nappa, 2015). According to Sorsamaki and Nappa (2015) settling is used for liquid-liquid (emulsion), liquid-gas (aerosol), solid-liquid (suspension) and solid-gas (smoke) separation, while sedimentation is used for solid-liquid separation (Coe, 1916; Concha & Bustos, 1991; Earle, 1983; Geankoplis, 1993; Heldman & Singh, 1981; Ortega-Rivas, 2012b; Parsapour et al., 2014; Saravacos & Kostaropoulos, 2002; Sorsamäki & Nappa, 2015; Talmage & Fitch, 1955). In addition, some other researchers have included both the force of gravity and centrifugal force in the definition of sedimentation (Heldman & Singh, 1981). In general, sedimentation could be defined as a unit operation used to separate solid or liquid droplets from a fluid, which can be gas or liquid, by the force of gravity. It is used in the food industry for cleaning/purification purposes, where particles are separated from a fluid, for thickening purposes, where particles, such as starch or sugar crystals are obtained as products, or for classification purposes, where particles are separated into fractions of different size or density (Geankoplis, 1993). Sedimentation could also be used for the solvent extraction in immiscible liquid-liquid systems to increase the mass transfer coefficient by the relative motion of the particles and liquid (Ortega-Rivas, 2012b).

Gravity sedimentation is extensively utilized in treating drinking water, industrial water, and wastewater (Green & Perry, 2007). Regarding the food and beverage industry; it can be used to obtain clear fruit juice by separating pulp (Téllez-Morales et al., 2021), to seperate cream from milk (Raikos et al., 2009), to remove sediment from butter (Patel et al., 2019), to clarify wine and beer by removing the suspended particles and yeast (Powell et al., 2004; Valdés et al., 2011; Vidgren & Londesborough, 2011), to remove impurities and sediment from vegetable oils (Hermann et al., 1999), from sugar syrups or from brines that are used in red meat and poultry production (Litchfield, 1982).

When used for cleaning/clarification or thickening purposes, a suspension is separated into two phases, a clear liquid and a thick slurry containing a high concentration of solids. Sedimentation is mainly influenced by two critical factors that depend on the concentration of the feed stream; the distance of a particle from the boundaries of the vessel and the distance from other particles. According to these parameters, sedimentation takes place as free or particulate settling and hindered or zone settling (Ortega-Rivas, 2012b).

Free settling occurs when a particle is sufficiently far from the walls of the vessel and other particles that the fall of the particle is not affected. Interference is reported to be less than 1% when the ratio of particle diameter to vessel diameter is less than 1:200 or the concentration of the solution is less than 0.2% (w/v). The settling velocity is only influenced by particle diameter, density, liquid viscosity, and gravitational force during a free settling. On the other hand, hindered settling occurs when particles settle at a lower rate when the concentration of the solution is high (Geankoplis, 1993). The particles fall together as a zone in this regime, resulting in a distinct clear water layer. Saravacos and Kostaropoulos (2002) described a third mechanism where particles are compressed by gravitational force generating a compressed bed at the bottom which is called compression regime.

The free settling approach tends to remain mostly theoretical because it cannot be used in real applications, such as industrial sedimentation, because of the effect of particles on each other. However, the approach could be used in the measurement of particle size distribution or particle classification (Ortega-Rivas, 2012b).

There are three forces acting on a free settling particle in a liquid: the downward force of gravity, the upward force of buoyancy due to the density difference between the liquid and the particle, and the upward force of friction acting in the opposite direction of particle motion (Geankoplis, 1993). The friction coefficient depends on the size of the particle, its shape (spherical or non-spherical) and whether it is rigid or non-rigid.

The fall of a rigid sphere consists of two periods, an accelerated fall and a fall at constant speed. The initial acceleration time is of the order of a tenth of a second, so the period of constant velocity fall, called terminal velocity, is important (Geankoplis, 1993). The equation for terminal velocity which is derivatized using the equations of acting forces is given below (Geankoplis, 1993):

$$v_t = \sqrt{\frac{4(\rho_P - \rho)gD_P}{3C_D\rho}} \tag{2.7}$$

where υ_t (m/s) is the terminal velocity, ρ_p (kg/m^3) is the density of particle, ρ is the density of liquid, g is (9.80665 m/s^2) gravitational constant, and D_p (m) is the diameter of the spherical particle and C_D is the drag coefficient. For turbulent flow, the drag coefficient is constant at $C_D = 0.44$. In the region of laminar flow or Stokes' law, the friction coefficient equation is substituted so that the terminal velocity becomes a function of the gravitational force, particle diameter and density, and the viscosity of the fluid, which is shown in the Eq. 2.1 (Geankoplis, 1993).

If the particles are a few μm in size or less, the Brownian motion is present, which is a random movement of the particle resulting from the collisions between the molecules of fluid surrounding the particle and particles, which tends to suppress the effect of gravity. In such a situation, the centrifugal force is utilized to precipitate the particles. When the particle diameter is larger, the walls of the vessel cause a retarding effect on settling and the terminal velocity is multiplied by the appropriate coefficients calculated using the diameters of the particle and the vessel, which are different for laminar and turbulent flows (Geankoplis, 1993).

In hindered settling, which occurs in suspensions consisting of a large number of particles, the surrounding particles interfere with the motion of individual particles (Geankoplis, 1993). In this regime, particles fall together leaving a distinct clear layer of water on the top (Saravacos & Kostaropoulos, 2002). The settling velocity is lower than that calculated by Stoke's law due to the effect of the true drag force caused by the particles in the suspension interfering each other. The effective viscosity μ_m is calculated by dividing the viscosity μ of the suspension by an empirical correction factor ψ_p which depends on ε, the volume fraction occupied by the liquid of the slurry mixture (Geankoplis, 1993). Earle (1983) indicated that for the highly concentrated suspensions, when the particle size range is not larger than 10:1, all particles settle at the same velocity. This velocity lies somewhere between the velocities of the smallest and largest particles in the suspension calculated by Stoke's law. In practice, when Stokes' law or its extensions cannot be applied, experiments are required to determine the velocity.

Sedimentation can be carried out in various equipment, such as thickeners, clarifiers, or classifiers (Ray, 1995). Thickeners and clarifiers are commonly employed for pre-concentration of liquid feeds before filtering due to their cost-effectiveness when dealing with large volumes of liquid. According to their mode of operation, these equipments can be divided into batch-operated settling tanks and continuously operated clarifiers and thickeners (Ortega-Rivas, 2012b). The use of batch tanks in food industry has declined considerably, but they are still used when small quantities of water or suspensions need to be treated (Ortega-Rivas, 2012b). In food processing plants, smaller metallic settling tanks, designed on similar principles as larger installations, are used for clarification or thickening of liquid suspensions. These cylindrical tanks have a conical bottom that are fed with a water suspension from the center, while sludge is removed from the bottom and clean water overflows from the top (Saravacos & Kostaropoulos, 2002). Generally, the feed concentration for a thickener is higher than that for a clarifier. Some types of equipment that can perform both thickening and clarification are designed to accomplish both operations in a single stage (Ortega-Rivas, 2012b). The circular pool continuous thickener, deep cone thickener and lamellar thickeners are some examples to alternative designs modified to increase the capacity of commercial thickeners (Ortega-Rivas, 2012b). A series of continuous thickeners are also used for a counter-current washing, where the slurry is continuously fed back to remove the mother liquor or to wash soluble materials from the solids (Ortega-Rivas, 2012b).

Ortega-Rivas (2012b) stated that a typical thickener has three operating layers that are clarification, settling and compression. The settling velocity of water suspensions is determined through laboratory tests in long or short cylindrical tubes. Earle (1983) explained the settling behavior observed in a cylinder as follows: The clarification zone appears as a clear liquid at the top. Below this zone is a region of almost constant composition, since the settling rate of particles of different sizes will be the same. If the particle size is large, this zone is replaced by a longer zone of variable composition. At the bottom is the sediment zone where Ortega-Rivas (2012b) defined as compression zone.

Sedimentation tanks should meet the qualifications in terms of both depth and width: The depth should be high enough to allow even the smallest particle to fall through. The width should also be large enough to ensure that the settling velocity of the particle is greater than the upward velocity of the liquid, which is important in hindered settling (Saravacos & Kostaropoulos, 2002). The cross-sectional area and diameter of the tank are calculated from the feed rate (kg/h) and the assumed load. Depending on the concentration of the suspension, the tank load can vary from 0.3 to 3 m^3/m^2h with a residence time of 10 hours (Saravacos & Kostaropoulos, 2002). Saravacos and Kostaropoulos (2002) reported that depths of the tanks are in the range of 3 to 5 m. Cylindrical tanks usually have diameters ranging from 10 to 50 m, while rectangular tanks can be 5 to 10 m wide and 80 m long (Saravacos & Kostaropoulos, 2002). The tanks usually also have a raking mechanism placed on a center column that moves the sediment with both vertical and horizontal movement and sends it to the center of the tank where it is pumped away (Ortega-Rivas, 2012b; Saravacos & Kostaropoulos, 2002).

Cross sectional area and the depth are important parameters of the tank due to requirements of three zones mentioned above. When the purpose is thickening, the area requirements of the settling and compression layers are important and the design is made according to which compression layer needs to be the largest. However, if the process also has a clarification purpose, the clear layer needs the largest area. On the other hand, the depth is only important for the compression layer as the static pressure and detention time will determine the concentration of the flow. The depth of the other two layers is only a matter of practical considerations (Ortega-Rivas, 2012b).

Given the large volumes of water involved, sedimentation tanks are designed based on laboratory tests and practical experience (Green & Perry, 2007). There are also some mathematical approaches which is used in designing the tanks. Coe-Clevenger and Talmage-Fitch approaches are the most commonly used approaches for designing thickeners. Coe (1916) introduced a method for determining the required area in settling-based thickeners by assuming that the settling velocity is only a function of the solid concentration. They established the following equation to determine the unit area of thickener:

$$U \cdot A = \left(\frac{1}{C} - \frac{1}{C-u}\right)\frac{1}{R \times \rho l} \tag{2.8}$$

This method is based on the hypothesis that there is a critical flux rate that determines the required area in concentration between the actual feed and the desired underflow concentrations. A series of tests are carried out at different concentrations and the results obtained are plotted against the underflow concentration, the largest area is used in the design of the thickener. The settling velocity is determined as the initial tangents of the curve of the height of the interface between the settled solids and clarified liquid versus time (Parsapour et al., 2014).

Talmage and Fitch (1955) applied Kynch's mathematical model (Kynch, 1952) for thickener design which is based on the free settling regime where flocs settle without receiving any impacts from other flocs. They used single batch settling test and observe settling velocity and local concentration form the gradient and y-intercept of the tangent constructed to the settling curve. The unit areas can be calculated using the Eq. 2.8 and the maximum unit area is chosen for designing the thickener. The Talmage-Fitch method may be more advantageous as the tests more closely resemble the actual thickening process in which flocs are formed (Parsapour et al., 2014).

Medina (2019) described Kynch model as a kinesthetic model applicable to ideal suspensions with homogeneous small particles and no mass transfer between particles. They also indicated that, the model assumes that the dispersion and cross-sectional area are constant and cannot accurately describe suspensions with flocculation (Concha & Bustos, 1991; Medina, 2019). Bürger et al. (2011) developed a mechanistic model consisting of a nonlinear partial differential equation to accurately describe the settling of sludge-like materials. The model includes one-dimensional, quadratic equation in which the solids concentration depends on depth and height.

Earle (1983) calculated the area required for a continuous thickener by equating the sedimentation rate in a given region to the reverse flow rate of the rising liquid as follows;

$$v_u = (F - L)(\mathrm{d}w/\mathrm{d}\theta)/A\rho \tag{2.9}$$

where υ_u is the upward flow rate of the liquid, F is the ratio of liquid mass to solid mass in the feed, L is the ratio of liquid mass to solid mass in the underflow liquid, $dw/d\theta$ is the mass rate of feed of the solids, ρ is teh density of the liquid, A is the settling area of the tank.

If the sedimentation rate is υ, then $\upsilon_u = \upsilon$ and, therefore

$$A = (F - L)(\mathrm{d}w/\mathrm{d}\theta)\nu\rho \tag{2.10}$$

To facilitate sedimentation of colloidal and challenging-to-settle particles in water, various flocculating agents, such as alum or some polymeric materials can be used to form fast settling agglomerations (Saravacos & Kostaropoulos, 2002). These substances neutralize the electric charges of the particles, preventing them from remaining dispersed by repelling each other (Sorsamäki & Nappa, 2015).

In summary, solid-liquid separation with sedimentation has a wide application area in food industry, such as clarification, thickening and classification. The settling velocity of a particle during sedimentation is influenced by many factors, such as particle concentration, particle shape, size of the vessel. The design of sedimentation tanks may require laboratory tests and mathematical modelling.

2.2.2 *Liquid/liquid separation with demulsification*

Demulsification is a technique in which a demulsifier is used to separate water from a crude oil emulsion by causing the water droplets of the emulsifier to agglomerate (Mohshim et al., 2023). Crude oil, a vital natural resource, originates from organic matter buried deep within the Earth's crust over millennia (Umar et al., 2018). With society's increasing demand for energy and fuel derivatives, crude oil exploration and production have thrived for over a century (Abdulredha et al., 2020; Abdulredha et al., 2020). Extraction involves a multi-stage process—primary, secondary, and tertiary—wherein initial recovery relies on underground pressure to bring crude oil to the surface (Alagorni et al., 2015). However, as underground pressure diminishes, recovery efficiency declines, necessitating secondary methods like injecting water and gases, such as CO_2, CH_4, or N_2 to enhance recovery rates (Pashakolaie et al., 2015).

These secondary techniques, improving recovery to approximately 50%, still leave a significant portion of crude oil unrecovered. Thus, tertiary methods, known as enhanced oil recovery (EOR), are employed to access remaining reserves (Zolfaghari et al., 2016). EOR encompasses thermal, miscible gas injection, and chemical methods, each tailored to different reservoir characteristics and crude oil types.

As reservoirs age, water content increases, resulting in the formation of crude oil emulsions—a costly challenge for the industry due to pipeline corrosion, increased viscosity, and catalyst poisoning (Abdulredha et al., 2020; Abdulredha et al., 2020). These emulsions, consisting of water-in-oil (W/O), or oil-in-water (O/W) configurations, must be treated to meet production standards and mitigate environmental risks.

Crude oil is seldom devoid of water, with the water cut (WC), defined as the volume of water divided by the total volume of fluids produced, often increasing over an oil well's lifespan, with mature wells reaching a WC of over 90% (Pal, 1994). Due to the surface-active nature of resins and asphaltenes in crude oil, water-oil mixtures typically manifest as water-in-oil (W/O) emulsions, where water forms the dispersed phase and oil the continuous phase. These emulsions may originate from the reservoir structure or processing after extraction, caused by turbulence and pressure drop (Sjöblom et al., 2003). To meet international standards, separating oil from water is crucial. Techniques, like freezing, hydrocyclones, and gravity settlers are employed, with the latter being the most common due to its simplicity and cost-effectiveness. Gravity settlers may incorporate enhancements like microwave, ultrasonic, electrostatic, chemical demulsifiers, mechanical barriers, or thermal methods (Chan & Chen, 2002; Cumming et al., 1999; Delfos et al., 2004; Eow & Ghadiri, 2002; Klasson et al., 2005; Li & Gu, 2005; Lorain et al., 2001; Meon & Blass, 1992; Rommel et al., 1992; Schlieper et al., 2004; Speth et al., 2002; Stack et al., 2005).

Gravity settlers operate on two primary mechanisms: droplet coalescence and droplet sedimentation. Coalescence involves droplet merging, while sedimentation refers to droplets settling under gravity. These processes are influenced by factors, such as droplet size, dispersed phase concentration, continuous phase viscosity, and presence of surfactants (Coons et al., 2003).

The phenomena observed during demulsification can be summarized as creaming or precipitation, flocculation, coalescence, and Ostwald ripening. The coalescence mechanism in liquid/liquid dispersions involves several critical steps, beginning with droplet approach and collision, followed by drainage of the interfacial film and film rupture. All these phenomena are briefly described below:

Creaming or sedimentation occurs due to the natural tendency of droplets to either rise or sink through a fluid under the influence of gravity or density differences. Creaming happens when the dispersed phase has a lower density than the continuous phase, causing the dispersed phase to rise to the top of the emulsion. This phenomenon is dependent on the size of the droplets; larger droplets will rise if their density is lower than the continuous phase or settle at the bottom if their density is higher. Creaming is also affected by factors, such as the rheology of the continuous phase, hydrodynamic and colloidal interactions between droplets, the electrical charge on the droplets, and the nature of the interfacial membrane or film (Jones et al., 1978).

Flocculation describes the phenomenon where the distance between dispersed phase droplets decreases, leading them to cluster together without rupturing, as the net attraction force (van der Waals) between them weakens. The rate of flocculation typically varies based on factors, such as the water cut, temperature, oil viscosity, and density difference between oil and water (Yonguep et al., 2022).

The process of droplet approach and collision in liquid/liquid dispersions involves multiple factors that influence the frequency and efficacy of collisions, which are crucial for coalescence. According to Chesters (1991), collision frequency is influenced by factors, such as the concentration of the dispersed phase and the level of turbulence. This notion is further explored by Coulaloglou and Tavlarides (1977) who propose an alternative expression that challenges the assumption that higher concentrations of dispersed phase necessarily increase collision frequency. In densely packed emulsions, where droplets are in constant contact, collision frequency often approaches, shifting the focus to drainage time as the critical limiting factor. Additionally, the force and duration of the collisions are vital, with viscous forces becoming predominant when the droplet size is smaller than the Kolmogorov length scale. The subsequent steps critical to the coalescence mechanism include the drainage of the interfacial film and the rupture of this film, which are central to the successful merging of droplets in such dispersions.

The drainage of the interfacial film is a pivotal process in determining the coalescence rate, especially in highly concentrated water/crude oil emulsions. Palermo (1991) defines drainage time as the period between the abrupt cessation of droplet movement and the rupture of the interfacial film. Complementing this, Yeo et al. (2003) proposed a model that simplifies drainage time by assuming immobile interfaces and disregarding the formation of dimples. More recent models have expanded on this understanding, providing a nuanced description of film drainage that incorporates factors, such as dimple formation and the instability of the film prior to rupture, offering deeper insights into the dynamics at play in the coalescence process.

Film rupture is a rapid process compared to the preceding stages of collision and drainage in liquid/liquid dispersions. Chesters (1991) notes that in the case of thick films, the intermolecular forces are primarily governed by interfacial tension, whereas in thin films, tangential forces that modify drainage need to be considered. Yeo et al. (2003) have further delineated this process into three distinct rupture regimes: nose rupture, film immobilization, and rim rupture, with the formation of dimples and nose rupture being most frequently discussed in the literature. These aspects of film rupture, alongside the drainage of the interfacial film, are critical in determining the coalescence rate and influencing the overall behavior of droplets within liquid/liquid dispersions.

Ostwald ripening refers to the phenomenon where larger droplets increase in size at the expense of smaller droplets. This occurs due to the mass transport of dispersed phase material between droplets, driven by differences in solubility within the continuous phase (Yonguep et al., 2022).

2.2.2.1 Thermal or conventional microwave demulsification

Thermal demulsification involves both microwave and conventional heating methods to separate water from oil in crude oil emulsions. Heating emulsions at high temperatures reduces the interfacial film's rigidity, promoting phase separation by enhancing settling rates and coalescence frequency (El-Ashtoukhy & Fouad, 2014). Microwave demulsification combines thermal and non-thermal effects, where high-frequency electromagnetic fields induce both thermal heating and droplet collision. Water molecules, with higher absorption capacity for microwaves, expand and weaken the interfacial film, aiding coalescence (Lv et al., 2020).

Microwave demulsification offers advantages like high efficiency, low energy consumption, and environmental friendliness (Mohammed & Mohammed, 2013). For instance, Mohammed and Mohammed (2013) achieved 85% demulsification efficiency using microwave irradiation, with optimal parameters of 900 W power, 50% water content, and 160 secs irradiation time. Kusumastuti et al. (2020) observed increased demulsification efficiency with prolonged irradiation and settling times.

Combining microwave and conventional heating improves efficiency, with microwave irradiation typically outperforming conventional methods (Abdurahman et al., 2017; Sun et al., 2020). Nour et al. (2012) found that microwave irradiation reduced emulsion viscosity more effectively than conventional heating, enhancing droplet coalescence. Lv et al. (2020) demonstrated that optimal microwave parameters improved oily sludge demulsification, with temperature and power influencing efficiency.

Kovaleva et al. (2021) compared microwave and radio-frequency demulsification, finding microwave more effective for high-water-content emulsions due to water molecule absorption. Ferreira et al., (2013) noted aging negatively affected demulsification efficiency, with microwave radiation less impacted due to its faster heating rate. While thermal demulsification effectively separates water and oil, excessive heat can degrade crude oil quality, making it unsuitable for all emulsions.

2.2.2.2 Chemical demulsification

Chemical demulsification involves the addition of demulsifying agents to promote the coalescence of emulsion droplets (Abed et al., 2019; Hassanshahi et al., 2020). These chemicals serve to destabilize the emulsion by neutralizing surfactants that maintain its stability, thus facilitating phase separation. Successful chemical demulsification requires careful selection of demulsifiers tailored to the specific characteristics of the emulsion, proper dosage, thorough mixing, and adequate retention time for effective phase separation. Table 2.1 provides an overview of recent significant studies on chemical demulsification of crude oil emulsions.

2.2.2.3 Demulsification by ultrasound

Ultrasound plays a crucial role in characterizing dispersions or emulsions, primarily by measuring droplet size distributions and monitoring their evolution over time. Recently, ultrasound technology has also been applied to assist in the separation of emulsions. Acoustic spectroscopy, which involves measuring ultrasound attenuation within a specific frequency range, and electroacoustic spectroscopy, which introduces an electrical field, are two prominent techniques in this field. Acoustic spectroscopy is primarily utilized for droplet size distribution measurement, while electroacoustic spectroscopy shows promise for measuring zeta potential (Frising et al., 2006).

In acoustic spectroscopy, sound pulses generated by an acoustic spectrometer pass through a sample and are then detected by a receiver. As the sound traverses the sample system, its intensity and phase change due to interactions with particles and liquid components, leading to energy losses (attenuation) and changes in sound speed. Acoustic spectrometers typically operate in the frequency range of 1–100 MHz. The effectiveness of emulsion separation using ultrasound relies on the principle that suspended droplets respond to the acoustic field, provided there is a nonzero acoustic contrast between the dispersed phase and the suspending fluid. This principle is utilized to organize droplets into thin parallel bands or to transport them using slowly moving pseudo-standing waves. Some methods induce droplet agglomeration, enabling recovery via conventional physical screening (Frising et al., 2006).

Despite its utility, ultrasound techniques face limitations, particularly in the presence of air or gas bubbles, which can significantly disrupt their reliability. This drawback may reduce their effectiveness in gravity settlers, where gaseous fractions are often present in abundance. Nonetheless, ultrasound remains a valuable tool in the characterization and separation of emulsions, with ongoing research aimed at overcoming these challenges (Yi et al., 2017).

In conclusion, the comprehensive discussion on liquid/liquid separation with a focus on demulsification elucidates the criticality of addressing emulsion challenges within the oil production sector. As crude oil extraction progresses to more complex stages, the increase in water cut and the formation of stable emulsions pose significant operational and environmental hurdles. These challenges necessitate advanced separation techniques to ensure efficient processing and compliance with environmental standards. The utilization of sedimentation and demulsification processes represents essential strategies for managing the inherent complexities of oil and water separation. Techniques, such as gravitational settling in thickeners and clarifiers, alongside more sophisticated methods like chemical and thermal demulsification, play pivotal roles in enhancing recovery rates and minimizing the environmental impact of oil production. The effectiveness of these methodologies is underscored by their ability to adapt to the varying characteristics of oil reservoirs and the evolving nature of extracted fluids. As such, demulsification not only improves the quality of crude oil, but also contributes to

Table 2.1 Overview of recent significant studies on chemical demulsification of crude oil emulsions.

Emulsion type	Demulsifier type	Demulsifier concentration	Demulsification efficiency	References
Crude Oil Emulsion	Carbonized Cotton-Coated Silica Nanoparticles ($CC@SiO_{fig}$)	400 mg/L	90.38% after 120 mins at 65°C	Yuan et al. (2021)
Water-in-Oil (W/O) Emulsion	Hydrothermally Treated Natural Lotus Leaf	1000 ppm	88.18% after 90 mins at 70°C	Ye et al. (2021)
Oil-in-Water (O/W) andand W/O Emulsions	Ethylene Oxide-Propylene Oxide (EO-PO) Block Copolymers	1, 2, and 5 ppm	Demulsification increased with dosage	Niu et al. (2019)
W/O from ASP Flooding	Novel Polyether-Polyquaternium (PPA) Copolymer	110 mg/L	80.6%	Sun et al. (2020)
W/O with Light, Medium, or Heavy Crude Oils	Alginite (Naturally Occurring Oil-Shale Rock)	0.0025 wt%	Less than 1.0% water remaining	Ahmed et al. (2021)
O/W from ASP Flooding	Silica-Supported Polyether Polysiloxane Quaternary Ammonium ($PPQA@SiO_2$)	300 mg/L	92% in 50 mins	Zhai et al. (2020)

(continued on next page)

Table 2.1 Overview of recent significant studies on chemical demulsification of crude oil emulsions—cont'd

Emulsion type	Demulsifier type	Demulsifier concentration	Demulsification efficiency	References
Crude Oil Emulsion	2-Amino-5-Dodecyl Benzene Sulfonic Acid (ADBSA) Surfactant	0.065–0.109 wt%	95%–98% at 60°C	Zargar et al. (2018)
Synthetic W/O using SW and DW	Ionic Liquids (IL) with Chloride, Decanoate, and Dicyanamide Ions	1000 ppm, 700 ppm	99% efficiency achieved	Adewunmi and Kamal (2019)
Water-in-Asphaltenes Solution Emulsion	Novel Aliphatic Alcohol Nonionic Polyether (MJTJU-2)	400 ppm of MJTJU-2	>97% after 15 mins at 60°C	Ma et al. (2020)
W/O Emulsion	SiO2 Nanoparticles (NP) and Graphene Oxide (GO-SiO_2) Nanocomposite	–	Complete (100%) demulsification achieved	Javadian and Sadrpoor (2020)
PS Produced Water Emulsion	Cationic Branched Quaternary Ammonium Chloride and Non-Ionic Copolymer of Propylene Oxide and Ethylene Oxide	350–500 mg/L	Cationic demulsifiers showed better performance	Chen et al. (2018)

the longevity and sustainability of oil production facilities. The strategic integration of mechanical, thermal, and chemical demulsification techniques offers a robust solution to the challenges posed by high water cuts and stable emulsions, thereby ensuring that oil production remains viable and environmentally responsible. Ultimately, the continuous refinement and integration of these separation technologies are vital for the industry's adaptation to the changing dynamics of oil reservoirs, and environmental regulations. This adaptive approach will facilitate more effective management of resource extraction and significantly mitigate the environmental footprint of the oil sector.

References

Abdulredha, M. M., Hussain, S. A., & Abdullah, L. C. (2020). Optimization of the demulsification of water in oil emulsion via non-ionic surfactant by the response surface methods. *Journal of Petroleum Science and Engineering, 184*. https://doi.org/10.1016/J.PETROL.2019.106463.

Abdulredha, M. M., Siti Aslina, H., & Luqman, C. A. (2020). Overview on petroleum emulsions, formation, influence and demulsification treatment techniques. *Arabian Journal of Chemistry, 13*(1), 3403–3428. https://doi.org/10.1016/J.ARABJC.2018.11.014.

Abdurahman, N. H., Yunus, R. M., Azhari, N. H., Said, N., & Hassan, Z. (2017). The potential of microwave heating in separating water-in-oil (w/o) emulsions. *Energy Procedia, 138*, 1023–1028. https://doi.org/10.1016/j.egypro.2017.10.123.

Abed, S. M., Abdurahman, N. H., Yunus, R. M., Abdulbari, H. A., & Akbari, S. (2019). Oil emulsions and the different recent demulsification techniques in the petroleum industry: A review. *IOP Conference Series: Materials Science and Engineering, 702*(1), 012060. https://doi.org/10.1088/1757-899x/702/1/012060.

Adewunmi, A. A., & Kamal, M. S. (2019). Demulsification of water-in-oil emulsions using ionic liquids: Effects of counterion and water type. *Journal of Molecular Liquids, 279*, 411–419. https://doi.org/10.1016/j.molliq.2019.02.008.

Ahmed, S. S., Hippmann, S., Roode-Gutzmer, Q. I., Fröhlich, P., & Bertau, M. (2021). Alginite rock as effective demulsifier to separate water from various crude oil emulsions. *Colloids and Surfaces A: Physicochemical and Engineering Aspects, 611*, 125830. https://doi.org/10.1016/j.colsurfa.2020.125830.

Alagorni, A. H., Yaacob, Z. B., & Nour, A. H. (2015). An overview of oil production stages: Enhanced oil recovery techniques and nitrogen injection. *International Journal of Environmental Science and Development, 6*(9), 693–701. https://doi.org/10.7763/ijesd.2015.v6.682.

Alam, R., & Shang, J. Q. (2017). Removal of bitumen from mature oil sands tailings slurries by electro-flotation. *Journal of Water Process Engineering, 15*, 116–123. https://doi.org/10.1016/j.jwpe.2016.10.002.

Ayalew, A., Kimanya, M., Matumba, L., Bandyopadhayay, R., Menkir, A., & Cotty, P. (2017). Controlling afl atoxins in maize in Africa: Strategies, challenges and opportunities for improvement. In D. Watson (Ed.), *Achieving sustainable cultivation of maize* (1st ed.) (pp. 371–394). Burleigh Dodds Science Publishing. https://doi.org/10.19103/as.2016.0002.23.

Basu, S., & Malpani, P. R. (2001). Removal of methyl orange and methylene blue dye from water using colloidal gas aphron: Effect of processes parameters. *Separation Science and Technology, 36*(13), 2997–3013. https://doi.org/10.1081/SS-100107642.

Belkacem, M., Khodir, M., & Abdelkrim, S. (2008). Treatment characteristics of textile wastewater and removal of heavy metals using the electroflotation technique. *Desalination, 228*(1–3), 245–254. https://doi.org/10.1016/j.desal.2007.10.013.

Bennett, G. F., & Peters, R. W. (1988). The removal of oil from wastewater by air flotation: A review. *Critical Reviews in Environmental Control, 18*(3), 189–253. https://doi.org/10.1080/10643388809388348.

Bhatia, D., Goel, G., Bhimania, S. K., & Bhaskarwar, A. N. (2005). Characterization and drainage kinetics of colloidal gas aphrons. *AIChE Journal, 51*(11), 3048–3058. https://doi.org/10.1002/AIC.10552.

Bi, P. Y., Chang, L., & Dong, H. R. (2011). Separation behavior of penicillin in aqueous two-phase flotation. *Fenxi Huaxue/Chinese Journal of Analytical Chemistry.* https://doi.org/10.1016/S1872-2040(10)60428-8.

Bjorndalen, H. N., Jossy, W. E., Alvarez, J. M., & Kuru, E. (2014). A laboratory investigation of the factors controlling the filtration loss when drilling with colloidal gas aphron (CGA) fluids. *Journal of Petroleum Science and Engineering, 117*, 1–7. https://doi.org/10.1016/j.petrol.2014.03.003.

Brookey, T., & House R. (2002). Aphron-containing well drilling and servicing fluids of enhanced stability. U.S. Patent 6422326, USA.

Bowser, T. J. (2012). Mechanical separation design. *Handbook of food process design* (pp. 811–833). Wiley-Blackwell. https://doi.org/10.1002/9781444398274.ch29.

Bulatovic, S. M. (2007). 3-Frothers. In S. M. Bulatovic (Ed.), *Handbook of flotation reagents* (pp. 43–51). Elsevier. https://doi.org/10.1016/B978-044453029-5/50012-5.

Bürger, R., Diehl, S., & Nopens, I. (2011). A consistent modelling methodology for secondary settling tanks in wastewater treatment. *Water Research, 45*(6), 2247–2260. https://doi.org/10.1016/j.watres.2011.01.020.

Caballero, M., Cela, R., & Perez-bustamente, J. A. (1989). Studies on the use of colloidal gas aphrons in co-flotation and colvent sublation processes. A comparison with the conventional technique. *Separation Science and Technology, 24*(9–10), 629–640. https://doi.org/10.1080/01496398908049797.

Casqueira, R. G., Torem, M. L., & Kohler, H. M. (2006). The removal of zinc from liquid streams by electroflotation. *Minerals Engineering, 19*(13), 1388–1392. https://doi.org/10.1016/j.mineng.2006.02.001.

Chan, C. C., & Chen, Y. C. (2002). Demulsification of W/O emulsions by microwave radiation. *Separation Science and Technology, 37*(15), 3407–3420. https://doi.org/10.1081/SS-120014434.

Chaphalkar, P. G., Valsaraj, K. T., & Roy, D. (1994). Flotation using microgas dispersions for the removal of pentachlorophenol from aqueous solutions. *Separation Science and Technology, 29*(7), 907–921. https://doi.org/10.1080/01496399408006634.

Chen, D., Li, F., Gao, Y., & Yang, M. (2018). Pilot performance of chemical demulsifier on the demulsification of produced water from polymer/surfactant flooding in the Xinjiang Oilfield. *Water, 10*(12), 1874. https://doi.org/10.3390/w10121874.

Chen, G. (2004). Electrochemical technologies in wastewater treatment. *Separation and Purification Technology, 38*(1), 11–41. https://doi.org/10.1016/j.seppur.2003.10.006.

Chen, X., Chen, G., & Yue, P. L. (2002). Novel electrode system for electroflotation of wastewater. *Environmental Science and Technology, 36*(4), 778–783. https://doi.org/10.1021/es011003u.

Chesters, A. K. (1991). The modelling of coalescence processes in fluid-liquid dispersions: A review of current understanding. *Chemical Engineering Research and Design, 69*(4), 259–270.

Cho, Y. S., & Laskowski, J. S. (2002). Effect of flotation frothers on bubble size and foam stability. *International Journal of Mineral Processing, 64*(2–3), 69–80. https://doi.org/10.1016/S0301-7516(01)00064-3.

Ciriello, S., Barnett, S. M., & Deluise, F. J. (1982). Removal of heavy metals from aqueous solutions using microgas dispersions. *Separation Science and Technology, 17*(4), 521–534. https://doi.org/10.1080/01496398208060256.

Coe, S. H. (1916). Methods for determining the capacities of slime settling tanks. *Transactions American Institute of Mining Engineerring, 55*. https://cir.nii.ac.jp/crid/1572543024126718336.

Colović, R., Puvača, N., Cheli, F., Avantaggiato, G., Greco, D., Đuragić, O., Kos, J., & Pinotti, L. (2019). Decontamination of mycotoxin-contaminated feedstuffs and compound feed. *Toxins, 11*(11). https://doi.org/10.3390/toxins11110617.

Concha, F., & Bustos, M. C. (1991). Settling velocities of particulate systems, 6. Kynch sedimentation processes: Batch settling. *International Journal of Mineral Processing, 32*(3), 193–212. https://doi.org/10.1016/0301-7516(91)90068-T.

Coons, J. E., Halley, P. J., McGlashan, S. A., & Tran-Cong, T. (2003). A review of drainage and spontaneous rupture in free standing thin films with tangentially immobile interfaces. *Advances in Colloid and Interface Science, 105*(1–3), 3–62. https://doi.org/10.1016/S0001-8686(03)00003-4.

Coulaloglou, C. A., & Tavlarides, L. L. (1977). Description of interaction processes in agitated liquid-liquid dispersions. *Chemical Engineering Science, 32*(11), 1289–1297. https://doi.org/10.1016/0009-2509(77)85023-9.

Cumming, I. W., Holdich, R. G., & Smith, I. D. (1999). The rejection of oil using an asymmetric metal microfilter to separate an oil in water dispersion. *Water Research, 33*(17), 3587–3594. https://doi.org/10.1016/S0043-1354(99)00085-8.

da Cruz, S. G., Dutra, A. J. B., & Monte, M. B. M. (2016). The influence of some parameters on bubble average diameter in an electroflotation cell by laser diffraction method. *Journal of Environmental Chemical Engineering, 4*(3), 3681–3687. https://doi.org/10.1016/j.jece.2016.05.017.

de Oliveira da Mota, I., de Castro, J. A., de Góes Casqueira, R., & de Oliveira Junior, A. G (2015). Study of electroflotation method for treatment of wastewater from washing soil contaminated by heavy metals. *Journal of Materials Research and Technology, 4*(2), 109–113. https://doi.org/10.1016/j.jmrt.2014.11.004.

De Witt, C. C. (1940). Froth flotation concentration. *Industrial and Engineering Chemistry, 32*(5), 652–658. https://doi.org/10.1021/IE50365A014.

Degremont, S. A. (2007). *Water treatment handbook* (7th ed.). Lavoisier.

Delfos, R., Murphy, S., Stanbridge, D., Olujić, Ž., & Jansens, P. J. (2004). A design tool for optimising axial liquid-liquid hydrocyclones. *Minerals Engineering, 17*(5), 721–731. https://doi.org/10.1016/J.MINENG.2004.01.012.

Drzymala, J., & Kowalczuk, P. B. (2018). Classification of flotation frothers. *Minerals, 8*(2). https://doi.org/10.3390/MIN8020053.

Earle, R. L. (1983). Mechanical separations. In R. L. Earle (Ed.), *Unit operations in food processing* (2nd ed.) (pp. 143–158). Pergamon https://doi.org/10.1016/B978-0-08-025536-1.50014-X.

El-Ashtoukhy, E. S. Z., & Fouad, Y. O. (2014). Oil removal from oil-water emulsion by electrocoagulation in a cell with rotating cylinder anode. *Electrochemistry, 82*(11), 974–978. https://doi.org/10.5796/ELECTROCHEMISTRY.82.974.

Eow, J. S., & Ghadiri, M. (2002). Electrostatic enhancement of coalescence of water droplets in oil: A review of the technology. *Chemical Engineering Journal, 85*(2–3), 357–368. https://doi.org/10.1016/S1385-8947(01)00250-9.

Farrokhpay, S., Filippov, L., & Fornasiero, D. (2021). Flotation of fine particles: A review. *Mineral Processing and Extractive Metallurgy Review, 42*(7), 473–483. https://doi.org/10.1080/08827508.2020.1793140.

Ferreira, B. M. S., Ramalho, J. B. V. S., & Lucas, E. F. (2013). Demulsification of eater-in-crude oil emulsions by microwave radiation: Effect of aging, demulsifier addition, and selective heating. *Energy and Fuels, 27*(2), 615–621. https://doi.org/10.1021/EF301110M.

Frising, T., Noïk, C., & Dalmazzone, C. (2006). The liquid/liquid sedimentation process: From droplet coalescence to technologically enhanced water/oil emulsion gravity separators: A review. *Journal of Dispersion Science and Technology, 27*(7), 1035–1057. https://doi.org/10.1080/01932690600767098.

Gallios, G. P., and Matis, K. A. (1992). *Flotation of salt-type minerals*. 357–382. https://doi.org/10.1007/978-94-011-2658-8_18.

Gao, Z., Wang, C., Sun, W., Gao, Y., & Kowalczuk, P. B. (2021). Froth flotation of fluorite: A review. *Advances in Colloid and Interface Science, 290*, 102382. https://doi.org/10.1016/j.cis.2021.102382.

García-Montoya, M. F., Gutiérrez-Granados, S., Alatorre-Ordaz, A., Galindo, R., Ornelas, R., & Peralta-Hernández, J. M. (2015). Application of electrochemical/BDD process for the treatment wastewater effluents containing pharmaceutical compounds. *Journal of Industrial and Engineering Chemistry, 31*, 238–243. https://doi.org/10.1016/j.jiec.2015.06.030.

Geankoplis, C. J. (1993). Transport processes and unit operations. In C. J. Geankoplis (Ed.), *Chemical engineering science* (3rd ed). Prentice-Hall.

Ghasemi Naghdi, F., & Schenk, P. M. (2016). Dissolved air flotation and centrifugation as methods for oil recovery from ruptured microalgal cells. *Bioresource Technology, 218*, 428–435. https://doi.org/10.1016/j.biortech.2016.06.093.

Ghernaout, D., Benblidia, C., & Khemici, F. (2014). Microalgae removal from Ghrib Dam (Ain Defla, Algeria) water by electroflotation using stainless steel electrodes. *Desalination and Water Treatment, 54*(12), 3328–3337. https://doi.org/10.1080/19443994.2014.907749.

Gonzales, L. G. V., Pino, G. A. H., & Torem, M. L. (2013). Electroflotation of cassiterite fines using a hydrophobic bacterium strain. *Rem: Revista Escola de Minas, 66*(4), 507–512. https://doi.org/10.1590/s0370-44672013000400016.

Green, D. W., & Perry, R. H. (2007). *Perry's chemical engineers' handbook* (8th ed.). McGraw Hill LLC.

Gupta, A. K., Banerjee, P. K., & Mishra, A. (2007a). Effect of frothers on foamability, foam stability, and bubble size. *Coal Preparation, 27*(1–3), 107–125. https://doi.org/10.1080/07349340701249794.

Gupta, A. K., Banerjee, P. K., Mishra, A., Satish, P., & Pradip (2007b). Effect of alcohol and polyglycol ether frothers on foam stability, bubble size and coal flotation. *International Journal of Mineral Processing, 82*(3), 126–137. https://doi.org/10.1016/j.minpro.2006.09.002.

Han, J., Wang, Y., Yu, C., Li, C., Yan, Y., Liu, Y., & Wang, L. (2011). Separation, concentration and determination of chloramphenicol in environment and food using an ionic liquid/salt aqueous two-phase flotation system coupled with high-performance liquid chromatography. *Analytica Chimica Acta*. https://doi.org/10.1016/j.aca.2010.11.033.

Hashim, M. A., Dey, A., Hasan, S., & Sen Gupta, B. (1999). Mass transfer correlation in flotation of palm oil by colloidal gas aphrons. *Bioprocess Engineering, 21*(5), 401–404. https://doi.org/10.1007/s004490050694.

Hashim, M. A., Gupta, B. Sen, Kumar, S. V., Lim, R., Lim, S. E., & Tan, C. C. (1998). Effect of air to solid ratio in the clarification of yeast by colloidal gas aphrons. *Journal of Chemical Technology and Biotechnology, 71*(4), 335–339. https://doi.org/10.1002/(SICI)1097-4660(199804)71:4<335::AID-JCTB849>3.0.CO;2-H.

Hashim, M. A., Mukhopadhyay, S., Gupta, B. Sen, & Sahu, J. N. (2012). Application of colloidal gas aphrons for pollution remediation. *Journal of Chemical Technology and Biotechnology, 87*(3), 305–324. https://doi.org/10.1002/JCTB.3691.

Hashim, M. A., Sen Gupta, B., & Vijaya Kumar, S. (1995a). Clarification of yeast by colloidal gas aphrons. *Biotechnology Techniques, 9*(6), 403–408. https://doi.org/10.1007/bf00160826.

Hashim, M. A., Sengupta, B., & Subramaniam, M. B. (1995b). Investigations on the flotation of yeast cells by colloidal gas aphron (CGA) dispersions. *Bioseparation, 5*(3), 167–173.

Hashim, Mohd. A., & Gupta, B. Sen (1998). The application of colloidal gas aphrons in the recovery of fine cellulose fibres from paper mill wastewater. *Bioresource Technology, 64*(3), 199–204. https://doi.org/10.1016/s0960-8524(97)00169-7.

Hassanshahi, N., Hu, G., & Li, J. (2020). Application of ionic liquids for chemical demulsification: A review. *Molecules (Basel, Switzerland), 25*(21), 4915. https://doi.org/10.3390/molecules25214915.

Heldman, D. R., & Singh, R. P. (1981). Mechanical separation processes. In D. R. Heldman, & R. P. Singh (Eds.), *Food process engineering* (pp. 364–384). Springer. https://doi.org/10.1007/978-94-010-9337-8_8.

Hermann, L., Mailer, R., & Robards, K. (1999). Sedimentation in canola oil: A review. *Australian Journal of Experimental Agriculture, 39*(1), 103–113. https://doi.org/10.1071/EA98091.

Hosny, A. Y. (1996). Separating oil from oil-water emulsions by electroflotation technique. *Separations Technology, 6*(1), 9–17. https://doi.org/10.1016/0956-9618(95)00136-0.

Huang, R., de Vries, D., & Chen, H. (2018). Strategies to enhance fresh produce decontamination using combined treatments of ultraviolet, washing and disinfectants. *International Journal of Food Microbiology*. https://doi.org/10.1016/j.ijfoodmicro.2018.06.014.

Jacob-Lopes, E., Maroneze, M. M., Queiroz, M. I., & Zepka, L. Q. (2020). *Handbook of microalgae-based processes and products: Fundamentals and advances in energy, food, feed, fertilizer, and bioactive compounds*. Academic Press.

Jakob, L., Heinzmann, M., & Nirschl, H. (2022). Development of a continuous aqueous two-phase flotation process for the downstream processing of biotechnological products. *Separation and Purification Technology, 278*, 119657. https://doi.org/10.1016/j.seppur.2021.119657.

Janssen, L. J. J., Sillen, C. W. M. P., Barendrecht, E., & van Stralen, S. J. D. (1984). Bubble behaviour during oxygen and hydrogen evolution at transparent electrodes in KOH solution. *Electrochimica Acta, 29*(5), 633–642. https://doi.org/10.1016/0013-4686(84)87122-4.

Jauregi, P., Gilmour, S., & Varley, J. (1997). Characterisation of colloidal gas aphrons for subsequent use for protein recovery. *Chemical Engineering Journal, 65*(1), 1–11. https://doi.org/10.1016/S1385-8947(96)03154-3.

Javadian, S., & Sadrpoor, S. M. (2020). Demulsification of water in oil emulsion by surface modified SiO2 nanoparticle. *Journal of Petroleum Science and Engineering, 184*, 106547. https://doi.org/10.1016/j.petrol.2019.106547.

Jiang, B., Na, J., Wang, L., Li, D., Liu, C., & Feng, Z. (2019). Reutilization of food waste: One-step extration, purification and characterization of ovalbumin from salted egg white by aqueous two-phase flotation. *Foods*. https://doi.org/10.3390/foods8080286.

Jiang, B., Wang, L., Na, J., Zhang, X., Yuan, Y., Liu, C., & Feng, Z. (2020). Environmentally-friendly strategy for separation of α-lactalbumin from whey by aqueous two phase flotation. *Arabian Journal of Chemistry*. https://doi.org/10.1016/j.arabjc.2018.11.013.

Jiang, B., Wang, L., Wang, M., Wu, S., Wang, X., Li, D., Liu, C., Feng, Z., & Chi, Y. (2021). Direct separation and purification of α-lactalbumin from cow milk whey by aqueous two-phase flotation of thermo-sensitive polymer/phosphate. *Journal of the Science of Food and Agriculture*. https://doi.org/10.1002/jsfa.11055.

Jones, T. J., Neustadter, E. L., & Whittingham, K. P. (1978). Water-in-crude oil emulsion stability and emulsion destabilization by chemical demulsifiers. *Journal of Canadian Petroleum Technology, 17*(02). https://doi.org/10.2118/78-02-08.

Jung, M. U., Kim, Y. C., Bournival, G., & Ata, S. (2023). Industrial application of microbubble generation methods for recovering fine particles through froth flotation: A review of the state-of-the-art and perspectives. *Advances in Colloid and Interface Science, 322*, 103047. https://doi.org/10.1016/j.cis.2023.103047.

Kastali, M., Mouhir, L., Saafadi, L., Yilmaz, L., & Souabi, S. (2021). Pretreatment of industrial wastewater by natural flotation: Application to pollution reduction from vegetable oil refinery wastewaters. *Environmental Science and Pollution Research*. https://doi.org/10.1007/s11356-021-12850-9.

Kee, P. E., Cheah, L. S., Wan, P. K., Show, P. L., Lan, J. C. W., Chow, Y. H., & Ng, H. S. (2021). Primary capture of Bacillus subtilis xylanase from crude feedstock using alcohol/salt liquid biphasic flotation. *Biochemical Engineering Journal*. https://doi.org/10.1016/j.bej.2020.107835.

Khelifa, A., Aoudj, S., Moulay, S., & De Petris-Wery, M. (2013). A one-step electrochlorination/electroflotation process for the treatment of heavy metals wastewater in presence of EDTA. *Chemical Engineering and Processing: Process Intensification, 70*, 110–116. https://doi.org/10.1016/j.cep.2013.04.013.

Khosla, N. K., Venkatachalam, S., & Somasundaran, P. (1991). Pulsed electrogeneration of bubbles for electroflotation. *Journal of Applied Electrochemistry, 21*(11), 986–990. https://doi.org/10.1007/bf01077584.

Klasson, K. T., Taylor, P. A., Walker, J. F., Jones, S. A., Cummins, R. L., & Richardson, S. A. (2005). Modification of a centrifugal separator for in-well oil-water separation. *Separation Science and Technology, 40*(1–3), 453–462. https://doi.org/10.1081/SS-200042503.

Kovaleva, L., Zinnatullin, R., Musin, A., Gabdrafikov, A., Sultanguzhin, R., & Kireev, V. (2021). Influence of radio-frequency and microwave electromagnetic treatment on water-in-oil emulsion separation. *Colloids and Surfaces A: Physicochemical and Engineering Aspects, 614*, 126081. https://doi.org/10.1016/j.colsurfa.2020.126081.

Kumar, P., Gupta, A., Mahato, D. K., Pandhi, S., Pandey, A. K., Kargwal, R., Mishra, S., Suhag, R., Sharma, N., Saurabh, V., Paul, V., Kumar, M., Selvakumar, R., Gamlath, S., Kamle, M., Enshasy, H. A. El, Mokhtar, J. A., & Harakeh, S. (2022). Aflatoxins in cereals and cereal-based products: Occurrence, toxicity, impact on human health, and their detoxification and management strategies. *Toxins, 14*(10), 687. https://doi.org/10.3390/toxins14100687.

Kusumastuti, A., Anis, S., Ahmad, A. L., Ooi, B. S., & Shah Buddin, M. M. H. (2020). Emulsion liquid membrane for heavy metals removal: Emulsion breaking study. *Jurnal Teknologi, 82*(5). https://doi.org/10.11113/jt.v82.14539.

Kynch, G. J. (1952). A theory of sedimentation. *Transactions of the Faraday Society, 48*, 166–176. https://doi.org/10.1039/tf9524800166.

Kyzas, G. Z., & Matis, K. A. (2014). Flotation of biological materials, *Processes* (2, pp. 293–310). MDPI AG. https://doi.org/10.3390/pr2010293.

Kyzas, G. Z., & Matis, K. A. (2016). Electroflotation process: A review. *Journal of Molecular Liquids, 220*, 657–664. https://doi.org/10.1016/j.molliq.2016.04.128.

Larmignat, S., Vanderpool, D., Lai, H. K., & Pilon, L. (2008). Rheology of colloidal gas aphrons (microfoams). *Colloids and Surfaces A: Physicochemical and Engineering Aspects, 322*(1–3), 199–210. https://doi.org/10.1016/j.colsurfa.2008.03.010.

Leong, H. Y., Ooi, C. W., Law, C. L., Julkifle, A. L., Katsuda, T., & Show, P. L. (2019). Integration process for betacyanins extraction from peel and flesh of Hylocereus polyrhizus using liquid biphasic electric flotation system and antioxidant activity evaluation. *Separation and Purification Technology*. https://doi.org/10.1016/j.seppur.2018.07.040.

Li, J., & Gu, Y. (2005). Coalescence of oil-in-water emulsions in fibrous and granular beds. *Separation and Purification Technology, 42*(1), 1–13. https://doi.org/10.1016/J.SEPPUR.2004.05.006.

Liakos, T. I., & Lazaridis, N. K. (2014). Melanoidins removal from simulated and real wastewaters by coagulation and electro-flotation. *Chemical Engineering Journal, 242*, 269–277. https://doi.org/10.1016/j.cej.2014.01.003.

Litchfield, J. H. (1982). Meat-, fish-, and poultry-processing wastes. *Journal (Water Pollution Control Federation), 54*(6), 688–692. http://www.jstor.org/stable/25041571.

Logan, M., Ravishankar, H., Tan, L. C., Lawrence, J., Fitzgerald, D., & Lens, P. N. L. (2021). Anaerobic digestion of dissolved air floatation slurries: Effect of substrate concentration and pH. *Environmental Technology and Innovation, 21*. https://doi.org/10.1016/j.eti.2020.101352.

Lorain, O., Thiebaud, P., Badorc, E., & Aurelle, Y. (2001). Potential of freezing in wastewater treatment: Soluble pollutant applications. *Water Research, 35*(2), 541–547. https://doi.org/10.1016/S0043-1354(00)00287-6.

Lv, X., Song, Z., Yu, J., Su, Y., Zhao, X., Sun, J., Mao, Y., & Wang, W. (2020). Study on the demulsification of refinery oily sludge enhanced by microwave irradiation. *Fuel, 279*, 118417. https://doi.org/10.1016/j.fuel.2020.118417.

Ma, J., Li, X., Zhang, X., Sui, H., He, L., & Wang, S. (2020). A novel oxygen-containing demulsifier for efficient breaking of water-in-oil emulsions. *Chemical Engineering Journal, 385*, 123826. https://doi.org/10.1016/j.cej.2019.123826.

Majlesi, M., Mohseny, S. M., Sardar, M., Golmohammadi, S., & Sheikhmohammadi, A. (2016). Improvement of aqueous nitrate removal by using continuous electrocoagulation/electroflotation unit with vertical monopolar electrodes. *Sustainable Environment Research, 26*(6), 287–290. https://doi.org/10.1016/j.serj.2016.09.002.

Maksimov, E. A., & Ostsemin, A. A. (2015). Intensifying the cleaning of emulsion- and oil-bearing waste water from rolled-product manufacturing by electroflotation. *Metallurgist, 58*(11–12), 945–949. https://doi.org/10.1007/s11015-015-0022-8.

Mansour, L. B., & Chalbi, S. (2006). Removal of oil from oil/water emulsions using electroflotation process. *Journal of Applied Electrochemistry, 36*(5), 577–581. https://doi.org/10.1007/s10800-005-9109-4.

Mansur, E. H. A., Wang, Y., & Dai, Y. (2006). Removal of suspensions of fine particles from water by colloidal gas aphrons (CGAs). *Separation and Purification Technology, 48*(1), 71–77. https://doi.org/10.1016/j.seppur.2005.07.022.

Marshall, H., Meneely, J. P., Quinn, B., Zhao, Y., Bourke, P., Gilmore, B. F., Zhang, G., & Elliott, C. T. (2020). Novel decontamination approaches and their potential application for post-harvest aflatoxin control. *Trends in Food Science and Technology, 106*, 489–496. https://doi.org/10.1016/j.tifs.2020.11.001.

Martínez-Huitle, C. A., Scialdone, O., & Rodrigo, M. A. (2018). Electrochemical water and wastewater treatment. *Electrochemical Water and Wastewater Treatment*, 1–553. https://doi.org/10.1016/C2016-0-04297-3.

Mathiazakan, P., Shing, S. Y., Ying, S. S., Kek, H. K., Tang, M. S. Y., Show, P. L., Ooi, C. W., & Ling, T. C. (2016). Pilot-scale aqueous two-phase floatation for direct recovery of lipase derived from Burkholderia cepacia strain ST8. *Separation and Purification Technology*. https://doi.org/10.1016/j.seppur.2016.07.017.

Matis, K. A., & Peleka, E. N. (2010). Alternative flotation techniques for wastewater treatment: Focus on electroflotation. *Separation Science and Technology, 45*(16), 2465–2474. https://doi.org/10.1080/01496395.2010.508065.

Matumba, L., Van Poucke, C., Njumbe Ediage, E., Jacobs, B., & De Saeger, S. (2015). Effectiveness of hand sorting, flotation/washing, dehulling and combinations thereof on the decontamination of mycotoxin-contaminated white maize. *Food Additives and Contaminants: Part A, 32*(6), 960–969. https://doi.org/10.1080/19440049.2015.1029535.

Medina, M. E. V. (2019). Secondary settling tanks modeling: study of the dynamics of activated sludge sedimentation by computational fluids dynamics. Doctoral dissertation, Université de Strasbourg.

Melo, F., & Laskowski, J. S. (2006). Fundamental properties of flotation frothers and their effect on flotation. *Minerals Engineering, 19*(6–8), 766–773. https://doi.org/10.1016/j.mineng.2005.09.031.

Meon, W., & Blass, E. (1992). Hydrodynamic modeling of droplet coalescence at liquid-liquid interfaces. *Separation Science and Technology, 27*(2), 129–159. https://doi.org/10.1080/01496399208018870.

Mesa, D., & Brito-Parada, P. R. (2019). Scale-up in froth flotation: A state-of-the-art review. *Separation and Purification Technology, 210*, 950–962. https://doi.org/10.1016/J.SEPPUR.2018.08.076.

Mohammed, S. A. M., & Mohammed, M. S. (2013). The application of microwave technology in demulsification of water-in-oil emulsion for missan oil fields. *Iraqi Journal of Chemical and Petroleum Engineering, 14*(2), 21–27. https://doi.org/10.31699/ijcpe.2013.2.4.

Mohshim, D. F., Nasir, R., Mannan, H. A., Abdulrahman, A., & Mukhtar, H. (2023). Challenges of industries in dealing with oily wastewater release and treatments. In A. Basile, A. Cassano, M. R. Rahimpour, & M. A. Makarem (Eds.), *Advanced technologies in wastewater treatment* (pp. 49–70). Elsevier. https://doi.org/10.1016/B978-0-323-99916-8.00009-2.

Mook, W. T., Aroua, M. K., & Issabayeva, G. (2014). Prospective applications of renewable energy based electrochemical systems in wastewater treatment: A review. *Renewable and Sustainable Energy Reviews, 38*, 36–46. https://doi.org/10.1016/j.rser.2014.05.042.

Nahui, F. N. B., Nascimento, M. R., Cavalcanti, E. B., & Vilar, E. O. (2008). Electroflotation of emulsified oil in industrial wastes evaluated with a full factorial design. *Brazilian Journal of Chemical Engineering, 25*(3), 435–442. https://doi.org/10.1590/s0104-66322008000300001.

Ndikubwimana, T., Chang, J., Xiao, Z., Shao, W., Zeng, X., Ng, I. S., & Lu, Y. (2016). Flotation: A promising microalgae harvesting and dewatering technology for biofuels production. *Biotechnology Journal, 11*(3), 315–326. https://doi.org/10.1002/biot.201500175.

Newcombe, G., & Dixon, D. (2006). *Interface science in drinking water treatment: Theory and applications*. Academic Press.

Nguyen, A. V. (2013). Froth flotation. *Reference Module in Chemistry, Molecular Sciences and Chemical Engineering*. https://doi.org/10.1016/B978-0-12-409547-2.04401-2.

Nguyen, T. L., Lee, D. J., Chang, J. S., & Liu, J. C. (2013). Effects of ozone and peroxone on algal separation via dispersed air flotation. *Colloids and Surfaces B: Biointerfaces, 105*, 246–250. https://doi.org/10.1016/j.colsurfb.2012.12.017.

Niu, Z., Yue, T., He, X., & Manica, R. (2019). Changing the interface between an asphaltene model compound and water by addition of an EO–PO demulsifier through adsorption competition or adsorption replacement. *Energy and Fuels, 33*(6), 5035–5042. https://doi.org/10.1021/acs.energyfuels.9b00741.

Noble, M., Brown, A., Jauregi, P., Kaul, A., & Varley, J. (1998). Protein recovery using gas-liquid dispersions. *Journal of Chromatography B: Biomedical Applications, 711*(1–2), 31–43. https://doi.org/10.1016/S0378-4347(98)00030-9.

Nour, A. H., Ilia Anisa, A. N., & Nour, A. H. (2012). Demulsification of water-in-oil (W/O) emulsion via microwave irradiation: An optimization. *Scientific Research and Essays, 7*(2), 231–243. https://doi.org/10.5897/SRE11.1657.

Nuorivaara, T., & Serna-Guerrero, R. (2020). Amphiphilic cellulose and surfactant mixtures as green frothers in mineral flotation. 2. Flotation of chalcopyrite and Cu-containing tailings. *Colloids and Surfaces A: Physicochemical and Engineering Aspects, 603*. https://doi.org/10.1016/J.COLSURFA.2020.125298.

Ortega-Rivas, E. (2012a). Common preliminary operations: Cleaning, sorting, grading. In E. Ortega-Rivas (Ed.), *Non-thermal food engineering operations* (pp. 11–25). Springer US. https://doi.org/10.1007/978-1-4614-2038-5_2.

Ortega-Rivas, E. (2012b). Separation techniques for solids and suspensions. In E. Ortega-Rivas (Ed.), *Non-thermal food engineering operations* (pp. 131–197). Springer US. https://doi.org/10.1007/978-1-4614-2038-5_7.

Pakhale, S. V., Vetal, M. D., & Rathod, V. K. (2013). Separation of bromelain by aqueous two phase flotation. *Separation Science and Technology, 48*(6), 984–989. https://doi.org/10.1080/01496395.2012.712596.

Pal, R. (1994). Techniques for measuring the composition (oil and water content) of emulsions: A state of the art review. *Colloids and Surfaces A: Physicochemical and Engineering Aspects, 84*(2–3), 141–193. https://doi.org/10.1016/0927-7757(93)02711-M.

Palermo, T. (1991). Le phénomène de coalescence. Etude bibliographique. *Revue de l'Institut Français Du Pétrole, 46*(3), 325–360. https://doi.org/10.2516/ogst:1991016.

Parsapour, G. A., Hossininasab, M., Yahyaei, M., & Banisi, S. (2014). Effect of settling test procedure on sizing thickeners. *Separation and Purification Technology, 122*, 87–95. https://doi.org/10.1016/j.seppur.2013.11.001.

Pascale, M., Logrieco, A. F., Graeber, M., Hirschberger, M., Reichel, M., Lippolis, V., De Girolamo, A., Lattanzio, V. M. T., & Slettengren, K. (2020). Aflatoxin reduction in maize by industrial-scale cleaning solutions. *Toxins, 12*(5), 331. https://doi.org/10.3390/toxins12050331.

Pashakolaie, V. G., Khaleghi, S., Mohammadi, T., & Khorsandi, M. (2015). Oil production cost function and oil recovery implementation-Evidence from an Iranian oil field. *Energy Exploration and Exploitation, 33*(4), 459–470. https://doi.org/10.1260/0144-5987.33.4.459.

Patel, A., Sartaj, K., Pruthi, P. A., Pruthi, V., & Matsakas, L. (2019). Utilization of clarified butter sediment waste as a feedstock for cost-effective production of biodiesel. *Foods, 8*(7). https://doi.org/10.3390/foods8070234.

Pattanaik, A., & Venugopal, R. (2018). Investigation of adsorption mechanism of reagents (surfactants) system and its applicability in iron ore flotation: An overview. *Colloid and Interface Science Communications, 25*, 41–65. https://doi.org/10.1016/j.colcom.2018.06.003.

Pawliszak, P., Bradshaw-Hajek, B. H., Skinner, W., Beattie, D. A., & Krasowska, M. (2024). Frothers in flotation: A review of performance and function in the context of chemical classification. *Minerals Engineering, 207*, 108567. https://doi.org/10.1016/J.MINENG.2023.108567.

Penaloza, I., Tita, A., McNew, E., & Chu, P. (2023). Barite resources, production and recovery using froth flotation: A review. *Minerals Engineering, 203*, 108327. https://doi.org/10.1016/J.MINENG.2023.108327.

Pereira, M. do S., Borges, A. C., Muniz, G. L., Heleno, F. F., & Faroni, L. R. D. A. (2020). Dissolved air flotation optimization for treatment of dairy effluents with organic coagulants. *Journal of Water Process Engineering, 36*. https://doi.org/10.1016/j.jwpe.2020.101270.

Perfil'eva, A. V., Brodskii, V. A., Il'in, V. I., & Kolesnikov, V. A. (2016). Effect of the composition of the medium and electroflotation processing parameters on extraction efficiency of chromium(III) dispersed phase from aqueous solutions. *Russian Journal of Applied Chemistry, 89*(3), 388–393. https://doi.org/10.1134/s1070427216030071.

Powell, C. D., Quain, D. E., & Smart, K. A. (2004). The impact of sedimentation on cone yeast heterogeneity. *Journal of the American Society of Brewing Chemists, 62*(1), 8–17. https://doi.org/10.1094/ASBCJ-62-0008.

Qin, X., Yang, B., Gao, F., & Chen, G. (2013). Treatment of restaurant wastewater by pilot-scale electrocoagulation-electroflotation: Optimization of operating conditions. *Journal of Environmental Engineering, 139*(7), 1004–1016. https://doi.org/10.1061/(asce)ee.1943-7870.0000687.

Quintanilla, P., Neethling, S. J., & Brito-Parada, P. R. (2021). Modelling for froth flotation control: A review. *Minerals Engineering, 162*. https://doi.org/10.1016/J.MINENG.2020.106718.

Raikos, V., Kapolos, J., Farmakis, L., Koliadima, A., & Karaiskakis, G. (2009). The use of sedimentation field-flow fractionation in the size characterization of bovine milk fat globules as affected by heat treatment. *Food Research International, 42*(5), 659–665. https://doi.org/10.1016/j.foodres.2009.02.001.

Rajendran, S., & Murty, Ch. V. G. K. (2023). *Mineral processing beneficiation operations and process optimization through modeling*. Elsevier Inc.

Ray, M. S. (1995). Chemical Process Design, by R. Smith, McGraw-Hill, Inc., New York, USA (1995). 460 pages. ISBN 0-07-059220-9. *Developments in Chemical Engineering and Mineral Processing, 3*(1), 56. https://doi.org/10.1002/apj.5500030107.

Ricordel, C., Darchen, A., & Hadjiev, D. (2010). Electrocoagulation–electroflotation as a surface water treatment for industrial uses. *Separation and Purification Technology, 74*(3), 342–347. https://doi.org/10.1016/j.seppur.2010.06.024.

Rommel, W., Meon, W., & Blass, E. (1992). Hydrodynamic modeling of droplet coalescence at liquid-liquid interfaces. *Separation Science and Technology, 27*(2), 129–159. https://doi.org/10.1080/01496399208018870.

Ross, A., Mason, R., Nottingham, S., Keilar, S., & Kowitz, T. (2002). The effect of water, brine and ethanol flotation on the quality and shelf life of macadamia kernels. I. Whole kernels. *Food Australia, 54*(9), 407–412.

Ross, C. C., Smith, B. M., & Valentine, G. E. (2000). Rethinking dissolved air flotation (DAF) design for industrial pretreatment. Industrial Wastes (IW) Conference 2000 (pp. 43–56). Water Environment Federation.

Roy, D., Valsaraj, K. T., & Kottai, S. A. (1992). Separation of organic dyes from wastewater by using colloidal gas aphrons. *Separation Science and Technology, 27*(5), 573–588. https://doi.org/10.1080/01496399208018903.

Rubio, J., Souza, M. L., & Smith, R. W. (2002). Overview of flotation as a wastewater treatment technique. *Minerals Engineering, 15*(3), 139–155. https://doi.org/10.1016/s0892-6875(01)00216-3.

Ryder, K., Ali, M. A., Billakanti, J., & Carne, A. (2018). Fundamental characterisation of caseins harvested by dissolved air flotation from dairy wastewater and comparison with skim milk powder. *International Dairy Journal, 78*, 112–121. https://doi.org/10.1016/j.idairyj.2017.11.007.

Sankaran, R., Manickam, S., Yap, Y. J., Ling, T. C., Chang, J. S., & Show, P. L. (2018a). Extraction of proteins from microalgae using integrated method of sugaring-out assisted liquid biphasic flotation (LBF) and ultrasound. *Ultrasonics Sonochemistry*. https://doi.org/10.1016/j.ultsonch.2018.06.002.

Sankaran, R., Parra Cruz, R. A., Show, P. L., Haw, C. Y., Lai, S. H., Ng, E. P., & Ling, T. C. (2019). Recent advances of aqueous two-phase flotation system for the recovery of biomolecules. *Fluid Phase Equilibria, 501*, 112271. https://doi.org/10.1016/j.fluid.2019.112271.

Sankaran, R., Show, P. L., Lee, S. Y., Yap, Y. J., & Ling, T. C. (2018b). Integration process of fermentation and liquid biphasic flotation for lipase separation from Burkholderia cepacia. *Bioresource Technology*. https://doi.org/10.1016/j.biortech.2017.11.050.

Sankaran, R., Show, P. L., Yap, Y. J., Lam, H. L., Ling, T. C., Pan, G. T., & Yang, T. C. K. (2018c). Sustainable approach in recycling of phase components of large scale aqueous two-phase flotation for lipase recovery. *Journal of Cleaner Production*. https://doi.org/10.1016/j.jclepro.2018.02.301.

Saravacos, G. D., & Kostaropoulos, A. E. (2002). Mechanical separation equipment. In G. D. Saravacos, & A. E. Kostaropoulos (Eds.), *Handbook of food processing equipment* (pp. 208–261). Springer US. https://doi.org/10.1007/978-1-4615-0725-3_5.

Sarkar, M. S. K. A., Donne, S. W., & Evans, G. M. (2011). Utilization of hydrogen in electroflotation of silica. *Advanced Powder Technology, 22*(4), 482–492. https://doi.org/10.1016/j.apt.2011.05.007.

Sarkar, Md. S. K. A., Evans, G. M., & Donne, S. W. (2010). Bubble size measurement in electroflotation. *Minerals Engineering, 23*(11–13), 1058–1065. https://doi.org/10.1016/j.mineng.2010.08.015.

Save, S. V., & Pangarkar, V. G. (1995). Harvesting of Saccharomyces cerevisiae using colloidal gas aphrons. *Journal of Chemical Technology and Biotechnology: International Research in Process, Environmental and Clean Technology, 62*(2), 192–199.

Schlieper, L., Chatterjee, M., Henschke, M., & Pfennig, A. (2004). Liquid–liquid phase separation in gravity settler with inclined plates. *AIChE Journal, 50*(4), 802–811. https://doi.org/10.1002/aic.10075.

Sebba, F. (1971). Microfoams-an unexploited colloid system. *Journal of Colloid And Interface Science, 35*(4), 643–646. https://doi.org/10.1016/0021-9797(71)90223-2.

Sebba, F. (1979). A surface-chemical basis for cell motility. *Journal of Theoretical Biology, 78*(3). https://doi.org/10.1016/0022-5193(79)90337-0.

Sebba, F. (1987). *Chapter 5: Colloidal gas aphrons.* Foams & Biliquid Foams-Aphrons (pp. 63–78). NY: J. Wiley & Sons.

Shammas, N. K., & Bennett, G. F. (2010). Principles of air flotation technology. *Flotation technology* (pp. 1–47). Humana Press https://doi.org/10.1007/978-1-60327-133-2_1.

Singh, A., Van Hamme, J. D., & Ward, O. P. (2007). Surfactants in microbiology and biotechnology: Part 2. Application aspects. *Biotechnology Advances, 25*(1), 99–121. https://doi.org/10.1016/j.biotechadv.2006.10.004.

Sjöblom, J., Aske, N., Auflem, I. H., Brandal, Ø., Havre, T. E., Sæther, Ø., Westvik, A., Johnsen, E. E., & Kallevik, H. (2003). Our current understanding of water-in-crude oil emulsions. Recent characterization techniques and high pressure performance. *Advances in Colloid and Interface Science, 100–102*, 399–473. https://doi.org/10.1016/S0001-8686(02)00066-0.

Skender, A., Moulai-Mostefa, N., & Tir, M. (2010). Effects of operational parameters on the removal efficiency of non-ionic surfactant by electroflotation. *Desalination and Water Treatment, 13*(1–3), 213–216. https://doi.org/10.5004/dwt.2010.992.

Sorsamäki, L., & Nappa, M. (2015). *Design and selection of separation processes*. Finland: VTT Technical Research Centre of Finland Ltd.

Speth, H., Pfennig, A., Chatterjee, M., & Franken, H. (2002). Coalescence of secondary dispersions in fiber beds. *Separation and Purification Technology, 29*(2), 113–119. https://doi.org/10.1016/S1383-5866(02)00067-9.

Stack, L. J., Carney, P. A., Malone, H. B., & Wessels, T. K. (2005). Factors influencing the ultrasonic separation of oil-in-water emulsions. *Ultrasonics Sonochemistry, 12*(3), 153–160. https://doi.org/10.1016/J.ULTSONCH.2003.10.008.

Subramaniam, M. B., Blakebrough, N., & Hashim, M. A. (1990). Clarification of suspensions by colloidal gas aphrons. *Journal of Chemical Technology andamp; Biotechnology, 48*(1), 41–60. https://doi.org/10.1002/jctb.280480105.

Sun, H., Wang, Q., Li, X., & He, X. (2020). Novel polyether-polyquaternium copolymer as an effective reverse demulsifier for O/W emulsions: Demulsification performance and mechanism. *Fuel, 263*, 116770. https://doi.org/10.1016/j.fuel.2019.116770.

Talmage, W. P., & Fitch, E. B. (1955). Determining thickener unit areas. *Industrial and Engineering Chemistry, 47*(1), 38–41. https://doi.org/10.1021/ie50541a022.

Tawfik, M. E., & Diez, F. J. (2014). On the relation between onset of bubble nucleation and gas supersaturation concentration. *Electrochimica Acta, 146*, 792–797. https://doi.org/10.1016/j.electacta.2014.08.147.

Téllez-Morales, J. A., Henández-Santos, B., & Rodríguez-Miranda, J. (2021). Impact of the melon juice clarification method on the physicochemical properties of an alcoholic beverage. *Journal of Food Processing and Preservation, 45*(5), e15427. https://doi.org/10.1111/jfpp.15427.

Tseng, H., Pilon, L., & Warrier, G. R. (2006). Rheology and convective heat transfer of colloidal gas aphrons in horizontal mini-channels. *International Journal of Heat and Fluid Flow, 27*(2), 298–310. https://doi.org/10.1016/j.ijheatfluidflow.2005.08.009.

Tzia, C., Sfakianakis, P., & Giannou, V. (2015). Raw materials of foods: Handling and management. *Handbook of food processing: Food safety, quality, and manufacturing processes*. https://doi.org/10.1201/b19398.

Umar, A. A., Saaid, I. B. M., Sulaimon, A. A., & Pilus, R. B. M. (2018). A review of petroleum emulsions and recent progress on water-in-crude oil emulsions stabilized by natural surfactants and solids. *Journal of Petroleum Science and Engineering, 165*, 673–690. https://doi.org/10.1016/J.PETROL.2018.03.014.

Valdés, E., Vilanova, M., Sabio, E., & Benalte, M. J. (2011). Clarifying agents effect on the nitrogen composition in must and wine during fermentation. *Food Chemistry, 125*(2), 430–437. https://doi.org/10.1016/j.foodchem.2010.09.024.

Venkatachalam, S. (1992). Electrogenerated gas bubbles in flotation. *Mineral Processing and Extractive Metallurgy Review, 8*(1–4), 47–55. https://doi.org/10.1080/08827509208952677.

Vidgren, V., & Londesborough, J. (2011). 125th anniversary review: Yeast flocculation and sedimentation in brewing. *Journal of the Institute of Brewing, 117*(4), 475–487. https://doi.org/10.1002/j.2050-0416.2011.tb00495.x.

Wang, C., Lü, Y., Song, C., Zhang, D., Rong, F., & He, L. (2022). Separation of emulsified crude oil from produced water by gas flotation: A review. *Science of the Total Environment, 845*. https://doi.org/10.1016/j.scitotenv.2022.157304.

Xu, Q., Nakajima, M., Ichikawa, S., Nakamura, N., & Shiina, T. (2008). A comparative study of microbubble generation by mechanical agitation and sonication. *Innovative Food Science and Emerging Technologies, 9*(4), 489–494. https://doi.org/10.1016/j.ifset.2008.03.003.

Ye, F., Mi, Y., Liu, H., Zeng, G., Shen, L., Feng, X., Yang, Y., Zhang, Z., Yuan, H., & Yan, X. (2021). Demulsification of water-in-crude oil emulsion using natural lotus leaf treated via a simple hydrothermal process. *Fuel, 295*, 120596. https://doi.org/10.1016/j.fuel.2021.120596.

Yeo, L. Y., Matar, O. K., Perez de Ortiz, E. S., & Hewitt, G. F. (2003). Film drainage between two surfactant-coated drops colliding at constant approach velocity. *Journal of Colloid and Interface Science, 257*(1), 93–107. https://doi.org/10.1016/s0021-9797(02)00033-4.

Yi, M., Huang, J., & Wang, L. (2017). Research on crude oil demulsification using the combined method of ultrasound and chemical demulsifier. *Journal of Chemistry, 2017*, 1–7. https://doi.org/10.1155/2017/9147926.

Yonguep, E., Kapiamba, K. F., Kabamba, K. J., & Chowdhury, M. (2022). Formation, stabilization and chemical demulsification of crude oil-in-water emulsions: A review. *Petroleum Research, 7*(4), 459–472. https://doi.org/10.1016/J.PTLRS.2022.01.007.

Yuan, H., Huang, Z., Shen, L., Xu, J., Feng, X., Yang, Y., Zhang, Z., Luo, Y., Yan, X., & Mi, Y. (2021). Demulsification of crude oil emulsion using carbonized cotton/silica composites. *Colloids and Surfaces A: Physicochemical and Engineering Aspects, 617*, 126421. https://doi.org/10.1016/j.colsurfa.2021.126421.

Zargar, G., Gheysari, R. G., Takassi, M. A., Rostami, A., & Zadehnazari, A. (2018). Evaluation of a sulfanilic acid based surfactant in crude oil demulsification: An experimental study. *Oil and Gas Sciences and Technology – Revue d'IFP Energies Nouvelles, 73*, 20. https://doi.org/10.2516/ogst/2018016.

Zarie, M., Jahedsaravani, A., & Massinaei, M. (2020). Flotation froth image classification using convolutional neural networks. *Minerals Engineering, 155*. https://doi.org/10.1016/J.MINENG.2020.106443.

Zhai, M., Wu, M., Wang, C., & Li, X. (2020). A novel silica-supported polyether polysiloxane quaternary ammonium demulsifier for highly efficient fine-sized oil droplet removal of oil-in-water emulsions. *RSC Advances, 10*(32), 18918–18926. https://doi.org/10.1039/d0ra01679a.

Zolfaghari, R., Fakhru'l-Razi, A., Abdullah, L. C., Elnashaie, S. S. E. H., & Pendashteh, A. (2016). Demulsification techniques of water-in-oil and oil-in-water emulsions in petroleum industry. *Separation and Purification Technology, 170*, 377–407. https://doi.org/10.1016/J.SEPPUR.2016.06.026.

Zouboulis, A. I., & Matis, K. A. (2012). Cadmium ion removal by electroflotation onto sewage sludge biomass. *International Journal of Environment and Waste Management, 9*(3/4), 245. https://doi.org/10.1504/ijewm.2012.046390.

Centrifugation in the food industry 3

Mondher Mzoughi and Ebru Firatligil
Department of Food Engineering, Faculty of Chemical & Metallurgical Engineering, Istanbul Technical University, Maslak, Istanbul, Türkiye

3.1 Introduction

Separation techniques are fundamental to the food industry, ensuring the removal of undesired components and the concentration of valuable constituents in various food products. These techniques are critical for achieving the desired quality, safety, and shelf-life of food items. Among the various separation methods available, centrifugation stands out due to its efficiency, versatility, and broad applicability (Brennan & Grandison, 2011).

Centrifugation utilizes centrifugal force to accelerate the separation process, significantly enhancing the speed and effectiveness compared to gravity-based methods (Olatunde et al., 2022). This technique is particularly advantageous in the food industry where rapid and precise separation of different phases like solid from liquid, or liquid from liquid is often required (Santhanam & Lipták, 2005). The benefits of centrifugation include high throughput, minimal labor, reduced processing time, and improved product consistency.

The diversity of centrifuges available allows for their use across a wide range of applications. From low-speed centrifuges used in routine sedimentation of particles to high-speed and ultracentrifuges designed for more complex biochemical separations, the variety of centrifuge designs, such as spintube, disk stack, tubular bowl, decanter centrifuges and others, responds to the specific needs of different food processing operations (Leung, 2020).

Applications of centrifugation in the food industry are extensive and varied. Centrifuges are employed in dairy processing for cream separation, in fruit juice production for pulp removal, in edible oil refining for degumming and neutralization, and in the fermentation industry for yeast and microbial cell separation. Additionally, centrifugation is used in the clarification of beverages, the concentration of protein solutions, and the separation of fine particles in various food suspensions (Berk, 2018).

The objective of this chapter is providing a comprehensive overview of the principles and forces underlying centrifugation, the different types of centrifuges and their specific designs, and the detailed applications of centrifugation within the food industry. By the end of this chapter, readers will have a thorough understanding of how centrifugation can be effectively applied to enhance food processing, improve product quality, and optimize operational efficiency.

Mechanical Separation Processes in the Food Industry: Unit Operations and Processing Equipment in the Food Industry.
DOI: https://doi.org/10.1016/B978-0-12-819523-9.00009-3

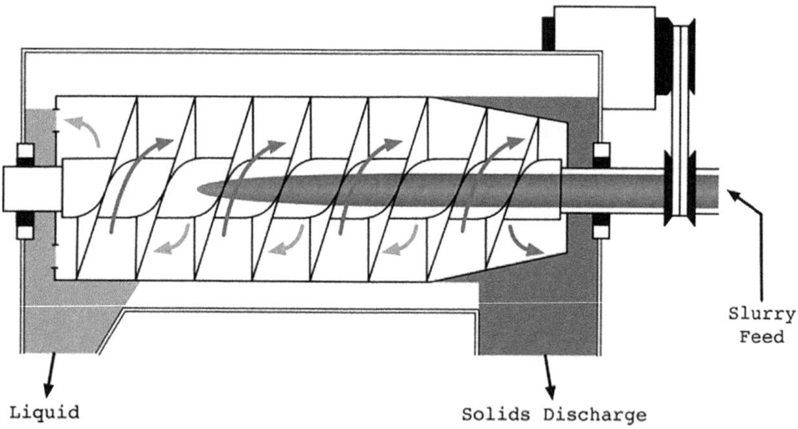

Figure 3.1 Schematic representation of a decanter centrifuge.
(From Todaro (2014)).

3.2 Principles of centrifugation

3.2.1 Basic principles of centrifugal separation

Centrifugal separation involves the mechanical division of a fluid mixture containing two or more immiscible phases with different densities by applying a centrifugal force. It refers to the separation process carried out in centrifuges that use power-driven rotating bowls, also known as rotating-boundary machines (McCarthy, 2011). Separation with centrifugation techniques can be utilized for two-phase systems where solid particles are dispersed in a liquid or where liquid droplets are emulsified within another liquid. In such systems, the particles or droplets constitute a discontinuous phase within the continuous liquid phase. Particles or droplets with a density greater than the surrounding liquid will settle downward (sedimentation), while those with a lower density will rise to the top, as illustrated in Fig. 3.1. The separation velocity is expressed by Stokes' law in the Eq. (3.1) below (Heymann, 2011). For effective separation at reasonable processing rates, there must be a density difference of at least 3% between the phases (McCarthy, 2011).

$$V_s = \frac{d^2 \times \Delta\rho \times g}{18 \times \eta} \tag{3.1}$$

where Vs represents the sedimentation velocity, d is the diameter of particle or droplet, $\Delta\rho$ is the difference of densities, g the acceleration caused by gravity, and η is the viscosity of the continuous phase.

Industrial centrifugation is classified into two types: sedimenting and filtering (Luckiram, 2022). Sedimenting centrifuges separate components based on density differences, using centrifugal forces up to 10,000 times greater than gravity to force denser particles to the bowl's periphery. This method is effective for separating solid-liquid and liquid-liquid mixtures with small density differences. Filtering centrifuges, on the other hand, use perforated bowls to separate solid-liquid mixtures. The liquid phase passes through the porous wall, retaining solids without relying on density differences. This method is suitable for particles larger than 10 microns and allows the separation of solids (Pouliot et al., 2014; Todaro, 2014).

3.2.2 Centrifugation equipment overview

The basic parts of a centrifuge include motor, rotor, chamber, lid, latch and the sample to be centrifuged (Tantray et al., 2023). According to Todaro (1996), a centrifuge comprises several essential components: the rotor (bowl or basket) that rotates to contain the product, a solids discharge system that may include ploughs, scrolls, inverting baskets, or nozzle systems for unloading solids, and a drive system that includes the main bearing shaft with seals and a motor (either electric or hydraulic) to rotate the bowl. The entire unit is supported by a frame and enclosed within a housing to contain the rotor (Todaro, 2014).

The categories of centrifuge rotors include fixed-angle, swinging-bucket, vertical-tube rotors and others depending on the centrifuge type concerned and the application of centrifugation. Rotors are typically categorized by their speed, design, as well as their material composition. Different centrifugal forces are generated during rotor spinning, depending on whether the centrifuge is high-speed, low-speed, or an ultra-centrifuge. Various materials can be utilized to produce these rotors: brass or steel for low-speed rotors, fiber-reinforced composites, titanium, or aluminum for high-speed rotors. Some rotors have protective coatings; for example, titanium alloy rotors used in ultracentrifugation are coated with polyurethane, while aluminum rotors have an aluminum oxide layer to prevent corrosion. It's crucial to handle rotors with care to avoid damaging these protective coatings (Olatunde et al., 2022).

3.2.3 Forces acting on particles in a centrifuge

The first force that act on particles during centrifugation is the gravitational force. Although much smaller compared to the centrifugal force, gravity still acts on the particles, pulling them downward. In the context of centrifugation, this force is often negligible compared to the centrifugal force generated at high speeds. The second force is g force. Along the centrifugation process, suspended particles and molecules are driven to the point furthest from the rotation center. This happens at a speed that is determined by the speed of the rotor, expressed in revolutions per minute (rpm). The force produced by the spinning rotor is known as the relative centrifugal force (RCF) or "g" force, which is proportional to the square of the rotor speed and the radial distance (the particle distance from the rotation axis) (Stephenson, 2016; Walker, 2012). The RCF can be calculated using Eq. (3.2).

$$\mathrm{RCF} = 11.18(\mathrm{r}_{\mathrm{cm}})\left(\frac{rpm}{1000}\right)^2 \tag{3.2}$$

where r (in centimeters) represents the radius from the centerline of the rotor to the point in the centrifuge tube where the RCF value is needed and rpm (in revolutions per minute) is equal to the rotor speed.

Since RCF rises with the square of the rotor speed, increasing rpm two times will result in increasing RCF four times. RCF is typically expressed in units of gravitational force, written as "$\times$ g" (times the gravity force). One g equals the gravity force, while two g is twice that force. An RCF of 12,500 means the force is 12,500 times greater than gravity, and it is denoted as 12,500 $\times$ g. Additionally, the centrifugal force increases as the distance from the rotor's center increases. Most protocols detailed in the *"Materials and Methods"* sections of journal articles describe centrifugation in terms of g force. This standardization enables other scientists to replicate the laboratory protocol, regardless of the type of centrifuge or rotor they are using (Stephenson, 2016). In order to find rpm when the g force (RCF) is given, Eq. (3.3) can be written as:

$$\text{rpm} = 1000\sqrt{\frac{RCF}{11.18\,(r_{cm})}} \tag{3.3}$$

The third force acting on particles during centrifugation is the frictional drag force. Molecules moving through a dense medium experience frictional drag, a force that opposes their motion. This frictional force is proportional to the velocity of particles and its friction coefficient, which depends on the shape and size of molecules. As these molecules travel to the bottom side of a centrifuge tube in fixed-angle or swing-out rotors, their velocity increases due to the rising of radial distance. Simultaneously, they encounter frictional drag that matches their velocity. The frictional force is the product of the frictional coefficient and the velocity and it is always acting in opposition to sedimentation (Olatunde et al., 2022).

The fourth force that acts on particles during centrifugation is the buoyant force. This force acts opposite to the direction of gravity and centrifugal force. It is the force exerted by the fluid medium on the particles and depends on the fluid density and the volume of the particles (Sood et al., 2023).

3.3 Types of centrifuges used in the food industry

According to the literature, various types of centrifuges are used for different applications in the field of food industry. Centrifuges can be classified based on their operation mode: batch or continuous. Batch processes are generally reserved for high-value products because they demand significant operator attention and result in higher operating costs (Vanyo, 1993).

Centrifuges can also be classified according to their speed. Low-speed centrifuges operate at speeds ranging from 4,000 rpm to 5,000 rpm and are typically utilized in laboratories for sedimentation of heavy particles. These centrifuges generally function at room temperature without any control of temperature. High-speed centrifuges, operating between 15,000 and 20,000 rpm, are employed for more advanced biochemical

Table 3.1 Advantages and disadvantages of different types of centrifuges.

Centrifuge type	Advantages	Disadvantages
Tubular bowl	- High *G* - Dewatering, drier cake - Simple dismantling of rotating assembly - Easy to clean - Plunger and plough designs, automatic cake removal	- Limited solids storage - Recovery of solids manually and difficult (loss of solid product) - Foaming unless centripetal pump used - Downtime during cake removal cleaning
Chamber bowl	- Clarification efficiency stays constant until solid holding space gets filled. - Large solids holding capacity - Good dewatering - Bowl cooling possible	- No solid discharge - Cleaning more difficult than tubular bowl - Recovery of solids manually and difficult (loss of solid product) - Downtime during cake removal and cleaning
Disk centrifuge	- Solids discharge possible - Liquid discharge under pressure eliminates foaming - Bowl cooling possible	- Poor dewatering - Difficult to clean
Decanter	- Continuous solids discharge (dry cake) - High feed solids concentration	- Lower *G* - Less surface area for clarification compared with disk stack

(From Leung (2020)).

applications and allow for manual control of speed and temperature to accommodate biological samples. Ultracentrifuges, which reach speeds of up to 65,000 rpm, are the most advanced type. The intense heat generated at such high speeds necessitates refrigeration and maintaining a high vacuum in the spinning chambers. Ultracentrifuges are used for both preparative and analytical works (Masoodi et al., 2021).

Another classification of centrifuges is according to their function or their structure. According to function, there are centrifuges used for solid–liquid separation named as clarifiers and there are those used for liquid-liquid separation named as separators. By structure, we distinguish tubular bowl, chamber bowl, decanter and disc stack centrifuges (Berk, 2018) and others. The advantages and disadvantages of some types of centrifuges are showed in Table 3.1. In this chapter, we will focus on the classification of centrifuges according to their structure. The following sections explains the characteristic of each type of centrifuge.

3.3.1 Hydrocyclone

The hydrocyclone, though not technically a centrifuge, is the simplest device that uses centrifugal force for separation. It achieves separation through the slurry motion, caused by the tangential entry of the feed material. This device operates based on the terminal settling velocity of a solid particle within a centrifugal field. The feed enters the hydrocyclone tangentially into its cylindrical section, generating a circular flow pattern with the fluid moving inward toward the vortex finder. High circulating velocities generate a centrifugal field much stronger than gravity, forming an air core along the axis. Heavier particles move outward to the walls of the cylindrical section and travel downward along the conical wall, while smaller particles are carried inward to the vortex finder. The separation happens as the suspension travels through hydrocyclone, resulting in a thickened slurry that exits continuously from the discharge nozzle (Majekodunmi, 2021).

Hydrocyclones are used in various processes, such as removing sand or grit from fruit mash streams before juice extraction, thereby protecting costly machines like decanter scrolls or disc stack desludging systems from premature wear. This type of centrifuge is also used for removing paper fibers from liquid streams in pulp mills. Additionally, hydrocyclones are extensively utilized to eliminate particles from gas streams, like flue gases, and can be applied for particle collection mechanisms in spray dryers to produce various powders in food industry. Although the medium is air rather than liquid, the principle remains the same. Hydrocyclones can also be used in conjunction with filtering centrifuges, due to their utility in feed concentration concentrate which can enhance the performance of the basket or filter centrifuge (Beveridge, 2000).

3.3.2 Tubular bowl centrifuges

The tubular centrifuge is also one of the most basic forms of mechanical centrifuges. It features a vertical tubular rotor with a feed inlet and separate outlets for light and heavy phases, all housed within a stationary casing. Primarily used as separators, tubular centrifuges continuously discharge the two liquid phases. Sludge (solid impurities) are retained in the bowl, and eliminated during cleaning when the device is halted. These machines operate at elevated speeds, generally around 15,000 rpm, generating centrifugal forces exceeding 10,000 g. This allows them to separate liquids that have small differences in density and capture very fine solid particles. However, their capacity is limited by their design (Berk, 2018). Since this type of system does not have a mechanism for ejecting solids, they can only be removed by stopping the machine, disassembling it, and manually scraping or flushing out the solids. Although tubular bowl centrifuges are capable of dewatering, they have a limited capacity for solids. Additionally, foaming can be an issue unless the system is equipped with special skimming or centripetal pumps (Majekodunmi, 2021).

Tubular bowl centrifuges are especially beneficial for capturing valuable particulates present in low concentrations that necessitate high relative centrifugal forces for separation. Another application is the final purification of a fluid stream, particularly using a solid-bowl machine. This is advantageous when the solids to be removed are highly abrasive, as they could cause significant wear on the nozzles or solid ejection ports of a disc stack device, or when high solid compaction is essential. This high compaction is particularly valuable in biotechnological industry where it is crucial to isolate costly precipitates (Beveridge, 2000).

3.3.3 Chamber bowl centrifuges

A chamber bowl centrifuge consists of multiple tubular bowls aligned coaxially. The device features a primary bowl with cylindrical inserts that divide its volume into a series of ring-shaped chambers, allowing sequential operation. The feed enters at the main bowl central part, and the chambers guide the suspension outward, increasing in distance from the axis. Particles settle on the outer walls of the chambers, while the clarified fluid overflows from the chamber with the largest diameter (Olatunde et al., 2022). This device also categorizes the suspended solids: coarse particles settle in the inner chamber, while progressively finer particles accumulate in the subsequent chambers. To remove the sedimented solids, the rotation must be halted for manual cleaning (Majekodunmi, 2021).

3.3.4 Decanter centrifuges

The decanter centrifuge operates on the principle of a clarifier or settling tank, where particles settle because of gravity. However, sedimentation driven only by earth gravity can be very slow. The decanter centrifuge accelerates this process by increasing the sedimentation rate by approximately 2000–5000 folds. Apart from majority of centrifuges, the decanter features a screw mounted concentrically in the bowl, which transports solid materials out of the unit. This design allows the centrifuge to continuously and mechanically separate solids and liquids with controlled efficiency (Letki, 2007; Leung, 2007). Fig. 3.1 shows a schematic representation of a decanter centrifugation device.

Decanter centrifuges, also known as conveyor bowl centrifugation machines, are utilized for solid-liquid separating tasks when feed contains a high concentration of solids and large particle sizes that disk bowl separators cannot handle. They are particularly effective when a high content in dry matter is needed in the separated solids phase (McCarthy, 2011). Decanters can be used for dewatering fine crystals and polymers, as well as for separating soft solids during the process of potable water and wastewater (domestic and industrial) (Letki, 2007).

The decanter centrifuge operates by using a rotating bowl to enhance separation. Inside the bowl, a screw conveyor is positioned concentrically, with the feed slurry introduced through a stationary feed tube into a designated feed zone. The centrifuge speeds up the slurry, forcing the solids to settle along the inner wall of the bowl. The conveyor and bowl work in similar direction but at different speeds, moving the solids to the conical end where they are consolidated and discharged. The liquid travels through the bowl, leaving solids along the way until it reaches adjustable dams or weirs, and is discharged at the opposite end. The liquid discharge may contain entrained gases, which can appear as foam in some applications (Brennan & Grandison, 2011; Letki, 2007).

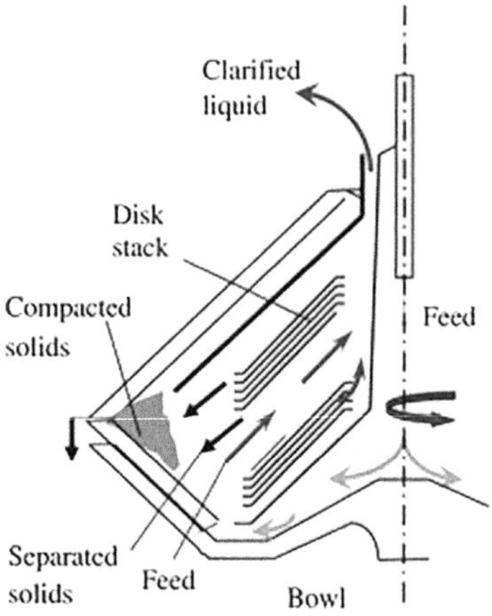

Figure 3.2 Schematic representation of disk stack centrifuge.
(From Leung (2020)).

3.3.5 Disk stack centrifuges

Separation efficiency of centrifugation can be enhanced via flow channeling through multiple narrow, parallel paths between conical discs. Disc-bowl centrifugation devices (or disc stack centrifuges), can operate as clarifiers or separators. The bowl (rotor), which contains a stack of discs, is mounted on a vertical spindle and spins within a fixed housing. These centrifuges have a capacity of up to 100 cubic meters per hour. A disk centrifuge can be used in similar applications as the continuous decanter, but it is generally better suited for situations with lower solids concentration and finer particles. There are various categories of disc-stack clarifiers that vaary in how they discharge the separated solids. These include centrifugation devices having a solid wall bowl, nozzle centrifuges and self-cleaning desludger centrifuges (Berk, 2018; Todaro, 2014). Fig. 3.2 gives a schematic representation of disk stack centrifuges.

In disk centrifuges having a solid wall bowl, liquid parts are continuously removed, but like the tubular centrifugation devices, they lack a sludge outlet. Collected solid fractions must be manually removed after stopping the machine and the bowl is disassembled for cleaning process. Consequently, these centrifuges are primarily used for separation processes, such as for separation of cream from skim milk, and are not suitable for products with elevated solid concentration. Nozzle centrifugation devices have nozzles that allow a continuous removal of solid materials while the centrifuge is working. They are used as clarifiers for products with moderate solid content (around 10% by volume). The bowl has a double-cone design, with solid particles gathering at the point with greatest diameter due to the highest centrifugal force. Narrow nozzles located at this point act as outlets for the solids. To prevent liquid from escaping, the pressure drops via the nozzles have to be elevated, requiring the sludge to be thick enough to ensure a controlled flow without clogging. The centrifuge can be adjusted

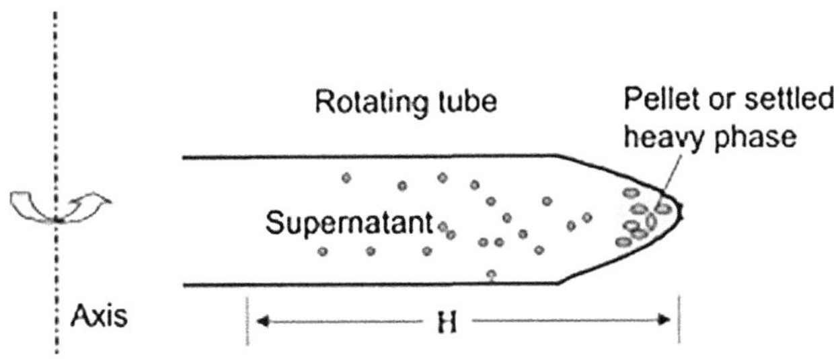

Figure 3.3 Schematic representation of a spintube centrifuge.
(From Leung (2020)).

to the solid content of the feed by using interchangeable nozzles with different hole sizes. Self-cleaning desludger centrifuges are designed to clarify suspensions with high solid content, around 30% to 40% by volume. In these centrifuges, collected sludge is eliminated periodically. The bowl consists of two different conical sections held one to the other with hydraulic pressure. During centrifugation process, solid particles collect in the zone that has the highest diameter, near the junction of the two cones. When this zone is full, the hydraulic system releases the bottom cone, creating an opening between the two parts. Solids are quickly expelled through this gap, after which the bowl closes and a new accumulation cycle begins. This process, which often includes a rinsing phase for the accumulation chamber, is usually automated (Berk, 2018).

3.3.6 Imperforate basket centrifuge

Imperforate basket centrifugation devices are used when the suspension contains a high concentration of solids. These centrifuges are characterized by a drum-shaped bowl or basket that typically rotates around a vertical axis. The centrifugal force causes the solids to accumulate and compress, but it does not remove water molecules. Once the rotation stops, the remaining liquid is drained, and the solid layers are manually dislodged by shoveling or scraping. Alternatively, a semiautomatic process can be employed: a skimmer pipe removes the residual liquid, and a knife blade cuts the solid from the main bowl, eliminating the need to turn off the centrifuge (Olatunde et al., 2022; Todaro, 1996).

3.3.7 Spintube (Bottle) centrifuge

Spintube centrifuges are utilized for laboratory trials and analyses. Samples are loaded in centrifuge tubes. The centrifuge works by spinning small tubes at high speeds around a vertical axis to separate particles based on their density. The centrifugal force pushes denser particles to the bottom side of the tube, forming a pellet, while lighter particles remain in the liquid above called supernatant (Leung, 2020). Fig. 3.3 shows a schematic representation of a spintube centrifuge.

Laboratory centrifuges are typically compact benchtop devices. Samples intended for centrifugation are loaded in tubes that are fitted into either fixed-angle or swing-out holders. These tubes typically hold volumes in the range of 1.5 ml and 15 ml, but there are specific tubes for volumes ranging from 0.5 ml to 100 ml. Rotor assemblies can accommodate between one and 24 tubes. The speed of centrifuge can be set at a constant 3,500 rpm or adjusted to reach 14,000 rpm. The maximum rotational centrifugal force (RCF) depends on the dimensions of the device and rpm, varying between 1,000 g and 16,000 g. More advanced units may include microprocessor control, refrigeration, and temperature regulation (Santhanam & Lipták, 2005).

3.4 Applications of centrifugation in food processing

3.4.1 Clarification of fruit juices and beverages

Juice extraction can be performed through several mechanical methods, including diffusion extraction, decanter centrifuges, screw-type juice extractors, fruit pulpers, and various types of presses (Sharma et al., 2016). Centrifugation plays a crucial role in industry of fruit juice for various purposes. Self-opening centrifuges help in pulp elimination and regulate its content in pineapple and citrus juices. In apple juice production, while tubular bowl centrifuges were initially used for clarification, more recent practices favor nozzle and self-opening machines. Hermetically sealed centrifuges are used to prevent excessive juice aeration. During the production of citrus oil, centrifugation occurs in two phases. Initially, the product generated by the extractor, containing an emulsion with an oil content of 0.5% to 3.0%, is concentrated to reach an oil content of 50% to 70% using a nozzle or self-opening centrifuge. Afterwards, a second centrifuge is used to process the concentrated emulsion in order to extract the citrus oil (Brennan & Grandison, 2011).

Ninga et al. (2022) investigated the effects of centrifugation parameters on the physical and chemical properties of guava juice and examined continuous disc-stack centrifuge operations based on laboratory centrifuge efficiency. The study explored how g-force (149 g to 3,731 g) and centrifugation time (10 to 40 minutes) influence juice properties such as protein, pectin, dry matter, sugars, pH, conductivity, clarity, and particle size. Results indicated a notable reduction in mean particle size and the contents of proteins and pectin, whereas juice clarity increased. Comparable clarification efficacy was achieved at g-forces exceeding 1,343 g. In the study of Biswas et al. (2016), centrifugation and microfiltration (MF) were utilized in the initial clarification of raw bottle gourd juice. Response surface methodology was employed to find optimum conditions for centrifugation (time and speed) were determined with and MF (transmembrane pressure drop (TMP) and cross-flow rate). The clarified juice was evaluated for clarity, total solids, protein, polyphenol, total soluble solids, sodium, and potassium. The quality of the clarified juice and power consumption were compared for both methods. Microfiltered juice exhibited higher clarity (97%) but had a lower polyphenol content (8 mg GAE/100 ml) in comparison with centrifuged juice (13 mg

GAE/100 ml). Centrifugation and microfiltration were also utilized and compared for the clarification process of stevia extract in the study of Chhaya et al. (2013) and for the clarification of jamun juice in the study of Ghosh et al. (2018).

Centrifugation can be used alone or in combination with other operations for juice clarification. Rai et al. (2007) used centrifugation and enzymatic treatment as clarification pretreatments during ultrafiltration of mosambi juice. De Oliveira et al. (2012) utilized centrifugation and enzymes as pretreatments before clarification of passion fruit juice using microfiltration. The study of Domingues et al. (2012) was about using chitosan for the clarification of passion fruit juice. The impacts of coagulation process parameters were evaluated and compared with centrifugation and enzymatic treatments. The optimal condition for chitosan treatment identified in these tests was compared with both centrifugation and enzymatic treatments. For the centrifugation process, two rotation speeds were tested: 4,000 rpm and 12,000 rpm. The enzymatic treatment involved using 1 ml/L of Pectinex 3X L (Novo Nordisk, Switzerland) for a period of 90 minutes at a temperature of 50°C. While the enzymatic treatment was effective solely in reducing viscosity, the chitosan treatment combined with mild centrifugation yielded the best results in the clarification of passion fruit juice. Singh and Das (2021), evaluated the use of centrifugation pretreatment of citrus fruit juice before clarification with membrane separation treatment. Results showed that juice clarity experienced sharp increase when rising device speed from 1,500 to 4,500 rpm. At the beginning, clarity improved from 0.6% to 85%, and viscosity decreased from 2.56 mPa.s to 1.42 mPa.s as the rpm increased from 0 to 6,000. After 4500 rpm, changes in clarity and viscosity were minimal. The optimal machine speed was determined to be 4,500 rpm. Meyer et al. (2001) utilized centrifugation as a pretreatment before clarification of cherry juice with enzymes. Centrifuging cherry juice at 10,000 g for 15 minutes before clarification significantly enhanced juice clarity.

3.4.2 Cream separation in dairy industry

Centrifugation serves multiple purposes in the dairy industry, including skimming hot or cold milk, whey skimming, and skimming buttermilk. Milk skimming is one of the most common and vital applications of separators today. Dairy regulations worldwide mandate the pasteurization of raw milk, and separators are a crucial component of pasteurizers, operating at temperatures between 50°C and 60°C. For milk skimming, there are three operating modes: full skimming (producing skimmed milk and cream), standardizing (producing standardized milk and cream), and back blending (producing whole milk). In addition to skimming, separators also clean the milk of solid impurities. In the standardizing mode, high skimming efficiency is not required, allowing the separator to operate at a higher capacity than in the full-skimming mode. Sometimes, skimming is done at cold temperatures (between 4°C and 15°C) to avoid thermal changes and eliminate the need for heating energy at that stage. However, cold skimming is less efficient than hot skimming and the viscosities of cream and skimmed milk are higher. Consequently, the flow paths must be significantly larger. Besides, to enhance performance, the temperature can be slowly raised to nearly 20°C. Unlike milk which has a fat content of (4%), whey has lesser fat content, approximately 0.4%. The main source of whey is cheese industry. Separators can be used for the recovery of the fat contained in cheese whey. Buttermilk, obtained after cream transformation into a concentrated form, contains certain amounts of fat. This remaining fat can be obtained by the use of separators (Brennan & Grandison, 2011; Heymann, 2011).

In the dairy processes, the frequently utilized centrifuges are disk bowl centrifuges and decanter centrifuges. Bowl centrifuges are used for milk skimming, partial homogenization, standardization, clarification from impurities and microorganisms as well as cream processing fresh cheese production by separating whey from curd after coagulation and whey processing to separate cheese fines and fat from it. Decanter centrifuges are generally employed to remove whey from casein curd. Initially, the mixture of curd and whey passes via a dewheying screen to lessen the hydraulic load on the decanter. Following a washing process, decanters are used to dewater the curd. The objective of this operation is to achieve the driest possible curd before proceeding to the drying stage. In addition, decanters are employed to extract lactose crystals from the crystal-mother liquor mixture formed during whey or whey permeate crystallization. This process involves two-step decantation, where wash water is added to the solid part from the first decanter. During the transformation of edible lactose to refined lactose, decanter centrifuges are also utilized for the recovery of lactose after recrystallization. Other applications of decanters include concentration of cheese fines issuing from disk bowl cheese whey separators and clarifiers. Furthermore, the sludge generated from flotation and sedimentation stages in dairy wastewater treatment can be concentrated using decanter centrifuges prior to disposal, thereby lowering disposal expenses (McCarthy, 2011).

3.4.3 Oil and fat extraction

In the initial stage of oil refining, named as degumming, water, dilute acid, or alkali are used to treat crude oil in order to remove phosphatides and mucilaginous materials. Nozzle or self-opening centrifuges are utilized for the extraction of gums after treatment. For acid-degumming, hot water is mostly used for oil washing. Then, washings are eliminated with centrifugation. The flowing stage is neutralization, where caustic soda is used to treat free fatty acids, phosphatides, and some pigments, forming soapstock. Afterwards, nozzle or self-opening centrifuges are utilized to separate this soapstock from the oil. At the end, the oil is washed again with hot water, and the washings are eliminated using centrifuges (Brennan & Grandison, 2011).

Centrifugation is widely used to extract olive oil (Alburquerque et al., 2004; Gila et al., 2022; Tamborrino et al., 2015). Klen and Vodopivec (2012) compared the traditional press with two- and three-phase centrifuge for the processing of olive oil. The two-phase centrifuge achieved the highest phenol transfer rate to oil (1.5%) and the greatest antioxidant potential. This was followed by a traditional press (1.2%), while a three-phase centrifuge had the lowest phenol transfer rate (0.5%), with over 30% of phenols being lost in the wastewater produced. In another research, two-phase and three-phase cetrifuging decanters were utilized in virgin olive oil production (Kalogeropoulos et al., 2014). Decanter centrifuge was also used to extract olive oil in the study of Altieri et al. (2013). In the study of Altieri et al. (2013), horizontal decanter centrifugation device equipped with a screw conveyor was used to extract olive oil and oil/water levels as well as differential speed were optimized.

Centrifugation can be utilized to extract coconut oil (Agarwal, 2017; Ng et al., 2021). In a research study, extraction of virgin coconut oil (VCO) using centrifuges was investigated, varying centrifugation speed, temperature, and time periods. Results indicated a VCO yield of 13.53% at 12,000 rpm over 120 minutes. The best yield, 13.80%, was achieved with a centrifugation temperature equal to 40°C. Lauric acid content in samples varied with centrifugation conditions, measuring 0.4543 μg, 6.2367 μg, and 6.4894 μg for different temperatures, speeds, and time intervals respectively (Wong & Hartina, 2014). In the study of Coulibaly et al. (2009), experiments were about extracting shea butter using a centrifugation machine. Laboratory tests with a small centrifuge examined the effects of kernel paste temperature, dilution rate, and centrifugation speed on extraction efficiency. High efficiencies, over 30%, were achieved with accelerations above 1,293 g, kernel paste temperatures above 90°C, and dilution rates over 50%.

Centrifugation was utilized to extract oil from groundnut and sesame seeds (Kehail et al., 2016). The oil extraction yield reached 45 % within 18 mins. The extraction yield increased with the increase of centrifugation speed. Pérez-Saucedo et al. (2021) combined centrifugation with ultrasound assisted extraction to recover oil from avocado. Results showed that centrifugation combined ultrasound-assisted extraction enhanced the physicochemical properties, including acidity, peroxide levels, and iodine values, compared to centrifugation extraction and the control (extraction with hexane). The combination also increased the yield, nutritional quality of fatty acids, oxidative stability, shelf life, and the omega-6 to omega-3 ratio comparing to other methods. Gupta and Muralidhara (2002) used centrifugation to efficiently separate lecithin from soybean oil. Moreau et al. (2010) recovered crude corn oil using centrifugation method. In another research, centrifugal force was used in combination with microwave heating to extract essential oils from orange peels (Angoy et al., 2020). This combination reduced the extraction time comparing to the hydrodistillation technique and offered the benefit of collecting both volatile and nonvolatile compounds in a single experiment.

3.4.4 Yeast separation during yeast production

Centrifugation can be utilized for harvesting and separating yeast after fermentation (Powell et al., 2003). Yeast can be processed using centrifugal machines that significantly reduce water content from fermentation media. The yeast is repeatedly washed and separated to eliminate extracellular materials, such as molasses remnants. After the first centrifugation, approximately 20% yeast solids are obtained, which contain about 50% cells by volume. The second centrifugation produces a cream-like yeast mass with 18% to 20% solid content, which can be pumped and further dewatered using a rotary filtering unit. Disk stack centrifuges are often used to concentrate yeast solids from fermentation and effectively separate big quantities of sludge. The centrifugal force pushes the yeast solids to the periphery of the bowl, where they sediment on the inclined surfaces of conical disks and are discharged through nozzles. The design of these conical elements and nozzles is crucial and affect the separation efficiency (Gélinas, 2017). Hydrocyclones can also be used for yeast separation (Bicalho et al., 2013). Lopes and Keshavarz-Moore (2012) dewatered and separated baker's yeast cells using tubular bowl and disk stack centrifuges.

The utilization of centrifuges to remove and collect yeast has become standard practice due to several benefits. These include reduced process time, cost savings (after initial capital investment) and enhanced productivity. However, it's crucial to ensure that centrifugation doesn't raise temperatures above 20°C. Centrifugation can sometimes damage yeast cells physically. This damage depends on the working parameters of the centrifugation device, like hydrodynamic forces, and yeast cell interactions within the disc stack centrifuge (Stewart, 2017).

3.4.5 Dewatering of food products

Centrifugation can be utilized for dewatering and concentration of food products and the production of high-quality food concentrates (Petzold & Aguilera, 2013). Santana et al. (2020) used centrifugation-assisted freeze concentration for concentrating blueberry juice. The study aimed to investigate how time and temperature parameters affect the centrifugation process utilized in assisting the first cycle of freeze concentration for blueberry juice. The solid content in the concentrate changed from the initial 18° Brix to nearly 60% in the first freeze concentration cycle. The combination between time and temperature significantly influenced the amounts of concentrate and solutes obtained, with an optimal concentration achieved at 15°C and a centrifugation period of 20 minutes. Orellana-Palma et al. (2017) utilized centrifugal freeze concentration to concentrate orange juice and increase its ascorbic acid retention. The process permitted to obtain a concentrate having high solid concentration and ascorbic acid retention. This technique was also utilized for the concentration of blueberry and pineapple juices (Petzold et al., 2015) and apple juice (Qin et al., 2021).

Dewatering of microalgae can be made using centrifugation. Centrifugation is one of the best techniques for fast harvesting of algae containing high-value compounds (Sharma et al., 2013). Centrifugation uses centrifugal force instead of gravity to separate microalgae from suspension. The efficiency of this process is determined according to the settling characteristics of the microalgae, the retention time of the slurry in the centrifuge (controlled by flow rate), and the settling depth, which can be minimized via centrifuge design. Common centrifuge types include disc stack, decanter, and hydrocyclones. While centrifugation can effectively harvest microalgae, it has several drawbacks. High gravitational and shear forces can affect the structure of cells. Processing large volumes of microalgae suspension can be time-consuming and may require expensive continuous centrifuge equipment. Additionally, energy consumption and capital costs for centrifugation are quite high (Chen et al., 2015).

3.4.6 Starch separation in starch industry

Many studies reported the use of centrifugation for separation purposes in starch production (Van Der Borght et al., 2005). Talab (2016) produced high-purity starch from green pea using centrifugation. Devereux et al. (2011) used centrifugation for separating starch from industrial wastewater. Results showed that centrifugation was demonstrated to be the best method of separation, with a high-quality starch obtained at the end. In another study, a conical basket centrifuge was used for obtaining tapioca starch from pulp (Saengchan et al., 2014). Impacts of rotational speed and liquid to solid (L/S) ratio on cake formation, and starch-pulp separation efficiency were evaluated. Results showed that raising the rotational speed and basket radius enhanced starch separation efficiency by boosting the centrifugal force, which created a significant pressure decrease through the filtering screen. This increased centrifugal force influenced the properties of the deposited cake, including its specific resistance, solidosity, permeability, and thickness. Saengchan et al. (2015) used centrifugation to enhance starch recovery from the pulp of cassava root. Results showed that as the revolution centrifugal force increased, some bound starch granules were released due to the force acting on the cassava pulp. Vernon-Carter et al. (2015) utilized centrifugation for the production of gelatinized corn starch dispersion fractions.

Density-gradient ultracentrifugation was used by Hong et al. (2005) for the molecular fractionation of corn and potato starches. The study highlighted the separation of amylose and amylopectin in corn and potato starches by centrifugation at 124,000 g for three hours to 72 hours at 40°C using a gradient media, Nycodenz. This separation was according to differences in their sedimentation rates. Amylopectin, being large and highly branched, migrated faster and reached its buoyant density (1.25 g/ml) quickly, forming a sharp carbohydrate peak. Amylose, a smaller and linear component, moved slower, covering a broader density range and eventually overlapping with amylopectin. After 3 hours of centrifugation, clear separation of amylose and amylopectin was achieved, revealing intermediate starch molecules in both corn (11.5%) and potato (7.7%) starches. Total carbohydrate analysis showed amylose contents of 22.6% for corn and 21.1% for potato. Under alkaline conditions, the sedimentation rate and buoyant density of amylopectin decreased, likely because of structural changes caused by the alkali.

3.4.7 Protein separation for food additives

Centrifugation is one among various techniques that can be used to extract and separate proteins from different sources (Novák & Havlíček, 2016; Sentís-Moré et al., 2022). Decanter and benchtop centrifuges were used during the extraction process of whole soybean proteins. Purschke et al. (2018) utilized centrifugal fractionation for the recovery and concentration of proteins from mealworm larvae. The study assessed a centrifugal fractionation method by examining the effects of solubilization pH, centrifugation speed/force, and centrifugation time on the protein extraction from mealworm larvae. Centrifugation force and solubilization pH were found to be the primary factors influencing protein amounts in the supernatant phase. Throughout experiments, protein concentration increased from an initial 39 g/100 g dry base (db) in the larvae to 56–65 g/100 g db in the extracts, with recovery rates ranging between 31% and 58%. These findings highlighted the critical factors affecting the performance of centrifugal separation for mealworm protein concentration and recovery, offering valuable data for designing industrial separation processes.

In the study of Haller and Kulozik (2019), continuous decanter centrifugation was used to separate whey protein aggregates. The study explored the use of decanter centrifugation device for efficient separation of whey proteins, focusing on two protein aggregates: α-lactalbumin (α-la) and β-lactoglobulin (β-lg). The α-la aggregates, stabilized non-covalently, showed enhanced sedimentation and high compressibility in scale-down experiments but faced particle breakdown and discharge issues in the decanter. In contrast, the β-lg aggregates, stabilized by disulfide bonds, formed rigid, low-compressibility aggregates that were successfully discharged as free-flowing powder. The experiments demonstrated the potential of decanters in whey protein separation and provided insights into the separability of protein aggregates using continuous centrifuges.

A decanter centrifuge was used in the study of Haller et al. (2021) to separate aggregated β-lactoglobulin. The study explored the thermal treatment for selective aggregation of β-lactoglobulin, which was then separated from the α-lactalbumin-enriched centrate using a decanter centrifugation device. The goal was to determine if pH-induced flocculation could enhance centrifugal separation compared to non-flocculated suspensions at pH 7.0. It was found that the most effective flocculation happened at pH 4.4, aiding the separation of aggregate fragments smaller than 20 μm. Ultimately, aggregate flocculation in the decanter achieved separation efficiencies of over 99% even at low centrifugal forces. Schubert et al. (2018) made casein fractionation with a decanter centrifuge.

3.4.8 Centrifugation in waste water treatment

In order to achieve the goal of sustainable production, many techniques can be used to recover and separate value-added compounds from food-processing wastewater (Lee & Stuckey, 2022). Centrifugation is one of these techniques. Centrifugation was used to recover proteins and other solids from stick-water generated during sardine fishmeal processing (Garcia-Sifuentes et al., 2009). In the study of Lateb et al. (2024), centrifugation was used for the treatment of olive mill wastewater and to separate it from the residual sludge. Results were compared with those of gravity decantation. Centrifugation significantly enhanced the treatment of olive mill wastewater and reduced the treatment duration from 4 h to 10 min comparing to gravity decantation. In one study, to reuse the waste water, centrifugation was utilized for effective removal of solids during wash-water treatment in the fresh-cut fruit and vegetable industry (Mundi & Zytner, 2015).

3.4.9 Laboratory applications and quality control

Centrifugation has many applications in quality control of food products. It can be utilized to separate and concentrate bacteria from complex sample matrices (Stevens & Jaykus, 2004). Brewster and Paul (2016) used centrifugation to isolate bacteria from raw milk in order to quantify *Salmonella* spp. With PCR detection method. This method allowed the recovery of bacteria without affecting its viability. In addition, centrifugation can be used for the characterization of food quality and structural stability as it was demonstrated by (Lerche, 2003). Choi et al. (2017) mentioned that centrifugation can be used for isolating and detecting pathogenic *Escherichia coli* in food products. Nishino et al. (2018) utilized density gradient centrifugation to separate viable probiotic lactic acid bacteria from fermented milk to evaluate its physiological state.

3.4.10 Separation of microorganisms and toxic compounds from food products

Centrifugation considered from the earliest methods used to effectively separate and concentrate microorganisms from aqueous broths and food products. When operated at high rotor speeds (around 15,000 × g), most microorganisms (>1 μm) will form a pellet from the suspension within 15 minutes. However, under these conditions, food particles and the microorganisms they carry will also pellet, making it difficult to separate microorganisms from food debris. By employing differential centrifugation, which involves alternating between low and high-speed centrifugation, microorganisms can be more effectively separated from larger food particles (Patel, 2000).

According to Gésan-Guiziou (2010), centrifugation can be used for removing bacteria, spores and somatic cells from milk. Ágoston et al. (2009) used continuous flow centrifugation to separate *Bacillus* and *Clostridium* spores from large milk quantities. Specifically, the research focused on concentrating around 10^3 bacterial spores in 3.7 liters of whole and skim milk using *Bacillus subtilis (B. subtilis), Bacillus atrophaeus (B. atrophaeus)*, and *Clostridium sporogenes* (*C. sporogenes*) spores. Standard plating techniques were used to enumerate the spores in concentrated samples. Conducted at 10,000 rpm with a 0.7 liter/min flow rate, three independent experiments revealed mean spore recoveries of 71.3% and 56.5% for *B. subtilis* in skim and whole milk, respectively; 55% and 59.3% for *B. atrophaeus*; and 88.2% and 78.6% for *C. sporogenes*. The results indicate that CFC is a promising method for rapidly concentrating bacterial spores from large milk volumes. Centrifugation was used among other method to remove aflatoxin M1 from milk (Kuharić et al., 2018). It was found that centrifugation improved the removal of aflatoxin from milk.

3.5 Conclusions

In this chapter, we have explored the fundamental principles and diverse uses of centrifugation in the field of food industry. Centrifugation, a pivotal separation technique, leverages centrifugal force in order to efficiently separate components according to their densities, enhancing the safety and quality of food products. The versatility of centrifuges, from batch to continuous operation and from low-speed to ultracentrifuges, underscores their indispensability in food processing. Different types of centrifuges, including disc stack, tubular bowl, chamber bowl, decanters and others, each serve unique roles in the industry, from clarifying juices to extracting valuable oils and removing unwanted particulates.

The advantages of centrifugation, such as high separation efficiency, scalability, and the capability to handle a wide range of particle sizes, make it a preferred choice in various applications of food processing. However, selecting the appropriate type of centrifuge for the appropriate food processing application is crucial to obtain an efficient separation process.

As food processing continues to evolve, the role of centrifugation will undoubtedly expand, driven by advancements in technology and a growing emphasis on product quality and process efficiency. This chapter aimed to provide a comprehensive overview of the mechanisms, equipment, and applications of centrifugation, offering valuable insights for professionals in the food industry. The knowledge gained here sets the stage for further exploration and innovation in the use of centrifugal techniques to meet the ever-changing demands of food production and safety.

References

Agarwal, R. K. (2017). Extraction processes of virgin coconut oil. *MOJ Food Processing & Technology, 4*(2). https://doi.org/10.15406/mojfpt.2017.04.00087.

Ágoston, R., Soni, K. A., McElhany, K., Cepeda, M. L., Udi, Z., Tzipori, S., Mohácsi-Farkas, C., & Pillai, S. D. (2009). Rapid concentration of Bacillus and Clostridium spores from large volumes of milk, using continuous flow centrifugation. *Journal of Food Protection, 72*(3), 666–668. https://doi.org/10.4315/0362-028x-72.3.666.

Alburquerque, J. A., Gonzálvez, J., García, D., & Cegarra, J. (2004). Agrochemical characterisation of "alperujo", a solid by-product of the two-phase centrifugation method for olive oil extraction. *Bioresource Technology, 91*(2), 195–200. https://doi.org/10.1016/S0960-8524(03)00177-9.

Altieri, G. (2010). Comparative trials and an empirical model to assess throughput indices in olive oil extraction by decanter centrifuge. *Journal of Food Engineering, 97*(1), 46–56. https://doi.org/10.1016/j.jfoodeng.2009.09.014.

Altieri, G., Di Renzo, G. C., & Genovese, F. (2013). Horizontal centrifuge with screw conveyor (decanter): Optimization of oil/water levels and differential speed during olive oil extraction. *Journal of Food Engineering, 119*(3), 561–572. https://doi.org/10.1016/j.jfoodeng.2013.06.033.

Angoy, A., Ginies, C., Goupy, P., Bornard, I., Ginisty, P., Sommier, A., Valat, M., & Chemat, F. (2020). Development of a green innovative semi-industrial scale pilot combined microwave heating and centrifugal force to extract essential oils and phenolic compounds from orange peels. *Innovative Food Science and Emerging Technologies, 61*, 102338. https://doi.org/10.1016/j.ifset.2020.102338.

Berk, Z. (2018). Centrifugation. *Food Process Engineering and Technology*, 243–259. doi:10.1016/B978-0-12-812018-7.00009-9.

Beveridge, T. (2000). Large-scale centrifugation. In I. D. Wilson, E. R. Adlard, M. Cooke, & C. F. Poole (Eds.), *Encyclopedia of Separation Science* (pp. 320–342). London: Academic Press.

Bicalho, I. C., Mognon, J. L., Shimoyama, J., Ataíde, C. H., & Duarte, C. R. (2013). Effects of operating variables on the yeast separation process in a hydrocyclone. *Separation Science and Technology (Philadelphia), 48*(6), 915–922. https://doi.org/10.1080/01496395.2012.712597.

Biswas, P. P., Mondal, M., & de, S. (2016). Comparison between centrifugation and microfiltration as primary clarification of bottle gourd (*Lagenaria siceraria*) juice. *Journal of Food Processing and Preservation, 40*(2), 226–238. https://doi.org/10.1111/jfpp.12599.

Brennan, J. G., & Grandison, A. S. (2011). Separations in food processing part 1. *Food processing handbook: 2nd ed.* https://doi.org/10.1002/9783527634361.ch11.

Brewster, J. D., & Paul, M. (2016). Short communication: Improved method for centrifugal recovery of bacteria from raw milk applied to sensitive real-time quantitative PCR detection of Salmonella spp. *Journal of Dairy Science, 99*(5), 3375–3379. https://doi.org/10.3168/jds.2015-9655.

Chen, C. L., Chang, J. S., & Lee, D. J. (2015). Dewatering and drying methods for microalgae. *Drying Technology, 33*(4), 443–454. https://doi.org/10.1080/07373937.2014.997881.

Chhaya, R., Majumdar, G. C., & De, S. (2013). Primary clarification of stevia extract: A comparison between centrifugation and microfiltration. *Separation Science and Technology (Philadelphia), 48*(1), 113–121. https://doi.org/10.1080/01496395.2012.674605.

Choi, Y., Lee, H., Lee, S., Kim, S., Lee, J., Ha, J., Oh, H., & Yoon, Y. (2017). Comparison of upgraded methods for detecting pathogenic Escherichia coli in foods using centrifugation or filtration. *Korean Journal for Food Science of Animal Resources, 37*(6), 799–803. https://doi.org/10.5851/kosfa.2017.37.6.799.

Coulibaly, Y., Ouédraogo, S., & Niculescu, N. (2009). Experimental study of shea butter extraction efficiency using a centrifugal process. *Journal of Engineering and Applied Sciences, 4*(6), 14–19.

De Oliveira, R. C., Docê, R. C., & De Barros, S. T. D. (2012). Clarification of passion fruit juice by microfiltration: Analyses of operating parameters, study of membrane fouling and juice quality. *Journal of Food Engineering, 111*(2), 432–439. https://doi.org/10.1016/j.jfoodeng.2012.01.021.

Devereux, S., Shuttleworth, P. S., Macquarrie, D. J., & Paradisi, F. (2011). Isolation and characterization of recovered starch from industrial wastewater. *Journal of Polymers and the Environment, 19*(4), 971–979. https://doi.org/10.1007/s10924-011-0342-0.

Domingues, R. C. C., Faria Junior, S. B., Silva, R. B., Cardoso, V. L., & Reis, M. H. M. (2012). Clarification of passion fruit juice with chitosan: Effects of coagulation process variables and comparison with centrifugation and enzymatic treatments. *Process Biochemistry, 47*(3), 467–471. https://doi.org/10.1016/j.procbio.2011.12.002.

Garcia-Sifuentes, C., Pacheco-Aguilar, R., Lugo-Sánchez, M., Garcia-Sánchez, G., Ramirez-Suarez, J. C., & Garcia-Carreño, F. (2009). Properties of recovered solids from stick-water treated by centrifugation and pH shift. *Food Chemistry, 114*(1), 197–203. https://doi.org/10.1016/j.foodchem.2008.09.064.

Gélinas, P. (2017). Compressed baker's yeast: Mapping patents on post-fermentation processes. *Comprehensive Reviews in Food Science and Food Safety, 16*(3), 456–476. https://doi.org/10.1111/1541-4337.12256.

Gésan-Guiziou, G. (2010). Removal of bacteria, spores and somatic cells from milk by centrifugation and microfiltration techniques. *Improving the Safety and Quality of Milk, 1*, 349–372. https://doi.org/10.1533/9781845699420.4.349.

Ghosh, P., Pradhan, R. C., & Mishra, S. (2018). Clarification of jamun juice by centrifugation and microfiltration: Analysis of quality parameters, operating conditions, and resistance. *Journal of Food Process Engineering, 41*(1). https://doi.org/10.1111/jfpe.12603.

Gila, A., Aguilera, M. P., Sánchez-Ortíz, A., Jiménez, A., & Beltrán, G. (2022). Effect of centrifugal froce (G) on stability of natural emulsions (water/oil) present in fresh virgin olive oils. *Journal of Food Engineering, 334*. https://doi.org/10.1016/j.jfoodeng.2022.111169.

Gupta, R., & Muralidhara, H. S. (2002). A novel approach to process crude oil membrane concentrate using a centrifuge. *JAOCS Journal of the American Oil Chemists' Society, 79*(5), 419–423. https://doi.org/10.1007/s11746-002-0499-1.

Haller, N., Greßlinger, A. S., & Kulozik, U. (2021). Separation of aggregated β-lactoglobulin with optimised yield in a decanter centrifuge. *International Dairy Journal, 114*, 104918. https://doi.org/10.1016/j.idairyj.2020.104918.

Haller, N., & Kulozik, U. (2019). Separation of whey protein aggregates by means of continuous centrifugation. *Food and Bioprocess Technology, 12*(6), 1052–1067. https://doi.org/10.1007/s11947-019-02275-1.

Heymann, B. (2011). Plant and equipment: Centrifuges and separators: Types and design. *Encyclopedia of Dairy Sciences: Second Edition*, 166–174. https://doi.org/10.1016/B978-0-12-374407-4.00406-4.

Hong, J. Y., Yoon, J.-W., & Lim, S. T. (2005). Molecular fractionation of starch in nycodenz-dimethyl sulfoxide by density-gradient ultracentrifugation. *Starch/Staerke, 57*(10), 488–493. https://doi.org/10.1002/star.200500405.

Jerman Klen, T., & Mozetič Vodopivec, B. (2012). The fate of olive fruit phenols during commercial olive oil processing: Traditional press versus continuous two- and three-phase centrifuge. *LWT, 49*(2), 267–274. https://doi.org/10.1016/j.lwt.2012.03.029.

Kalogeropoulos, N., Kaliora, A. C., Artemiou, A., & Giogios, I. (2014). Composition, volatile profiles and functional properties of virgin olive oils produced by two-phase vs three-phase centrifugal decanters. *LWT, 58*(1), 272–279. https://doi.org/10.1016/j.lwt.2014.02.052.

Kehail, M. A., Mohamed, R. A., Abdelrahim, Y. M., & Ibrahim, N. A. (2016). Extraction rates of oils of some seeds by using different speeds of the centrifuge. *Asian Journal of Applied Sciences, 4*(6), 2–7.

Kuharić, Ž., Jakopović, Ž., Čanak, I., Frece, J., Bošnir, J., Pavlek, Ž., Ivešić, M., & Markov, K. (2018). Removing aflatoxin M1 from milk with native lactic acid bacteria, centrifugation, and filtration. *Arhiv Za Higijenu Rada i Toksikologiju, 69*(4), 334–339. https://doi.org/10.2478/aiht-2018-69-3160.

Lateb, Y., Moussaoui, R., Elias, A., Sharrock, P., & Amrouche, T. (2024). Centrifugation effect on olive mill wastewater treatment by coagulation-flocculation using $Ca(OH)_2$-pectin system. *Waste and Biomass Valorization, 15*(7), 4051–4060. https://doi.org/10.1007/s12649-024-02418-2.

Lee, S. Y., & Stuckey, D. C. (2022). Separation and biosynthesis of value-added compounds from food-processing wastewater: Towards sustainable wastewater resource recovery. *Journal of Cleaner Production, 357*, 131975. https://doi.org/10.1016/j.jclepro.2022.131975.

Lerche, D. (2003). *Characterisation of food qualityand structural stability by analytical centrifugation characterization of dispersions view project*. 149–153. https://www.researchgate.net/publication/242574741.

Letki, A. G. (2007). Centrifugation: Decanters. *Encyclopedia of Separation Science*, 1–10. https://doi.org/10.1016/b978-012226770-3/10688-0.

Leung, W. (2007). Decanter centrifuge. *Centrifugal Separations in Biotechnology*. https://doi.org/10.1016/B978-185617477-0/50005-7.

Leung, W. W.-F. (2020). Selection and sizing of centrifuges. *Centrifugal Separations in Biotechnology*, 203–227. https://doi.org/10.1016/b978-0-08-102634-2.00009-x.

Lopes, A. G., & Keshavarz-Moore, E. (2012). Prediction and verification of centrifugal dewatering of P. pastoris fermentation cultures using an ultra scale-down approach. *Biotechnology and Bioengineering, 109*(8), 2039–2047. https://doi.org/10.1002/bit.24478.

Luckiram, B. (2022). Centrifugation. *Integration and optimization of unit operations: Review of unit operations from R&D to production: Impacts of upstream and downstream process decisions*. Elsevier Inc. https://doi.org/10.1016/B978-0-12-823502-7.00014-1.

Majekodunmi, S. O. (2021). A review on centrifugation in the pharmaceutical industry. *International Journal of Pharmaceutical Research, 13*(3), 67–78. https://doi.org/10.31838/ijpr/2021.13.03.240.

Masoodi, K. Z., Lone, S. M., & Rasool, R. S. (2021). Study of principle of centrifugation. *Advanced Methods in Molecular Biology and Biotechnology*, 133–137. https://doi.org/10.1016/b978-0-12-824449-4.00023-2.

McCarthy, O. J. (2011). Plant and equipment: Centrifuges and separators: Applications in the dairy industry. *Encyclopedia of Dairy Sciences: Second Edition*, 175–183. https://doi.org/10.1016/B978-0-12-374407-4.00463-5.

Meyer, A. S., Köser, C., & Adler-Nissen, J. (2001). Efficiency of enzymatic and other alternative clarification and fining treatments on turbidity and haze in cherry juice. *Journal of Agricultural and Food Chemistry, 49*(8), 3644–3650. https://doi.org/10.1021/jf001297n.

Moreau, R. A., Hicks, K. B., Johnston, D. B., & Laun, N. P. (2010). The composition of crude corn oil recovered after fermentation via centrifugation from a commercial dry grind ethanol process. *Journal of the American Oil Chemists' Society, 87*(8), 895–902. https://doi.org/10.1007/s11746-010-1568-z.

Mundi, G. S., & Zytner, R. G. (2015). Effective solid removal technologies for wash-water treatment to allow water reuse in the fresh-cut fruit and vegetable industry. *Journal of Agricultural Science and Technology A, 5*(6), 396–407. https://doi.org/10.17265/2161-6256/2015.06.003.

Ng, Y. J., Tham, P. E., Khoo, K. S., Cheng, C. K., Chew, K. W., & Show, P. L. (2021). A comprehensive review on the techniques for coconut oil extraction and its application. *Bioprocess and Biosystems Engineering, 44*(9), 1807–1818. https://doi.org/10.1007/s00449-021-02577-9.

Ninga, K. A., Desobgo, S. C. Z., Nso, E. J., & Kayem, J. (2022). White-flesh guava juice clarification by a fixed-angle conical rotor centrifuge laboratory and characterization of continuous disk stack centrifuges. *Heliyon, 8*(11), e11606. https://doi.org/10.1016/j.heliyon.2022.e11606.

Nishino, T., Matsuda, Y., & Yamazaki, Y. (2018). Separation of viable lactic acid bacteria from fermented milk. *Heliyon, 4*(4), e00597. https://doi.org/10.1016/j.heliyon.2018.e00597.

Novák, P., & Havlíček, V. (2016). Protein extraction and precipitation. *Proteomic Profiling and Analytical Chemistry: The Crossroads: Second Edition*, 52–62. https://doi.org/10.1016/B978-0-444-63688-1.00004-5.

Olatunde, A., Obidola, M. S., & Tijjani, H. (2022). Centrifugation techniques. *Analytical techniques in biosciences: From basics to applications* https://doi.org/10.1016/B978-0-12-822654-4.00008-7.

Orellana-Palma, P., Petzold, G., Andana, I., Torres, N., & Cuevas, C. (2017). Retention of ascorbic acid and solid concentration via centrifugal freeze concentration of orange juice. *Journal of Food Quality, 2017*. https://doi.org/10.1155/2017/5214909.

Patel, P. (2000). A review of analytical separation, concentration and segregation techniques in microbiology. *Journal of Rapid Methods and Automation in Microbiology, 8*(4), 227–248. https://doi.org/10.1111/j.1745-4581.2000.tb00326.x.

Pérez-Saucedo, M. R., Jiménez-Ruiz, E. I., Rodríguez-Carpena, J. G., Ragazzo-Sánchez, J. A., Ulloa, J. A., Ramírez-Ramírez, J. C., Gastón-Peña, C. R., & Bautista-Rosales, P. U. (2021). Properties of the avocado oil extracted using centrifugation and ultrasound-assisted methods. *Food Science and Biotechnology, 30*(8), 1051–1061. https://doi.org/10.1007/s10068-021-00940-w.

Petzold, G., & Aguilera, J. M. (2013). Centrifugal freeze concentration. *Innovative Food Science and Emerging Technologies, 20*, 253–258. https://doi.org/10.1016/j.ifset.2013.05.010.

Petzold, G., Moreno, J., Lastra, P., Rojas, K., & Orellana, P. (2015). Block freeze concentration assisted by centrifugation applied to blueberry and pineapple juices. *Innovative Food Science and Emerging Technologies, 30*, 192–197. https://doi.org/10.1016/j.ifset.2015.03.007.

Pouliot, Y., Conway, V., & Leclerc, P. (2014). Separation and concentration technologies in food processing. *Food Processing: Principles and Applications*, 33–60. doi:10.1002/9781118846315.ch3.

Powell, C. D., Quain, D. E., & Smart, K. A. (2003). The impact of brewing yeast cell age on fermentation performance, attenuation and flocculation. *FEMS Yeast Research, 3*(2), 149–157. https://doi.org/10.1016/S1567-1356(03)00002-3.

Purschke, B., Mendez Sanchez, Y. D., & Jäger, H. (2018). Centrifugal fractionation of mealworm larvae (Tenebrio molitor, L.) for protein recovery and concentration. *LWT, 89*, 224–228. https://doi.org/10.1016/j.lwt.2017.10.057.

Qin, F. G. F., Ding, Z., Peng, K., Yuan, J., Huang, S., Jiang, R., & Shao, Y. (2021). Freeze concentration of apple juice followed by centrifugation of ice packed bed. *Journal of Food Engineering, 291*, 110270. https://doi.org/10.1016/j.jfoodeng.2020.110270.

Rai, P., Majumdar, G. C., Das Gupta, S., & De, S. (2007). Effect of various pretreatment methods on permeate flux and quality during ultrafiltration of mosambi juice. *Journal of Food Engineering, 78*(2), 561–568. https://doi.org/10.1016/j.jfoodeng.2005.10.024.

Saengchan, K., Nopharatana, M., Lerdlattaporn, R., & Songkasiri, W. (2015). Enhancement of starch-pulp separation in centrifugal-filtration process: Effects of particle size and variety of cassava root on free starch granule separation. *Food and Bioproducts Processing, 95*, 208–217. https://doi.org/10.1016/j.fbp.2015.05.008.

Saengchan, K., Nopharatana, M., & Songkasiri, W. (2014). Recovery of tapioca starch from pulp in a conical basket centrifuge: Effects of rotational speed and liquid to solid (L/S) ratio on cake formation and starch-pulp separation efficiency. *Separation and Purification Technology, 127*, 192–201. https://doi.org/10.1016/j.seppur.2014.02.026.

Santana, T., Moreno, J., Petzold, G., Santana, R., & Sáez-Trautmann, G. (2020). Evaluation of the temperature and time in centrifugation-assisted freeze concentration. *Applied Sciences (Switzerland), 10*(24), 1–11. https://doi.org/10.3390/app10249130.

Santhanam, C. J., & Lipták, B. G. (2005). Centrifuge controls. *Instrument Engineers Handbook, Fourth Edition: Process Control and Optimization, 2*, 1632–1639. https://doi.org/10.1016/b978-0-7506-2255-4.50116-1.

Schubert, T., Meric, A., Boom, R., Hinrichs, J., & Atamer, Z. (2018). Application of a decanter centrifuge for casein fractionation on pilot scale: Effect of operational parameters on total solid, purity and yield in solid discharge. *International Dairy Journal, 84*, 6–14. https://doi.org/10.1016/j.idairyj.2018.04.002.

Sentís-Moré, P., Ortega-Olivé, N., Mas-Capdevila, A., & Romero-Fabregat, M. P. (2022). Impact of centrifugation and vacuum filtration step on the yield and molecular weight distribution of protein hydrolysates from rapeseed and sunflower meals. *LWT, 165*. https://doi.org/10.1016/j.lwt.2022.113741.

Sharma, H. P., Patel, H., & Sharma, S. (2016). Enzymatic extraction and clarification of juice from various fruits. *Trends in Post Harvest Technology Journal, 2*(1). www.jakraya.com/journal/tpht.

Sharma, K. K., Garg, S., Li, Y., Malekizadeh, A., & Schenk, P. M. (2013). Critical analysis of current microalgae dewatering techniques. *Biofuels, 4*(4), 397–407. https://doi.org/10.4155/bfs.13.25.

Singh, V., & Das, C. (2021). Clarification of citrus fruit (Mosambi) juice by hybrid (Pretreatment and Membrane) process. *Materials Today: Proceedings, 47*, 1384–1388. https://doi.org/10.1016/j.matpr.2021.02.435.

Sood, G., Sharma, M., & Kaushal, R. (2023). Centrifugation: Basic principle, types. *Basic biotechniques for bioprocess and bioentrepreneurship*. Elsevier Inc. https://doi.org/10.1016/B978-0-12-816109-8.00007-6.

Stephenson, F. H. (2016). Chapter 12: Centrifugation. *Calculations for molecular biology and biotechnology* (3rd ed., pp. 431–438.) https://doi.org/10.1016/B978-0-12-802211-5.00012-6.

Stevens, K. A., & Jaykus, L. A. (2004). Bacterial separation and concentration from complex sample matrices: A review. *Critical Reviews in Microbiology, 30*(1), 7–24. https://doi.org/10.1080/10408410490266410.

Stewart, G. G. (2017). Brewing and distilling yeasts. *Brewing and distilling yeasts* https://doi.org/10.1007/978-3-319-69126-8.

Talab, H. A. (2016). Starch-protein extraction and separation it from green pea. *European Online Journal of Natural and Social Sciences, 5*(4), 1012–1017. http://www.european-science.com.

Tamborrino, A., Leone, A., Romaniello, R., Catalano, P., & Bianchi, B. (2015). Comparative experiments to assess the performance of an innovative horizontal centrifuge working in a continuous olive oil plant. *Biosystems Engineering, 129*, 160–168. https://doi.org/10.1016/j.biosystemseng.2014.10.005.

Tantray, J. A., Mansoor, S., Wani, R. F. C., & Nissa, N. U. (2023). Principle and working of centrifuge. *Basic Life Science Methods*, 15–20. https://doi.org/10.1016/b978-0-443-19174-9.00005-2.

Todaro, C. L. (1996). Centrifugation. *Fermentation and Biochemical Engineering Handbook: Principles, Process Design and Equipment: Second Edition*, 558–589. https://doi.org/10.1016/B978-081551407-7.50015-5.

Todaro, C. M. (2014). Centrifugation. *Fermentation and biochemical engineering handbook: Principles, process design, and equipment* (3rd ed.). Elsevier Inc. https://doi.org/10.1016/B978-1-4557-2553-3.00013-1.

Van Der Borght, A., Goesaert, H., Veraverbeke, W. S., & Delcour, J. A. (2005). Fractionation of wheat and wheat flour into starch and gluten: Overview of the main processes and the factors involved. *Journal of Cereal Science, 41*(3), 221–237. https://doi.org/10.1016/j.jcs.2004.09.008.

Vanyo, J. P. (1993). Rotors and centrifuges. *Rotating Fluids in Engineering and Science, 2*, 251–264. https://doi.org/10.1016/b978-0-7506-9261-8.50018-1.

Vernon-Carter, E. J., Hernandez-Jaimes, C., Meraz, M., Lara, V. H., Lobato-Calleros, C., & Alvarez-Ramirez, J. (2015). Physico-chemical characterization and in vitro digestibility of gelatinized corn starch dispersion fractions obtained by centrifugation. *Starch/Staerke, 67*(7–8), 701–708. https://doi.org/10.1002/star.201500074.

Walker, G. S. (2012). Sample preparation for chromatographic applications in forensic science. *Comprehensive Sampling and Sample Preparation: Analytical Techniques for Scientists*: 3. Elsevier. https://doi.org/10.1016/B978-0-12-381373-2.00120-4.

Wong, Y. C., & Hartina, H. (2014). Virgin coconut oil production by centrifugation method. *Oriental Journal of Chemistry, 30*(1), 237–245. https://doi.org/10.13005/ojc/300129.

Filtration and clarification

4

Bharati Kumari and Nandan Sit
Department of Food Engineering and Technology, Tezpur University, Assam, India

4.1 Introduction

Filtration is a mechanical separation process that plays a crucial role in the food industry, helping to ensure the quality and safety of food products. It is used for a solid-liquid system to separate solids from liquids or gases (Tien, 2012). In the filtration process, a solid-liquid mixture is forced to flow through a porous media, mesh, or cloth. The solids accumulated on the filter media in porous media are known as filter cake. The fluid flow through porous media is known as filtrate. Either the cake or filtrate can be desired, depending on the intended use. Based on the retention of particles on porous media, the filtration process is classified into two forms: surface filtration and depth filtration. In surface filtration, the suspension is forced through the filter media, which holds the particles on the surface. In depth filtration, the suspension is forced to flow through a thick, porous media, and particles are retained within the media through the adsorption process (Berk, 2018). Surface filtration is also known as cake filtration, while depth filtration is known as deep bed filtration. The suspension with high solids content is treated by surface filtration, whereas depth filtration is used to treat suspension that has lower solids content. Filter paper, cellulose filter papers, muslin cloths, perforated plates, etc. are examples of surface filter media, and granular materials (e.g., sand, glass wool, and rock wool or anthracite) are examples of depth filter media Berk (2018); Pouliot et al. (2014). In the surface filtration process, solution feeds through porous media and as a result liquid passes through filter media, whereas solids are accumulated on the surface of filter media. The separation of particles is based on the particle size density. The three main categories of surface filtration are sieving filtration, cross-flow filtration, and cake filtration. In the sieve filtration method, low-concentration particles are moving towards the filter media and fouling individual pores, which is known as blocking or sieve filtering. The filter medium pores need to be smaller than all the particles, which need to be fully separated. In cake filtration, the direction of feed (suspension) flow coincides with that of the filtrate flow, and cake growth takes place along the opposite direction. For economic purposes, the slurry concentration and particle size must be sufficiently large. Particle sizes in between micrometers to millimeters. In cross-flow filtration, filtrate flow is normal to that of suspension flow. In crossflow filters, the filtrate, or permeate, is released through the membrane, while a low-concentrated slurry of microscopic particles (smaller than roughly 10 μm) is pumped across a microporous membrane and subsequently concentrated. As much as feasible, the crossflow prevents the creation of a highly impermeable particle layer (filter cake), which should permanently wash away the deposited particles. Crossflow filters can separate a wide range of particles,

Mechanical Separation Processes in the Food Industry: Unit Operations and Processing Equipment in the Food Industry.
DOI: https://doi.org/10.1016/B978-0-12-819523-9.00008-1

from some micrometers to tiny molecules like Na+ or Cl-ions. The driving force needed for the flow of suspension may be generated due to gravity, vacuum, pressure, centrifugation, etc. (Tien, 2012). Regarding the filtration technique, it includes conventional filtration, mechanical pressing for juice extraction, centrifugation, and membrane technologies (Pouliot et al., 2014). Membrane technology plays primary role in various industries, such as food, pharmaceutical, chemical, and many more. Recently, this filtration process has gained popularity due to its higher efficiency compared to other separation methods (Schroën & de Aguiar, 2023). It uses the semi-permeable membrane to separate fluid suspensions. There are two ways to perform membrane filtration: cross-flow and dead-end. Based on the pore size, membrane filtration could be classified into four types: microfiltration (MF), ultrafiltration (UF), nanofiltration (NF), and reverse osmosis (RO). The membrane filtration method is used for the filtration and clarification of beverages in various industries. Recently, the clarification of beverages has gained popularity globally. A clarification process is used to remove suspended solids, which cause turbidity in beverages. This process improved the clarity of beverages and produced appealing beverages, as well as increased the shelf-life of the beverages. Thus, clarification of juices is considered an essential step in ensuring the quality, safety, and marketability of food products in the processing industry.

4.2 Principle of filtration

Filtration is solely based on particle or droplet size, bigger particles whose size greater than the filter pores opening might accumulate on filter media, whereas particles less than a given size will flow through the filter media. A pressure difference across the filter medium is typically required for the functioning of the filter. Then, the flow of filtrate and accumulation of cake on the surface can be calculated. Assume the flow of filtrate was laminar, then Poiseuille's law can be used to describe the filtrate flow and can be expressed as

$$\frac{\Delta P}{L} = \frac{32\mu V}{D^2} \tag{4.1}$$

where,

ΔP = Pressure drop (Pa), L = length of cake accumulated on surface, V = velocity of filtrate (m/s), μ = Viscosity of filtrate in pascal second (Pa.S), D = diameter of filtarate (m).

4.2.1 Theory of filtration

As filtration proceeds, a porous cake of solid particles accumulates on a porous filter medium and as a result, a compressible cake is formed that deforms itself under pressure with time. The cake porosity (ε) and permeability (κ) are decreased with time

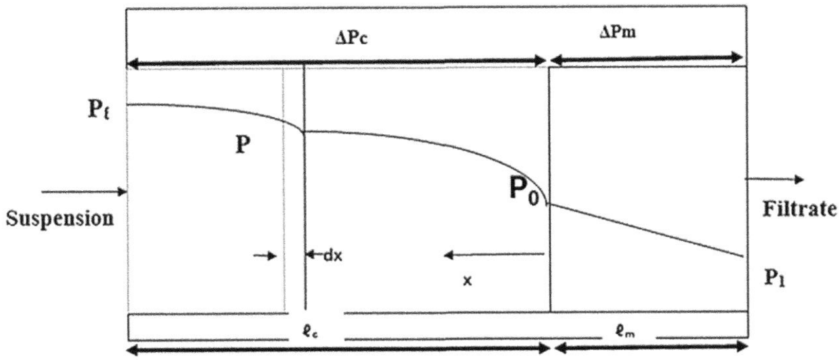

Figure 4.1 Schematic diagram of cake filtration.

having minimum value near filter media. The variation in cake resistance is carried out with changing pressure drop.

- In filtration theory, some assumptions can be followed, deformation of the compressible cake occurs in the plane.
- The fluid flow is one-directional and follows the behavior of an incompressible Newtonian fluid.
- Darcyan flow (low fluid velocity).
- The internal velocity of cake were substantially higher than its solid velocities.
- The acceleration, gravity, and viscous forces are very less than compared to the pressure force and the force that the fluids impart on the solids.
- Cake porosity and permeability depend on the pressure exerted on the solids Ps, which is the difference between the pressure at the cake surface and at the given point, (Ps = Pt - P).

Fig. 4.1 illustrates the process of cake filtration in detail. Where ΔPc and ΔPm represent the pressure drops across the cake and filter medium respectively. The variables ℓm and ℓc correspond to the thicknesses of the filter medium and the cake. The filtration area is denoted by A and dx is the thin layer of cake.

The rate of filtration could be calculated using this equation:

$$\text{Rate of Filtration} = \frac{\text{Driving Force}}{\text{Cake resistance}} \tag{4.2}$$

The flow rate of filtrate will be determined by applying Darcy's law:

$$Q = \frac{dV}{dt} = \frac{A\Delta P}{\mu R} \tag{4.3}$$

$$\Delta P_m = \frac{\mu R_m}{A}\frac{dV}{dt} \tag{4.4}$$

Darcy's Law's

$$\frac{dPc}{dx} = \frac{\mu q}{k} \tag{4.5}$$

The mass of solids dm in the cake slice of thickness dx is given below

$$\mathrm{dm} = \rho_s(1-\varepsilon)\mathrm{Adx} \tag{4.6}$$

$$\mathrm{dx} = \mathrm{dm}/\rho_s(1-\varepsilon)\mathrm{A} \tag{4.7}$$

By putting the value of dx Eq. (4.7) in Eq. (4.5), then we will get

$$\mathrm{d}Pc = \frac{\mu q\, dm}{k\rho s(1-\varepsilon)A} \tag{4.8}$$

$$\Delta \mathrm{Pc} = \frac{\alpha\ \mu qm}{A} \tag{4.9}$$

Where specific cake resistant $(\alpha) = \frac{1}{k\rho s(1-\varepsilon)}$ is given
Suppose m $\cong \rho$cV

$$\Delta \mathrm{Pc} = \frac{\alpha\ \mu q\rho cV}{A} \tag{4.10}$$

$$\Delta P_c = \frac{\mu V\alpha C}{A^2}\frac{dV}{dt} \tag{4.11}$$

$$\Delta P = \Delta P_c + \Delta P_f \tag{4.12}$$

By putting Eqs. (4.4) and (4.11) in Eq. 4.2, then we will get,

$$\Delta P = \frac{\mu V\alpha_{av}\, C}{A^2}\frac{dV}{dt} + \frac{\mu R_m}{A}\frac{dV}{dt} \tag{4.13}$$

$$\Delta P = \left(\frac{\mu V\alpha_{av}\, C}{A^2} + \frac{\mu R_m}{A}\right)\frac{dV}{dt} \tag{4.14}$$

$$\frac{dV}{dt} = \frac{\Delta P}{\left(\frac{\mu V\alpha_{av}\, C}{A^2} + \frac{\mu R_m}{A}\right)} \tag{4.15}$$

where *Rm* is the resistance of the filter medium to filtrate flow in the filtration method

4.2.2 Constant pressure filtration

Constant pressure filtration: The initial flow rate is the highest in this filtration process due to the lowest cake thickness., If the pressure is maintained consistently over the duration of the filtration process, the flow rate will gradually decrease with increasing time of filtration because of the growing cake thickness. The initial flux through the filter is relatively high, resulting in severe fouling. Thus, the overall capacity of the filter reduced with time and the cake increased.

In constant pressure filtration,

Pressure (ΔP) is constant, the Eq. (4.15) can be integrated with boundary conditions, assuming α_{av} is constant, t=0, v=0 and t=t and V=V (KHEAN, 2003).

$$\frac{V}{t} = \frac{\Delta P}{\left(\frac{\mu V \alpha_{av} C}{A^2} + \frac{\mu R_m}{A}\right)} \tag{4.16}$$

4.2.3 Constant rate filtration

Constant rate filtration allows the fixed liquid flow rate across the filter medium. It is commonly used in industry due to numerous advantages, such as working under low pressure at the initial stage of filtration cycle. In this filtration process, both the cake concentration and the pressure drop across the filter cake will change considerably over time. A positive displacement pump is used to feed pressure across filter to maintain constant filtrate rate until back flashing pressure is applied.

In constant rate filtration,

Flow rate of filtrate is constant, we will get,

$$\Delta P = \frac{\mu q^2 \alpha_{av} C}{A^2} + \frac{\mu q R_m}{A} \tag{4.17}$$

It is simple to determine the maximum and minimum limits of fluctuation in pressure throughout the filter cake. Industries use this filtration due to higher filtrate amounts in a shorter time and better particle packing structure regardless of its thickness. However, increasing pressure has an extensive effect on soft and flocculated materials, leading to highly compressible filter cake. This process is used in the pharmaceutical industry to enhance purity of product, undesired solvents and dissolved contaminants are removed from the filter cake containing active pharmaceutical ingredient (Girard & Fukumoto, 2000).

4.2.4 Filter media and filter aids

4.2.4.1 Filter media

Selecting the appropriate filter medium is often crucial for ensuring efficient filter operation. The medium must effectively retain the solids being separated from the liquid while offering a suitable lifespan. The selection of a filter medium typically involves the consideration of the following criteria: ion in cake resistance is carried out with changing pressure drop.

- The solids should not clog the filter openings.
- Ability to quickly form solids layers its pores after feeding is initiated.
- Low resistance to fluid flow through the filter medium.
- Resistance to chemical corrosion or damage
- Adequate durability against mechanical abrasion.
- Sufficient strength to and withstand the filtration pressure.
- The cake may be discharged efficiently and effortlessly.
- Reasonably priced.

Textiles made of cotton or synthetic fibers are used to make filter media. For filtering fine crystals metal fabric with 400 mesh apertures may also be used.

4.2.4.2 Filter aids

Filtration of certain solids can be hindered due to a slow filtration rate or inadequate filter clarity. This could be increased using filter aids, which is a granular or fibrous material that forms a highly permeable filter cake, trapping the very fine solids. An effective filter aid should be light-weight, porous, and chemically inert to the filtrate. Common filter aids include diatomite, a highly porous form of silica, and expanded perlite, a porous alkali aluminosilicate. Their bulk density ranges from 200 to 400 kg/m^3. Filter aids can be applied in two ways: (1) as a precoat to protect the filter medium and prevent fine particles from escaping into the filtrate (applied at around 0.5 kg/m^2 of filter area), or (2) mixed with the slurry to capture difficult-to-filter particles in a permeable cake.

4.3 Types of filters

4.3.1 Plate and frame filter presses

Plate and frame filter is the oldest batch type filtration and commonly pressure-driven equipment for filtration of food juices, and wider application in various industries. It is a sophisticated and extensively used solid-liquid separation machinery employed in industries, such as mining, metallurgy, petroleum, chemicals, pharmaceuticals, textiles, printing, paper production, food, alcohol, inorganic salt, coal washing, dyeing, and wastewater treatment industries. It is an important and economical process for recovering solid particles from liquid-solid suspensions. A pressure displacement pump is maintain pressure throughout the filtration. In this case, filtration rate decreased with increasing time. It is assumed that the flow resistance between plate and frame is negligible. The constant pressure drop must be maintained throughout the process.

4.3.1.1 Working of plate and frame filter

Plates and frames are fitted in alternation, with an absorbent material encompassing both sides of the plates, to form these filters. Clear filtrate fluids could transport down each of the plates through sections that have been carved into the plates. The feed slurry

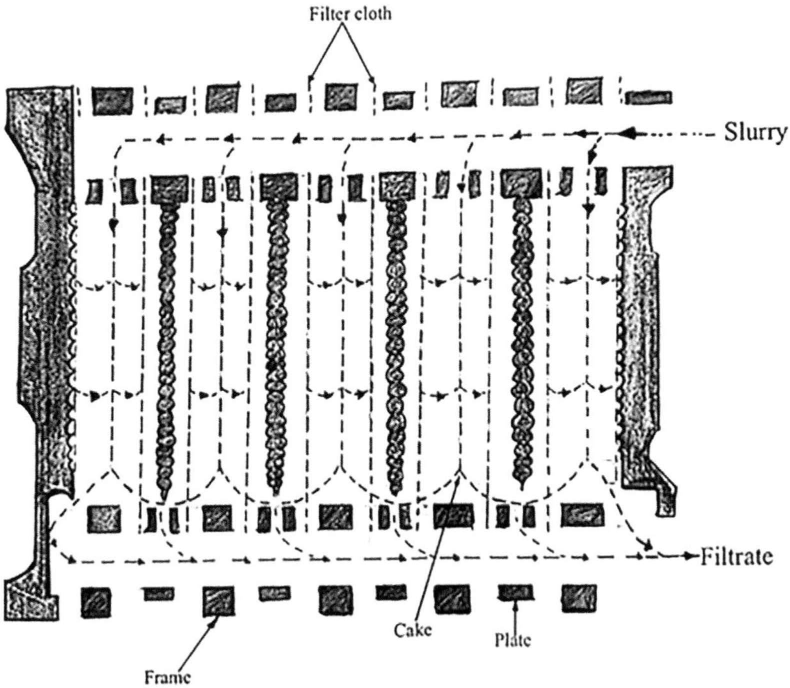

Figure 4.2 Plate and frame filter.

fed into the press, filling the open frames with slurry as it proceeds down the pipe. The solids accumulate as a cake on the frame side of the filter cloth when the filtrate passes through it. Filtration continues until all solid particulates have been collected on the frames. Each discharge outlet connects to a main header Fig 4.2 (Mann, 1980). The frames and plates are disassembled immediately as soon as the frames are full. The cycle is then carried out once again when the filter is reassembled. **Performance efficiency of filter:** The press performance is evaluated using final cake solids and filtrate volume produced per unit of time.

4.3.1.2 Advantages

- Plate and frame filtration provide a few key benefits, including the ease of separating particles from suspension.
- Rotating the plate and frame in some cross-flow plate and frame systems increases shearing forces and decreases fouling.
- Plate and frame filters lack feed spacers, there is less chance of clogging, and they can usually manage high solids concentrations.

4.3.1.3 Disadvantages

- The drawbacks typical of batch operations affect the plate-and-frame presses.
- Low packing density is one major problem for plate and frame filtration.

- A considerable portion of the overall running cost may come from labor expenses associated with taking out cake and reassembling the parts of equipment as well as fixed charges for maintenance.

4.3.1.4 Recent advancements in plate and frame filter

A rotating shaft is used to attach duplicate sets of frames in some of the more recent models of plate-and-frame presses. To reduce maintenance and labor expenses, semi-automated filters, such as filter press, are used. It is simple to operate at high temperature, is versatile, and flexible in nature. The principle of the new advanced plate-and-frame-filter press remains the same, but some improvements have been made in this updated machine. Earlier, filter press was made by using cast iron and wood, but nowadays, alloys such as aluminum, bronze, lead, and stainless steel are used to make filter press. A specific type of plate and frame was designed to withstand higher pressures of up to 1,000 Psi, and features a leakproof filtration system with proper drainage system and sealing. For achieving this requirement, gasket and grooved-type plate and frame filter were developed, as well as a filter press with hollow plate and frames for the circulation of steam oil or brine (Kriegel, 1938). These specific types of filters have been treated to more automation advancements used in food industry, chemical industry, and other industries to filtration of beverages, non-beverages, alcohol, and other solids matters from suspension. Although they cannot be used for high throughput operations. Ito Mfg. Co., based in Setocity, Japan, developed an automatic plate-and-frame-type filter press for automatic discharge of cake from filter press (Kurita et al., 2010).

4.3.2 Leaf filter

Leaf filter is batch-type filter commonly used in oil refinery industry. The filter system consists of a horizontal cylindrical steel pressure tank with a retractable carriage that supports several vertical filter leaves. These leaves are mounted parallel and stacked on a hollow shaft within a filter vessel in Fig 4.3 (Grimm, 1956). The suspension enters the pressure tank and is forced to flow through the filter leaves. The filtrate enters the hollow framework and exits through a header, while the cake is deposited on the outside of the filter leaf. The paths of the clear liquid and suspension flow in opposite directions. The shell is opened to collect the cake, and sometimes water jets can be used to discard cake if solid cake is not useful.

4.3.2.1 Advantages

- Dry cake (low moisture content) is produced.
- Extremely small particles (0.5 to 1 micron) can be filtered through this filter, because of a precoated layer.
- This filter is easily available.
- Precoated cloth of this filter is easily available at a cheaper cost.

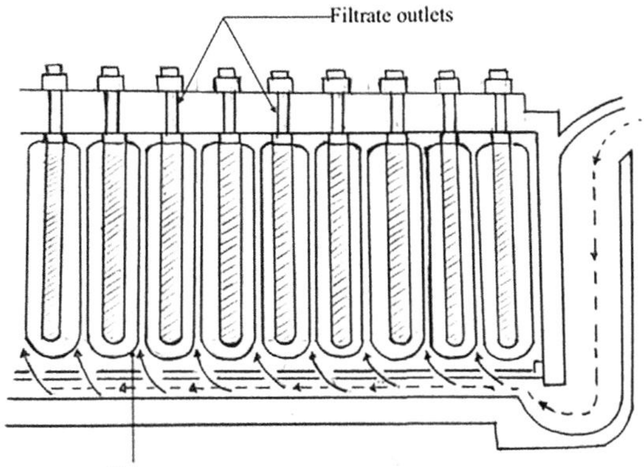

Figure 4.3 Leaf filter.

4.3.2.2 Disadvantages

- Only batch operations, and diluted liquids (less than 5% solids) are appropriate uses for pressure leaf filters.
- High headroom is needed to disassemble pressure leaf filters on vertical vessels.
- The mesh or filter cloth of a pressure leaf filter may need to be changed regularly.
- If the precoat layer is not adequately maintained, the pressure leaf filter may clog more quickly.

4.3.2.3 Application

- Extraction of pectin from citrus peels using pressure leaf-filter (Bähner et al., 2017).
- Filtration of beer using the horizontal leaf filter (Gehring et al., 1997).
- Leaf filter was used to remove foot from Crude Oil (Campbell, 1921).

4.3.2.4 Advancement in leaf filter

Although the concept of a leaf-filter system has been established, but use of this filter was limited due to some shortcomings such as lower efficiency and higher cleaning cost. To overcome this shortcoming, some advancements had been made in this filter, For example, Kelly leaf filter and Sweetland leaf filter (Jayaprahash, Lakshmana, & Kingston, 2020). In this modern system, construction is designed suchthat leaves can be completely exposed for cleaning using an automatic opening device, and the cake can be automatically discharged.

4.3.3 Bag filter

Bag filters are a kind of filtering technology that filters solids and unwanted substances from a liquid or gas by using a filter a bag. A porous material bag encased in a metal or plastic frame makes up the bag filter. The pollutants are held in the bag as the liquid or gas to be filtered flows through the porous material. When the bag filter blocked with particles, it is simple to remove and replace. Polypropylene is widely used in bag filters due to their smooth surface, strong resistance to acids and alkalis, and limited moisture absorption capacity (Lozano-Sánchez et al., 2012). Although woven fabric filter media is commonly used in liquid bag filters, non-woven composite materials are also used in certain applications. The teabag made with wet lay Manila hemp (abaca) fibers is a popular type of bag filter.

4.3.3.1 Advantages

- It is a cost-effective filtration method.
- Construction and operation of bag filter is relatively simple.
- Bag filters provide an environmentally friendly solution to filtration needs.
- They can handle a variety of liquids and gases and can be customized to meet specific filtration requirements.

4.3.3.2 Disadvantages

- It requires special types of fabrics to filter higher temperature contaminants.
- It requires high power consumption and ongoing maintenance to function at peak efficiency.

4.3.3.3 Application

The application of bag filters is versatile in food and other industries. Industrial bag filters are often used in the food and beverages industries, oil industry, and water-treatment industry, due to their low cost and a high degree of reliability. For example, the polypropylene filter bag, which makes up the filtering apparatus, is directly filled with olive oil. The suspended solids are eliminated as the fluid flows through the filter bag (Jin et al., 2021). A bag filter is also used with other equipment due to the simplicity of this filter. Filter bags are encased with homogenization bag to improve the use of homogenization bags (Lozano-Sánchez et al., 2012).

4.3.4 Cartridge filter

The working principle of a cartridge filter involves the separation of solids particles (solute) from liquids or gases (solvent) by passing them through a porous medium, which is made from paper, cloth, or synthetic fibers. It is generally used for the removal of dust or unwater material from the water and juices. It is installed before membrane filtration to increase microfiltration efficiency. Cartridge filter is divided in various types of filters such as Membrane cartridge filters, pressure cartridge filter and membrane cartridge filter. Membrane cartridge filters are mostly used in pharmaceutical and biotech processing. Materials for membrane filters include PTFE, nylon, and polypropylene.

4.3.4.1 Working principle of pressure cartridge filter

It consists of many more cylindrical or pleated filter elements stacked in a single house with inlet, outlet, and nozzle system. A pressure tank is used to provide pressure during filtration. The pulse-jet system is used for cleaning of membranes of filter (Farhat et al., 2020). For many cartridge filter applications, the common filter media is pleated resin-treated filter paper. The suspension is fed from the inlet valve. The solids particles (particle size > the size of opening of filter) are retained on the first filter, and clear liquid passes through all filter, while solids are retained on the surface of membrane (Beer, 1989). Cartridge filters are usually used to remove particles smaller than approximately 50 microns. Pressure is applied to compel the suspension to flow through the filters. The clear liquids enter from the outlet. Cartridge filters are usually designed for depth filtration. But an increasing number of these filters are staked in single house, functioning as surface filtration systems (Beer, 1989). Filters with pleats have a surface filtering method, whereas string wound, resin-impregnated, and molded cartridges have a depth filtration mechanism. Surface filters had been slightly more costly than depth filters. Usually, cartridge filters are put before reverse osmosis units. There are several types of cartridge filters: metal, carbon, nonwoven, membrane, and string-wrapped.

4.3.4.2 *Working principle of vacuum cartridge filter*

The cartridge vacuum filter is a batch-type filter made of sintered porous materials. Each filter element has an average filtration area of 0.6 m^2. The receiving bins' lower section contains the horizontally stacked filter elements, which are spaced 30 mm apart from the bottom. On one end of the cylindrical filter element is connected by a branch pipe to the vacuum system manifold, while the other end is blanked off. This filter is used for removing the moisture from powder.

4.3.4.3 *Advantages*

- Depth cartridges provide the benefit of increased thickness and depth, which enables them to trap more material in the filter.
- They are also convenient to replace and are designed for single use.
- Smaller particles can easily be removed due to manifolds of cartridges filter in one housing.
- It is an economical method due to its low maintenance costs.
- It is a safer filtration method because the membranes do not react with suspension. Thus, the chemistry of liquid will not be changed after filtration.

4.3.4.4 *Disadvantages*

- Fouling is the primary problem of this filter.
- Cartridge filter must be removed or disposed of when it is blocked. The cost of removing this filter is high.
- The use of coagulant in cartridge filter was not permitted.
- The system area and size limited the use of this filter.

4.3.4.5 *Application*

Cartridge filters are extensively utilized in various industries, such as food and beverages, water purification, pharmaceuticals, natural gas transportation, power generating, cement making, and tunnel driving, owing to its high dust removal effectiveness, modest floor space occupation, and zero water consumption. The cartridge filter was tailored to filter wine, beer, mineral water, and non-alcoholic beverage.

- It is used for sterile filtration of beer (Beer, 1989).
- It is used for cleaning bulk grain loading (Jin et al., 2021).
- This filter is combined with a cyclone separator for dust removal (Xie et al., 2018).

4.3.4.6 Recent advancements in this filter

The solid particles are continuously collected on filter surface during filtration, a cartridge filter system will not be able to efficiently clean out the many pollutants. Thus, the cartridge filter must be removed from the system. Recently, there has been a growing interest in developing designs of filter cartridges with lower environmental impact during use and disposal. This is because of their lighter weight, the incorporation of alternative substances for construction, and lower expenses for disposal (Rosas et al., 2013). To enhance the cleaning efficiency of cartridge filter, modify the pulse-jet cleaning systems structure, including the pleated structure, nozzle diameter, nozzle type, injection distance, and built-in rotary (Jin et al., 2021).

Continuous rotatory filtration: All batch filters have some drawbacks such as they are time consuming and only suitable for the smaller capacity, which limits their use in industries at higher scale. To mitigate this problem continuous filtration methods were introduced. Different types of continuous rotatory filtration were discussed below:

4.3.5 Continuous pressure drum filter

A revolving drum housed inside a pressure vessel is known as a pressurized drum filter as shown in Fig 4.4. It consists/ of a pressure let-down rotary valve, wash bars, an agitator, a slurry trough, and a filter drum of area up to 120 m^2 enclosed within a sealed housing. The particle size and feed concentration are approximately 1–100 μm and 5–30% w/w (Tarleton & Wakeman, 2006) in range. Pressuring the pressure vessel with compressed gas of up to 800 kPa facilitates constant pressure filtration at the outer drum surfaces. The suspension feed pipe fills the filter trough. To maintain the solids in suspension, the agitator is then turned on. To release the cake, a gas blast expands the filter medium. The cake is directed into the chute by a scraper-deflector plate. For continuous cake discharge, the rotary valve is attached to the chute. Applications for the pressure drum filter include deep cakes of up to six inches, and enormous filter areas; however, the larger the area, greater the amount gas is consumed (Perlmutter, 2015).

4.3.5.1 Advantages

- The rotary vacuum drum filter operates continuously and automatically, resulting in low operating costs.
- It is an affordable and environmentally-friendly process.

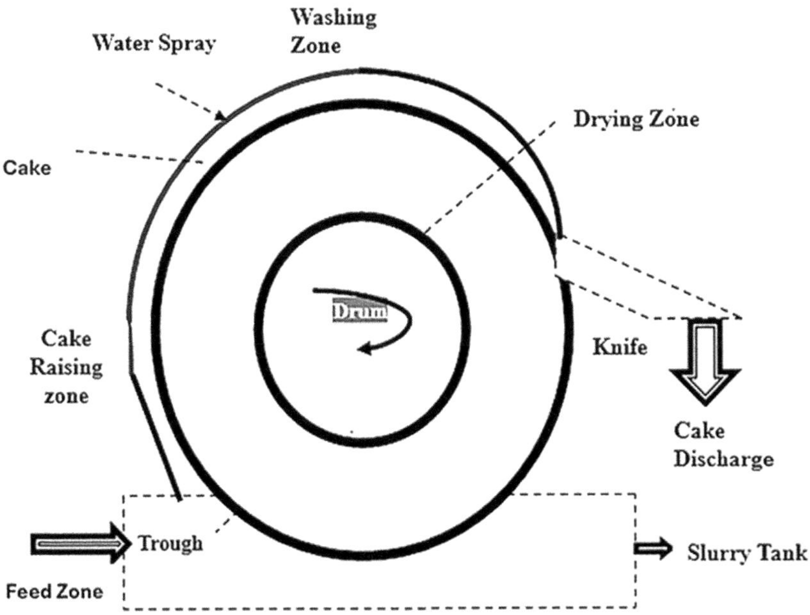

Figure 4.4 Drum filter.

- The process can be easily adapted by incorporating a pre-coating filter process.
- Additionally, adding a showering device allows it to produce relatively clean products.

4.3.5.2 Disadvantages

- The pressure difference in the system is theoretically limited to atmospheric pressure (1 bar) and is usually slightly lower in practice due to its structure.
- Additional equipment like agitators, vacuum pumps, vacuum receivers, and slurry pumps are needed alongside the drum.
- The discharge cake retains residual moisture.
- The cake is prone to cracking as air is drawn through by the vacuum system, reducing the efficiency of washing and drying.
- The vacuum pump consumes a significant amount of energy.

4.3.5.3 Application

The drum filter is the oldest and most widely used filter for liquid-solid separation, ranking as the top choice in the mechanical separation process. It is the cost-effective and demanding for many sectors listed below:

- Aquaculture
- Wastewater treatment plants.
- Pre-treatment before biological treatment
- Municipal and industrial wastewater treatment

4.3.5.4 Recent advancements

The drum filter is one of the key equipment used to maintain water quality in recirculating aquaculture systems (RAS). Recently, research has made significant advancements in drum filters, leading to the design of micro-screen drum filter. Micro-screen drum filters a simple, efficient, and reliable due to their overall design and operation, as well as low maintenance costs. Rotating micro-screen filters are an advanced version of micro-screen drum filters, which have become increasingly popular on farms in the recent years. These typically include a fine mesh screen with a pore size ranging from (60–200 mm) arranged in rotating drum through which effluent is transported. Particles trapped on the mesh are either backwashed or scraped off into a waste collection trough. Rotating micro-screens are well suited for application prone to blockage, such as fish farms, because of the huge flow of wastewater that must pass through the screen and the small screen pore size required to separate the solids (Ali, 2013). To measure a drum filters performance, its mechanical efficiency is crucial. The amount of material removed by the filter in relation to the total amount of material contained in the influent waters is measured by mechanical efficiency. To choose an optimal configured micro-screen drum filter would be considered the micro-screen rating (μm), particle size distribution of the suspended solids in the RAS waters and the required water quality.

4.3.6 Continuous vacuum disc filter

4.3.6.1 Principle

It is a continuous process. The working principle of disc filters is the same as that of rotary drum filters. The disc filter is made up of steel sectors covered in cloth, which are fitted on a central shaft. The shaft also connects a portion of the sectors to vacuum. As the disc rotates, solids form a cake on both sides of the disc. The disc then lifts the cake above the slurry level in the trough, where it is suction-dried and removed by a whirling air blow, assisted by a scraper (Wills & Finch, 2015).

4.3.6.2 Construction and working of disc filter

The disc filter functions similarly to a drum filter that is fed from below. The compartments that arise on both faces of vertical discs are the main variations to be noticed. These consist of eight or more segments, each of which is attached to the shaft of equipment. As seen in Fig 4.5, the discs are spaced approximately 30 cm apart on the shaft to maximize filtering area per unit area, which is far more than can be achieved with a drum-style filter. Eight discs, each about 2.5 m in diameter, can be placed in the same area as a 2.5 m diameter $\times$ 2.5 m drum filter, which would have a filtering area of 20 m^2, forming a disc filter area of around 80 m^2. Two rotary valves are usually installed, one at each end of the main shaft, due to the enormous areas and the resulting large liquor and air flows (Richardson & Harker, 2002). A pipeline transports the slurry to a bath containing sunken discs, which are hollow chambers wrapped with a filter material. The remaining slurry is directed into a chute to collect the overflowing slurry. Scrapers are employed to remove the filter cake. The purpose of a distributor head is

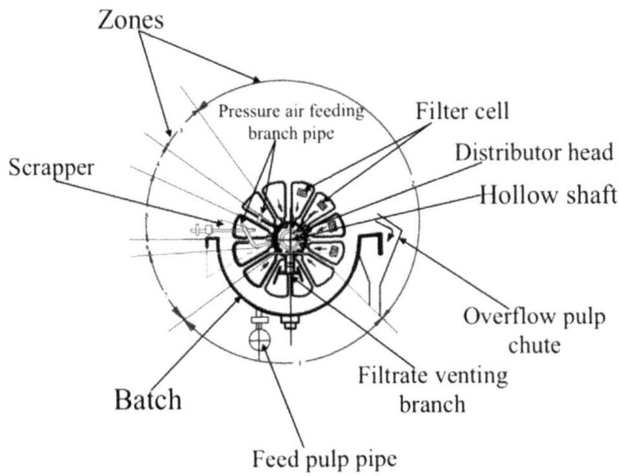

Figure 4.5 Vacuum disc filter (Neikov et al., 2009).

to supply air to remove the filter cake layer while regulating the filtrate and exhaust air away (Neikov et al., 2009). The filtrate and air passed through the filter media, into the drainage section, and then through rotary shaft passageways to a valve at either or both ends.

4.3.6.3 Advantages

- The trough can be split into two sections to allow different slurries to be handled simultaneously in the same machine.
- Small floor space allows for extensive filtration areas at a low cost per unit.
- It is easy to clean and maintenance.

4.3.6.4 Disadvantages

- These devices have the drawback of requiring bulky filter cloths, which have a high propensity to "blind."
- It is difficult to maintain the homogenous condition throughout the filtration process.

4.3.7 Continuous belt filter

4.3.7.1 Working principle of belt filter

The continuous-belt filter press has been used recently. It is made up of several pulleys that remove moisture from the slurry. The feed is fed in a feed tank before being pumped to the filter at low pressure. One or two static mixing vessels, positioned in the pump line, are used to introduce flocculent to the feed. After the raw feed is fed into the press at atmospheric pressure, the slurry settles on the filter cloth, rapidly thickening and becoming flocculated. After being first made into a very wet cake, it is caught between a second filter cloth and squeezed through a sequence of pulleys to remove filtrate and create the cake. The degree of dewatering achieved will depend on the number of pulleys the twin cloths surrounding the cake travel over and the pressure applied. Lastly, the fabrics are divided, and the completed cake is released from the end of belt.

4.3.7.2 Advantages

- Belt filter press can process special sludge.
- It is a continuous process.
- Low energy costs and easy operation.
- Small space occupation.
- Stable operation and long service life.

4.3.7.3 Disadvantages

- It's suitable for high sludge moisture content.
- Maintenance is difficult.

Continuous belt filter is divided into different types, listed below:

1. Horizontal belt filter
2. Gravity belt filter
3. Rubber belt filter

4.3.7.4 Horizontal belt filter

When it is essential to clean the cake for food and beverage filtration, horizontal belt filters are employed. In the dry zone, the horizontal belt filter may remove soluble materials from the filter cake, and drastically cut down on residual moisture.

4.3.7.4.1 Working principle of horizontal belt filter

The horizontal belt filter cycle involves driving an endless cloth past a series of evacuated suction boxes, across two rollers on a perforated belt as seen in Fig 4.6. The top-fed filtration principle allows for gravity-directed filtration, which results in a layer of coarse material settling over the filter cloth to facilitate cake release, and enhance filtrate clarity (Osborne, 2013). The feed suspension is injected towards one end of the filter and processed at a constant vacuum to form a cake, which is then theoretically gas-dispersed at the right vacuum levels, and washed and/or dried multiple times as needed. Though concessions may be necessary due to space or budgetary limits, the filter's length can, in theory, be as long as needed to meet the process requirements. As it goes over the second roller, the last cake is released (Tarleton & Wakeman, 2006).

4.3.7.4.2 Advantages and application

In several sectors, horizontal belt filters are becoming more and more popular as effective solid-liquid separation methods. The horizontal layout of these filters enables high throughput and continuous filtration. They have excellent cake washing and drying

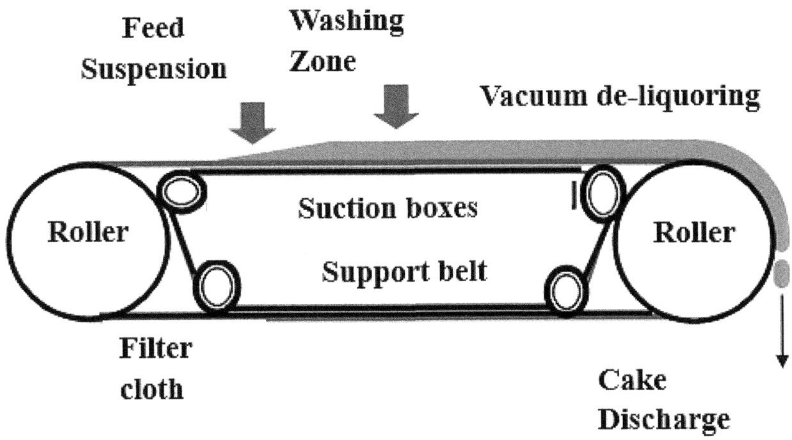

Figure 4.6 Continuous horizontal belt filter (Tarleton & Wakeman, 2006).

capabilities, and have a wide range of applications, such as in the food processing, chemical, and mining industries. Their small size maximizes the use of available space, and their automated controls make them easier to use. Their low care requirements and flexibility to adapt to different feed conditions are other factors in their popularity. Horizontal belt filters are a key technology, that improve process efficiency and product quality as companies look for sustainable and affordable filtration solutions. When compared to traditional filter presses, these devices have the advantages of a continuous process, and a much smaller plant footprint needed for dewatering. The horizontal belt filter is applied in food and beverages industry, such as beverages industry, alcoholic beverage industry, and manufacturing of high fructose corn syrup.

4.3.7.5 Rubber belt filter

The rubber belt filter is comprised of a rubber belt with grooves that move constantly, functioning as a conveyor belt for suspension. These grooves drain through holes, typically aligned along the centre-aline of the belt into a vacuum box that is forced against the underside of the belt. For chemical compatibility and withstanding ability, the belt can be designed from a wide variety of rubber materials. The slurry can be placed upon the continuously moving belt filter at a constant rate. To obtain a good solids distribution, it is typical to direct this slurry flow against the direction of the belt travel. Cake discharge is a continuous process and can be pulled off with a scraper. This type of filter is especially suited for slurries, that are quickly settled. The slurry can be divided into factors, like coarser and finer. The coarser particles could feed first and finer particles in last to collect clearer filtrate.

4.3.7.5.1 Advantages and disadvantages

- It is a continuous process.
- Filter cloth can be automatically washed and regenerated and produced higher efficiency.

- Good corrosion resistance.
- Low energy consumption.
- Not suitable for longer duration processing.

4.3.7.5.2 Application

Rubber belt vacuum filter has reached the international level and is widely used in electric power environmental protection, mining, chemical industry, metallurgy, paper making, food, pharmaceuticals, and coal industries. This filter is used for oil filtration in the oil industry.

4.3.7.6 The gravity belt filter

The gravity belt filter is a rather straightforward apparatus that resembles the horizontal belt filter in basic form, with the exception that separation is entirely driven by gravity. The diluted and flocculated feed is introduced at one end of a semi-permeable belt that moves continually. The solids that are discharged, typically have the consistency of a pumpable slurry. To aid in the promotion of liquid evacuation, more advanced versions gently agitate the thickened solids down the length of the belt. The most common application of a gravity belt filter is pre-concentration of a suspension to reduce the downstream separator's load and consequently size required. Depending on the type of feed, flocculant usage can be minimal. The throughput range for individual units is between 10 and 200 m^3 per hour.

4.3.7.6.1 Application of gravity belt filter

A gravity belt filter is used for the thickening and concentration of solids by allowing the water to drain from the sludge under gravity through a porous moving belt medium on which the sludge sits. It is applied in several industries including:

- Marine
- Oil and gas
- Chemicals
- Medical

4.3.8 Membrane separation

In the food industry, membrane technology is used as a new technology, and as an alternative to conventional filtration method. Membrane processes were developed to separate a dispersed phase from a continuous phase. The portion of the feed that passes through the membrane is called the permeate stream, and it contains particles smaller than the membrane pore size. The portion that accumulates in the membrane is known as concentrate stream, and it mainly consists of particles that are larger than membrane pore size (Daneluz et al., 2023). This process separated ions, macromolecules, and tiny molecules from complex mixtures. Membrane acts as a heart of the membrane process, and function as a permeable, semi-permeable, and selectively-permeable

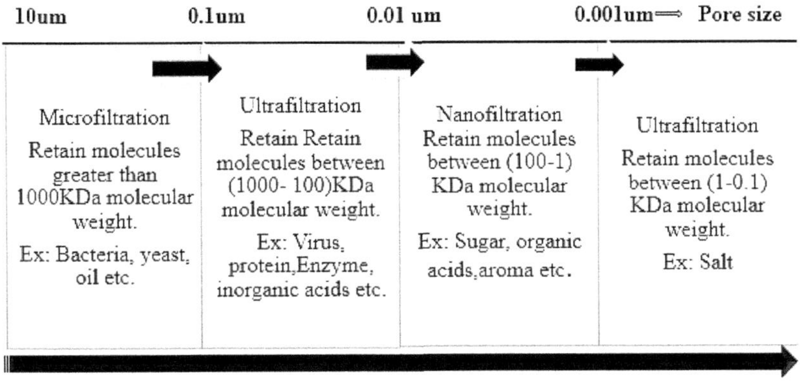

Figure 4.7 Pore sizes of membranes in different membrane proceses (Dornier et al., 2018).

barrier to separate one or more solutes in a liquid or gas (Girard & Fukumoto, 2000). Pressure differences are the primary driving force underlying separation in membrane technology. Membrane filtration uses a semi-permeable membrane as a filter, just like traditional filtration, but its pores are significantly smaller (Pouliot et al., 2014). A membrane is a thin, unique interface that regulates how quickly chemical species that come into touch with it can permeate through.

There are different types of membranes, such as microporous membrane, non-porous or dense membranes, electrically charged membranes, and metal and ceramic membranes. For the concentration, fractionation, and purification of liquid solutions, as well as the treatment of water, pressure-driven membrane processes, such as reverse osmosis (RO), microfiltration (MF), ultrafiltration (UF), and nanofiltration (NF) have long been established (Dornier et al., 2018). Based on the pore size and molar mass of retentate membrane, filtration could be classified into four types: microfiltration (MF), ultrafiltration (UF), nanofiltration (NF), and reverse osmosis (RO). The pore size of MF membranes is usually between 0.1–10 μm. UF membranes have pore sizes between 0.01 to 0.1 μm, and may hold 300–10^6 Da molecular weight particles. The pore size of NF and RO range from 0.01–0.001, and less than 0.001 respectively, and have retained a molar mass between 1,000 and 3,000 Da, and less than 1,000 Da respectively (Charcosset, 2021). Membrane technology is a cost-effective, energy-efficient, and eco-friendly filtration process. Moreover, membrane technology has been used in many manufacturing industries, including the fruit juice industry. In fruit juice processing, reverse osmosis (RO) can be applied to concentration, UF, and MF for clarification. RO is used for rejuvenation and dealcoholization of wine, whereas MF/UF is employed for wine clarification during production. Membrane filtration processes are typically categorized based on the size range of solutes that they are designed to separate. The porous membranes use a different range of pore diameters as shown in Fig. 4.7. Microfiltration has the following advantages over conventional methods: (1) separation can be carried out without altering the pH level and temperature of solution, and (2) chemical additives are not needed. The drawbacks of membrane processes include: (1) relatively high investment costs, and (2) occurrence of concentration polarization phenomena. Despite these drawbacks, membrane technology has emerged as viable alternative to traditional thermal techniques for fruit juice clarification and concentration, which were previously common in the dairy and beverage industries.

4.3.8.1 Basic theory of membrane filtration

There are two ways to perform membrane filtration: cross-flow and dead-end shown in Fig 4.5. In dead-end filtration, the feed stream is forced to flow perpendicularly to a semi-permeable membrane under pressure. The particulate matter (retentate) accumulates on the membrane surface, and the permeate passed through the membrane can be collected on the opposite side of the membrane (Girard & Fukumoto, 2000). The buildup of a cake layer on the membrane barrier will restrict the filtration rate or flux. Dead-end filtering occurs when there is more convective movement of particles towards the filter than there is back-transport of particles. As a result, particles that are retained on the membrane create numerous cake layers. In crossflow filtration, the feed stream passes parallel or tangential through the membrane due to pressure difference. The buildup of the cake layers can be somewhat suppressed due to pressure differences in crossflow (Nazir et al., 2019).

The retention coefficient: The retention of a component on the membrane can be defined by the retention coefficient (R) and is calculated as:

$$\mathrm{R} = 1 - \frac{C_P}{C_R} \tag{4.18}$$

where $\mathrm{C_P}$ is the concentration of the component in the permeate and $\mathrm{C_R}$ is the concentration in the retentate.

Permeate flux: The permeate flux is defined the amount of the permeate produced during membrane separation per unit of time and membrane area. It is measured in units of volume per square meters per hour.

$$J = \frac{\text{Permeate flow rate}}{\text{Membrane area}} \tag{4.19}$$

Recovery rate: The recovery rate refers to the fraction of the feed flow that passes through the membrane during filtration. It is typically expressed as a percentage:

$$Q_r = \frac{Q_P}{Q_f} \tag{4.20}$$

Q_r: recovery rate, Q_p: permeate flow rate; and Q_f: feed-flow rate.

The primary benefit of employing membranes in separation procedures is their ease of use. It is a cost-effective, environment-friendly, and green technology with minimal waste generation. Moreover, it is a lesser time- and energy-consuming process. The primary goal in industrial applications is to keep the permeate flux high. Furthermore, permeability and selectivity are the two main characteristics that define a membrane's performance.

For industrial applications, permeability and selectivity are the two main characteristics that define a membrane performance. Selectivity is the capacity of the membrane to hold particles on the surface of membrane, whereas permeability is related to permeate flux.

The liquid flow through a pore (q) is calculated using Poiseuille's Law model:

$$q = \frac{\Pi d^4}{128\mu\ell} \times \Delta P \tag{4.21}$$

Where $\Delta p =$ The pressure difference across the pore, μ the viscosity of liquid, and $\ell =$ The pore length.

The permeate flux (flow per unit membrane area represents the total flows passing through the individual pores and can be expressed as:

$$J = N \times \frac{\Pi d^4}{128\mu\ell} \times \Delta P \tag{4.22}$$

Where J = permeate flux, N = the number of pores per square centimeter of membrane

For membranes of equal pore area and porosity (ε), the number of pores per square centimeter is proportional to the inverse square of the pore diameter. That is,

$$N = \varepsilon \frac{4}{\Pi d^2} \tag{4.23}$$

It follows that the flux through membranes can be calculated by giving Eq. (4.23) in Eq. (4.22) is

$$J = \frac{\Delta P \varepsilon}{32\mu\ell} \times d^2 \tag{4.24}$$

From Eq. (4.24), it was clearly shown that the permeate flux is directly proportional to the square of the pore diameter. Because of that, microfiltration membranes have significantly higher permeance compared to ultrafiltration membranes, which in turn have much higher than reverse osmosis membranes. These variations have a greater influence on the operating pressure and industrial applications of these membranes (Saleh & Gupta, 2016).

4.3.8.2 *Application of membrane separation*

Membrane systems have been extensively employed for several decades in many different aspects of food production. In cases where the retained material is valuable, such as in cheese making, they help increase product yield and concentrate liquid streams, thereby enhancing product value. UF membranes are commonly employed in the fruit juice and syrup sector to remove pollutants such as yeasts, moulds, germs, pyrogens, colloids, and solids from different fruit juice,s such as apples, pear, orange, and many more. Furthermore, the ongoing expulsion of proteins, polyphenols, tannins, and polysaccharides improves quality of final juices. As a result, the juices can be concentrated into syrups with significantly increased viscosity using RO membranes.

4.3.8.3 Classification of membrane filtration

Membrane separation processes can be categorized based on the size of the membrane pores and the particles that are retained in the membrane shown in Fig 4.7:

- Microfiltration: A membrane with pore sizes ranging from 10 to 0.1 μm was employed.
- Ultrafiltration: The pore size of membranes varies between 0.1 to 0.01 μm.
- Nanofiltration: The membrane size for NF is between 0.01 and 0.001 μm.
- Reverse Osmosis: Membrane thickness of RO is less than 0.001 μm. Compared to the other membrane technologies, RO membranes merely allowed water molecules to travel through extremely high pressure drops (10–100 MPa).

4.3.8.4 Microfiltration

The first commercial MF membrane had been developed in the year 1960. Since then, MF membranes have found extensive use in juice processing, dairy industry, egg processing, and wastewater treatment to eliminate germs, bacteria, colloids, fat globules, and sterilize beer and wine (Younas & Rezakazemi, 2022). Recently, the implementation of microfiltration technology has been increased in wastewater treatment, clarification of juices, wine, and beer. Membranes could extract different types of impurities from water based on their characteristics and pore size such as suspended particles, oils and greases, proteins, microbes, colloids, dissolved salts, and organic acids. In the process of microfiltration (MF), proteins flow across the membrane easily, whereas collides, suspended particles are retained on the membrane surface as seen in Fig 4.8 (Urošević et al., 2017). It is a pressure-driven membrane process, functioning at low transmembrane pressures ($<$ 50 Psi or 3.4 bar or 0.35 MPa), and very high permeation fluxes (Belfort et al., 1994). It is operated in both ways: crossflow and dead-end filtration method. The introduction of crossflow microfiltration (CMF) in the food industry brings a massive change in this sector. This process not only clarifies the juices, but also removes the microorganisms from the juices, leading to an increase in the shelf-life of juices. Fruit and vegetable juices have seen an increased use of MF to eliminate bacteria and protozoa. The permeate flux is governed by Darcy's law:

$$\mathrm{J} = \frac{\Delta P - \Delta \pi}{\mu R_h} \tag{4.25}$$

Where J ($m^3/m^2/s$) is the permeate flux, ΔP (Pa) the applied pressure, $\Delta\pi$ the osmotic pressure, (Pa s), μ=solvent viscosity and R_h (m^{-1}) the hydraulic resistance. The (CMF)

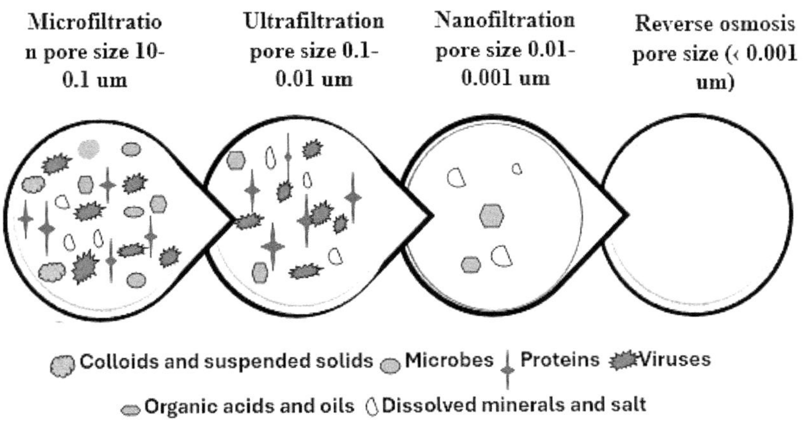

Figure 4.8 A schematic diagram of retentate and permeate from different membrane filtration processes.

membranes retain only colloids and particles allowing solutes and salts to pass through the membrane. As a result, $\Delta\pi$ would be considered negligible in this case and Darcy's law can be expressed

$$J = \frac{\Delta P}{\mu R_h} \tag{4.26}$$

MF is a powerful and versatile instrument for the dairy sector to remove fat globules, bacteria, and casein micelles from the milk. The main applications of MF in cheese-making are bacteria removal, whey defatting, and micellar casein enrichment. Additional uses include the fractionation of whey or milk protein, the separation of milk fat, and the selective extraction of somatic cells from raw whole milk. MF can be used to isolate heterologous proteins from transgenic milk. It has been demonstrated that MF works incredibly well to remove somatic cells, bacteria, and spores from skim milk. As a non-thermal technique for eliminating germs from milk to maintain natural nutritional and sensory qualities (Chew et al., 2020).

4.3.8.5 Ultrafiltration

Ultrafiltration process is used to clarify, separate, and concentrate the solution. The pore size of UF is in the range of 0.01–0.1μm, operated at pressure 0.1–0.8MPa, to retain 0.05–10μm diameter components with 500~500000 Da molecular weight substance. The common polymers used to prepare the UF membranes are polysulfone, polyethersulfone, polyacrylonitrile, etc. (Mondal et al., 2016). During the UF process, all low molecular weight components readily flow through the membrane while high molecular weight components—like protein and suspended solids, inorganic solids—are retained on the membrane surface seen in Fig 4.8. Moreover, sodium hydroxide, mono-and di-saccharides, salts, amino acids, and organic acids are also retained to some extent (Urošević et al., 2017). There are three main mechanisms of ultrafiltration membrane retention operates through three primary mechanisms: adsorption onto the membrane surface and within the pores, retention within the pores, and removal of particles by mechanical filtration on the surface of the membrane. It can effectively remove pathogenic microorganisms. The turbidity level is high and the filtration precision is excellent, ensuring that the quality of the treated water remains consistent and dependable (Li et al., 2018).

4.3.8.5.1 Application

- Ultrafiltration process is widely used to remove pathogenic bacteria from water to make safe drinking water. It is also used as a pretreatment of RO and has a wide range of applications in the field of wastewater reuse and seawater desalination (Belfort et al., 1994).
- It is used in beverages industries to clarify juices, because macromolecules of raw juices settled during storage, leading to destroying the shelf-life, deteriorating the quality and appearance of the juice.
- It is used as a pre-treatment process before concentration of juice using reverse-osmosis, or other membrane technology.

4.3.8.6 Nanofiltration

The nanofiltration membrane was first introduced in the food sector in 1976 to1977. A membrane manufacturer company created a unique softening membrane that had a rejection rate of roughly 47% for monovalent ions. This softening membrane was named nanofiltration membrane owing to its pore sizes. NF is a pressure-driven separation method, widely used for the concentration of juices. The effectiveness of filtration method is dependent on the charge and steric effects (Warczok et al., 2004). The characteristics of nanofiltration membranes can be classified between those of reverse osmosis (RO) and ultrafiltration (UF) membranes, with pore sizes of less than one nm and molecular weight cut-offs of 300–500 Da as seen in Fig 4.8. NF membranes can retain most organic compounds, viruses, a variety of salts, and inorganic ions. It is a great choice for the food industry as it offers a better separation procedure at minimal loss of nutritional characteristics (Yadav et al., 2022). The surface of NF membranes is negatively charged. It illustrates anion repulsion, which is the primary factor behind the rejection of the species. Salts containing monovalent anions are seen to have low rejection rates. Nonetheless, salts containing divalent and multivalent anions exhibit significant rejection. It was observed that NF is widely used for eliminating divalent ions from hard water to soften it. NF has several advantages over RO, including the ability to operate at low pressure, high permeate flow, retention of organic and multivalent salts, and lower investment, operating, and maintenance expenses (Younas & Rezakazemi, 2022).

4.3.8.7 Reverse osmosis

It helps to understand the first osmosis principle to comprehend the basic principles underlying RO. When two salt solutions with differing concentrations are separated

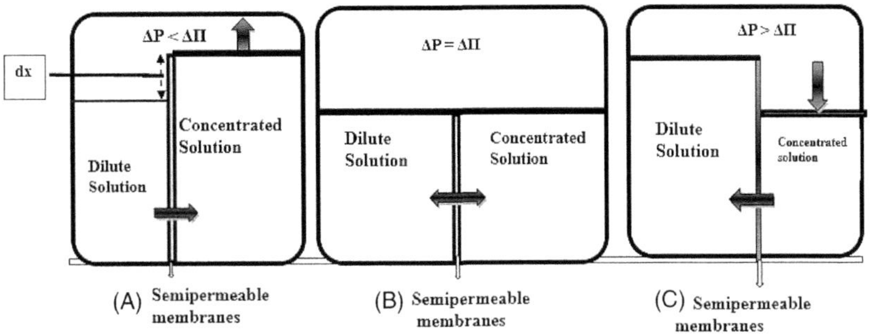

Figure 4.9 Osmotic phenomena: (A) osmosis, (B) osmotic equilibrium, and (C) reverse osmosis (Younas & Rezakazemi, 2022).

through a semi-permeable membrane, water migrates from the weaker solution to the stronger solution. This process is known as osmosis. Salts migrate until the concentrations of both solutions reach osmotic equilibrium. In RO, the two solutions are separated through a semi-permeable membrane, in which water flows from the higher concentrated solution to the lower concentration due to applied pressure. Water flow stops when applied pressure equal to the osmotic pressure difference between two solutions. Water flows from the concentrated solution into the diluted solution when the applied pressure is higher than the osmotic pressure ($\Delta P > \Delta\pi$). The concept of osmosis and RO is described schematically in Fig 4.9 where, (a) represents the osmosis (b) osmosis equilibrium and (c) represents the RO.

Among the membranes, RO has the smallest pore size of approximately $0.0001\mu m$It has the ability to retain all dissolved and suspended species, as well as monovalent salts, but only allows water to flow through membrane. To desalinate the saline water, RO eliminates monovalent ions. NF and RO are referred to as dense membrane separation processes. RO systems are generally employed as the last stage in reclamation of wastewater to remove total organic carbon. It has been demonstrated that RO effectively removes microorganisms, neutral base compounds, and dissolved species (Younas & Rezakazemi, 2022). Reverse osmosis is a pressure-driven membrane process operated at room temperature that can be extensively used for juice concentration. Fruit juices are concentrated to save costs associated with distribution, storage, and transportation. As compared to traditional thermal method, it is a cost and energy-effective process. Moreover, RO concentration is the attainment of high-quality products due to low temperature operation, resulting in the maintenance of nutritional, aroma, and flavor of final product. The primary drawback is its elevated viscosity, and high osmotic pressure, which restrict the process efficiency (Echavarría et al., 2012).

Jesus et al. (2007) observed concentrated orange juice, and preserved nutritional characteristics using RO at different transmembrane pressure. The results showed that the soluble solids content, vitamin C, permeate flux increased with increasing pressure. Echavarría et al. (2012) used the reverse osmosis for concentration of four fruit juices (peach, pear, apple, and mandarin). The osmotic pressure, viscosity, red color, and soluble solids of final product was increased whereas, permeate flux decreased as the concentration factor increased, which can be due to concentration polarization and fouling. RO is used to preconcentrate the clarified apple juice and increased 25°Brix soluble solids concentration (Alvarez et al., 2000). It was observed that RO is a promising process for concentration of juices over the thermal processing methods. This process not only concentrates the solution but also maintains the nutritional value of juices. It is also used as a pre-concentration process before pervaporation process.

4.4 Clarification

The consumption of fruit juices has been increasing globally. Freshly pressed juice contains suspended solids (polysaccharides, protein, and polyphenols, especially tannins) causes turbidity and a cloudy appearance in juices during storage. This cloudy appearance negatively affects consumer attitude towards consumption of juice because consumers liked clear juice with an appealing color. Before filtration, the suspended solids are effectively precipitated, a process known as clarification. The clarification process is used to remove suspended solids (polysaccharide, pectin, protein, and polyphenols, especially tannins), which causes turbidity during storage and debris from the juice (Shirvani et al., 2023). By clarifying fruit juices, producers can enhance their appearance, taste, and overall quality, making them more appealing to consumers and extending their marketability. It is typically accomplished by microfiltration, enzymatic treatment, and the use of common clarifying agents such as polyvinyl pyrrolidone, silica sol, gelatin, bentonite (Chatterjee et al., 2004).

4.4.1 Clarification of beverages using polysaccharide-based agents (Chitosan and xanthon gum)

Clarification is an essential step in the processing of fruit juice, which is typically accomplished by microfiltration, enzymatic treatment, and the use of common clarifying agents, such as polyvinyl pyrrolidone, silica sol, gelatin, bentonite (Chatterjee et al., 2004). Chitosan is a linear positively-charged polymer with a pH of less than 6.5. It consists of β-1,4 linked D-glucosamine and *N*-acetyl-glucosamine due to deacetylation. Unlike chitin, which is extremely hydrophobic, chitosan is soluble in acidic solutions depending on its degree of deacetylation (Colangelo et al., 2018). Due to its polycationic character, chitosan (deacetylated chitin) has been shown to be a useful coagulating agent for assisting in the removal of suspended particles from liquids. Furthermore, since chitosan is biodegradable and harmless, it can be utilized as a clarifying agent in the purification of fruit juices (Chatterjee et al., 2004). The clarification effect of polysaccharide-based agents may occur through two possible mechanisms: (1) the formation of water-soluble polysaccharide-phenolic protein complex, and (2) the removal of proteins (Erkan-Koç et al., 2015). Apple, grape, lemon, and orange juices were clarified using chitosan. The results showed that lowered the protein content of the juice, which causes turbidity of juices whereas, the minimal decrease in titratable acidity and total soluble solids of juices. It was observed that chitosan can be used as a clarifying agent in juice processing, but that has minimal effect on the biochemical parameters of the juices (Chatterjee et al., 2004). Chokeberry juice had lower turbidity and higher antioxidant activity and polyphenol content after clarification of juice using chitosan and xanthan gum. The color of juices was more appealing red color without browning (Lachowicz et al., 2018). The turbid substances (polyphenol, polysaccharides, and proteins) lowered the commercial value of wine. Thus, several studies have been carried out to clarify the variables impacting protein instability and haze formation in wine. The effect of chitosan on wine was investigated. The results revealed that tartrate stability improved, while no change was observed in the polyphenols' content (Colangelo et al., 2018). During pressing of pomegranates, tannins hydrolyzed and formed a cloudy appearance in juice. Moreover, pectin and soluble protein also caused turbidity in juices. Chitosan and xanthan gum caused lower turbidity and higher antioxidant activity whereas, these clarifying agents had no effect on the phenolic compounds of pomegranates juices (Erkan-Koç et al., 2015). The strawberry juice was clarified using chitosan, the clarified juice showed lowered turbidity, phenolic content, flavonoids, protein, and antioxidant activity, while there was no effect on soluble solids, titratable acidity, pH level, folate, vitamin C, anthocyanin (Heidarizadeh et al., 2023).

4.4.2 Clarification of beverages using enzymes

Raw extracted juices are cloudy and dense due to the presence of pectic substances in fruits. Consumers prefer juice, that is low in viscosity, crystal clear, and high nutrient content. As a result, treating pectinaceous compounds is essential from a marketing perspective. The use of pectinase for juice clarification offers several benefits, including improved pulp fluidification, increased juice yield, enhanced clarification, reduced sugar content, decreased turbidity and viscosity, and increased total soluble solids.

Pectic compounds (pectinic acid and pectin) are a group of complex colloidal plant carbohydrates made up primarily of (1,4)-a-D-polygalacturonicacid. Pectin is a mixture of water-soluble pectinic acid that can form gels under appropriate conditions, typically involving high sugar, and acid concentration (Kilara & Van Buren, 1989). Pectin is a high-molecular weight polysaccharide consisting of linear chains of α-1, 4-linked D-galacturonic acid residues present in middle lamella of plant cell wall. It is extensively utilized as a gelling agent, thickener, emulsifier, or stabilizer in the food sector. However, the fruit juices became hazy and dense due to the presence of pectin. The settlement of suspended solids during storage is also an issue for industries. To mitigate this problem, most industries are showing interest in enzyme-based degradation of pectin. The uses of enzymes in fruit juice processing have been demonstrated be a cost-effective and efficient method. Recently, pectinases, cellulases, and hemicellulases enzymes have been used for clarification of fruit and vegetable juices. Pectinolytic enzymes, or pectinases, act in different forms on their substrate, the pectin. Temperature, incubation time, pH, and concentration are the main variables affecting enzyme activity. These specifications change depending on the fruit (Singh & Singh, 2015). Acidic pectinases are utilized in juice processing and winemaking industries to break down pectic substances (Patel et al., 2022).

The cherry juice was clarified using pectinase, the results revealed that turbidity was lowered during storage (Pinelo et al., 2010). Clarification of apple, butia palm fruit, blueberry, and grape juices using pectinase enzyme at different temperature and time. The results showed that clarification of juices increased with increased time, while there was no direct relation established with temperature. Butia palm and grape juices showed more clear and opaque juice than other juices (Sandri et al., 2011). Pectolytic enzymes are generally used in the production of clear apple juices before filtration. The application of enzymes lowered the viscosity of the juice as a result of depolymerizing the juice pectin. Additionally, depolymerization minimizes the protective colloid function of pectin, allowing suspended particles to coagulate and settle quickly (Kilara & Van Buren, 1989). This improves the performance of filtration methods. Cheirsilp and Umsakul (2008) reported that enzyme-treated banana showed lower viscosity during banana wine production, resulting in improved clarification of the wine without altering nutritional properties.

Cellulose is an abundantly available polysaccharide. It is a crystalline polymer of D-glucose residues linked with β-1, 4 glucosidic linkages. It is a stiff polymer in a highly ordered manner due to inter- and intrachain hydrogen bonds (Singh & Singh, 2015). Thus, breakdown of cellulose is challenging and requires extra efforts. Cellulases are the enzymes used for degradation of cellulose hydrolyze cell wall polysaccharides, and are responsible for converting cellulose into glucose. They break the glycosidic bonds of cellulose microfibrils, releasing oligosaccharides, cellobiose, and glucose. Cellulases are used commercially in extraction and clarification of fruits and vegetable juices for production of nectars and purees (Sharma et al., 2017). The cellulase treated juice showed clear and haze free than raw juice. The apple juice was clarified using pectinase, amylase, and cellulase at a standard concentration and temperature. It was observed that all enzyme-treated juices showed lower viscosity than raw juices whereas, pectinase-treated apple juice showed highest reduction in viscosity compared to others (Padma et al., 2017). This might be due to the different activity of enzymes on different substances. Amylase, pectinase, and cellulase enzymes degrade starch, pectin, and cellulose polysaccharides respectively (Ozyilmaz & Gunay, 2023). There is a point to consider before clarification that the select the targeted substances to be clarified. According to targeted substances, we can use enzymes to clarify the juices. Then it will prove to be a cost-effective method.

4.4.3 Clarifying of beverages using gelatin and bentonite

The turbidity or haze development in beverages is caused by the protein-polyphenol complex, must be eliminated through the clarifying process by employing clarifying agents such as gelatin, bentonite, silica, and polyvinylpyrrolidone, etc. These are the most popular and economical clarifying agents. To stop haze from forming after bottling during cold storage, gelatin, and bentonite absorb suspended particles from the raw juice that contain proteins, hydrophobic polyphenols, and cellulose sludge. The mechanism of clarifying agents during the clarification process relies on the differences in ionic charges of protein, polyphenols, and the fining agents. This includes flocculation and sedimentation, leading to the removal of these haze-forming substances from the juices (Jafari et al., 2024). Jafari et al. (2024) used gelatin and bentonite to clarify gooseberry juice, the results exhibited higher light transmittance total phenolic content, vitamin C, and antioxidant activity as compared to raw sample. Apple juice processing industry uses gelatin and bentonite as a standard commercial procedure for clarifying juice. The fining treatment using gelatin, silica sol, and bentonite likely functions by enabling non-specific binding of haze-active polyphenols and proteins during the formation of colloid. Tannins, which are primarily negatively charged, are drawn to colloidal gelatin because of its positive charge (Awe, 2018). Furthermore, bentonite has negative charge attracted positively charged proteins. There are plenty of attachment locations due to the large interior surface area between the various alumina-silica plates. Bentonite particles break into sheets of alumina-silicate after swelling easily in water. An enormous surface area is made possible by the sheets' separation, which allows for the interchange of cations, as well as adsorption and hydrogen bonding (Valdhuber & Pulko, 2022). As the particles bond to each other, their overall weight rises, causing settling to take place due to gravity. It takes additional filtration steps to get a crystal-clear product. These possible haze precursors are removed from solutions because of flocculation and sedimentation caused by variations in the ionic charges of proteins, polyphenols, and fining agents. The results of clarification of apple juice using bentonite and gelatin, showed that juice was haze-free and less turbid after treatment (Benitez & Lozano, 2007). Clarification of wine prepared from locally available fruits (pawpaw, pineapple, cashew, and banana) resulted in a haze-free, turbid-free wine with low astringency. Gelatin is not only used for clarification, but also to reduce astringency and improve clarity before fermentation (Awe, 2018).

4.4.4 Clarification of beverages using immobilized enzyme

Fruit juice clarity is an important stage in the production process that affects the appearance, sensory attributes, and stability during storage. Most of the tannins and phenolic compounds in juice are eliminated by adsorption. Gelatin, bentonite, (PVPP) polyvinyl pyrrolidone are widely employed for adsorption clarifying, but it is a time-consuming process (Wang et al., 2023). Moreover, juices are clarified using pectinolytic enzymes that break the polysaccharide chain of pectic substances and solubilizes them in the bulk liquid. It reduces the viscosity and the haziness of juice. Enzymatic treatment has the advantages of high selectivity, sustainable procedures, and delicate operating conditions. However, the use of biocatalyst in juice processing is limited due to its low stability, non-reusability, recovery problem, and high cost (Patel et al., 2022). To mitigate these issues, enzyme immobilization technology has been introduced. Immobilization is the process which changes the homogeneous (free) enzyme state into heterogenous (immobilized enzyme) state. So that enzymes can be easily recycled and reused. Proper immobilization can also improve the stability, activity, purity, selectivity, and specificity of an enzyme. The ability of immobilized enzymes to continue performing an enzymatic process after immobilization is one of its most significant benefits (Hosseini et al., 2021). However, some limitations persist, including challenges in scaling up immobilization for industrial applications, the high cost of biocatalysts, and potential negative effect on biocatalyst performance

(Wang et al., 2023). Oliveira et al. (2024) used immobilized enzyme for clarification of wine. The results showed that complete clarification was done within two to three hours. The reusability of the immobilized pectinase conjugates was easier than free enzyme. The xylanase immobilized on functionalized magnetic particle exhibited excellent catalytic activity at pH level of 6.5 and 65°C. It significantly improved juice clarity and showed remarkable stability retaining about 55% of its initial activity even after ten reaction cycles in the juice clarification process (Shahrestani et al., 2016). Preparing or building a bioreactor is the initial stage in carrying out such a continuous enzymatic process. The continuous operation utilizes bioreactor configurations, like fixed bed, packed bed, fluidized bed, and membrane reactor. The barberry juice was clarified using an immobilized pectinase enzyme in both batch and continuous process, resulting in significant enhancement of color and antioxidant properties Whereas, turbidity was reduced after clarification (Hosseini et al., 2021).

4.4.5 *Clarification of beverages using microfiltration and ultrafiltration*

Earlier, bentonite and diatomaceous earth, gelatin and tannin, were commonly used for clarifying fruit juices. However, in recent years, this traditional clarification method has increasingly been replaced by membrane processes such ass microfiltration and ultrafiltration. These methods allow for simultaneous clarification and filtration. To the extent of membrane activity, juices are treated with enzymes immediately before filtration (Urošević et al., 2017).

4.4.5.1 *Clarification of beverages using microfiltration*

Regarding beverages, MF is used in the processing of juices and beer. After fermentation is complete, most beers are treated in the beer-making process to reduce the overall count of microorganisms and to clarify them. The presence of plant residue, which includes fiber, cellulose, hemicellulose, starch, and fat, as well as colloid macromolecules like pectin, protein, and polyphenols causes fruit juices to become cloudy. These residues must be removed to produce clear juice. Crossflow MF is used to recover beer from the yeast-beer mixture. Furthermore, MF is currently used to clarify beer—that is, to get rid of microbes, yeasts, and haze— without having a negative impact on the beer taste or aroma. Wine is cloudy because of the presence of macromolecules (phenolic compounds, polysaccharides, and proteins). When a deposit forms at bottom of bottles, it is often seen as an indication of product deterioration. In fact, clarity is the primary visual quality that consumers expect from wine (El Rayess et al., 2011).

4.4.5.2 *Clarification of beverages using ultrafiltration*

Juice extracted from fruits and vegetables after pressing is typically turbid, brown, very viscous, and tends to settle during storage. Therefore, it is essential to clarify the colloids present in the fruit juices before commercialization. MF and UF are extensively used for clarification of fruit juices. Moreover, MF with an average pore diameter of 0.2 microns is not adequate for eliminating microorganisms completely. UF is a promising choice to obtain a final clarified juice that is free from haze, suspended particulates, and high molecular weight solutes. This process could clarify juices and preserve all their natural flavor, aroma, and nutritional value as well as enhance shelf-life by eliminating debris and bacteria.

Ultrafiltration of bottle gourd juice was done using hollow fibers membrane. The results showed that the nutritional value of juice was maintained, as well as shelf-life of juice was increased eight weeks at refrigerated temperature. It was used to clarify raw juices of peach, pear, apple, and mandarin to eliminate microorganisms from the juices (Mondal et al., 2016).

Alvarez et al. (1998) used the ultrafiltration process to remove pectinase from apple juice. The results reported that permeate flux was much higher, whereas pectinase and viscosity of juice was much lower. Furthermore, clarification of vinegar employing UF not only provides wider vinegar range with same sensorial attributes without turbidity, but it also minimizes the number of processing steps and reduced storage time.

4.5 Conclusion and future scope

Mechanical filtration plays a crucial role in various processes within the food industry. Here are some key applications: water treatment, fruit juice filtration, wine and beer industry and dairy industry. Fruit juice industry used filtration method not only for filtration but also for clarification of beverages. Filtration is a mechanical separation process used to separate solids from fluids by applying pressure and vacuum. Filtration could be done using two ways: surface filtration and deep filtration; in both filtration processes, fluid mixture is forced to pass through filtering media recover liquid or solids according to the purpose. Surface filtration retained the sample on the surface of filtering media, while in deep filtration process the particles could be retained inside the pores. Filtration can be divided into various types, such as plate and frame filter, leaf filter, and continuous filter like rotatory disc and rotary drum filter, etc. However, some disadvantages limit the application of filter in various industries. To mitigate this problem, the recent advancements has been made to improve the application of filter.

Membrane also plays a crucial role in filtration of water, juices, beer, and wine in the food industry. Membrane filtration is classified into four types: microfiltration, ultrafiltration, nanofiltration and reverse osmosis, according to the pore sizes of membranes. Microfiltration only retained the collides and debris, ultrafiltration retained the protein and higher molecular weight substances, nanofiltration retain salts and protein, and reverse osmosis retain all soluble substances while only water passed through this membrane. Clarification is a process to clarify the juices and beer in beverages industry. This process helps in achieving clarity and purity in the final product. This process decreased the turbidity and improved the shelf-life of the juices, as well as the provided an appealing color that attracts the consumer. Thus, clarification process has become a necessity in the beverage industry. It is done using chitosan, enzymes, gelatin, etc. This process is used as a pre-treatment in membrane separation to reduce the fouling in membrane process and as a post-treatment to produce haze-free beverages. Clarification and filtration both are necessary steps in food industry and can be done in continuous process. This will help in reducing the cost and time of the process. The process becomes more energy-efficient. Clarifying agents have different characteristics and play differently with different substances, for example, pectolytic enzymes clarify pectin but have minimal effect on polyphenols, while amylase, pectinase, and cellulase enzymes degrade starch, pectin, and cellulose polysaccharides respectively. Since we can used clarifying agents according to the targeted substance provide better results.

References

Ali, S. A. (2013). Design and evaluate a drum screen filter driven by undershot waterwheel for aquaculture recirculating systems. *Aquacultural Engineering, 54*, 38–44.

Alvarez, S., Alvarez, R., Riera, F. A., & Coca, J. (1998). Influence of depectinization on apple juice ultrafiltration. *Colloids and Surfaces A: Physicochemical and Engineering Aspects, 138*(2-3), 377–382.

Alvarez, S., Riera, F. A., Alvarez, R., Coca, J., Cuperus, F. P., Bouwer, S. T., & Panyor, L. (2000). A new integrated membrane process for producing clarified apple juice and apple juice aroma concentrate. *Journal of Food Engineering, 46*(2), 109–125.

Awe, S. (2018). Effect of clarifying agents (gelatin and kaolin) on fruit wine production. *International Journal of Agriculture Innovations and Research, 6*(4), 130–132.

Bähner, F. D., Santacoloma, P. A., Abildskov, J., & Huusom, J. K. (2017). Analysis and modelling of an industrial pressure filtration using process data. *IFAC-PapersOnLine, 50*(1), 12137–12142.

Beer, C. (1989). Sterile filtration of beer. *Master Brewers Association of the Americas-Technical Quarterly, 26*(3), 89–93.

Belfort, G., Davis, R. H., & Zydney, A. L. (1994). The behavior of suspensions and macromolecular solutions in crossflow microfiltration. *Journal of Membrane Science, 96*(1-2), 1–58.

Benitez, E. I., & Lozano, J. E. (2007). Effect of gelatin on apple juice turbidity. *Latin American applied research, 37*(4), 261–266.

Berk, Z. (2018). *Food process engineering and technology*. Academic press.

Campbell, R. C. (1921). Modern leaf type filters. *Industrial & Engineering Chemistry, 13*(11), 1002–1004.

Charcosset, C. (2021). Classical and recent applications of membrane processes in the food industry. *Food Engineering Reviews, 13*(2), 322–343.

Chatterjee, S., Chatterjee, S., Chatterjee, B. P., & Guha, A. K. (2004). Clarification of fruit juice with chitosan. *Process Biochemistry, 39*(12), 2229–2232.

Cheirsilp, B., & Umsakul, K. (2008). Processing of banana-based wine product using pectinase and α-amylase. *Journal of Food Process Engineering, 31*(1), 78–90.

Chew, J. W., Kilduff, J., & Belfort, G. (2020). The behavior of suspensions and macromolecular solutions in crossflow microfiltration: An update. *Journal of Membrane Science, 601*, 117865.

Colangelo, D., Torchio, F., De Faveri, D. M., & Lambri, M. (2018). The use of chitosan as alternative to bentonite for wine fining: Effects on heat-stability, proteins, organic acids, colour, and volatile compounds in an aromatic white wine. *Food Chemistry, 264*, 301–309.

Daneluz, J., da Silva, G. F., Duarte, J., Turossi, T. C., dos Santos, V., Baldasso, C., & Daneluz, A. C. (2023). Characterization of microfiltration and ultrafiltration membranes for application in kombucha filtration. *Journal of Industrial and Engineering Chemistry, 126*, 264–269.

Dornier, M., Belleville, M. P., & Vaillant, F. (2018). Membrane technologies for fruit juice processing. In A. Rosenthal, R. Deliza, J. Welti-Chanes, & G. Barbosa-Cánovas (Eds.), *Fruit preservation. Food engineering series* (pp. 211–248). New York: Springer.

Echavarría, A. P., Falguera, V., Torras, C., Berdún, C., Pagán, J., & Ibarz, A. (2012). Ultrafiltration and reverse osmosis for clarification and concentration of fruit juices at pilot plant scale. *LWT-Food Science and Technology, 46*(1), 189–195.

El Rayess, Y., Albasi, C., Bacchin, P., Taillandier, P., Raynal, J., Mietton-Peuchot, M., & Devatine, A. (2011). Cross-flow microfiltration applied to oenology: A review. *Journal of Membrane Science, 382*(1-2), 1–19.

Erkan-Koç, B., Türkyılmaz, M., Yemiş, O., & Özkan, M. (2015). Effects of various protein- and polysaccharide-based clarification agents on antioxidative compounds and colour of pomegranate juice. *Food Chemistry, 184*, 37–45.

Farhat, N. M., Christodoulou, C., Placotas, P., Blankert, B., Sallangos, O., & Vrouwenvelder, J. S. (2020). Cartridge filter selection and replacement: Optimization of produced water quantity, quality, and cost. *Desalination, 473*, 114172.

Gehring, B., Oechsle, D., & Kottke, V. (1997). Parameterisation of horizontal leaf filters and optimisation for the filtration of beer. *Filtration & separation, 34*(1), 67–72.

Girard, B., & Fukumoto, L. R. (2000). Membrane processing of fruit juices and beverages: A review. *Critical Reviews in Food Science Nutrition, 40*(2), 91–157.

Grimm, R. T. (1956). Leaf filter for foots removal from crude oil reduces labor and eliminates filter cloth and paper. *Journal of the American Oil Chemists' Society, 33*, 437–439.

Heidarizadeh, F., Kolahi, M., & Akhond, M. (2023). Enhancing the qualitative and nutritional properties of strawberry juice through chitosan treatment. *International Journal of Food Science & Technology, 58*(10), 5214–5226.

Hosseini, S. S., Khodaiyan, F., Mousavi, S. M., & Azimi, S. Z. (2021). Continuous clarification of barberry juice with pectinase immobilised by oxidized polysaccharides. *Food Technology and Biotechnology, 59*(2), 174–184.

Jafari, S., Shiekh, K. A., Mishra, D. K., Kijpatanasilp, I., & Assatarakul, K. (2024). Combined effects of clarifying agents improve physicochemical, microbial and sensorial qualities of fresh Indian gooseberry (*Phyllanthus emblica* L.) Juice during refrigerated storage. *Foods, 13*(2), 290.

Jayaprahash, C., Lakshmana, J. H., & Kingston, J. J. (2020). Self-contained food sample homogenisation filter bag for microbial analysis. *Defence Life Science Journal, 5*(4), 241–247.

Jesus, D. F., Leite, M. F., Silva, L. F. M., Modesta, R. D., Matta, V. M., & Cabral, L. M. C. (2007). Orange (*Citrus sinensis*) juice concentration by reverse osmosis. *Journal of Food Engineering, 81*(2), 287–291.

Jin, H., Hu, S., Xie, B., Yan, Y., Yang, M., & Zhou, F. (2021). Experimental optimization for cleaning parameters and field application of cartridge filter in bulk grain loading. *Powder technology, 378*, 421–429.

Khean, T. S. (2003). Studies in filter cake characterisation and modelling.

Kilara, A., & Van Buren, J., P. (1989). Clarification of Apple Juice. In D. Downing, L. (Ed.), *Processed Apple Products* (pp. 83–96). Springer.

Kriegel, P. (1938). Plate and frame filter press. *Industrial & Engineering Chemistry, 30*(11), 1211–1213.

Kurita, Y., Suwa, S., & Murata, S. (2010). Filter presses: A review of developments in automatic filter presses. *Filtration & Separation, 47*(3), 32–35.

Lachowicz, S., Oszmiański, J., & Kalisz, S. (2018). Effects of various polysaccharide clarification agents and reaction time on content of polyphenolic compound, antioxidant activity, turbidity and colour of chokeberry juice. *LWT, 92*, 347–360.

Li, X., Jiang, L., & Li, H. (2018). Application of ultrafiltration technology in water treatment. *IOP Conference Series: Earth and Environmental Science*: 186. IOP Publishing.

Lozano-Sánchez, J., Cerretani, L., Bendini, A., Gallina-Toschi, T., Segura-Carretero, A., & Fernández-Gutiérrez, A. (2012). New filtration systems for extra-virgin olive oil: Effect on antioxidant compounds, oxidative stability, and physicochemical and sensory properties. *Journal of Agricultural and Food Chemistry, 60*(14), 3754–3762.

Mann, J. (1978). *Transport processes and unit operations: By Christie J* p. 650. Geankoplis Published by Allyn and Bacon Inc. 1980.

Mondal, M., Biswas, P. P., & De, S. (2016). Clarification and storage study of bottle gourd (*Lagenaria siceraria*) juice by hollow fiber ultrafiltration. *Food and Bioproducts Processing, 100*, 1–15.

Nazir, A., Khan, K., Maan, A., Zia, R., Giorno, L., & Schroën, K. (2019). Membrane separation technology for the recovery of nutraceuticals from food industrial streams. *Trends in food science & technology, 86*, 426–438.

Neikov, O. D., Yefimov, N. A., & Naboychenko, S. (2009). *Handbook of non-ferrous metal powders: Technologies and applications*. Elsevier.

Oliveira, S. C., Dencheva, N. V., & Denchev, Z. Z. (2024). Immobilization of enological pectinase on magnetic sensitive polyamide microparticles for wine clarification. *Foods, 13*(3), 420.

Osborne, D. G. (2013). *The coal handbook: Towards cleaner production: Coal production*: 1. Woodhead Publishing Limited.

Ozyilmaz, G., & Gunay, E. (2023). Clarification of apple, grape and pear juices by co-immobilized amylase, pectinase and cellulase. *Food Chemistry, 398*, 133900.

Padma, P. N., Sravani, P., Mishra, P., Sneha, N., & Anuradha, K. (2017). Synergistic effect of multiple enzymes on apple juice clarification. *Indian Journal of Science and Technology, 10*(10), 1–5. doi:10.17485/ijst/2017/v10i10/107716.

Patel, V. B., Chatterjee, S., & Dhoble, A. S. (2022). A review on pectinase properties, application in juice clarification, and membranes as immobilization support. *Journal of Food Science, 87*(8), 3338–3354.

Perlmutter, B. A. (2015). *Solid-liquid filtration: Practical guides in chemical engineering*. Butterworth-Heinemann.

Pinelo, M., Zeuner, B., & Meyer, A. S. (2010). Juice clarification by protease and pectinase treatments indicates new roles of pectin and protein in cherry juice turbidity. *Food and bioproducts processing, 88*(2-3), 259–265.

Pouliot, Y., Conway, V., & Leclerc, P.-L. (2014). Separation and concentration technologies in food processing. In: S. Clark, S. Jung & B. Lamsal (Eds.). *Food processing: Principles and applications* (2nd, pp. 33–60). New Jersey: John Wiley & Sons Ltd.

Richardson, J. F., & Harker, J. H. (2002). *Chemical engineering-volume 2: Particle technology and separation processes*. Butterworth-Heinemann.

Rosas, I., Collado, S., Fernández, I., Gutierrez-Lavin, A., & Diaz, M. (2013). Fouling in pleated microfiltration cartridges caused by Pseudomonas putida. *Journal of Membrane Science, 443*, 107–114.

Saleh, T. A., & Gupta, V. K. (2016). *Nanomaterial and polymer membranes: Synthesis, characterization, and applications*. Elsevier.

Sandri, I. G., Fontana, R. C., Barfknecht, D. M., & da Silveira, M. M. (2011). Clarification of fruit juices by fungal pectinases. *LWT-food Science and Technology, 44*(10), 2217–2222.

Schroën, K., & de Aguiar, I. B. (2023). The under-recognized fundamental effects of component interaction and deformability for membrane processes for food. *Separation and Purification Technology, 332*, 125844.

Shahrestani, H., Taheri-Kafrani, A., Soozanipour, A., & Tavakoli, O. (2016). Enzymatic clarification of fruit juices using xylanase immobilized on 1, 3, 5-triazine-functionalized silica-encapsulated magnetic nanoparticles. *Biochemical Engineering Journal, 109*, 51–58.

Sharma, H. P., Patel, H., & Sugandha. (2017). Enzymatic added extraction and clarification of fruit juices: A review. *Critical Reviews in Food Science and Nutrition, 57*(6), 1215–1227.

Shirvani, A., Mirzaaghaei, M., & Goli, S. A. H. (2023). Application of natural fining agents to clarify fruit juices. *Comprehensive Reviews in Food Science and Food Safety, 22*(6), 4190–4216.

Singh, R., & Singh, R. K. (2015). *Role of enzymes in fruit juices clarification during processing: A review International Journal of Biological Technology, 6*, 114–124.

Tarleton, S., & Wakeman, R. (2006). *Solid/liquid separation: Equipment selection and process design*. Elsevier.

Tien, C. (2012). *Principles of filtration*. Elsevier.

Urošević, T., Povrenović, D., Vukosavljević, P., Urošević, I., & Stevanović, S. (2017). Recent developments in microfiltration and ultrafiltration of fruit juices. *Food and Bioproducts Processing, 106*, 147–161.

Valdhuber, J., & Pulko, B. (2022). The impact of bentonite fining on apple juice clarity and sodium content. *Agricultura Scientia, 19*(2), 21–28.

Wang, F., Xu, H., Wang, M., Yu, X., Cui, Y., Xu, L., & Qian, J. (2023). Application of immobilized enzymes in juice clarification. *Foods, 12*(23), 4258.

Warczok, J., Ferrando, M., Lopez, F., & Güell, C. (2004). Concentration of apple and pear juices by nanofiltration at low pressures. *Journal of Food Engineering, 63*(1), 63–70.

Wills, B. A., & Finch, J. (2015). *Wills' mineral processing technology: An introduction to the practical aspects of ore treatment and mineral recovery*. Butterworth-heinemann.

Xie, B., Li, S., Jin, H., Hu, S., Wang, F., & Zhou, F. (2018). Analysis of the performance of a novel dust collector combining cyclone separator and cartridge filter. *Powder Technology, 339*, 695–701.

Yadav, D., Karki, S., & Ingole, P. G. (2022). Nanofiltration (NF) membrane processing in the food industry. *Food Engineering Reviews, 14*(4), 579–595.

Younas, M., & Rezakazemi, M. (Eds.). (2022). *Membrane contactor technology: Water treatment, food processing, gas separation, and carbon capture*. John Wiley & Sons.

Sieving, sifters, and entoleters

5

Humeyra Cavdar Dincturk, Rukiye Gundogan and Esra Capanoglu
Department of Food Engineering, Faculty of Chemical and Metallurgical Engineering, Istanbul Technical University, Maslak, Istanbul, Türkiye

5.1 Introduction

The food and agricultural industries rely on various processes to obtain the desired functionality, safety, and quality in specific products. Some processes produce homogeneous and pure products by separating materials based on size, while others eliminate undesirable particles to obtain clean and safer products. One of the most commonly utilized processes is separation. In the case of the separation of particles or impurities, sieving, sifters, and entoleters hold significant importance to achieve different purposes.

Sieving is a separation-based method used to differentiate particles based on their size. Different sieving methods include gravitational, centrifugal, vibratory, and rotary sieving. Each method has various advantages and application areas. Gravitational sieving is the preferred method for separating large quantities of grain or flour. In this method, particles fall through a mesh by gravitational force. Centrifugal sieving works based on centrifugation force and provides obtaining fine powders. Vibratory sieving moves particles with vibration in the sieve and tries to achieve maximum separation yield for desired particles. Rotary sieving acts with rotating drums or cylinders with mesh surfaces. Cylinders turn continuously and provide a high amount of separation (Sayyad et al., 2022). Sifters are equipment that works on the sieving and screening principle. They classify particles into different fractions based on their size. Sifters are crucial in food and agriculture to improve the quality of various products. Different types of sifters are preferred to achieve the desired final product properties. Vibratory sifters are commonly used equipment that use vibration energy to separate particles. They can provide fine particles, making them suitable equipment for flour milling and spice processing (Sanchez-Suarez et al., 2022). Other than vibratory sifters, rotary or centrifugal sifters also have a significant place in separation processes. Rotary or centrifugal sifters consist of rotating paddles that move materials through the sieve to increase the screening efficiency of particles (Cauvain, 2015). Gyratory sifters provide a gentle circular motion that separates sensitive seeds (Fišteš, 2021). Recently, air-jet sifters have been utilized to separate fine and light materials. The use of air streams to disperse particles is the main principle of this sifter type (Doddabematti Prakash et al., 2024; Lee et al., 2020). Each sifter has a vital role in the purity and consistency of products by manipulating the texture of food products (Bala et al., 2023). Additionally, entoleters are separation equipment that is used for the sanitation of products. They are technological devices that apply high-speed impact to the products and destroy insects, eggs, and larvae. The destruction of insects resulted from the collision of impact plates. This impact-type technology sanitizes products and supports food quality and safety

Mechanical Separation Processes in the Food Industry: Unit Operations and Processing Equipment in the Food Industry.
DOI: https://doi.org/10.1016/B978-0-128-19523-9.00001-9

(Bhattacharyya et al., 2023). The integration of all these techniques and equipment in food and agricultural processes is critical to ensure food safety, prevent foodborne illnesses, increase the quality of products, and improve the efficiency of systems by increasing yield and decreasing effort.

This chapter explores the fundamental principles and advantages of sieving, sifters, and, entoleters by investigating their use in food, food, and agricultural processes. This comprehensive overview covers the functions and benefits of these techniques and equipment to inform industry professionals about their compartmentalization of these into several processes. Additionally, this chapter provides outcomes of future innovations and trends in these technologies, highlighting how they are improved to meet the requirements of food and agricultural industries and specific products.

5.2 Fundamentals of sieving

Sieving is used to achieve specific particle sizes without requiring grinding or crushing. During sieving, materials are placed on a vibrating mesh screen. Activation methods, such as ultrasound, vibration, and manual shaking, which directs smaller particles downward to pass through the screen (undersize), while larger particles accumulate on the surface and are discharged as an oversize product. This process is utilized to characterize many products on the market that have been milled and sieved into distinct fractions (Saha et al., 2022).

Sieving employs screens with consistent mesh sizes to separate materials based on their size. Larger particles are retained on the screen until they exit, while smaller particles pass through. Fig. 5.1 represents the fundamental particle separation principle of the sieving process. The effectiveness of sieving is measured by comparing the amount of collected undersized particles to those smaller than the mesh openings in the original material. Smaller particles generally pass through more readily than those near the size of the mesh openings (Zhang et al., 2016). The size of sieves varies depending on the products and reaches 25 mm. Sieves with a size of 20 μm may be used for powdered-form products (Lewis, 2023). A sieve is defined as an open container that contains square openings that are identical to each other. "screen aperture" refers to the gap between the wires in a wire mesh screen, and "mesh number" denotes the number of wires per inch. Although these terms have been removed from the latest British standards, they are still commonly used (Lewis, 2023). Particle classification depends on physical properties such as shape, size, and density. In a vacuum, solid particles accelerate indefinitely regardless of size or density. In air or water, viscosity creates resistance that increases with velocity until particles settle at their terminal velocity, where gravitational and fluid forces balance. This settling allows particle separation based on their different sizes and densities (Modrzewski et al., 2022).

5.2.1 Types of sieves used in the food and agricultural industry

Sieving is a fundamental process in the food and agricultural industries that separates particles based on size. Various types of sieving techniques are employed

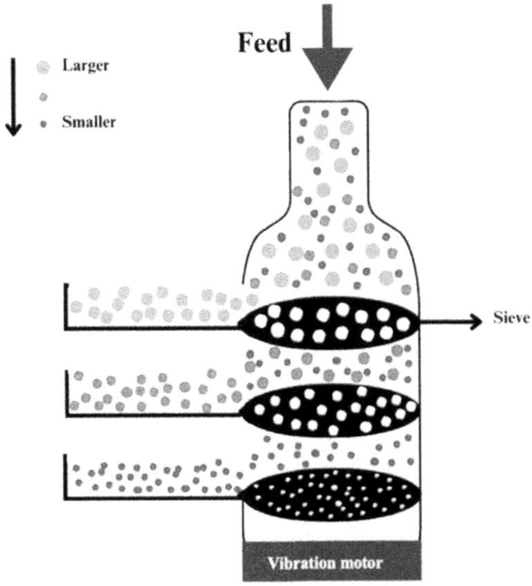

Figure 5.1 Vibratory sieving of the mixture by size separation (Nagarajan et al., 2016; Sanchez-Suarez et al., 2022).

depending on the material and the desired outcome. Gravitational sieving operates on the principle of the fall of particles through a mesh that separates large amounts of grains or flours. Centrifugal sieving utilizes centrifugal force to push materials through a mesh, efficiently separating fine powders and flours (Al-Mogahwi and Baker, 2005). Vibratory sieving incorporates vibration to enhance the movement of particles through the sieve, increasing the separation efficiency. Rotary sieving involves rotating drums or cylinders with mesh surfaces that facilitate continuous processing and high-capacity screening. This method is a part of centrifugal sifters with internal paddle systems that enable the separation of fine and coarse fractions (Cappelli et al., 2020b). Oscillating sieving is a roller-milling process; a purification step is applied to remove by-products. For this process, purifiers consisting of oscillating sieves are used. Airflow falls heavier particles from sieves (Cappelli et al., 2020a). Each sieving method has its characteristics and capabilities. However, apart from their differences, they effectively achieve uniform and high-quality products without impurities.

5.2.2 Parameters affecting sieving efficiency

Sieving efficiency is influenced by several factors that play a critical role in the proper and efficient separation of particles. The shape of particles is the key determinant of the sieving efficiency. Generally, irregularly shaped particles can clog the sieve surface (Ulusoy, 2023). Also, moisture content, density, and surface properties affect passage through the mesh (Shaaya et al., 2016; Yu et al., 2021). High moisture content can lead to agglomeration, resulting in an increase in overall particle size. In the case of process parameters, the feed rate must be controlled within the optimum range, as excessive flow rate decreases sieving efficiency with high pressure. Specifically, for each type of sieving process, parameters must be controlled, such as vibration amplitude (Yu et al., 2022). Finally, maintaining the cleanliness of the sieves is an easy way to prevent undesirable situations during the screening of particles. As a result, the optimization of each factor is required to achieve high sieving efficiency and product quality.

5.2.3 Applications in the food and agricultural industry

Separating particles based on their size can provide different functional and physicochemical properties. Therefore, investigation of the sieving process in this context is critical in manipulating the various properties of particles. Table 5.1 presents the findings of recent studies about the effects of sieving on particle properties.

As reported by Cheng et al. (2023), their study explored the impacts of differential sieving on the physicochemical properties of pea, lentil, barley, and oat flours. After sieving, coarse and fine flour fractions were obtained for each product. Across these four groups, the barley presented the highest yield of coarse flour (63.1%) but the lowest yield of fine flour (35.6%). This was identified as resulting from the largest particle size of the whole barley flour. According to the chemical composition analysis, the highest starch content (53.1-73.3%) was achieved in fine flours due to the requirement for higher mechanical force to break the cotyledon structure. With this, fine flours have higher L* values (84.7-88.5) since they contain more starch. Also, the fine flour showed stronger gel than coarse flour, as the starch was the main contributor to the gel structure. The coarse flours had the highest protein (12.1-24.7%) and dietary fiber content (21.0-36.5%), which provided higher water holding capacities (WHC) (1.51-2.30%) than fine flours (1.12-1.23%). For all crops, the foaming capacity (FC) of fine and coarse flours was close to each other as the same protein content is responsible for FC. Pea flours did not significantly differ between the emulsifying activity index (EAI) and emulsifying stability (ESI) values. However, the ESI of whole lentil flour (65.4%) was higher than that of coarse (56.1%) and fine flour (55.0%) (Nagarajan et al., 2016; Sanchez-Suarez et al., 2022). Similarly, the higher starch content in smaller particle-sized flours was attributed to starch degradation due to increased shear stress caused by sieves that enhance starch extraction (Joseph et al., 2024). These findings were consistent with the study of Gu et al. (2022). In this study, starch granules were commonly found in smaller particle-size fractions. Also, starch, amylose, amylopectin, and damaged starch contents of oat flours increased with decreasing particle size, approximately 35, 6, 29, and 2%, respectively. The highest amylopectin content was reported in fractions with a particle size range of 100–132 μm that provided the highest water absorption index (WAI) of 4.02 g/g. Also, the low amount of proteins and dietary fibers provided a higher water solubility index. Proteins and dietary fibers were the main constituents of the large particle-size fractions. However, the amino acid content of large particle

Table 5.1 Effect of sieving process on functional and physicochemical properties of particles.

Sieving	Product(s)	Pre-treatment	Sieve specifications	Findings	References
Differential sieving	Pea, lentil, barley, oat	Milling into whole flour	0.15 mm	• The highest yield of coarse flour (63.1%) and the lowest yield of fine flour (35.6%) for barley. • The highest starch content (53.1-73.3%) and lowest protein (11.0-25.1%) and dietary fiber content (8.8-19.1%) with fine flours. • Higher ESI (65.4%) of whole lentil flour than that of coarse (56.1%) and fine flour (55.0%).	Cheng et al. (2023)
Electrical sieving	Oat	-	180, 150, 132, 100, and 74 μm	• Starch, amylose, amylopectin and damages starch contents decreased with decreasing particle size, approximately 35, 6, 29 and 2%, respectively. • The highest highest water absorption index (4.02 g/g) fraction with particle size range of 100–132 μm.	Gu et al. (2022)
Sieving	Wheat	Milling into wheat flour	80 and 120 mesh	• Better breakage effect for coarser particle size with sieving	Tian et al. (2022)
Sieving	Wheat	Milling into wheat flour	315 μm and 32 grid gauze	• Higher starch content in lower particle size fractions (40.7-43.5%)	Joseph et al. (2024)
Plan sifter sieving	Wheat	Milling into wheat flour	100-300 μm	• Higher total phenolic content (TPC) (231 and 261 mg/kg) and higher dietary fiber content (1.70, 2.20 g/100 g) in higher particle sized-fractions.	Cappelli et al. (2020b)

(continued on next page)

Table 5.1 Effect of sieving process on functional and physicochemical properties of particles—cont'd

Sieving	Product(s)	Pre-treatment	Sieve specifications	Findings	References
Square-hole sieving	Alfaalfa leaf	Crushing	0.063-0.250 mm	• Increase in angiotension-converting enzyme (ACE) inhibitory activity from 9.63% to 54.59% with decrease in particle size. • Higher tyrosinase inhibitory activity values approximately 70-85% in lower particle sized flours.	Lai et al. (2024)
Sieving	Sweet potato	Milling into sweet potato flour	250 and 500 μm	• Higher water absorption capacity (WAC) with particle size of 500 μm flour than 250 μm. • The highest water solubility with particle size of 250 μm for peeled (5-40%) and unpeeled (5-15%) flours	Chikpah et al. (2020)
Laboratory sieving	Oilseed cake	Flax, hemp, milk thistle, poppy, pumpkin, rapeseed, safflower, sunflower	250 μm	• The highest TPC of flax (7.42 mg GAE/g), hemp (2.99 mg GAE/g) and milk thistle (40.89 mg GAE/g) flour with higher particle sizes (>250 μm). • The highest DPPH radical scavenging activity in sunflower (15.49 mg AAE/g). • No significant difference between DPPH radical scavenging activity of flours with different particle sizes.	Bárta et al. (2021)

AAE: Ascorbic acid equivalent, ESI: Emulsion stability index, GAE: gallic acid equivalent.

fractions (>150 μm) was lower than that of small fractions. The content of cysteine amino acid decreased from 2.51% to 1.97% with decreasing particle size, especially since it is located near the oat cortex. Also, the lipid content of flours decreased with decreasing particle sizes. This was likely a result of the presence of mostly in the outer layer of the kernel. The thermal property analysis revealed that the highest onset temperature (T_0=62.36°C) and the lowest gelatinization enthalpy (ΔT=16.19°C) were achieved in sieve fractions of 74–100 μm. The higher T_0, the temperature limit that presents the beginning of the fluid-structure and crystallization of molecules, represents a higher temperature requirement for smaller sieve fractions. The high amount of starch granules in smaller sieve fractions is the reason for lower ΔT (Gu et al., 2022). In another study, the higher total phenolic content (TPC) (231 and 261 mg/kg) and higher dietary fiber content (1.70, 2.20 g/100 g) in higher particle sized-fractions was associated with the locations of dietary fibers with phenolics in the non-endospermic kernel (Cappelli et al., 2020). Lai et al. (2024) researched to explore the effect of sieving on the functional properties of alfalfa leaf powder. The research indicated a decrease in particle size increased angiotensin-converting enzyme (ACE) inhibitory activity, responsible for the antihypertensive properties, from 9.63% to 54.59%. They pointed to increased protein solubility and availability of hydrophobic amino acid residues as the reason for his finding. Also, higher tyrosinase inhibitory activity, responsible for the prevention of enzymatic browning and color changes, values are approximately 70-85% in lower particle-sized flours. This result was ascribed to the increased flavonoid solubility, which interacts with the active site of the tyrosinase enzyme (Lai et al., 2024).

5.3 Design and operation of sifters

5.3.1 *Types of sifters*

Various sifters are employed in the industry to assist in processing raw materials, such as removing impurities and achieving homogenization. Producers can standardize product sizes by selecting sieves that best fit their processes and capacity requirements. Efficiency and capacity, measured by the mass of material processed per unit of time and surface area, are crucial factors in industrial sieves (Sanchez-Suarez et al., 2022). A commonly used sifter for separating stocks from break and reduction rolls is the plansifter. The plansifter employs a set of around 30 sieves in compartments arranged in parallel, gyrating horizontally to classify particles by size (Serna-Saldivar, 2010). In contrast, vibrating sifters, which use either vibrating motors or eccentric shafts, are frequently used in industries like food processing, pharmaceuticals, and chemicals to separate particles by size. The vibrating machines described used a traditional mechanism to convert rotary and reciprocating motions. Vibrating sifters, powered by vibrating motors or eccentric shafts, are widely utilized in food processing, pharmaceuticals, and chemical industries to sort particles by size. This drive design resulted in higher metal consumption and increased energy losses, particularly when inertial flywheels were added to reduce periodic force pulses (Modrzewski et al., 2022).

During the sieving process, grains pass through a series of screens to eliminate stones, dust, and other unwanted materials. The vibrating screen is commonly used for this purpose and typically includes components like a drive engine, feed hopper, sieves, dust removal system, and cleaner. In mechanical systems, both aspiration and sieving are often combined into a single unit of equipment. The sieves are arranged in layers with varying slopes to effectively sort out large, medium, and small-sized impurities. The screen structure enabled the vibrators to be adjusted by symmetrically moving them away from the center of the sieve, usually in opposite directions. Adjusting the spacing of the vibrators changed the oscillation paths of the screen. The screen could also control the excitation force of each vibrator by modifying the position of the unbalanced masses on its shaft. These masses, positioned at both ends of the shaft, each consisted of two identical parts. Adjusting the excitation force involved shifting one mass relative to the other by a specific angle. Inverters controlled the motors' rotational speed, allowing for various combinations of vibrational frequencies (Modrzewski et al., 2022). Rotary sifters have a rotating cylindrical drum that sifts and separates particles. Smaller particles pass through the sieve as the drum rotates, while larger particles are retained (Liu, 2009). The sieving surface can be constructed from wire mesh or a perforated plate. Sieving is primarily utilized after grinding and drying processes to standardize products, ensuring high quality for customers (Prasad, 2024). Centrifugal sifters use centrifugal force to separate particles, employing a cylindrical screen or sieve basket with high-speed rotating blades that force particles through the screen. Tumbler sieves, on the other hand, use a gentle tumbling or gyrating motion for separation, making them ideal for delicate or fragile materials. These are commonly used in industries such as cosmetics and fine chemicals (Liu, 2009). The air-jet sifters apparatus is tailored for fine powders smaller than 50 μm. It features a single sieve with a tightly sealed lid on the instrument's body. Connected to a vacuum cleaner, it draws powder through the sieve while air circulates up through a rotating radial arm to maintain sieve cleanliness. The powder collected is stored in a cyclone or the vacuum cleaner's filter. The amount retained on the sieve is quantified by weighing, with some models equipped with an integrated system to monitor gradual weight loss over time (Dodds, 2013).

5.3.2 Design approach

Sifters are essential in many industrial and agricultural applications for the efficient separation, screening, and processing of materials. When designing sifters, it is essential to consider several factors, which can be classified into three categories, each affecting particle movement and process efficiency. The first category includes the properties of the raw material, such as particle size distribution, shape, and moisture content. The second category involves the design of the equipment, including the shape and size of screen openings, type and material, screen angle, vibration mode, direction, amplitude, and frequency. The third category pertains to the sieving process itself, including mass flow (for continuous processes), material feed amount (for batch processes), sieving duration, and particle layer thickness (Jezsó & Peciar, 2022).

The capacity of a sifter plays a pivotal role in its design, directly influencing the throughput and productivity of the system. Key factors in the design process include determining the size and configuration of the sifting mechanism, optimizing the material feeding rate, and understanding the specific characteristics of the materials being processed. These considerations are essential for maximizing the efficiency and effectiveness of the sifting operation. Sieve blinding is particularly significant as it directly controls the process. It occurs when particles block and stick in the sieve mesh, reducing the effective screening area. This results in lower sieving rates (capacity) and decreased precision in particle separation (efficiency) (Liu, 2009). Sanitation is paramount in industries like food processing and pharmaceuticals, where maintaining a contaminant-free environment is crucial. In flour milling, for example, the final stage involves sieving the flour through fine mesh screens to eliminate all insect stages. To prevent insects from bypassing damaged screens, an impact device is used to crush any remaining insects. Regular visual inspections of the material separated during sieving enable ongoing process monitoring and facilitate the detection of potential contaminants (Liu, 2009).

5.3.3 Advantages and limitations

Sieving is widely utilized due to its numerous advantages over traditional filtration methods. Some of these benefits include its speed, which exceeds that of conventional filtration methods. Sieving is a straightforward technique that does not demand specialized skills to execute. Moreover, it is cost-effective as it does not necessitate expensive equipment (Prasad, 2024). However, sieving areas are often dusty, which may lead to cross-contamination risks, particularly with products like flour. It is best to conduct sieving in a separate area or use dedicated equipment. Some suppliers use screens and effective extraction to manage this. If handling small amounts of allergens, the same equipment can be used if thoroughly cleaned and validated, with allergens sieved last in the day (Stein, 2015). Sieving is not suitable for separating mixtures with identical particle sizes. Additionally, applying force during sieving can break apart mixture particles, making it inefficient for filtering brittle materials. Moreover, sieving is restricted to dry particles, as wet particles cannot be effectively separated using this method (Prasad, 2024).

5.3.4 Applications in food industry

Sieving is one of the oldest and most commonly utilized techniques for separating solid particles. In flour milling, particle breakdown followed by separation, with sieving being a critical operation in this process (Sultanbawa et al., 2001). Previous studies indicate that milling and sieving result in flours with varying particle sizes. This is a crucial parameter that affects the proximate compositions and techno-functional properties of the flours. Dhen et al. (2016) separated soy flour into three fractions using a sifter: coarsest (> 156 μm), intermediate (132–156 μm), and finest (<132 μm). Their findings showed that as particle size decreased, the protein, lipid, and ash content in the soy flour increased. Cheng et al. (2023) examined pulse (pea, lentil) and cereal (barley, oats) seeds ground into whole flours, and then divided into coarse and fine fractions. Across all crops, fine fractions consistently had higher starch and damaged starch contents compared to whole and coarse fractions. Ash and total dietary fiber contents showed the opposite trend. Techno-functional properties like brightness starch gelatinization enthalpy, and gel hardness also followed a fine>whole>coarse pattern. Protein content varied minimally in peas and lentils, while barley and oats exhibited a coarse> whole>fine trend.

After the initial cleaning stage, rice is subjected to additional purification using either cylindrical or flat sieves to eliminate larger foreign materials, broken kernels, and contaminants. Cylindrical sieves are aligned in a series, while standard sifter tables usually incorporate a minimum of two sieves, sometimes with an integrated air aspiration system. Subsequently, cleaned rough rice is subjected to final purification on gravity tables designed to separate stones and materials with varying densities. This meticulous process ensures the removal of damaged kernels and stones that differ in density from sound paddy rice. Integrated magnets are employed to capture ferrous metals, safeguard milling equipment, and maintain the quality of the final products (Serna-Saldivar, 2010). The grinding process enhances the physicochemical and functional qualities of spice powders, making them applicable in the food, pharmaceutical, and medical industries. These improved characteristics support the development of new formulations of functional foods rich in bioactive compounds and innovative composite or functional materials for various related sectors (Poliński et al., 2022). The recent research related to food applications of sifters is given in Table 5.2.

5.4 Entoleter technology

Sustainable agriculture is a term that one of the main purposes is to crop yield and wealth of farmers by preventing damage and waste of products due to insects. Integrated pest management (IPM) is one of the key strategies to handle pesticide attacks on crops. Recently, some eco-friendly techniques have been preferred instead of traditional pesticides due to possible health problems and environmental deterioration. One of the well-known IPM approaches is mechanical control. This method applies direct mechanical force and energy to destroy insects. Entoleter is an impacting machine that breaks down the kernel of grains through de-hulling and kills pests by the effect of centrifugation forces (Bhattacharyya et al., 2023; Farid et al., 2024). Also, it provides the removal of friable parts of damaged wheat since they are easily broken and removed by aspiration (Doddabematti Prakash et al., 2024). Spinning disks are the main constituents of entoleters that flour and grains are retained to the periphery disk and striked to the pegs. This provides the destruction of insects. The entoleter machines are used the milling process of flour. Firstly, the flour is entered into the entoleter and it is blown and thrown by centrifugal force between two disks. This mechanism turns at a very high speed (Shaaya et al., 2016). However, entoleters can break down grain kernels. Therefore, the entoleter technology must be improved to prevent loss of yield (Thorpe, 2022).

Table 5.2 Recent studies about different food applications of sifters.

Sample	Procedure/relevance	References
Soy flour	Soy flour was sifted into three fractions based on particle size: coarsest (>156 μm), intermediate (132–156 μm), and finest (<132 μm). Smaller particle sizes correlated with higher protein, lipid, and ash content in the soy flours.	Dhen et al. (2016)
Pulse (pea, lentil) and cereal (barley, oats) flour	Across all four crops, the flour streams ranked fine > whole > coarse for starch and damaged-starch contents but reversed this order for ash and total dietary fiber contents.	Cheng et al. (2023)
Yellow pea flour	The authors used sifters of different sizes (1.27, 1.00, 0.79, and 0.5 mm) to mill yellow peas into flours with varying particle sizes. They found that finer pea flour contained more starch, while protein content across the flours was similar.	Bourré et al. (2019)
Teff (Eragrostis tef (Zucc.) Trotter) flour	Three varieties of teff grains were milled and sifted to achieve three different flour fractions categorized by particle sizes (360-250 μm, 250-160 μm, and 160-90 μm). Teff flours showed increased water and oil absorption capacity, swelling power, and solubility index with smaller particle sizes. However, differences in oil absorption capacity and dispersibility due to variety were not significant.	Boka et al. (2023)
Whole seed flour and kernel flour of Jack bean	The crystallinity, swelling power, solubility, and pasting properties of WJB and KJB flours showed a tendency to increase as particle size decreased, correlating with higher starch content. Smaller sieve apertures selectively eliminate non-starch components, enabling customized flour properties.	Anuntagool and Soonthonsun (2023)
Yellow pea, lentil, and chickpea flour	Small, medium, and large-sized flours were obtained using bottom sieves of 75 μm, 150 μm, and 200 μm, respectively. Chickpea flours had the lowest bulk density and highest aeration ratio, making them more compressible and cohesive compared to other pulse and wheat flours. Small-sized pulse flours were generally more cohesive than larger ones.	Nkurikiye et al. (2023)
Gorgon nut flour	The flour underwent sieving into five different size ranges (250-75 micrometers). Across all these size fractions, various characteristics of gorgon nut flour, including peak viscosity, onset temperature, enthalpy, peak, and amylose content, were found to exhibit an inverse correlation with setback, breakdown, relative crystallinity, pasting temperature, and pasting time.	Sharma et al. (2023)
Flax, hemp, milk thistle, poppy, pumpkin, rapeseed, safflower, sunflower) cake flours	Oilseed cake flours were sieved into fractions. The flours from hemp, milk thistle, and safflower contained elevated levels of carbohydrates, especially noticeable in fractions above. This fraction also demonstrated a significant water-holding capacity in flax, poppy, pumpkin, and sunflower flours.	Bárta et al. (2021)

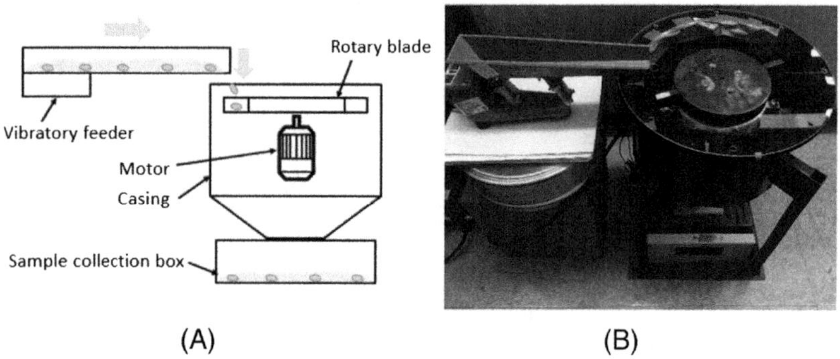

Figure 5.2 Rotary blade impact machine (A) Schematic representation of impact machine (B) Photograph of impact machine.
(Adapted from Chen et al. (2021)

The study led by Brabec et al. (2015) provided insights into the detection of insects in wheat using a laboratory entoleter machine. Laboratory-scale impact machine is a simple method that works on rotary systems. It counts larvae, pupae, and adults in wheat products. Fig. 5.2 from Chen et al. (2021) presents a rotary-type impact machine. The grain hardness and moisture content determine the entoleter impeller speed. The outcomes revealed that approximately 70-90% of the grains released insects after impact milling since entoleter is an effective tool in cleaning insects (Brabec et al., 2015). Also, it may be used for different purposes in the cereal industry. In another study, an entoleter machine was preferred as a de-huller during oat flour production (Cheng et al., 2023). Singh and Johnston (2004) revealed that in the wet-milling process of the corn, the entoleter impact milling increases the released starch amount. Also, it provides lower fiber disintegration in corn slurry, which is crucial for different functions (Singh & Johnston, 2004).

5.5 Innovations and future trends

The sieving, sifters, and entoleters field is progressing remarkably due to technological innovations and changing industry demands. These advancements enhance efficiency, accuracy, and safety in particle separation processes, particularly within the food, pharmaceutical, and chemical sectors. Additionally, advanced sorting methods that enhance resource recovery are increasingly advocated to minimize harm, and promote a circular economy. Contemporary advancements in sieving include vibratory sieves, rotary sifters, and ultrasonic sieving machines, all designed to improve sieving efficiency. Vibratory sieves utilize vibration to aid material movement through the sieve, while ultrasonic sieves employ high-frequency sound waves to prevent clogging and improve particle separation. These technologies have demonstrated significant enhancements in sieving speed, precision, and throughput (Mowade et al., 2024).

Recent developments in mechanical separation techniques within the food industry are setting the stage for upcoming innovations. One exciting development involves combining intelligent sensors and artificial intelligence to boost the efficiency and accuracy of separation techniques within the food industry. Sensor technologies have transformed sorting by swiftly detecting, classifying, and accurately separating materials. Challenges include complex material compositions and stricter purity requirements. Ongoing research aims to enhance sensors and data analysis to meet these demands. Despite challenges, sensors promise to significantly enhance sorting (Chen et al., 2024). Over the past two decades, advancements in sensor technologies like laser emitters, dual-energy X-ray transmission (DE-XRT), and engine management system (EMS) sensors, and hyperspectral cameras, have led to a revolution in sorting techniques. By integrating sensors with sophisticated algorithms and employing industrial robots, these systems can deliver highly efficient performance in diverse operational environments, often outperforming human capabilities in numerous aspects. (Zhao & Li, 2022). Furthermore, advancements in materials science have the potential to result in more durable and lightweight components for sieves, thereby enhancing the performance and longevity of sieving machines. Collaborative initiatives involving food scientists, engineers, and technologists are anticipated to propel advancements in mechanical separation techniques, thereby enhancing the safety and efficiency of food processing operations in the coming years. (Mowade et al., 2024).

5.6 Conclusions

Sieving is a widely used separation process in the food and agriculture industries, chosen on the basis of the desired product characteristics. The sieving process ensures that the particles are separated depending on their size, resulting in a uniform product. Sieving has effects on the functional and physicochemical properties of particles. In that case, examining the changes caused by each process in different products provides the necessary information infrastructure to obtain products with the desired properties. Sifters are the equipment used in the sieving process, and the other particles are separated depending on their size. Commonly preferred sifters are vibratory, rotary, gyratory, and air-jet sifters for refining products. Entoleters, like sifters, are equipment used for separation but are selected for different purposes. They are among the equipment used to sterilize flour or grain group products, destroy insect eggs and larvae, and ensure food safety thanks to high-speed impact technology. Each technique and equipment provides different benefits in the industry, increasing the quality and safety of the products. Therefore, their detailed investigation and inclusion in the processing of food and agricultural products can not only improve product quality and safety but also increase productivity, which is essential in terms of cost. In addition, the latest developments and innovations in screening, sieves, and entoleters require changes that will create solutions to the problems encountered in the industry. Especially nowadays, the increasing demand for plant-based products and nutrition will trigger innovations in this sector. Accordingly, it will be valuable for consumers to reach safer, higher-quality products. Businesses will also benefit in many ways by following and implementing such innovations. By understanding and improving the specific capabilities of these systems, manufacturers can optimize their processes, prevent potential problems, and ensure safe and high-quality food products are delivered to consumers.

References

Al-Mogahwi, H. W. H., & Baker, C. G. J. (2005). Performance evaluation of mills and separators in a commercial flour mill. *Food and Bioproducts Processing, 83*(1), 25–35. https://doi.org/10.1205/fbp.04172.

Anuntagool, J., & Soonthonsun, S. (2023). Effect of particle size classification on properties of flour from jack bean: An under-utilized high protein legumes. *LWT, 189*, 115418. https://doi.org/10.1016/j.lwt.2023.115418.

Bala, M., Tushir, S., Garg, M., Meenu, M., Kaur, S., Sharma, S., & Mann, S. (2023). Wheat milling and recent processing technologies: Effect on nutritional properties, challenges, and strategies. In O. P. Gupta, S. Kumar, A. Pandey, M. K. Khan, S. K. Singh, & G. P. Singh (Eds.), *Wheat science* (pp. 219–256). CRC Press.

Bárta, J., Bártová, V., Jarošová, M., Švajner, J., Smetana, P., Kadlec, J., Filip, V., Kyselka, J., Berčíková, M., Zdráhal, Z., Bjelková, M., & Kozak, M. (2021). Oilseed cake flour composition, functional properties and antioxidant potential as effects of sieving and species differences. *Foods, 10*(11), 11. https://doi.org/10.3390/foods10112766.

Bhattacharyya, B., Gogoi, I., Das, P. P. G., & Kalita, B. (2023). Chapter 7—Management of agricultural insect pests for sustainable agriculture and environment. In M. Farooq, N. Gogoi, & M. Pisante (Eds.), *Sustainable agriculture and the environment* (pp. 161–193). Academic Press https://doi.org/10.1016/B978-0-323-90500-8.00001-4.

Boka, A. L., Tolesa, G. N., & Abera, S. (2023). Effects of variety and particles size on functional properties of Teff (Eragrostis tef (Zucc.) Trotter) flour. *Cogent Food & Agriculture, 9*(1), 2242635. https://doi.org/10.1080/23311932.2023.2242635.

Bourré, L., Frohlich, P., Young, G., Borsuk, Y., Sopiwnyk, E., Sarkar, A., Nickerson, M. T., Ai, Y., Dyck, A., & Malcolmson, L. (2019). Influence of particle size on flour and baking properties of yellow pea, navy bean, and red lentil flours. *Cereal Chemistry, 96*(4), 655–667. https://doi.org/10.1002/cche.10161.

Brabec, D. L., Pearson, T. C., Maghirang, E. B., & Flinn, P. W. (2015). Detection of fragments from internal insects in wheat samples using a laboratory entoleter. *Cereal Chemistry, 92*(1), 8–13. https://doi.org/10.1094/CCHEM-08-13-0173-R.

Cappelli, A., Mugnaini, M., & Cini, E. (2020). Improving roller milling technology using the break, sizing, and reduction systems for flour differentiation. *LWT, 133*, 110067. https://doi.org/10.1016/j.lwt.2020.110067.

Cappelli, A., Oliva, N., & Cini, E. (2020). Stone milling versus roller milling: A systematic review of the effects on wheat flour quality, dough rheology, and bread characteristics. *Trends in Food Science & Technology, 97*, 147–155. https://doi.org/10.1016/j.tifs.2020.01.008.

Cauvain, S. (2015). Wheat milling and flour testing. In S. Cauvain (Ed.), *Technology of breadmaking* (pp. 339–375). Springer International Publishing https://doi.org/10.1007/978-3-319-14687-4_12.

Chen, X., Kroell, N., Feil, A., & Greiff, K. (2024). Chapter 9—Sensor-based sorting. In C. Meskers, E. Worrell, & M. A. Reuter (Eds.), *Handbook of recycling* (2nd ed.) (pp. 145–159). Elsevier https://doi.org/10.1016/B978-0-323-85514-3.00028-2.

Chen, Z., Wassgren, C., & Ambrose, R. K. (2021). Measured damage resistance of corn and wheat kernels to compression, friction, and repeated impacts. *Powder Technology, 380*, 638–648. https://doi.org/10.1016/j.powtec.2020.11.012.

Cheng, F., Ding, K., Yin, H., Tulbek, M., Chigwedere, C. M., & Ai, Y. (2023). Milling and differential sieving to diversify flour functionality: A comparison between pulses and cereals. *Food Research International, 163*, 112223. https://doi.org/10.1016/j.foodres.2022.112223.

Chikpah, S. K., Korese, J. K., Hensel, O., & Sturm, B. (2020). Effect of sieve particle size and blend proportion on the quality properties of peeled and unpeeled orange fleshed sweet potato composite flours. *Foods, 9*(6), 6. https://doi.org/10.3390/foods9060740.

Dhen, N., Román, L., Ben Rejeb, I., Martínez, M. M., Garogouri, M., & Gómez, M. (2016). Particle size distribution of soy flour affecting the quality of enriched gluten-free cakes. *LWT - Food Science and Technology, 66*, 179–185. https://doi.org/10.1016/j.lwt.2015.10.032.

Doddabematti Prakash, S., Rivera, J., & Siliveru, K (2024). 10—Unit operations in wheat processing. In C. K. Sunil, K. A. Athmaselvi, N. Venkatachalapathy, C. Anandharamakrishnan, & V. M. Balasubramaniam (Eds.), *Unit operations in food grain processing* (pp. 289–329). Academic Press https://doi.org/10.1016/B978-0-443-18965-4.00010-8.

Dodds, J. (2024). Chapter 15—Techniques to analyse particle size of food powders. In B. Bhandari, N. Bansal, M. Zhang, & P. Schuck (Eds.), *Handbook of food powders* (2nd ed.) (pp. 249–268). Woodhead Publishing https://doi.org/10.1016/B978-0-323-98820-9.00070-3.

Farid, A., Hayat, M., Ghazanfar, S., & Javed, R. (2024). Manufacturing of starch from rice, corn, potato, and wheat and its industrial applications. In M. Z. Hashmi, A. Saeed, S. G. Musharraf, & W. Shuhong (Eds.), *Recent advances in industrial biochemistry* (pp. 225–249). Springer International Publishing https://doi.org/10.1007/978-3-031-50989-6_10.

Fišteš, A. (2021). The evolution of milling process. In F. Boukid (Ed.), *Cereal-based foodstuffs: The backbone of Mediterranean cuisine* (pp. 19–45). Springer International Publishing https://doi.org/10.1007/978-3-030-69228-5_2.

Gu, Y., Qian, X., Sun, B., Ma, S., Tian, X., & Wang, X. (2022). Nutritional composition and physicochemical properties of oat flour sieving fractions with different particle size. *LWT, 154*, 112757. https://doi.org/10.1016/j.lwt.2021.112757.

Jezsó, K., & Peciar, P. (2022). Influence of the selected sieving parameters on the sieving efficiency of material MCC Avicel PH102. *Strojnícky Časopis - Journal of Mechanical Engineering, 72*(1), 77–88. https://doi.org/10.2478/scjme-2022-0008.

Joseph, M., Alavi, S., Adedeji, A. A., Zhu, L., Gwirtz, J., & Thiele, S. (2024). Adaptation of conventional wheat flour mill to refine sorghum, corn, and cowpea. *AgriEngineering, 6*(3), 3. https://doi.org/10.3390/agriengineering6030114.

Lai, S., Cui, Q., Sun, Y., Liu, R., & Niu, Y. (2024). Effects of particle size distribution on the physicochemical, functional, and structural properties of alfalfa leaf powder. *Agriculture, 14*(4), 4. https://doi.org/10.3390/agriculture14040634.

Lee, D., Kim, M. J., Kwak, H. S., & Kim, S. S. (2020). Characteristics of bread made of various substitution ratios of bran pulverized by Hammer Mill or Jet Mill. *Foods, 9*(1), 1. https://doi.org/10.3390/foods9010048.

Lewis, M. J. (1996). Chapter 9—Solids separation processes. In A. S. Grandison, & M. J. Lewis (Eds.), *Separation processes in the food and biotechnology industries* (pp. 243–286). Woodhead Publishing https://doi.org/10.1533/9781855736573.243.

Liu, K. (2009). Some factors affecting sieving performance and efficiency. *Powder Technology, 193*(2), 208–213. https://doi.org/10.1016/j.powtec.2009.03.027.

Modrzewski, R., Obraniak, A., Rylski, A., & Siczek, K. (2022). A study on the dynamic behavior of a sieve in an industrial sifter. *Applied Sciences, 12*(17), 17. https://doi.org/10.3390/app12178590.

Mowade, S. M. (2024). Multipurpose sieving and screening machine. *Interantional Journal of Scientific Research in Engineering and Management, 8*, 1–5. https://doi.org/10.55041/IJSREM34665.

Nagarajan, V., Mohanty, A., & Misra, M. (2016). Biocomposites with size-fractionated biocarbon: Influence of the microstructure on macroscopic properties. *ACS Omega, 1*, 636–647. https://doi.org/10.1021/acsomega.6b00175.

Nkurikiye, E., Pulivarthi, M. K., Bhatt, A., Siliveru, K., & Li, Y. (2023). Bulk and flow characteristics of pulse flours: A comparative study of yellow pea, lentil, and chickpea flours of varying particle sizes. *Journal of Food Engineering, 357*, 111647. https://doi.org/10.1016/j.jfoodeng.2023.111647.

Poliński, S., Topka, P., Tańska, M., Kowalska, S., Czaplicki, S., Szydłowska-Czerniak, A., et al. (2022). Effect of grinding process parameters and storage time on extraction of antioxidants from ginger and nutmeg. *Molecules, 27*(21), 7395. doi:10.3390/molecules27217395.

Prasad, S. (2024). Sieving and its application in dairy and food industry. *Agriculture & Food, 6*, 397–399.

Saha, N., Goates, C., Hernandez, S., Jin, W., Westover, T., & Klinger, J. (2022). Characterization of particle size and moisture content effects on mechanical and feeding behavior of milled corn (*Zea mays L.*) stover. *Powder Technology, 405*, 117535. https://doi.org/10.1016/j.powtec.2022.117535.

Sanchez-Suarez, N., Orozco-Mendoza, G. L., Zartha-Sossa, J. W., Gafaro-Garcés, D. C., Melchor-Cahuana, L. G., & Gonzalez-Tovar, C. (2022). Trends in sieving and its applications in cereals. A literature review. *Frontiers in Sustainable Food Systems, 6*. https://doi.org/10.3389/fsufs.2022.902147.

Sayyad, F. G., Sharma, H. K., Kumar, N., Sharma, H. K., & Kumar, N. (2022). Cleaning and separation. *Agro-processing and food engineering: Operational and application aspects* (pp. 307–352). Springer https://doi.org/10.1007/978-981-16-7289-7_8.

Serna-Saldivar, S. O. (2022). Production of cereal-based raw materials for the snack industry. In S. O. Serna-Saldivar (Ed.), *Snack foods*. CRC Press.

Shaaya, E., Kostyukovsky, M., & Quinn, E. (2016). Chapter 12—Improving the control of insects in food processing. In H. Lelieveld, J. Holah, & D. Gabrić (Eds.), *Handbook of hygiene control in the food industry* (2nd ed.) (pp. 179–189). Woodhead Publishing https://doi.org/10.1016/B978-0-08-100155-4.00012-1.

Sharma, R., Kotwaliwale, N., Jindal, N., & Saxena, D. C. (2023). Principal component analysis for particle size-based classification of gorgon nut flour and their morphological, structural, thermal and pasting characteristics. *International Journal of Food Science & Technology, 58*(6), 3144–3153. https://doi.org/10.1111/ijfs.16442.

Singh, V., & Johnston, D. (2004). An enzymatic process for corn wet milling, *Advances in food and nutrition research* (48, pp. 151–171). Academic Press https://doi.org/10.1016/S1043-4526(04)48003-6.

Stein, K. (2015). 6—Effective allergen management practices to reduce allergens in food. In S. Flanagan (Ed.), *Handbook of food allergen detection and control* (pp. 103–131). Woodhead Publishing https://doi.org/10.1533/9781782420217.1.103.

Sultanbawa, F. M., Owens, W. G., & Pandiella, S. S. (2001). A new approach to the prediction of particle separation by sieving in flour milling. *Food and Bioproducts Processing, 79*(4), 211–218. https://doi.org/10.1205/096030801753252270.

Thorpe, G. (2022). Chapter 4—Alternative and emerging storage practices and technologies. In K. A. Rosentrater (Ed.), *Storage of cereal grains and their products* (5th ed.) (pp. 81–111). Woodhead Publishing https://doi.org/10.1016/B978-0-12-812758-2.00011-8.

Tian, X., Wang, X., Ma, S., Sun, B., Li, L., & Wang, Z. (2022). Study of the ball milling condition effect on physicochemical and structural characteristics of wheat flour. *Journal of Food Processing and Preservation, 46*(1), e16203. https://doi.org/10.1111/jfpp.16203.

Ulusoy, U. (2023). A review of particle shape effects on material properties for various engineering applications: From macro to nanoscale. *Minerals, 13*(1), 1. https://doi.org/10.3390/min13010091.

Yu, C., Geng, R., & Wang, X. (2021). A numerical study of separation performance of vibrating flip-flow screens for cohesive particles. *Minerals, 11*(6), 6. https://doi.org/10.3390/min11060631.

Yu, C., Pu, K., Geng, R., Qiao, D., Lin, D., Xu, N., Wang, X., Li, J., Gong, S., & Zhou, Q. (2022). Comparison of flip-flow screen and circular vibrating screen vibratory sieving processes for sticky fine particles. *Minerals Engineering, 187*, 107791. https://doi.org/10.1016/j.mineng.2022.107791.

Zhang, B., Gong, J., Yuan, W., Fu, J., & Huang, Y. (2016). Intelligent prediction of sieving efficiency in vibrating screens. *Shock and Vibration, 2016*(1), 9175417. https://doi.org/10.1155/2016/9175417.

Zhao, Y., & Li, J. (2022). Sensor-based technologies in effective solid waste sorting: Successful applications, sensor combination, and future directions. *Environmental Science & Technology, 56*(24), 17531–17544. https://doi.org/10.1021/acs.est.2c05874.

Magnetic and metal detectors

6

Sanket Balasaheb Kokane and Vinkel Kumar Arora
Department of Food Engineering, National Institute of Food Technology Entrepreneurship and Management-Kundli, Sonipat, Haryana, India

6.1 Introduction

Metal detectors are crucial in ensuring food safety by metallic contaminants in the food products. Magnetic separators are commonly used for ferromagnetic impurties, while eddy current metal detectors are suitable for non-ferromagnetic impurites. These detectors helps precisely, sort metals in food products. Advanced metal detection devices can adjust the detection frequency to effectively detect metals of different sizes. In pet food industry, metal detectors with rejecting devices ensure continuous operation and improve efficiency by promptly removing metallic impurities. Additionally, innovative designs, like metal detectors with silica-gel plates and slide bars, further improve the stability and accuracy of metal detection in food processing applications.

It protect food products from any foriegn metal inclusion and hence, ensure thefood safety. It is essential to have access to the appropriate procedures and equipment for metal fragment control. For years, the focus has primarily been on ensuring the safety and security of machinery through asceptic separation and metal detection. However, today the emphasis on food safety has become the paramount. Ensuring the safety and integrity of products, consumers, and brands has never been more crucial than in todays scnerio. (Yang et al., 2025). The combination of high-grade rare Earth magnets and efficient metal detectors is crucial for meeting the product security demands of the food industry. These two components work together to effectively minimize metal risks. Various companies utilize different methods for detection, such as magnets or metal detection. Both aspects carries equal significance. Thus, it is imperative to direct our attention toward them.

6.1.1 Why do food manufacturers use metal detectors?

Metal detectors for food are typically used to ensure the safety of consumers. Even with utmost caution, it is quite difficult to eliminate the risk of metallic contamination in food products during the production process. It is important to understand that the presence of metal particles in the product, during the production or are already present in the raw materials? can pose a significant risk of injury to consumers. The impact on the food processing companies are extensive and significant, encompassing compensation claims and costly recalls. Metallic impurities contaminated food products can cause significant harm by tarnishing a brand's reputation and eroding consumer trust, leading to long-term consequences (Sapp, 2024).

Mechanical Separation Processes in the Food Industry: Unit Operations and Processing Equipment in the Food Industry.
DOI: https://doi.org/10.1016/B978-0-12-819523-9.00011-1

Metal detectors offer reliable protection against both ferrous and non-ferrous metals, such as Aluminum and Stainless steel. They can be fitted in every step of the manufacturing process. It has a wide range of applications, such as inspecting from raw mwrial cleaning, sorting and grading lines, bread and bakery products, meat and sausage, fruit, vegetables, dairy products, spices, sugar etc. Apart from safeguarding consumers health and companies reputation, metal detectors are employed to ensure the safety of machinery as well. Even a small metal part has the potential to cause machinery breakdowns, leading to costly repairs and downtime, which can negatively impact production and revenue.

6.2 Sources of magnetic and metallic contamination

Metal and magnetic contamisnation in the food industry can originate from various sources throughout the production process, ranging from farm to plate. They are unique in virtually every stage of food production and processing. Several factors contribute to the contamination of the product, including farming instruments, packaging materials like stapler pins and screws, washers, nuts, bolts, broken machinery parts, damaged sieves or screw or chain conveyors, human interference etc. These contaminants pose significant health risks to consumers and can affect the quality and safety of food products. Here are some familiar sources of such contamination:

6.2.1 Raw materials

The origin of all food materials, whether derived from plants or animals, is deeply rooted in fundamental agricultural methods. Throughout the agricultural process, from field preparation to harvesting and transportation, there is potential for contamination from various sources such as soil, pesticides, fertilizers, and farming practices. These fertilizers, for instance, may inadvertently introduce heavy metals like arsenic, lead, and cadmium into the food chain.Furthermore, raw agricultural produce may inherently contain magnetic minerals in the soil, or these minerals may be introduced during harvesting. Water used for irrigation or processing may also carry dissolved metals from natural sources or pollution. As the journey continues from harvesting to transportation and through primary processing at the industrial level, there remains a risk of contamination for all unprocessed raw materials managed along the way. However the seprating and removal of heavy metals and metallic pollutants in soil and water are very chanllenging issues. The raw material such as cereals and grains poses the threta of metallic contamination during the proceesing like cleaning, sorting and grading. this may occur largely due to damaged sieves, falling of loosened screws/nuts due to the vibration of the moving parts in the machines. So, its is very important to install the magnetic seperator in the raw material lines and ensure proper and timely cleaning schedules.

6.2.2 Processing equipment

The machinery and equipment utilized in food processing are crafted from food-grade metals to ensure safety and quality. However, due inevitable wear and tear pose a significant challenge, particularly in heavy processing tasks like cutting, shearing, material handling, grinding, mixing, heat treatment, and packaging. These operations can release metal fragments into the food, compromising its integrity. Additionally, corrosion and rust on metal surfaces in wet processing or beverage processing, further exacerbate this risk, particularly in environments with high moisture or acidity levels. Magnetic components of food processing equipment, such as motors, conveyors, and buckets, hammers, rolling stones, rollers, agitators, are susceptible to degradation and damage, causing metallic particles to be shed into the food product. Furthermore, constant metal-to-metal contact between various machinery parts can also generate small matellic fragments, adding to the contamination concerns.

6.2.3 *Packaging, storage, and transportation*

The machinery that handles and prepares packaging materials is predominantly crafted from metals, increasing the risk of metal contamination. Specific metal containers utilized for packaging food items have been found to leach metals into the contents. Additionally, metallic coatings applied to packaging materials contribute to this contamination. The Tin coated MS cans, aluminum cans are largely used in the fruits and vegetables processing. The can prepartion invovles rolling, forming and seaming operation during the canning process. The operations poses the treat of metallic inclusions during the processing. Furthermore, metal fragments from transportation equipment, including trucks and railcars, also pose a contamination risk.

6.2.4 *Human error*

Neglecting proper maintenance and handling techniques can lead to the misplacement of metal components or insufficient sanitation, allowing metal pieces to contaminate food. Occasionally, items like tools or personal accessories such as jewelry, and utensils inadvertently find their way into food items.

6.3 Types of metal and magnetic contaminants in food

Innovations in metal detection technology have significantly enhanced the capabilities of these devices, enabling them to detect minuscule amounts of metal with remarkable precision. These detectors function by detecting either metal particles' magnetic permeability or conductivity. Metal contaminants are categorized into three categories according to their distinct properties.

6.3.1 *Ferrous (Mild steel/cast iron)*

The metallic compound known as ferrous material consists of iron. With inherent characteristics such as magnetism and conductivity, it stands out as one of the most easily detectable types of metals. Metal detectors capitalize on their magnetic nature of the material, swiftly identifying them by sensing the magnetic field emanating from even minute traces of iron. While some ferrous metals may exhibit limited conductivity, their magnetic attributes remain prominent. Notable examples include iron oxide, commonly known as rust, which manifests in the material of contact as fragmented mild steel or mineral form like hematite (Fe_2O_3). Despite their diminished conductivity, these compounds readily magnetize upon exposure to an external magnetic force.

6.3.2 Non-ferrous (Copper/brass/aluminum)

This substance category is deficient in iron despite its lack of magnetic properties. However, non-ferrous metals excel as a conductors, rendering them conveniently locatable. Furthermore, their electromagnetic resonances facilitate their detection. Unlike ferrous, stainless steel, non-ferrous materials like silver or copper generate more significant "eddy currents," which further simplify their identification.

6.3.3 Stainless Steel (SS)

While stainless steel does contain iron, it lacks significant magnetic characteristics. Its conductivity is exceptionally low, making it challenging to detect using metal detection technology. Moreover, environmental factors like moisture and high salinity can further hinder its detection.

6.4 Principles of magnetic and metal detection

6.4.1 Magnetic detection

Since the 1930s, magnetic separators have been employed in the food industry to eliminate undesirable ferrous materials and minerals by exploiting variances in magnetic properties. Magnetic detection operates on the fundamental principles of magnetism, whereby ferrous particles are drawn towards magnetic fields (Fig. 6.1). Such fields originate from permanent magnets or electric currents flowing through circuits. They exert a magnetic force, either attractive or repulsive, on materials. Materials with strong magnetism are easily attracted to magnets, but only a few have enough susceptibility to be drawn to commercially available magnets. Substances that include iron are drawn towards these magnets, making it easier to eliminate undesirable ferrous items from food-related substances. Magnets provide a direct and economical solution (Payne et al., 2023).

The strength and gradient of the magnetic field determine the force exerted on a ferrous particle by a magnet. A proficient magnetic separator must possess the capability to produce a potent magnetic field to magnetize the ferrous particle. Additionally, it should have a magnetic gradient that can attract the particle toward the magnet as it traverses through the processed material. Foods containing ferrous materials has a diverse array of magnetic characteristics, leading to varying magnetic susceptibilities. The magnetic susceptibility of a material can vary, as it is defined by the ratio of its magnetization intensity to the strength of the applied magnetic field. Rust and

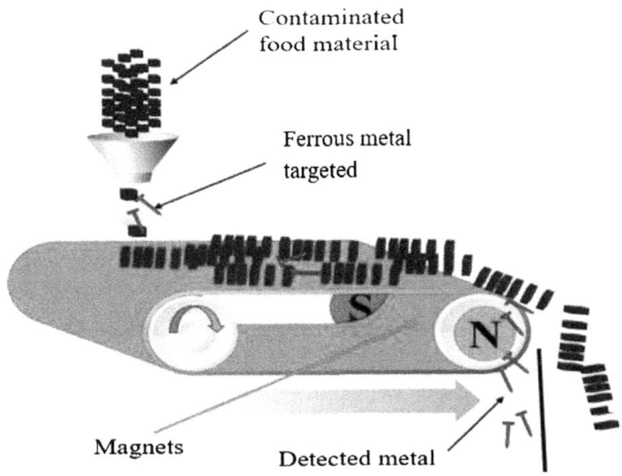

Figure 6.1 Principle of magnetic detection.

scale, which are composed of iron oxides, exhibit a reduced magnetic susceptibility, necessitating a more intense magnetic field for their magnetization. Stainless steels, a variant of steel with varying chromium content, exhibit a diverse array of magnetic characteristics based on the grade. Although certain types of stainless steels exhibit magnetic properties and can be influenced by weak magnetic fields, most stainless steels are non-magnetic and remain unaffected even when exposed to strong magnetic fields in typical situations.

To effectively separate unwanted ferrous fragments, it is not enough to magnetize them. It is essential to divert them away from the current flow and into a specific point where they can be gathered or removed. Establishing a magnetic gradient that directs magnetized particles towards the collection point is essential for guiding their movement in the desired direction. This gradient assures that the magnetic force is sufficiently powerful to magnetize any ferrous contaminant. Even at the most remote location from the magnet where it is prone to occur. Furthermore, the magnetism must surpass the resistance resulting from the motion of food material and the gravitational force, which is dependent upon its direction of motion. The magnetic power diminishes by the inverse square law when the distance between the magnet and the material is increased. Consequently, there is typically enough force to draw and retain the ferrous impurity once it is near the magnet (Apoussidis & Wells, 2004).

6.4.2 Metal detection

Metal detection, a fascinating technology, employs electromagnetic fields to discern between ferrous and non-ferrous metal impurities in various food items. These detectors consist of coils that generate electromagnetic fields. Whenever a metal impurity traverses this field, it disturbs the electromagnetic equilibrium, prompting

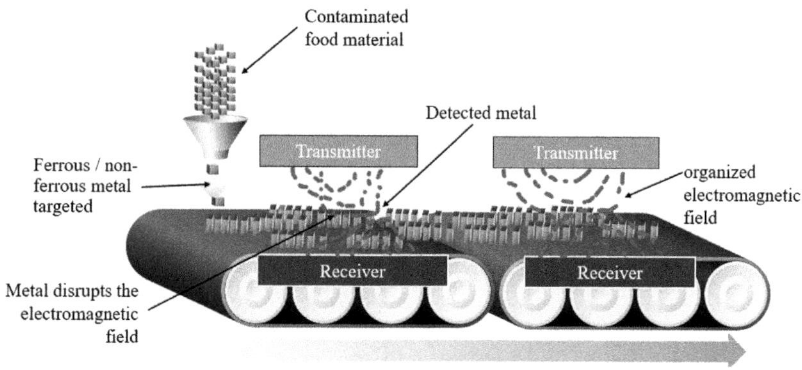

Figure 6.2 Principle of magnetic detection.

Table 6.1 Difference between magnetic and metal detectors (Magnattackglobal, 2018).

Magnetic detector	Metal detector
It Extracts contaminants from the flow of product	It identifies/rejects the contaminated product
Highly Magnetic, Ferrous (fine, medium, and Large)	Ferrous (Mostly for -2 mm to +2mm)
Weak Magnetics & Work Hardened stainless steel (-3 mm to +6 mm)	Non-ferrous metals stainless steel (-3 mm to +3 mm)
Suitable for stone (-3 mm, most)	For stone (3 mm, little)
Installed at Intake, processing, packing	Installed permanently after scalping magnet, preferably after packaging

an alert. Innovations in metal detection include integrating multi-frequency technology and digital signal processing to heighten sensitivity and minimize erroneous alerts. It is fascinating how diverse types of metal detectors work based on their unique detection mechanisms. The technique most frequently employed in the food processing industry is typically the transmitter-receiver strategy. Food sector metal detectors are fitted with a configuration consisting of a transmitter coil and two receiving coils. The transmitter coil produces a uniform electromagnetic field. Each time a metal object traverses the detector, it disrupts the field, leading to the detection of a signal by the receiver coils. Subsequently, the electronic component within the metal detector scrutinizes and interprets this signal, identifying any metal impurities present (Fig. 6.2). These detectors feature automatic reject mechanisms, ensuring swift removal of any contaminated product directly from the production line. The technical difference between magnetic and metal detectors are reported in Table 6.1.

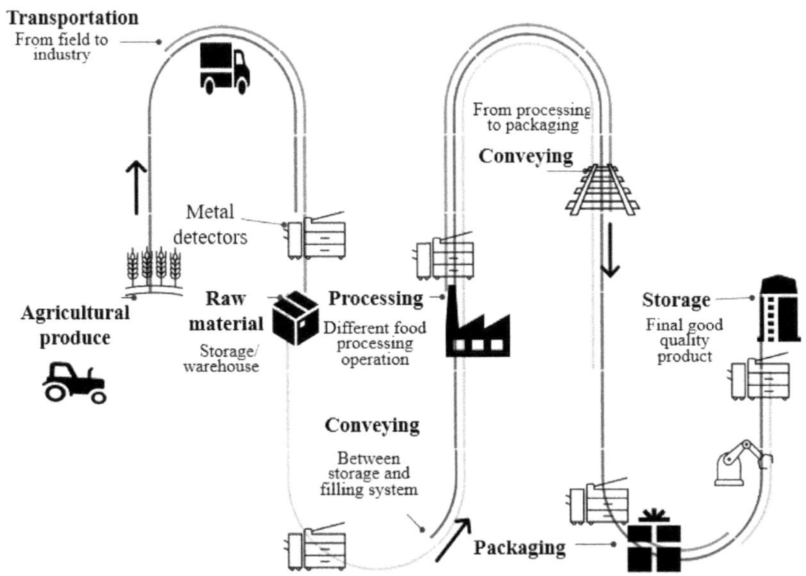

Figure 6.3 Flow diagram for different locations os metal detection in processing.

6.5 Where do food manufacturers deploy detectors?

Implementing metal and magnetic detectors holds significant importance in eliminating metallic contaminants. The number of detectors needed hinges on several factors, such as operational phases, equipment utilized, characteristics of the food substance, and its properties. Safeguarding consumers within the food industry is paramount; thus, multiple installation points prove advantageous. Moreover, specific detectors are specifically installed to protect machinery rather than for consumer safety. Metal detectors can be installed at various points on the production line (Smith, 2023). Illustrated in Fig. 6.3. These detectors detect and capture metallic/magnetic impurities at various critical points. A well-defined cleaning schedule is essential for maintaining the efficiency and accuracy of metal detectors used in food processing. The regular cleaning ensures their maximum performance and prevents further contamination. The schedule should include the timely removal of detected impurities, surface cleaning, and routine calibration checks to maintain detection accuracy. Depending on operational needs, cleaning should performed at scheduled intervals, such as daily or per shift. Additionally, keeping a record of cleaning activities, including date, time, and personnel involved, helps ensure compliance with food safety standards and enhances overall detector reliability.

6.5.1 Inspection of incoming raw materials

Examining raw materials upon arrival provides the advantage of isolating metal fragments before they undergo further processing, averting their fragmentation into numerous smaller pieces that could pose challenges in detection and separation. Diligent scrutiny of raw materials shields machinery from additional harm and diminishes the risk of wear and tear caused by metallic impurities, thereby mitigating metal contamination. Consequently, installing a detector just before the storage or warehouse area for incoming raw materials is prudent. This precaution guarantees the identification and elimination of contaminated materials before infiltrating the production process, thus safeguarding the integrity of finished goods.

6.5.2 In-Line between storage and filling systems

Detecting metallic or magnetic impurities along the production line, from storage to filling, is crucial to pinpoint any lingering contaminants from earlier stages.

6.5.3 Within the production process

Conducting inspections at crucial checkpoints (HACCPs) throughout the manufacturing process is advised to promptly identify equipment malfunctions (like damaged blades). To ensure the timely detection and resolution of any issues, magnetic and metal detectors are seamlessly incorporated into production lines. These detectors, strategically placed along the assembly line, vigilantly scrutinize food items for any potential contaminants, thereby safeguarding the quality and safety of the end products.

6.5.4 Conveying system

Following the processing and manufacturing of food products, as they journey through a network of conveyors toward packaging, it becomes imperative to implement metal detection systems at both the inception and culmination of these conveyor routes.

6.5.5 Packaging process

In the culinary journey, this penultimate stage marks the moment when ingredients are lovingly embraced by packaging materials, ready to embark on their storage adventure. As various machines whirl and hum, each deploying its unique packaging prowess, there is a crucial task: ensuring the contents' purity. Detecting stray metallic intruders during this delicate packaging ballet is essential, safeguarding against contamination before the seal is set.

6.5.6 As a final checkpoint

As a precautionary measure, food items are carefully examined after packaging to ensure their purity. This examination involves the use of either a metal detector or an X-ray system to eliminate any potential contamination. It is highly advisable for the optimal safeguarding of the products to conduct inspections not only on the final output but also on the raw materials and at various crucial checkpoints throughout the production process.

6.6 Food industry magnetic and metallic detectors

6.6.1 Metallic detectors

Metals possess specific characteristics that can impact a transmitted signal due to their conductive and magnetic properties. When it comes to food processing, metal detectors play a crucial role in analyzing the received signal and comparing it to the expected signal. This helps identify any discrepancies that could indicate the presence of contaminants. Currently, three diverse types of detection systems are being utilized, with one type being the most widely used, making up over 90% of current applications. These systems utilize pulse induction technology, ferrous foil detection, and the balanced three-coil system. All-purpose metal detectors operate on the same principles, although it is important to choose a metal detector specifically designed for your intended use to achieve the best performance. The construction techniques ensure that the search head's components cannot move independently, and that water and dirt cannot enter.

6.6.1.1 Pulse induction technology (PIT)

Pulse Induction Technology (PIT) is a primarily handheld/door frame metal detection. Technological innovation is crucial in handheld metal detectors to guarantee precise and effective detection of metallic objects. One of the sophisticated technologies that enable these gadgets is PIT. The PIT system overcomes the constraints of conventional metal detecting techniques, rendering it an essential element of contemporary security solutions. (Madavha et al., 2020). PIT, in its essence, produces rapid and accurately timed electromagnetic pulses. The pulses generate a magnetic field that interacts with metallic objects, enabling the detector to detect their presence. Pulse Induction Transmitter (PIT) differs from previous detection systems by employing a technique of broadcasting and receiving signals alternately. This unique approach allows PIT to effectively overcome obstacles like mineralization, conductivity, and depth problems.

PIT works in four stages, as mentioned below, and the working principle depicted in Fig. 6.4 (Team, 2023):

1) Transmitting pulse: A powerful electromagnetic pulse is emitted by the PI metal detector while it is in the transmitting phase. A magnetic field is produced by this pulse, and it can penetrate the area that is going to be scanned.
2) Eddy currents and disruption: Whenever a magnetic field meets a metallic object, it results in the formation of circular electric currents within the metal. These currents are referred to as eddy currents. Eddy currents are responsible for a momentary disruption in the magnetic field's capability to maintain its stability.
3) Receiving pulse: Following this, the detector returns to the receiving mode. As the device enters this phase, it will stop generating pulses and instead concentrate on detecting any changes in the magnetic field brought about by eddy currents in the metal object.
4) Signal analysis and alert/display: The portable metal detector examines changes in the magnetic field and measures the duration for the field to revert to its equilibrium state. The duration of this decay, referred to as the "ringing down/flyback" time, offers insights into the type, size, and distance of the detected metal object.

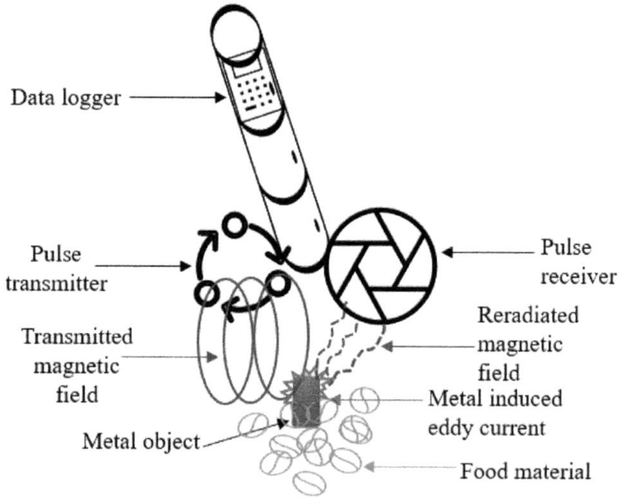

Figure 6.4 Working principle PIT technology.

Advantages: PI technology can detect metallic objects at different depths without experiencing a major decrease in sensitivity, unlike conventional continuous wave detectors. PIT is especially advantageous for doing scans across various surfaces and settings. Pulse Induction (PI) detectors are less susceptible to interference from mineralized soils or settings with high conductivity, making them well-suited for use in difficult conditions, such as food with a high mineral content. PI technology is highly effective in detecting many metals, particularly non-ferrous metals such as copper, brass, and aluminum, which might provide difficulties for conventional detection technologies. PI detectors are less vulnerable to interference from nearby electronic equipment or electromagnetic sources since they operate by transmitting and receiving in brief bursts.

To summarize, PIT plays a crucial role in portable metal detectors by facilitating precise and effective identification of metallic objects, regardless of their depth, mineralization, or conductivity difficulties. The device's capacity to produce accurate electromagnetic pulses, examine variations in the field of magnetism and offer valuable observations about identified items renders it an essential instrument in the fields of safety, exploration, and restoration.

6.6.1.2 Ferrous in foil technology

Ferrous foil detectors are specifically designed to identify ferrous metal contaminants in products that are packaged in foil trays, such as pies, quiches, and casseroles. However, their applications are limited. These detectors are utilized in situations where the amount of metal passing through the detection system is substantial, rendering other detection methods insufficiently sensitive. They are also employed when the signal from the packaging is stronger than the signal from the contaminant. This detection

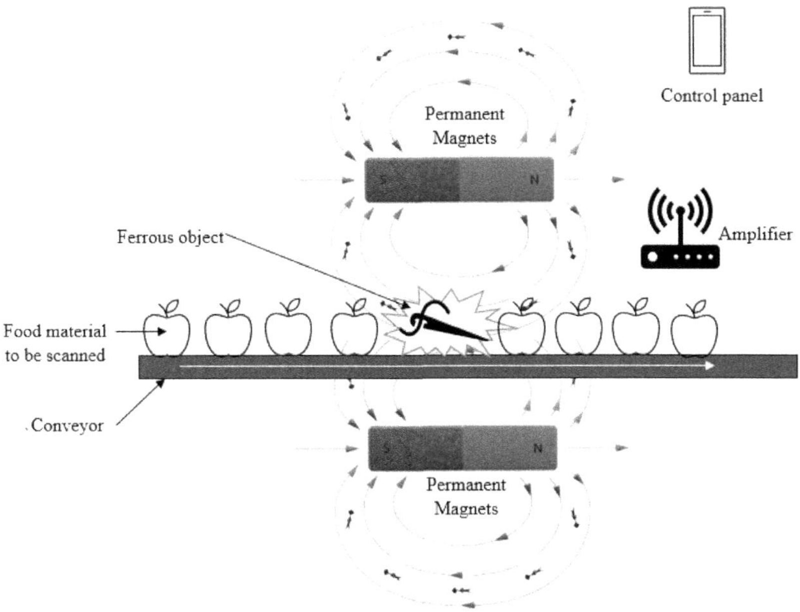

Figure 6.5 Working principle of ferrous in foil technology.

method is based on magnetism and can detect metals that contain ferrous material. The system consists of upper and lower discrete permanent magnets, which generate a powerful magnetic field. Coils are strategically placed at the center to enhance their effectiveness (Fig. 6.5). When a ferrous metal enters the magnetic field, it becomes magnetized. As the metal moves through or under the coils, a small current is generated. The coils are connected to amplifiers that generate an output signal to a control device, which in turn activates a reject system. These detectors are specifically designed to identify ferrous metals in fresh or frozen products packaged in foil wrapping (Craig, 2004).

Application: The height of the aperture between magnets can influence the sensitivity in detecting small metal fragments. In general, a 100 mm high aperture can detect a 1.0 mm piece of ferrous metal. The maximum effective aperture height is typically considered to be 200 mm. During a time when machinery was primarily constructed from mild steel, ferrous foil detection systems proved to be advantageous. Nevertheless, these systems are losing their significance in the industry due to the transition to 300-grade stainless steel with minimal magnetic properties.

6.6.1.3 Balanced three-coil system

Today's often used detection system is the balanced three-coil system, which accounts for around 90 to 95% of all current systems. The reason for its extensive usage is its exceptional sensitivity. This system analyses the disparities in signals detected by

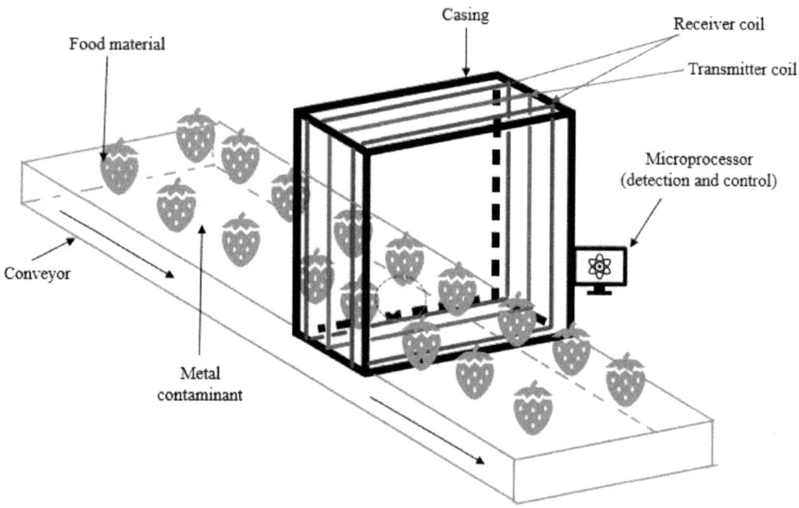

Figure 6.6 Working principles of a balanced three-coil system.

two receiver coils positioned on opposite sides of a transmitter coil along the length of the conveyor (Craig, 2004). The receiver coil was positioned on either side of the transmitter coil. When metal passes through the signal fields, it causes a disturbance in the normal pattern of the electrical fields. Magnetic metals cause an elevation in the induced voltage, whereas non-magnetic metals cause a reduction. As the metal fragment passes through a coil, the change in voltage causes the signals in each receiver coil to fluctuate. The resulting signal imbalance in the system reveals the existence of metal contamination.

As illustrated in Fig. 6.6, When a magnetic metal object is moved through the coils, the magnetic field is attracted to the metal, increasing voltage. An object made of a material that is neither magnetic nor conductive will cause the magnetic fields to change direction, resulting in a decrease in the induced voltage. The ratio between the aperture and product size is essential for achieving optimal performance. The sensitivity of the detector is evaluated at the geometric centre of the aperture, which corresponds to the location of minimum sensitivity. The sensitivity is inversely proportional to the size of the aperture, particularly its smaller dimension.

6.6.2 Magnetic detectors

Magnetic tubes, magnetic plates, magnetic pipeline traps, magnetic drum separators, and magnetic pulleys are the five basic types of magnetic separators that are utilized by the food industry to remove ferrous impurities.

6.6.2.1 Magnetic tube

A tube magnet is comprised of cylindrical permanent magnets enclosed within a slender stainless-steel tube, usually with a 25 mm diameter. The length of the tube

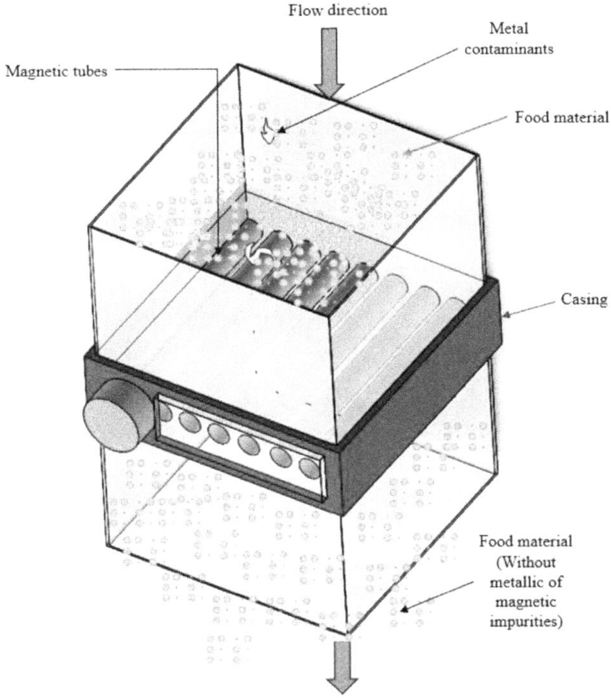

Figure 6.7 Construction and working of magnetic tube type detector.

is tailored to accommodate the precise location of the installation site. These magnets have the flexibility to be utilized either individually or in groups inside the processing system. Additionally, they may be seamlessly incorporated into grates, enabling the smooth flow of food ingredients (Fig. 6.7).

When dealing with powders that have low flow characteristics, such as ground spices, there is a possibility of bridging occurring between the tubes that are next to each other, leading to clogs. To enhance the probability of catching ferrous impurities, it is advisable to augment the space between tubes and incorporate supplementary rows of staggered magnets. Some manufacturers recommend using vibrations to enhance the movement of food materials via the magnetic system. Nevertheless, this can gradually diminish the magnetic potency and should only be applied as a final option. In such instances, it is imperative to constantly assess the magnetic intensity and promptly replace tubes when there is a decline of 20% in potency. To improve the extraction of ferrous impurities from materials that flow freely, the spacing between tubes can be reduced to increase the effectiveness of removal. Self-cleaning tube assemblies are suggested for challenging-to-reach areas and ongoing processes that cannot be stopped for cleaning.

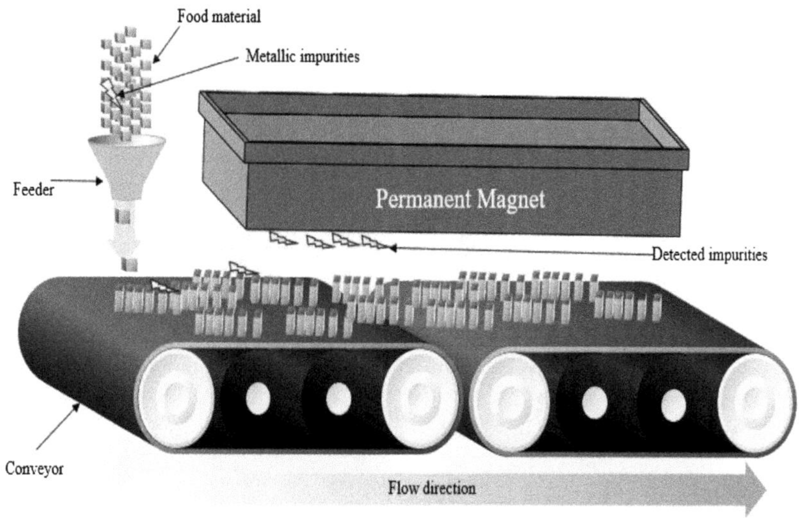

Figure 6.8 Construction and working of magnetic plate type detector.

6.6.2.2 *Magnetic plates*

A plate magnet is a rectangular enclosure that contains permanent magnets. These magnets are strategically placed either above conveyor belts or at the bottom of chutes, ducts, or pipelines to effectively eliminate bigger fragments of unwanted iron, typically with ferrite magnets. To remove tiny ferrous particles or manage deeper material layers, more powerful versions equipped with rare earth magnets are used. For installations beyond a height of 150 mm, it is advisable to use suspension magnets, which are larger variants of plate magnets (Fig. 6.8). Plate magnets are available in a range of sizes and magnetic strengths to suit different configurations. They can be stacked sequentially to improve the collection of ferrous impurities and prolong the duration between cleanings.

6.6.2.3 *Magnetic pipeline trap*

For liquid form food material like juices, soups, and sauces transported through pipelines, magnetic traps with transverse tube magnets are used to capture ferrous contaminants. These traps come in many sizes to fit different pipe diameters (Fig. 6.9). The magnets are affixed to a detachable plate to facilitate cleaning and are specifically engineered to retain impurities downstream, thus preventing further contamination. The housing design guarantees unrestricted circulation around the magnets. It is advisable to arrange the trap in a manner that allows for the removal of magnets from above to prevent any leakage. For heat-sensitive liquids like chocolate and honey, a steam-jacketed trap prevents solidification. Corrosive liquids, such as tomato sauce, require traps made from resistant materials. Soups and stews with solids need traps with external plate magnets to avoid flow impediments, using a rectangular cross-section to

Figure 6.9 Construction and working of magnetic trap for pipe type system.

maintain effectiveness. Traps are often set in series to improve capture rates and reduce cleaning frequency, and parallel lines with valves allow continuous operation during cleaning. The pressure drop in these traps is like a 90° bend in the pipeline.

6.6.2.4 Magnetic drum separator

A magnetic drum separator consists of a drum that rotates and has a stationary permanent magnet covering approximately 180° of its surfaces. The material is introduced into the drum through a chute or a vibratory feeder. The presence of a magnet causes the iron particles to be drawn towards it and dragged along until they reach the non-magnetic area. At this point, the centrifugal force causes the iron particles to be forcefully ejected. A splitter beneath the drum separates the clean product from the ferrous contaminants. The work illustrated in Fig. 6.1 is identical to the magnetic drum separator; instead of a pully, a cylindrical drum is there instead of a pully in the figure. These separators are ideal for removing iron from dry, free-flowing goods like grain, tea, rice, and sugar. Larger ferrous pieces require ferrite magnets, while smaller particles need high-intensity rare earth magnets. A dust-tight drum with extraction ports is essential for dusty materials like tea. The motor, placed outside the housing, prevents contamination from grease or oil.

6.6.2.5 Magnetic pulley separator

A magnetic pulley consists of a shaft-mounted pulley with permanent magnets around its edge, which rotate together, unlike a drum magnetic separator. Positioned at the conveyor's head, the magnetic pulley attracts ferrous contaminants from the food product. As depicted in Fig. 6.1, the pulley spins, the product is flung off by centrifugal force, while metal contaminants stick to the pulley. The magnetic force diminishes as the belt moves further from the pulley, resulting in the metal falling off. A separator located below the pulley diverts the purified product and impurities into separate channels.

6.6.3 Advanced technologies for metal and magnetic detector

6.6.3.1 Metal detection robot

Technology's surge in our lives brings both perks and pitfalls. To truly amplify its benefits and minimize downsides, it should be harnessed to uplift humanity and society. Researchers are fascinated by creating resilient, collaborative robots that tackle challenging tasks autonomously. Among these innovations, the metal detector robot stands out for its efficiency, surpassing the sluggish, manual methods of the past. The detector is designed to uncover hidden metal surfaces in specific navigation areas. The robots consist of hardware and software components that have been chosen based on their cost-effectiveness, usability, regional availability, and ease of programming. The hardware is crucial for the implementation and testing of the control application, utilizing Arduino Uno microcontroller boards as the central processing unit of the system. Multiple sensors, such as metal detecting and ultrasonic sensors, are used to identify metals and promptly navigate around obstructions. The Arduino UNO, which functions as a microcontroller, serves as the central processing unit, overseeing the execution of orders and the interpretation of outcomes. Bluetooth serves as a wireless connection that links the computer and the robotic circuit, enabling smooth communication between peripheral devices. Meanwhile, the robot utilized an ultrasonic sensor to create sound waves, allowing it to identify and avoid obstructions. The ultrasonic sensor, controlled by Arduino, measures the distance to obstacles ahead of a robot by calculating the time difference between sending and receiving ultrasound waves. This helps the robot choose the best path by identifying and navigating around obstacles. Using an intelligent algorithm, the robot detects obstacles ahead and chooses the path with the fewest barriers. It then sends this data to the computer for display and analysis. The goal is to use a budget-friendly automated metal detector to replace human or automatic or semiautomatic detectors. This detector will wirelessly send data, including the location and images of discovered metal objects, to a server for analysis and storage (Kadhim et al., 2019).

6.6.3.2 Optical detection strategies for microfluidics

In recent times, microfluidic technology has made significant strides, enhancing its potential for applications in water quality and environmental monitoring. Research groups globally have developed multiple microfluidic detecting devices utilizing optical and electrochemical principles. Several of these technologies exhibits remarkable analytical capabilities and demonstrate encouraging compatibility with conventional laboratory detection methods. Nevertheless, microfluidic technology has not yet solidified its position as a dependable approach for regular water monitoring, nor has it fully achieved its potential in the field of environmental monitoring (Hu et al., 2023). Only a restricted quantity of microfluidic sensors has effectively progressed to completely automated, field-ready devices that possess the capability of uninterrupted operation. To tackle this issue, it is imperative to progress in wireless networking and ensure the broad adoption of IoT technology. Collaboration among interdisciplinary scientists is also vital for efficient system design. While initial costs hinder commercial-scale production, the accessibility of microfluidic component manufacturing is steadily improving (Lace & Cleary, 2021).

6.6.3.2.1 Absorbance-based detection

UV-vis spectroscopy is preferred due to its high sensitivity and straightforwardness. Shorter optical path lengths in microfluidic chips result in decreased sensitivity. However, altering channel shapes, adding mirrors, and integrating waveguide optics effectively enhance the optical length and improve measurements. For absorbance detection in microfluidic systems, a widespread practice is to utilize LEDs combined with fiber optics to serve as the light source. LEDs produce a focused range of wavelengths, which removes the necessity for optical couplers or monochromators. Due to their exceptional longevity, they are ideally suited for challenging environments such as high humidity and vibrations. A twin LED optical detection setup employs one LED as a lighting source and another LED, operating under reverse bias, as the sensor. The widespread use of digital imaging for colorimetric detection of analytes, such as heavy metals, has been facilitated by the development of high-resolution smartphone cameras and advanced image analysis algorithms (Lace & Cleary, 2021).

6.6.3.2.2 Fluorescence detection

Fluorescence detection is widespread for its exceptional selectivity and sensitivity. Familiar excitation sources include LEDs, lasers, mercury, and xenon arc lamps. LEDs possess a broad variety of wavelengths, making them adaptable for various purposes. They are particularly suitable for integrating into microfluidic chips because of their effective emission of light and minimal energy usage. Laser-induced fluorescence exhibits exceptional sensitivity and is compatible with point detectors such as photomultiplier tubes. It is particularly well-suited for analyzing small sample volumes in microfluidic applications. High-pressure vapor discharge lamps, with their broad spectrum, are frequently used in fluorescence microscopy (Lace & Cleary, 2021).

6.6.3.2.3 Chemiluminescence detection

Chemiluminescent detection methods offer exceptional sensitivity and straightforward equipment needs compared to other optical methods. Moreover, they do not rely on an external light source, cutting out interference, and making the analysis more budget-friendly. However, they are restricted to specific reagents like luminol, and factors like temperature and pH can affect their reliability, potentially leading to inconsistent results (Lace & Cleary, 2021).

6.6.3.2.4 Surface plasmon resonance (SPR)

The technology of SPR is phenomenally successful in enhancing the degree of sensitivity and precision of chemical detection by accurately recording interactions between molecules. This method generates plasmonic waves by inducing an interaction between a layer of metals and dielectric material. The simplicity, enhanced sensitivity, and cost-effectiveness of SPR make it a preferred method for detecting a wide range of analytes. The gadget uses a microfluidic configuration on a smartphone to identify the existence of arsenic and mercury by utilizing the SPR of gold nanoparticles. Their method caused a shift in color by promoting the formation of groups of exceedingly small particles when gold particles react with dithiothreitol-10,12 pentacosadiynoic acid and lysine in an environment of quicksilver and arsenic trioxide. Their method exhibited a linear spectrum of detection from 710 to 1278 μg L^{-1} for quicksilver and arsenic trioxide, with detection limits fluctuating between 10.77 to 53.86 μg L^{-1}, respectively (Lace & Cleary, 2021).

6.6.3.3 *Electrochemical detection for microfluidics*

Electrochemical detection involves the interaction of an analyte of interest with electrodes, leading to distinct electrical signals that can be utilized for quantitative assessment. The use of electrodes in microfluidic detection systems provides a straightforward, energy-efficient, and economical method of detection that enhances sensitivity. Although electrochemical detection offers benefits, it can be limited by the brief lifespan of electrodes. However, due to its great sensitivity, quick response time, and smooth integration into microfluidic devices, it is often favored for detecting heavy metals (Lace & Cleary, 2021).

6.6.3.4 *Quartz crystal microbalance (QCM)*

The Quartz Crystal Microbalance (QCM) harnesses the piezoelectric effect, detecting subtle changes in mass on its surface by monitoring the crystal's resonant frequency. Renowned for its sensitivity, durability, and cost-effectiveness, QCM enables seamless, real-time measurements. Its adaptable substrates offer boundless opportunities for tailored sensing applications (Lace & Cleary, 2021).

6.7 Factors affecting the performance of magnetic and metallic detection

This section explores critical factors for achieving top results in real-world applications. When it comes to the installation of a metal detecting system, it is recommended that collaborate with a provider who is skilled in delivering holistic detection services. Given the numerous interconnected factors in a working environment, leveraging expert knowledge over component-specific expertise is beneficial. The following sections highlight crucial considerations for specifying and installing a metal detection system. Nevertheless, contemporary metal and magnetic detectors cannot detect every metal particle that passes through them. The inherent physical principles that the technology relies on limit the instrument's maximum effectiveness. Thus, like any measuring device, metal detectors have limitations in their precision. These limitations differ based on the application, with the primary factor being the metal particle size that can be detected. Despite these constraints, metal detectors are crucial and indispensable for maintaining process quality control (Herson, 2024).

The detection capability of a metal detecting mechanism is determined by the height of the aperture through which the product travels, especially when the width is smaller than 500 mm. Therefore, it is crucial to consider the dimensions of current and future products during the design phase. Sensitivity spikes dramatically within about 10 mm of the casing, so this must be factored in to prevent the product effect signal from overpowering other signals and compromising the system's ability to detect contaminants. Additionally, the system's capacity to filter out product signals is vital. While two machines might show similar sensitivity in the open air, the one better at phasing out product signals will perform significantly better in real-world conditions. The metal detector is a delicate device that must be shielded from external disturbances like vibrations and electro-mechanical influences, which can alter its output signals. The primary disruption is caused by the conveyor for transporting goods through the detector. The detector's sensitivity surpasses the boundaries of its housing, necessitating a metal-free space at both ends around one-point times the height of the opening. The establishment of this zone is difficult since the detector is highly susceptible to static electric charge, which restricts the usage of specific polymers.

Furthermore, it is crucial to manage the phenomenon known as the "loop effect" caused by conveyor roller bearings, which have the potential to imitate electrical impulses. This impact can be noticed within a range five times the height of the aperture. Although it may not be immediately apparent after installation, the issue can arise as the bearings get less tight. Any substantial alterations in the environment can have an impact on the area devoid of metals or trigger the initiation or cessation of a loop effect, necessitating meticulous deliberation. Besides these characteristics, some of the characteristics of materials from which impurities are removed also affect the detector's performance. Temperature: When magnetic materials are subjected to elevated temperatures, they lose their magnetism, a loss that cannot be restored by merely cooling them down. Flow Characteristics: The moisture content in many food products can alter their flow properties, impacting how they move through the magnetic separator. Product Characteristics: Understanding the specific traits of the material processed is essential for choosing the appropriate magnetic separator (Mythri, 2024). Metal detectors can be misled by the magnetic and electrical properties of contaminants and by materials rich in water or salt that disrupt the signal. Freezing a "wet" product alters its conductivity, simplifying metal detection. For instance, a fresh chicken breast loaded with water hampers metal detection. However, once flash-frozen, its altered conductivity aids in spotting metal fragments.

6.8 Industrial metal detectors applied in the food industry

In the selection of metallic and magnetic detectors for the food industries, several parameters must be considered to assure efficiency and precision. One key factor is the availability of customized detectors, which fulfill the specific production requirements. Customization allows the selection of detectors according to the specific type and form of food material being processed, whether solid, liquid, or powder. The quantity of food material influences the requirements of the detector, as greater quantities could require more sensitive or durable devices. The method of food material flow, whether gravity-driven or forced circulation, can affect the selection of the detector, as these systems must be compatible with various flow patterns to ensure detection precision. The product's packaging, whether bundles, cardboard, cartons, plastic, or unpackaged., greatly affects the design and sensitivity specifications of the detectors. Unpacked products can require an alternative scanning process to efficiently identify impurities. The installation location is a critical aspect, as detectors may be located at multiple locations throughout the production line, such as raw material input, during processing, before or following packaging, or even post-storage. Every stage requires a different detection method reliant upon the risk of contamination at that stage. Moreover, the detectors may differ in mobility, being either stationary or portable, and in orientation, such as horizontal or vertical. These variations facilitate the detection of impurities in large quantities or the handling of various food product sizes, particularly in continuous or batch processes. The type of magnet used for magnetic detectors is another vital consideration. Different magnets, such as rare-earth, plate, rotary, inline, grate, and ceramic magnets, provide distinct levels of detecting sensitivity and strength, crucial for identifying ferrous impurities.

Finally, considerations such as the operational environment, temperature, and integration ease into the current production line are essential for selecting the most appropriate metallic or magnetic detector for a particular food processing product. By evaluating these many elements, food manufacturers can ensure effective contamination control and maintain superior standards of food safety and quality. Table 6.2 explores some examples of metallic and magnetic detectors employed in the food industry. It details their specifications according to parameters such as food material type, flow technique, product form, installation location, detector mobility, orientation, product volume, process type, and magnet type.

6.9 Standards use in metallic and magnetic detectors

The food industry employs various international standards for metallic and magnetic detectors to ensure product safety and quality. These standards outline specifications for detection efficiency, sensitivity, and operational protocols. Below is a table summarizing key international standards relevant to metallic and magnetic detectors in the food sector.

Table 6.2 Industrial metal/magnetic detectors used in the food industry.

Sr. no.	Metal/magnetic detectors	Specifications
1	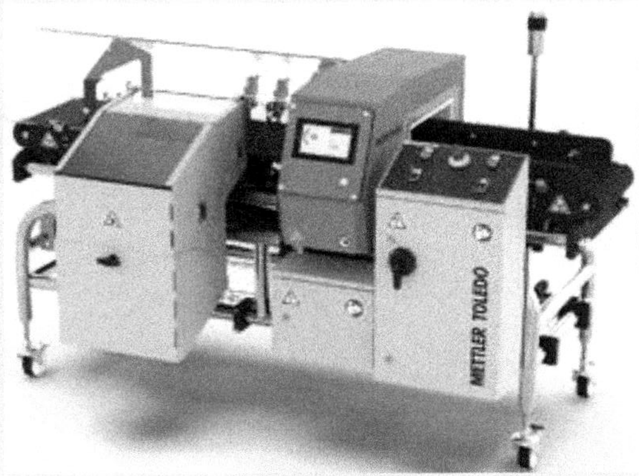	**Courtsey_Company name:* METTLER TOLEDO ***Model:*** Metal Detector M33R ***IP Rating:*** IP69k ***Compatible software:*** ProdX ***Max Product Width***: 1900 mm ***Max Product height***: 1000 mm ***Packaging Type***: Bundles, cardboard, cartons, plastic, and unpackaged. ***Product flow***: Conveyor ***Frequencies***: Up to 600 kHz ***Industry segment***: Bakery and Confectionery, Beer, Wine and Other Alcoholic/Non-Alcoholic Beverages, Cosmetics, Dairy, Fine and specialty chemicals, Food Additives, Meat & Poultry & Seafood, Nutraceuticals, Oils and Fats, Pet Food and Animal Feed, Ready Meals, Snack Foods, Sugar Starch and Yeast, Water.

(continued on next page)

Table 6.2 Industrial metal/magnetic detectors used in the food industry—cont'd

Sr. no.	Metal/magnetic detectors	Specifications
	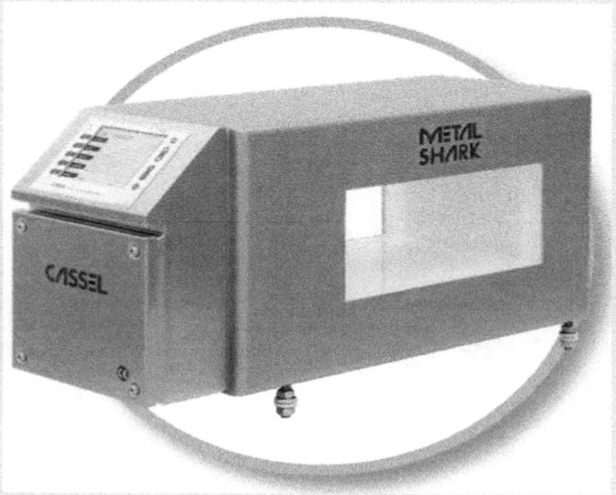	***Model:*** METAL SHARK® BD ***IP Rating:*** IP65 ***Compatible software:*** ProdX ***Product temperature:*** Standard: -22°F to 104°F/ optional: -40°F to 140°F ***Ambience***: Standard: 14°F to 104°F, Relative humidity of 30% up to 80% ***Product memory***: 250 storage spaces ***Electronics***: Digital signal processor, noise filter, digital balance control, digital frequency control. ***Metal Detection***: Ferrous, non-ferrous (e.g. aluminium or copper) and stainless steel ***Measuring Method***: Inductive, multi-channel operation, balanced receiver coils ***Power Supply***: Single phase 110-240 VAC, 50/60 Hz ***Industry segment***: Food, petfood ***Packaging Type***: Paper, cardboard, packaging

(*continued on next page*)

2		***Courtsey_Company name:** BUNTING ***Model:*** pipeLINE™ ***Product flow***: Conveyor ***Manufacturing***: All components are made of stainless steel or food-safe plastic ***Industry segment***: Ground meat, Liquids or pastes moving through pressurized conveying lines. ***Other Features***: 1) Custom designed system fits any pipeline application requirements. Built to your specifications and comes ready for installation. 2) 3-way separation mechanism is suitable for both pneumatically-driven cleaning (PIG) and Clean in Place (CIP). 3) Features 05 or 07 controls, the most advanced metal detection controls, delivering superior results with ease of use. 4) Horizontal or vertical applications. 5) Fixed base standard or available with mobile base.

(continued on next page)

Table 6.2 Industrial metal/magnetic detectors used in the food industry—cont'd

Sr. no.	Metal/magnetic detectors	Specifications
		***Model:*** X5C - Compact X-ray Inspection System ***IP Rating:*** IP66 ***Compatible software:*** ProdX ***Max Product Width***: 280 mm ***Max Product height***: 110 mm ***Product Length***: 280 mm ***Finish***: 304 stainless steels ***Weight on Belt***: Up to 6 kg ***Speed Range***: 10 - 50 m/min ***Power Requirements***: 220-230 V, 50Hz, Single Phase ***Operating Environment***: 2 to 10°C with cabinet warm up,10 to 35°C with Vortex. ***X-Ray Output***: 80W ***X-Ray Safety***: Leakage less than 1 μSv/h ***Packaging Type***: Retail-packed convenience food, ready meals, and small end of line packaged goods ***Detection Algorithms:*** 2 x contaminant, 3 x object checking, 2 x image optimization. ***Metal Detection:*** All metals, bone, glass and dense plastics, even in foil trays or metalized film. ***Industry segment***: Any food industry.

(*continued on next page*)

3	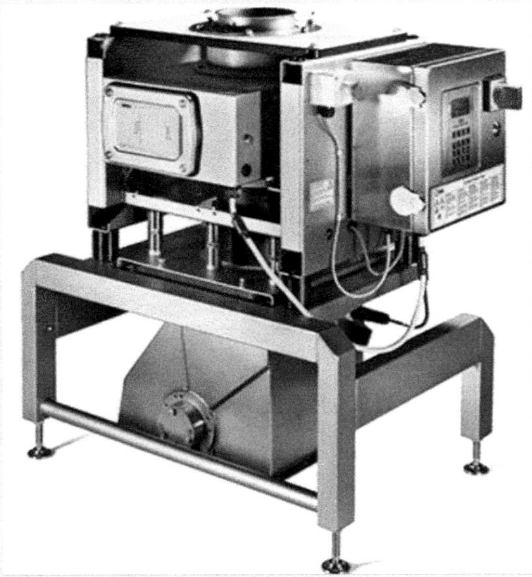	**Courtsey_Company name:* COSTRUZIONI ELETTRONICHE INDUSTRIALI AUTOMATISMI ***Model:*** THS/FFV21 series ***Detection sensitivity***: High ***Immunity to environmental interference***: High ***Construction compliant***: 21 CFR Parts 210 and 211 Inspected/contaminated product counter ***Certification***: ISO 9001 certified manufacturer I/O interfaces: RS232, Auxiliary RS232, ***Protection degree***: IP65 ***Electrical insulation***: Galvanic isolation ***Pollution degree (IEC61010-1)***: Class 2 ***Power Supply***: 100-240 VAC, 50/60 Hz - single phase ***Environmental conditions***: Temperature -10 to +55°C, Relative humidity of 5 – 90 %. ***Maximum temperature***: 70°C (158°F)

(continued on next page)

Table 6.2 Industrial metal/magnetic detectors used in the food industry—cont'd

Sr. no.	Metal/magnetic detectors	Specifications
4		*Courtsey_***Company name:*** ECLIPSE MAGNETICS ***Model:*** Housed Easy Clean Grid ***Protection:*** Secondary ***Feed type***: Gravity ***Cleaning type***: Manual ***Manufacturing***: 316 grade stainless steel ***Magnetic Material type***: NdFeB ***Magnetic performance***: 7000, 9000,10000 ***Magnet Grade***: N35 & N45 ***Temperature***: -20°C / +90°C ***Pressure***: ± 0.2 Bar ***Suitable Products***: Dry powders and granulates ***Suitable Location***: Inlet/outlet points, pre-/post-silo and machinery points ***Other Features:*** 1) Stainless Steel Toggle Clamps 2) Pharmaceutical specification 3) Larger sizes available 4) Safety relay switches 5) Metal Detectable Seal 6) Higher temperature versions available.

(*continued on next page*)

		***Model:*** Magnetic Filter for liquid/slurry form ***Feed type***: Force/pipe ***Manufacturing***: Stainless Steel 304 / 316 ***Cleaning type***: Manual ***Magnetic Material type***: NdFeB Magnet– 11000 gauss ***Max Operating Temperature***: 80°C ***Suitable Products***: Liquid, paste, gel, etc. ***Other Features:*** 1) Suitable for fine ferrous particle as 10 microns. 2) The magnets are installed in polished condition for the higher life of the rods and ease of cleaning. 3) For different flow rates, different sizes of magnetic filters are available. 4) Normally these filters are designed to take up pressures 5kg/sq cm. 5) They can be designed to withstand higher pressure

(continued on next page)

Table 6.2 Industrial metal/magnetic detectors used in the food industry—cont'd

Sr. no.	Metal/magnetic detectors	Specifications
	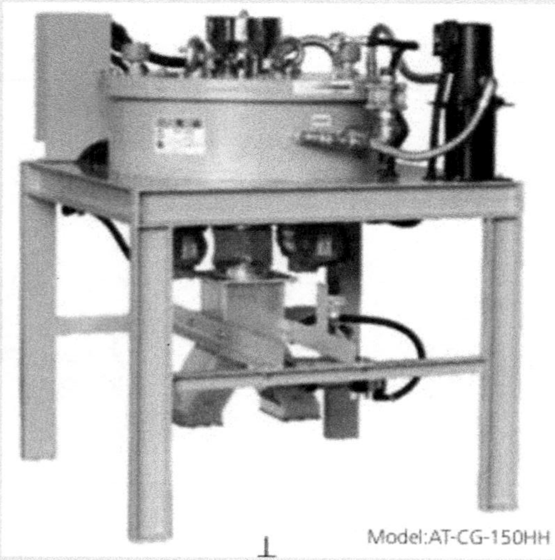	***Model:*** AT-CG-150HH ***Protection:*** Secondary ***Feed type***: feeding device ***Cleaning type***: Automatic ***Manufacturing***: Stainless steel Magnetic Power: 4.12 KW Flux Density: 2,400 GAUSS (not screen) Number of screens: 18 ***Magnet***: Electro-magnetized screens ***Suitable Products***: Dry powders of grain size less than 0.1mm ***Other Features:*** 1) Electro-magnetized screens capture fine iron particles in the micron range 2) Vibration motors ensure effective and smooth flow of material through the screens. 3) The rate of feed material is controlled.

* Source: The images are the courtesy of the respective company

Standard	Purpose	Specifications
ISO 22000:2018	Ensures food safety management systems include metal detection as part of hazard analysis.	Outlines critical control points (CCPs) for metal detection, ensuring food products meet safety standards.
BRCGS Global Standard for Food Safety (Issue 9)	Provides guidelines for effective metal detection systems in food processing environments.	Recommends sensitivity, calibration, and validation practices for metal detectors to detect ferrous, non-ferrous, and stainless steel contaminants.
Codex Alimentarius General Principles of Food Hygiene (CXC 1-1969, Rev. 2020)	Ensures preventive controls for physical contaminants in food.	Recommends integration of metal detection systems at CCPs in Hazard Analysis and Critical Control Points (HACCP) plans.
FDA Food Safety Modernization Act (FSMA)	Mandates preventive controls for identifying and mitigating hazards, including metal contamination.	Requires the use of detection technology capable of identifying metals in food manufacturing and processing environments to comply with Preventive Controls Rule.
IFS Food Standard (Version 7)	Ensures food safety and quality through effective detection systems.	Specifies operational requirements for metallic detectors to detect all contaminant types, emphasizing equipment validation and sensitivity tests.
European Commission Regulation (EC) No. 852/2004	Establishes food hygiene requirements, including detection of physical hazards like metal.	Requires detection and prevention of physical contaminants at every stage of the production process, emphasizing compliance with HACCP systems.
GFSI Benchmarking Requirements	Standardizes metal detection systems across GFSI-certified schemes (e.g., SQF, BRCGS).	Requires systems to meet industry-standard detection thresholds and adhere to routine testing, calibration, and documentation.
ISO/IEC 17025:2017	Ensures testing and calibration of metal detection equipment meets recognized quality standards.	Establishes guidelines for accuracy, repeatability, and precision of metal detectors used in food safety testing laboratories.

While these standards provide a robust framework for ensuring food safety, some argue that reliance on technology alone may not address all contamination risks, emphasizing the need for comprehensive training and human oversight in detection processes.

6.10 Conclusion

Magnetic and metal detectors are vital in upholding the safety and quality of food products within the food industry. Utilizing the forces of magnetism and electromagnetic fields, these technologies efficiently identify and eliminate contaminants, protecting consumers from health hazards. With technological advancements, incorporating sophisticated sensors and automation will elevate the effectiveness of these detection systems, solidifying their essential role in preserving food safety standards.

References

Apoussidis, E., & Wells, I. (2004). Magnets. In M. Edwards (Ed.), *Detecting Foreign Bodies in Food* (pp. 63–85). Elsevier.

Craig, J. P. (2004). Metal detection. *Detecting foreign bodies in food* (pp. 47–62). Edited by M. Edwards: Elsevier.

Herson, M. (2024). "A guide to metal detection in the food manufacturing industry." Loma Systems. Accessed on 28/05/2024.

Hu, T., Lai, Q., Fan, W., Zhang, Y., & Liu, Z. (2023). Advances in portable heavy metal ion sensors. *Sensors, 23*(8), 4125–4151. https://doi.org/10.3390/s23084125.

Kadhim, H. A., Ali, N. S., & Abdulsahib, D. M. (2019). Management and archiving system for metal detection robot using wireless-based technology and online database registry. *International Journal of Power Electronics and Drive Systems (IJPEDS), 10*(1), 219–229. https://doi.org/10.11591/ijpeds.v10.i1.pp219-229.

Lace, A., & Cleary, J. (2021). A review of microfluidic detection strategies for heavy metals in water. *Chemosensors, 9*(4), 60–85. https://doi.org/10.3390/chemosensors9040060.

Madavha, L., Laseinde, T., Daniyan, I., & Mpofu, K. (2020). Functional design and performance evaluation of a metal handheld detector for land mines detection. *Procedia CIRP, 91*, 696–703. https://doi.org/10.1016/j.procir.2020.04.134.

Magnattackglobal. (2018). Magnetic separators & metal detectors in the food industry: Magnattack™ global blog. *Magnattackglobal.Com.* https://www.magnattackglobal.com/blog/2018/magnetic-separators-metal-detectors/. (Accessed 24 May 2024).

Mythri, R. (2024). Magnets in food industry. *Food Safety Works.* https://foodsafetyworks.com/insights/magnets-in-food-industry/. (Accessed 24 May 2024).

Payne, K., O'Bryan, C. A., Marcy, J. A., & Crandall, P. G. (2023). Detection and prevention of foreign material in food: A review. *Heliyon, 9*(9), e19574–e19587. https://doi.org/10.1016/j.heliyon.2023.e19574.

Sapp, J. G. (2024). Application of rapid detection technology in food heavy metals. *Journal of Trends in Life Sciences, 2*(1), 23–28. https://doi.org/10.61784/jtls240129.

Smith, J. (2023). Our complete guide to metal detectors for food processing. *Metal Detectors For Food Processing.* https://www.eclipsemagnetics.com/resources/guides/metal-detectors-for-food-processing/. (Accessed 24 May 2024).

Team, Hawkberg Product. (2023). Pulse Induction (PI) technology for enhancing handheld metal detection. *X Ray Baggage Scanners UVSS Walkthrough Gates Metal Detectors Baggage Scanners Walkthrough Gates Metal Detectors.* https://www.hawkberg.com/pulse-induction-pi-technology-in-handheld-metal-detectors/. (Accessed 24 May 2024).

Yang, H., Liu, L., Shu, Z., Zhang, W., Huang, C., Zhu, Y., Li, S., Wang, W., Li, G., Zhang, Q., Liu, Q., & Jiang, G. (2025). Magnetic iron oxide nanoparticles: An emerging threat for the environment and human health. *Journal of Environmental Sciences, 152*, 188–202. https://doi.org/10.1016/j.jes.2024.04.045.

Cyclone separators for the food industry

7

Christian O. Díaz-Ovalle
Tecnológico Nacional de México/I.T. Roque, Departamento de Ingenierías, Celaya, Guanajuato, México

7.1 Introduction

Cyclone separators recollect, totally or partially, the powder contained in a fluid stream by taking advantage of the centrifugal force and gravity during flowthrough. This mechanic unit aids in recovering powder or dust, and is common in the industry to handle, store, manipulate, operate, and modify any solid, whether it is a raw material, intermediate, or a final product. Thus, the application of cyclone separators is extensive, and herein this is focused on the food industry that reclaims particular attention. Around the world, most of the food industry relates to solid products, such as dairy powders, sugar, and flavor powders, as well as raw materials, such as grains, and protein powders (Bhandari et al., 2013).

The cyclone separators are constrained to reach a high level of efficiency to avoid environmental and monetary losses. The literature presents a huge list of studies to improve the separation efficiency of cyclones where the geometric design is the main element. Hoffmann and Stein (2008) compiled all the required elements to guarantee an adequate design in cyclone separators and indicated the need to understand the hydrodynamic behavior. In principle, the cyclone gas flow pattern is a descending-ascending swirling flow composed of tangential, axial, and radial velocities, where the tangential velocity influences more the transporting solids. In addition, the pressure drop is associated with the frictional loss and is a suitable operational condition to evaluate the cyclone performance, which depends on the geometrical design. Nowadays, the theoretical study of cyclone separators is possible by numerical techniques, such as the computational fluid dynamics (CFD), which is discussed herein.

The cyclone gas patterns are complicated, and the real scenario is a multi-phase system that includes the solids as particles. Michaelides et al. (2017) described the importance of the fluid-solid and solid-solid interactions (one-way and two-way coupled flows) that influence the flow through the cyclone separator. Furthermore, the particle shape and size distribution make particular cases for the separation efficiency (Fayed & Otten, 1997). For the food industry, Onwulata (2006), Tamime (2009) and Bhandari et al. (2013) provide complete guidelines for handling powders, which reclaims the use of cyclone separators.

Mechanical Separation Processes in the Food Industry: Unit Operations and Processing Equipment in the Food Industry.
DOI: https://doi.org/10.1016/B978-0-128-19523-9.00005-6

7.2 Basics of food powder technology

Powders and dust are defined as particulate solid materials of various sizes and shapes that are either bulked or suspended into fluid, depending on their concentration (Bhandari et al., 2013). This description indicates an unpredictable system, where the nature of the particle (size, shape, density, etc.) makes a dispersed or discrete phase. Nowadays, the understanding of powder and dust products considers a variety of properties, classified as (Bhandari et al., 2013): (a) fundamental properties (moisture, shape, size, structure, density, etc.), (b) functional properties (flowability), and (c) defective properties (dustiness, stickiness, etc.). Other characteristics of powders arise from their dispersibility and the possible interactions, which are summarized by the cohesive forces, for instance: liquid bridges (humidity dynamic behavior), solid bridges (results of dissolved solids), and van der Waals forces (suspended dust, $<1\,\mu m$).

The dispersibility of powders yields a void volume, void fraction, or porosity, ε, occupied by either air, water, or air-water vapor. Porosity exists between particles and within the particles themselves (internal porosity), and it is an important factor in determining the density of solids in the dispersed phase. Besides, the density might be obtained experimentally by the standards, either EN ISO 60 (1999), EN 1236 (1995), or ISO 697 (1981) (Lanzerstorfer, 2020). Fig. 7.1 depicts the main concepts about density proposed by Barbosa-Cánovas and Juliano (2005):

a) True particle density (ρ_S) relates to the solid mass and its volume without pores.
b) Apparent particle density (ρ_A) relates to the solid mass and its volume without external pores.
c) Bulk particle density (ρ_B) involves the ρ_S and the porous density, ρ_G, gas density, or air density as $\rho_B = \rho_S(1-\varepsilon_p) + \varepsilon_p \bullet \rho_G$.

Some examples of this densities are (kg/m^3): cocoa powder ($\rho_S=1420$, $\rho_B=500$), flour ($\rho_S=1540$, $\rho_B\approx600$), starch ($\rho_S\approx1540$, $\rho_B\approx460$), and milk powder ($\rho_S\approx1360$, $\rho_B\approx480$), and the reader might consult Lanzerstorfer (2020) and Pugliese et al. (2017) for further data. In fact, the mixing of powders is required to produce the desired foodstuff, where the density is affected (Shenoy et al., 2015).

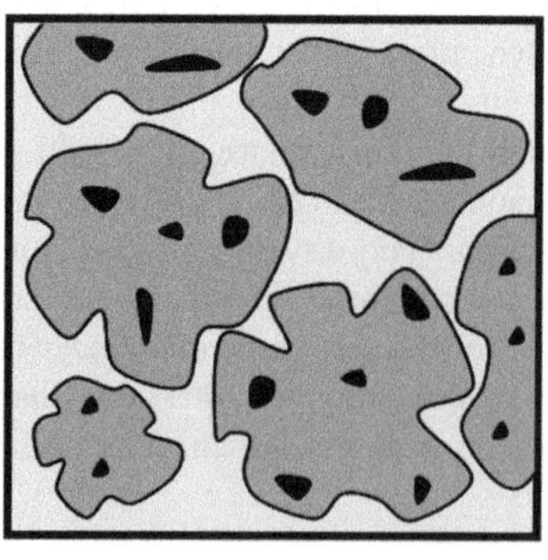

$$\rho_S = \frac{m_S}{V_S} = \frac{m_S}{V_T(1-\varepsilon_p)} = \frac{m_S}{V_T - V_E - V_I}$$

$$\rho_A = \frac{m_S}{V_S + V_I} = \frac{m_S}{V_T - V_E}$$

External pores with volume V_E
Solid with volume V_S and mass m_S
Internal pores with volume V_I
Sample with volume V_T

Figure 7.1 Representation of the particle densities and some mathematical equivalencies.

7.3 Applications of cyclone separators in the food industry

The high level of adaptability characterizes cyclone separators, which is demonstrated by operating at high and low temperatures (Yamasaki et al., 2020), handling a variety of fluid flows, and handling a wide range of particulate sizes (Leith & Jones, 1997). This point describes the importance of cyclone separators in industries worldwide. Particularly, the food industry has used cyclone separators since the handling of grains, and this requirement has been increased with the spray drying technology. Thus, the covering of applications of cyclone separators is transcendental for food processes, and the fundamentals of powder products are essential to understand the purpose and function of cyclone separators.

The food industry has extended the versatility of its products due to innovative processes, where the powders are outstanding. The powder might be identified as either a product, raw material, or subproduct, and the common powder products are milk powder, coffee, cocoa, sugar, egg, tea, whey proteins, casein, starch, encapsulated butter oil, etc. (Bhandari et al., 2013). In the food industry, the powders are classified according to their origins (Cuq et al., 2011): (a) natural powders, such as flours, salts, etc., (b) transformed and formulated powders, such as instant coffee, milk powder, etc., and (c) functional powders, such as additives and spices. Most of these products result from unit operations, such as spray drying, freeze drying, crystallization, milling, granulation, crushing, etc. In Fig. 7.2, Cuq et al. (2011) have defined the main unit operations in the transformation of food powders from agricultural raw materials, where the drying and milling are elements into some processes.

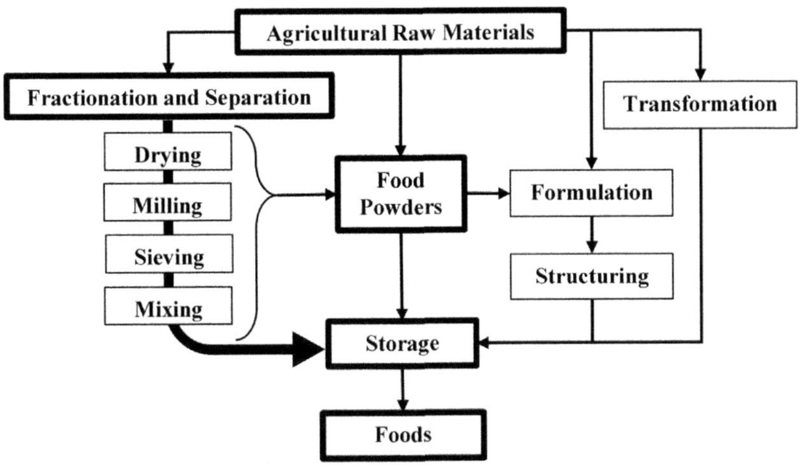

Figure 7.2 Representation of the unit operations involved in the food powder processes. (Modified from Cuq et al. (2011)).

Table 7.1 Identification of some uses of cyclone separators in the food industry.

Process goal	Food industry	Product	Unit operation
Powder production	Dairy powder Cane sugar Instant beverages Egg and derivatives	Milk powder Instant coffee and tea Powder egg Powder proteins Powder vegetables	Spray drying Crystallization Freeze drying Grinding Bag filter
Grain process	Grain cleaning Flour Coffee industry Cocoa industry Brewering and malting	Flour Roasted bean Rice, wheat, barley, malt, corn, oats	Milling Classifying Bowers Conveyor Storage equipment

The cyclone separators aim to accomplish fundamental goals, such as the in cases like: (a) a collecting system in dryers (the spray drying is the most remarkable one), (b) a suction system for avoiding dust accumulation during the mechanical operation of solids and grains (handling, milling, crushing, classification, etc.), and (c) a recovering element at the end of the powder transport system. Table 7.1 identifies examples of cyclone separators used in the food industry, based on the process goal, products, and related unit operations. For instance, the instant beverage industry produces instant coffee (powder product) through freeze-drying, where the cyclone separator collects the product.

The mechanical operations of grain were among the first food processes to require the use of cyclone separators. Dust from grains results from harvesting, handling, and storage, with some particles being starch granules formed by friction (such as grain hull, grain brush, and small pieces of broken grain). Additional elements from the environment during the recollection (insects, pesticide residues, ground dust, stalk, etc.) (Rosentrater & Bucklin, 2022). Nowadays, coming in contact with dust grain is a health and safety problem (Perry, 2008), whereby it is mandatory to collect it. The treatment of dust in grain processes is possible by suction systems that collect the dust in each operation (transport, milling, cleaning, classification, etc.) by low-pressure vacuum from an extractor (centrifugal fan); and the final element is an either a cyclone separator or bag filter (Williams & Rosentrater, 2022). Fig. 7.3 a) depicts a dust suction system where the cyclone separator is located before the bag filter, and Fig. 7.3 b) presents a typical cyclone separator of the dust suction system.

Powdered products are important foodstuffs for the economy due to their long shelf life and ease of handling and transportation. Spray dryers yield most of these products and operate in a manner: (a) a hot air stream enters into a chamber to evaporate liquid droplets by direct contact, (b) this evaporation yields suddenly a solid (powder product) that falls through the chamber and is collected at the bottom, and (c) a suction system extracts the humid air with powder upon a multi-cyclone separator arrangement to recover the dust (Bhandari et al., 2008; Woo & Bhandari, 2013). Fig. 7.4 a) shows the 3D representation of spray drying equipment, and Fig. 7.4 b) presents a multi-cyclone separator arrangement connected to a suction system. The dairy industry leads

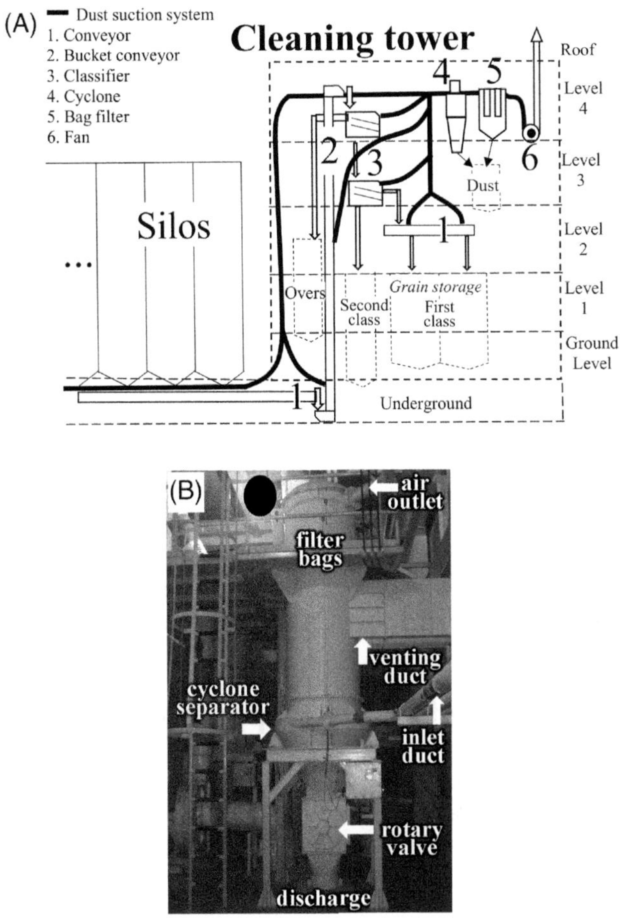

Figure 7.3 Application of a cyclone separator in the cleaning grain process: A) dust suction system diagram, and B) typical cyclone separator in a dust suction system.

the production of powders through spray drying, which requires cyclone separators to guarantee a profitable production by avoiding loss of product (Tamime, 2009; Westergaard, 1994). Gamiño-Tovar et al. (2018) have analyzed the importance of the cyclone separator design in the spray drying process and its impact on particle agglomeration.

The trends in encapsulated and powder products respond to the increasing consumption of bioactive food ingredients, where nanotechnology provides benefits for the food industry, particularly in the production of functional and nutraceutical delivery systems (Jafari et al., 2021). A nano spray drying system is suitable to yield these products, but the conventional cyclone separators are not able to collect nanoparticles with size lower than 5 μm. Nevertheless, Wang et al. (2012) declared the possibility to apply the

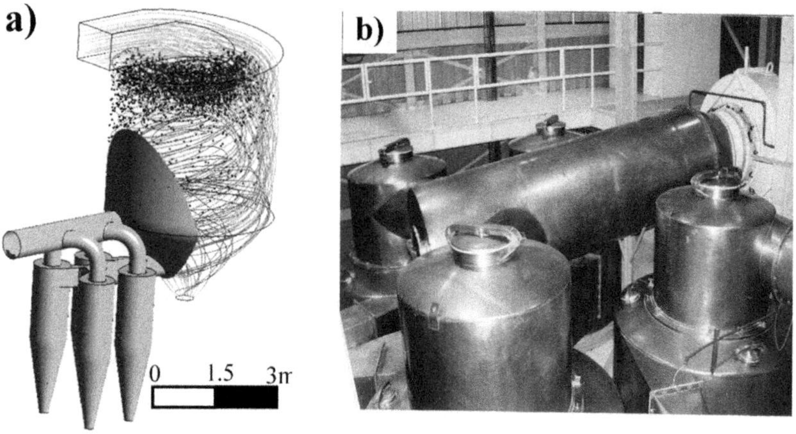

Figure 7.4 Application of cyclone separators in the spray drying process: A) 3D representation and B) upper view of a multi-cyclone arrangement.

fish-hook phenomenon in cyclone separators with nanoparticles. These enounces new tendencies of scaling for cyclone separators in the food industry.

7.4 Fundamentals of cyclone separators

The understanding of cyclone separators is supported by the hydrodynamic behavior inside the equipment. Leith and Jones (1997), Cortés and Gil (2007), and Hoffmann and Stein (2008) have reported theoretical and experimental explanations on this topic, and foregrounded the importance of the velocity profile and pressure drop. The cyclone separator presents a swirling descendent flow near the walls and a swirling ascendent flow in the cyclone core. In detail, an inlet duct conducts the flow to the wall, forms a tangential effect that produces a descent by encircling the exit duct (vortex finder), and the fluid goes down inside a conical section, where the flow results in ascending upon the vortex finder due to the pressure field. For the particles, this behavior equivalates to a centrifugal force on the wall that reduces its kinetic energy and the gravitational effect during descent makes possible its collection. Fig. 7.5 depicts the flowthrough inside a cyclone separator (streamlines from a CFD simulation) with the main elements of cyclone separators.

The velocity is a vector conformed by three components, and this number is representative of a fluid flow in a 3D analysis. The velocity components adapt the flowthrough of the fluid to the geometrical system and are standardized by coordinates, such as cartesian, cylindrical, and spherical coordinates. The flow inside a cyclone separator falls into the cylindrical coordinates (r, θ, z), where the velocity has the axial, radial, and angular components, v_z, v_r, and v_θ, respectively, considering $v_\theta = \omega r$ with ω as the angular velocity and r as the radial position. Besides, the exerted centrifugal force makes a tangential effect on the wall that corresponds to a tangential velocity, v_t.

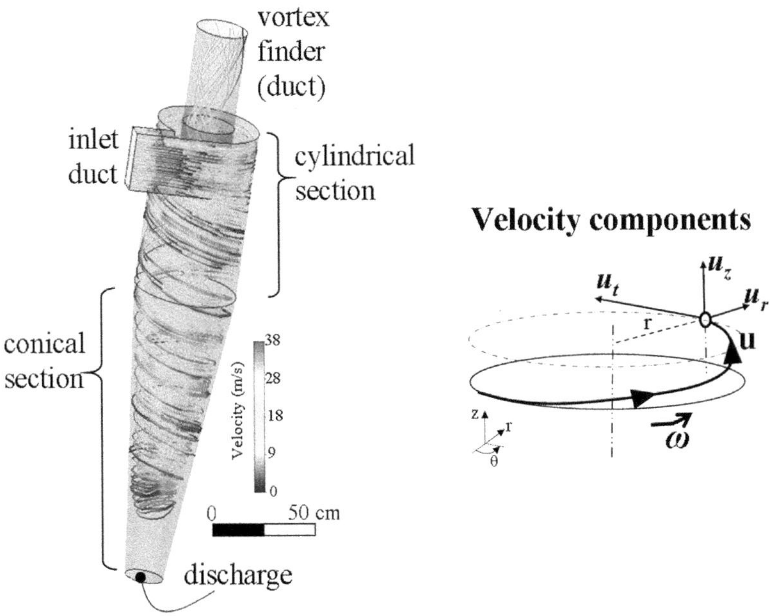

Figure 7.5 Internal flowthrough for a cyclone separator and velocity components.

7.4.1 Flow patterns inside cyclone separators

Shepherd and Lapple (1939), Ogawa (1997), Cortés and Gil (2007), and Hoffmann and Stein (2008) have succinctly described the flow patterns for cyclone separators. They concluded a particular hydrodynamic inside due to the upper cylindrical and lower conical shape. The experimental tests have identified velocity and pressure fields as observable parameters. Particularly, the velocity field includes its components: tangential, axial, and radial velocities. Fig. 7.6 depicts 2D profiles from a center plane for these elements, whose descriptions are:

- Total velocity is shown in Fig. 7.5 with acceleration (increment) inside the cyclone body and reduction in the conical section. Fig. 7.6 demonstrates this behavior at the center plane. A transversal profile is similar to the letter M (known as Rankine vortex), and a vertical profile corresponds to a reduction of the total velocity.
- The tangential velocity, v_t, is the more influential velocity component. Fig. 7.6 shows the low values of v_t in the cyclone core and close the walls. This is a Rankine vortex and approaches a swirling flow. Mathematically, the Rankine vortex comes from the mass and momentum conservation equations for the angular component of velocity in steady state and Newtonian fluid, such as: $v_\theta = \mathrm{r}C_1$ (forced vortex) and $v_\theta = C_2/\mathrm{r}$ (free vortex).
- The axial velocity follows a down-upward pattern, as Fig. 7.6 presents. In the core zone, the axial velocity reaches similar scale values to the tangential velocity.
- The radial velocity has not been experimentally studied, its values are the lowest of the velocity components and are a result of secondary inner flows that produce a double-eddy motion (Shepherd & Lapple, 1939). Ogawa (1997) and Wang et al. (2023) reviewed these

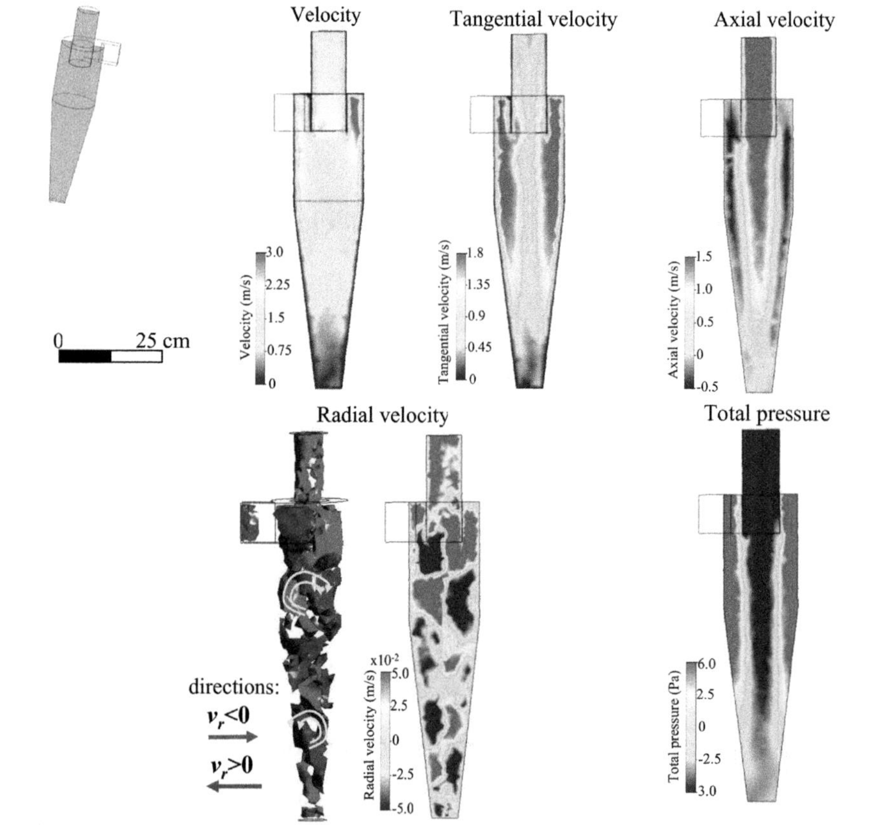

Figure 7.6 2D profiles from a center plane with a 3D detail of the radial velocity.

low instabilities and negative values in the cyclone core. Fig. 7.6 depicts an alternation of radial velocity directions (positive and negative) and the inner secondary flows.

- The pressure indicates the friction level on the cyclone walls. Fig. 7.6 observes the distribution of pressure that relates to the centrifugal force and the existence of secondary flows. Furthermore, Hoffmann and Stein (2008) have explained that the low vertical change of pressure coincides with the low effect of the hydrostatic pressure.

7.4.2 Geometrical aspects of cyclone separators

A cyclone separator geometry is particular to reach its characteristic and desirable flow behavior, as Fig. 7.5 depicts. In detail, the cyclone separator presents a set of geometry elements cataloged by inlet, cyclone body, and vortex finder. The inlet duct supplies the gas with particles to the cyclone and might present either a circular or rectangular

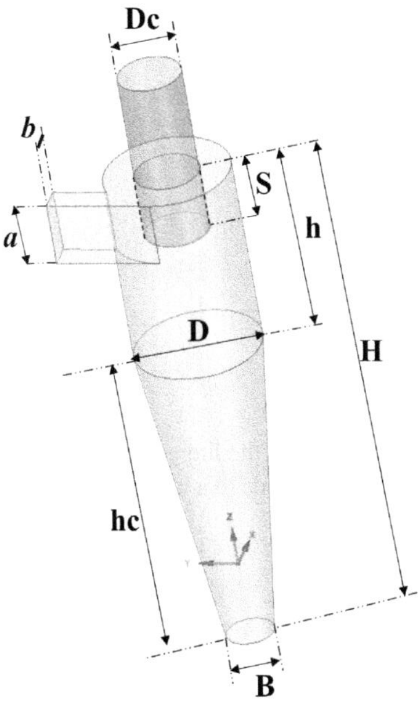

Figure 7.7 The main geometrical elements of cyclone separators in a 3D view. (From Guo et al. (2024)).

transversal shape. In most designs, the rectangular transversal shape has demonstrated appropriate results, with the parameters height, *a*, and width, *b*. The cyclone body contains cylindrical and conical shapes, upper and lower zones respectively, that are united with the same diameter, D. The conical shape is an inverse frustum with height, hc, and tip diameter, B; the cylindrical shape is an annular region with height, h, this shape results from the immersed vortex finder. For each cyclone shape, the height measures from the central union (sharing diameter), and the sum of both is the total height of the cyclone, H=h+hc. Finally, the vortex finder is an immersed exit duct with diameter, Dc, and height inside the cyclone, S. Fig. 7.7 shows all the cyclone elements in a 3D representation.

The inlet duct design influences the cyclone separator performance (Hoffmann & Stein, 2008; Ogawa, 1997), and three designs are used: (a) tangential inlet, (b) descending inlet (see Vibco from Fig. 7.8), and (c) wrap-around or ter Lynden.

The geometry and size of cyclone separators rule the flow patterns inside, which affects the dust motion and its collection. The change of these elements will produce different operative conditions and might benefit the dust collection based on its characteristics (density, concentration, and size). These changes have yielded several standard cyclone separators, which have been cataloged based on geometrical ratios with respect to the diameter, for instance Dc/D, S/D, etc. Table 7.2 summarizes the

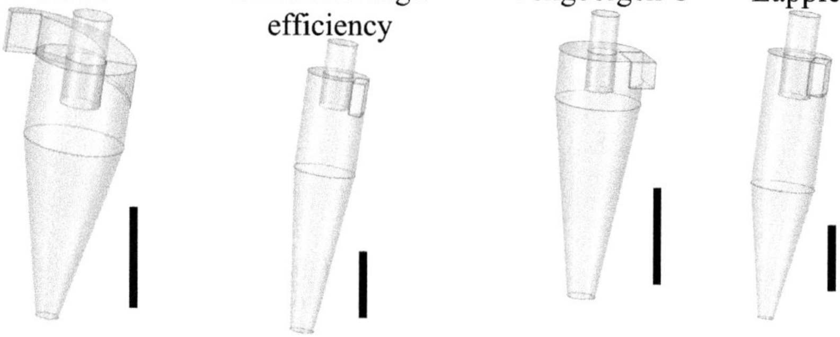

Figure 7.8 Examples of cyclone separator geometries reported in the literature. The black line indicates the diameter scale, whereby the shorter length corresponds to a larger cyclone separator.
(From Guo et al. (2024)).

Table 7.2 Geometrical ratios of standard cyclone separators in respect to Dc.

	Name	D	Dc	S	H	h	*a*	*b*	B
1	Muschelknautz E	1	0.25	0.457	1.374	0.254	0.254	0.085	0.335
2	Muschelknautz D	1	0.333	0.89	2.417	0.734	0.524	0.15	0.546
3	Storch 4	1	0.45	0.677	6.215	3.496	1	0.146	0.35
4	Storch 3	1	0.557	1.042	4.276	2.406	0.87	0.313	0.48
5	Storch 2	1	0.48	1.062	4.876	2.062	0.836	0.236	0.373
6	Storch 1	1	0.337	0.39	5.323	1.5	0.274	0.274	0.175
7	Tengbergen C	1	0.332	0.43	2.76	0.555	0.297	0.297	0.332
8	Tengbergen B	1	0.533	1.067	2.876	1.543	0.852	0.267	0.533
9	Tengbergen A	1	0.404	0.567	2.336	0.65	0.487	0.267	0.73
10	TSN-11	1	0.39	0.7	2.756	0.63	0.529	0.155	0.443
11	TSN-15	1	0.594	1.316	4.226	2.214	0.624	0.226	0.447
12	Stairmand high efficiency	1	0.5	0.5	4	1.5	0.5	0.2	0.377
13	Stairmand high flowrate	1	0.742	0.87	3.974	1.49	0.742	0.374	0.374
14	VanTongeren AC	1	0.31	1	3.79	1.342	0.46	0.206	0.4
15	Vibco	1	0.39	0.434	2.517	0.797	0.39	0.315	0.23
16	Lapple GP	1	0.5	0.625	4	2	0.5	0.25	0.25

Modified from Cortés and Gil (2007).

cyclone separation ratios for the extended in the literature, and Fig. 7.8 schematizes some examples.

The cyclone separator geometry has been studied to enhance the cyclone performance, and demands change in size, shape, and addition of devices. The literature presents some theoretical and practical proposals that Guo et al. (2024) have cataloged based on the reason of the geometrical change:

a) Basic cyclone design: a standard cyclone shape presented a difference in the geometrical ratios. Table 7.2 contains some important cases.

b) Modification of the geometrical elements: the standard cyclones might suffer modifications in the geometrical ratios in order to enhance the cyclone performance. The most influencing geometrical elements are:

- Inlet duct dimensions (*a* and *b*) with inverse effect on the cyclone performance due to the applied centrifugal force on the particles to the walls.
- Vortex finder diameter (Dc) with inverse effect on the cyclone performance due to the change of the axial velocity profile.
- Cylinder and cone height (H and hc) with a direct effect on the cyclone performance due to the augmentation in the separation length and particle pathway.

c) Remarkable geometrical modifications: the geometrical elements are prone to innovative modifications in order to improve the cyclone separator performance. Some examples are: the vortex finder modifications (shape and eccentricity), and insertions of devices (filters, spiral vanes, etc.).

7.5 Elements of design

The cyclone separators focus on reaching a high particle separation or collection efficiency. As already mentioned, the geometrical shape influences the particle separation by taking in advance the swirling flow, particle density, and particle size. The literature has demonstrated how the geometrical aspects enhance the separation efficiency. This section provides the calculation of the steering elements of design cyclone separators: pressure drop, and collecting efficiency; and includes a basic description of the particle tracking analysis.

7.5.1 Pressure drop

The literature has described the direct proportion between this parameter and the collecting efficiency (Demir, 2014; Guo et al., 2024), in a range value from 0.5 to 8 in. of water (Leith & Jones, 1997). In principle, the pressure drop is the difference of static pressure between the inlet and the outlet (Chen & Shi, 2007) and it results from the frictional losses of the fluid and the particles, and decreases in proportion to the particle load concentration (Shepherd & Lapple, 1939), frictional loss and cyclone height (Hoffmann & Stein, 2008). Chen and Shi (2007) have proposed the components of the pressure drop: (a) local losses (expansion loss at the inlet and contraction loss at the outlet), and (b) frictional losses (swirling loss by friction to the walls and dissipation loss by dynamic energy change in the outlet).

The computing of the pressure drop is relevant in the cyclone performance and is a dimensionless form, such as Euler number (Eu) or number of velocity heads (N_H). The Eu number expresses the flow losses in a pressure drop and relates the pressure forces/inertial forces: Eu=$\Delta P/(\frac{1}{2}\rho_G v^2)$, where Eu→0 indicates a perfect flow without losses. The Eu number is considered a clean pressure drop without dust loading; otherwise, it uses a function of the concentration value as a corrector parameter. Some mathematical models for Eu or N_H have differed from the researcher's perspective of cyclone separator losses. Table 7.3 contains eight models reported in the literature, and the reader might consult them in Leith and Jones (1997), Cortés and Gil (2007), Hoffmann and Stein (2008), and Demir (2014). For all cases, the geometrical parameters *a, b*, and Dc are fundamental elements for the internal swirling flow losses and are grouped by $Z=(ab)/Dc^2$. Moreover, some models present groups of terms in order to reduce the mathematical expression.

7.5.1.1 *Case study to compute the pressure drop*

The Tengbergen C cyclone separator is recommended for treating the milk powder after the spray drying process (Gamiño-Tovar et al., 2018). A conventional cyclone size is around 1.5 m in diameter to process an air stream (2 m^3s^{-1}) at 90°C and 1 atm, for the air: ρ_G=0.9718 kgm^{-3} and μ=2.135 × 10^{-5} $kg{\bullet}m^{-1}s^{-1}$ from Çengel and Cimbala (2017). Determine the pressure drop for this case by using the Chen and Shi (2007) model.

Solution:

This problem illustrates a complicated model from Table 7.3. The reader can observe the systematic and numerical solution of the case by following the calculation of each parameter. Basically, all calculations for the Eu number are dimensionless, except for some parameters from the models proposed by Stairmand (1949) and Alexander (1949). For this case study, the geometrical sizes correspond to a diameter of 1.5 m by multiplying each geometrical element in Table 7.2. Table 7.4 contains these results.

The initial step computes the fundamental parameters from the Chen and Shi (2007) model, such as:

$$Q = \frac{\mathrm{Dc}}{\mathrm{D}} = \frac{0.5\mathrm{m}}{1.5\mathrm{m}} = 0.333$$

$$M = 0.38(0.333) + 0.5(0.333)^2 = 0.1815$$

$$\xi = \frac{\pi \mathrm{D}^2}{4ab} = \frac{\pi(1.5\mathrm{m})^2}{4(0.445\mathrm{m})(0.445\mathrm{m})} = 8.92$$

Secondly, the intermediate parameters are determined:

Inlet velocity:

$$\mathrm{v} = \frac{F}{ab} = \frac{2\frac{\mathrm{m}^3}{\mathrm{s}}}{(0.445\mathrm{m})(0.445\mathrm{m})} = 10.1\frac{\mathrm{m}}{\mathrm{s}}$$

Table 7.3 Models to compute pressure drop in cyclone separators (Part 1).

Model	Equation	Parameters
Shepherd and Lapple (1939)		
	$\text{Eu} = 16\text{Z}$	
Stairmand (1949)		
	$\text{Eu} = 1 + 2\xi^2\left(\frac{1}{M} - 1\right) + 2\left(\frac{4\text{Z}}{\pi}\right)^2$	$M = \frac{\text{Dc}}{2(\text{D} - b)}$ $Q = \frac{2fA}{ab}, f = 0.005$ $\xi = \frac{-\sqrt{M} + \sqrt{M + 2Q}}{Q}$ Total area of the cyclone walls: $A = \frac{\pi}{4}(\text{D}^2 - \text{Dc}^2) +$ $\pi\text{Dh} + \pi\text{DcS} + \frac{\pi}{2}(\text{D} + \text{B})\sqrt{\text{hc}^2 + \left(\frac{\text{D} - \text{B}}{2}\right)^2}$
Alexander (1949)		
	$\text{Eu} =$ $4.62\frac{Z}{\text{D} \cdot \text{Dc}}((M - 1)Q + \xi M)$	$M = \left(\frac{\text{D}}{\text{Dc}}\right)^{2n}$ $Q = \left(\frac{1 - n}{n}\right)$ $\xi = 0.8\left(\frac{1}{n^2 Q}\left(\frac{4 - 2^{2n}}{3}\right) - Q\right) +$ $0.2\left((2^n - 1)Q + 1.5(2^{2n})\right)$ Thermal correction $n = 1 - \left(\frac{T\rvert_K}{283K}\right)^{0.3}\left(1 - 0.67(\text{D}\rvert_{meters})^{0.14}\right)$
First (1950)		
	$\text{Eu} = 24\text{Z}\sqrt[3]{\frac{\text{D}^2}{\text{h} \cdot \text{hc}}}$	
Barth (1956)		
	$\text{Eu} = \left(\frac{4Z\theta}{\pi}\right)^2(\xi_B + \xi_{VF})$	$\theta = \frac{\pi\text{Dc}(\text{D} - b)}{4\text{ZDc}^2\alpha + 2\pi(\text{D} - b)(\text{H} - \text{S})f}$ $\alpha = 1 - \frac{1.2b}{\text{D}}$ Pressure losses at the cyclone body: $\xi_B = \frac{\text{Dc}}{\text{D}}\left(\frac{\text{Dc}^2}{(\text{Dc} - 2\theta(\text{H} - \text{S})f)^2} - 1\right), f =$ 0.02 Pressure losses at the vortex finder: $\xi_{VF} = 4.4\sqrt[3]{\theta^2} + 1$
Casal and Martinez-Benet (1983)		
	$\text{Eu} = 11.3\text{Z}^2 + 2.33$	

(*continued on next page*)

Table 7.3 Models to compute pressure drop in cyclone separators (Part 1)—cont'd

Chen and Shi (2007)	
$\mathrm{Eu}=\left(1-\frac{2b}{\mathrm{D}+1.33b-\mathrm{Dc}}\right)^2$ $+1.11f\xi A_S v_t^{\,3}Q^{-1.5k}+4.5\frac{(1-3Q^2)}{\xi^2}$ $+v_t^{\,2}(MQ)^{-k}+\xi^{-2}(Q^2-M^2)^{-2}$	$M=0.38Q+0.5Q^2$ $Q=\frac{\mathrm{Dc}}{\mathrm{D}}$ $\xi=\frac{\pi\mathrm{D}^2}{4ab}$ Inlet velocity: $\mathrm{v}=\frac{F}{ab}$; F : volumetricflow Reynolds number: $\mathrm{Re}=\frac{\rho_G \mathrm{vD}}{\mu_G\xi Q}$ Swirl exponent: $k=1-\exp\left(-0.26\mathrm{Re}^{0.12}\left(1+\left\|\frac{\mathrm{S}-a}{b}\right\|\right)^{-0.5}\right)$ Dimensionless area of contact surface: $A_C=1-Q^2+4\left(\frac{\mathrm{S}\cdot Q}{\mathrm{D}}+\frac{\mathrm{h}}{\mathrm{D}}\right)+$ $\left(1+\frac{\mathrm{B}}{\mathrm{D}}\right)\sqrt{\frac{4\mathrm{h}}{\mathrm{D}}+\left(1-\frac{\mathrm{B}}{\mathrm{D}}\right)^2}$ Tangential velocity (m/s): $v_t=\frac{1.11Q^{0.16}\mathrm{Re}^{0.06}\xi^{-0.21}}{1+fA_C\sqrt{\xi Q}}$, $f=0.02$
(Demir, 2014)	
$\mathrm{Eu}=14.8587\mathrm{Z}-0.6984\frac{\mathrm{h}}{\mathrm{D}}-0.4841\frac{\mathrm{hc}}{\mathrm{D}}+0.5738\frac{\mathrm{S}}{\mathrm{D}}$	

From Demir, (2014).

Table 7.4 Dimension for the cyclone separator of case study 7.5.1.1.

Geometrical elements	D	Dc	S	H	h	*a*	*b*	B
size (m)	1.5	0.5	0.645	4.14	0.832	0.445	0.445	0.5

Dimensionless area of contact surface:

$$A_C=1-Q^2+4\left(\frac{\mathrm{S}\cdot Q}{\mathrm{D}}+\frac{\mathrm{h}}{\mathrm{D}}\right)+\left(1+\frac{\mathrm{B}}{\mathrm{D}}\right)\sqrt{\frac{4\mathrm{h}}{\mathrm{D}}+\left(1-\frac{\mathrm{B}}{\mathrm{D}}\right)^2}$$

$$A_C=1-0.333^2+4\left(\frac{(0.645\mathrm{m})(0.333)}{1.5\mathrm{m}}+\frac{0.832\mathrm{m}}{1.5\mathrm{m}}\right)$$

$$+\left(1+\frac{0.5\mathrm{m}}{1.5\mathrm{m}}\right)\sqrt{\frac{4(0.832\mathrm{m})}{1.5\mathrm{m}}+\left(1-\frac{0.5\mathrm{m}}{1.5\mathrm{m}}\right)^2}$$

$$A_C=1-0.111+4(0.6978)+(1.333)\sqrt{2.6635}=5.856$$

Reynolds number:

$$\mathrm{Re} = \frac{\rho_G \mathrm{vD}}{\mu_G \xi Q} = \frac{\left(0.9718\,\frac{\mathrm{kg}}{\mathrm{m}^3}\right)\left(10.1\,\frac{\mathrm{m}}{\mathrm{s}}\right)(1.5\mathrm{m})}{\left(2.139\times10^{-5}\,\frac{\mathrm{kg}}{\mathrm{ms}}\right)(8.92)(0.333)} = 231723$$

Swirl exponent:

$$\begin{aligned} k &= 1 - \exp\left(-0.26\mathrm{Re}^{0.12}\left(1 + \left|\frac{\mathrm{S}-a}{b}\right|\right)^{-0.5}\right) \\ &= 1 - \exp\left(-0.26(231723)^{0.12}\left(1 + \left|\frac{0.645\mathrm{m} - 0.445\mathrm{m}}{0.445\mathrm{m}}\right|\right)^{-0.5}\right) \\ k &= 1 - \exp\left((-1.145)(0.8306)\right) = 0.6137 \end{aligned}$$

Tangential velocity:

$$\begin{aligned} v_t &= \frac{1.11Q^{0.16}\mathrm{Re}^{0.06}\xi^{-0.21}}{1+0.02A_C\sqrt{\xi Q}} = \frac{1.11(0.333)^{0.16}(231723)^{0.06}(8.92)^{-0.21}}{1+0.02(5.856)\sqrt{(8.92)(0.333)}} \\ &= \frac{1.2338}{1.2019} = 1.0265\,\frac{\mathrm{m}}{\mathrm{s}} \end{aligned}$$

Finally, the ΔP from the Eu number calculation:
Euler number:

$$\begin{aligned} \mathrm{Eu} &= \left(1 - \frac{2b}{\mathrm{D}+1.33b-\mathrm{Dc}}\right)^2 + 1.11f\xi A_S {V_T}^3 Q^{-1.5k} + 4.5\frac{\left(1-3Q^2\right)}{\xi^2} \\ &\quad + {V_T}^2(MQ)^{-k} + \xi^{-2}\left(Q^2 - M^2\right)^{-2} \\ \mathrm{Eu} &= \left(1 - \frac{2(0.445\mathrm{m})}{1.5\mathrm{m}+1.33(0.445\mathrm{m})-0.5\mathrm{m}}\right)^2 \\ &\quad + 1.11(0.02)(8.92)(5.856)\left(1.0265\,\frac{\mathrm{m}}{\mathrm{s}}\right)^3(0.333)^{-1.5(0.6137)} \\ &\quad + 4.5\frac{\left(1-3(0.333)^2\right)}{(8.92)^2} + \left(1.0265\,\frac{\mathrm{m}}{\mathrm{s}}\right)^2\left((0.1815)(0.333)\right)^{-0.6137} \\ &\quad + (8.92)^{-2}\left((0.333)^2 - (0.1815)^2\right)^{-2} \end{aligned}$$

$$\begin{aligned} \mathrm{Eu} &= (0.4409)^2 + 1.11(3.1094) + 4.5(0.008387) + \left(1.0265\,\frac{\mathrm{m}}{\mathrm{s}}\right)^2(5.5963) \\ &\quad + (8.92)^{-2}(164.59) \\ \mathrm{Eu} &= 0.1944 + 3.4515 + 0.03774 + 5.8968 + 2.068 = 11.648 \end{aligned}$$

Pressure drop in the Tengbergen C cyclone separator:

$$\begin{aligned} \Delta\mathrm{P} &= \frac{1}{2}\rho_G \mathrm{v}^2 \mathrm{Eu} = \frac{1}{2}\left(0.9718\,\frac{\mathrm{kg}}{\mathrm{m}^3}\right)\left(10.1\,\frac{\mathrm{m}}{\mathrm{s}}\right)^2(11.648) \\ &= 577.37\mathrm{Pa}\left|\frac{1\,\mathrm{in.ofWater}}{249.084\mathrm{Pa}}\right| = 2.378\,\mathrm{in.ofwater} \end{aligned}$$

Table 7.5 Comparison of the pressure drop models for the case study 7.5.1.1.

Models sorted by Demir's study	Values of pressure drop for this case study (in. of water)
Demir (2014)	2.1167
Alexander (1949)	2.9406
First (1950)	3.5598
Barth (1956)	2.9346
Shepherd and Lapple (1939)	2.5382
Casal & Martinez-Benet (1983)	1.8927
Chen and Shi (2007)	2.3253
Stairmand (1949) (not evaluated)	2.0593

The pressure drop value falls between 0.5 to 8 in. of water as Leith and Jones (1997) have indicated for an acceptable performance. However, a diversity of models generates suspicion about the best prediction. Demir (2014) compared all the models with experimental data and determined the models with low observed error. For this case study, all the models computed the pressure drop similarly to Chen and Shi (2007) model in a Tengbergen C cyclone separator. Table 7.5 presents the pressure drop values for the models from Table 7.3, sorted based on the lowest observed error from Demir's study, which did not include the Stairmand (1949) model. By following Demir's comparison, it is not possible to guarantee a more exact calculation as there is no correspondence between the similar values of pressure drop and the sorting list. This observation suggests that the best model for pressure drop would only be determined in particular cases with validation in experimental data.

In addition, the set of geometries makes it possible to compare the pressure drop per each cyclone geometry by using each pressure drop model. For this step, a Visual Basic code in Microsoft Excel was used to compute the pressure drop. Although Table 7.2 indicates the same diameter size, the geometry parameters have to resize to guarantee the same inlet mass flow rate based on the inlet area, $a{\cdot}b$, as Guo et al. (2024) suggested. Herein, an area conversion parameter for a j-cyclone, ϕ_j, is determined from the maximum inlet proportion area (Stairmand high flow cyclone, j=13) as: $\phi_j=(a_jb_j/(a_{13}b_{13}))^{\frac{1}{2}}$, now the geometrical rations from Table 7.2 are resized by multiplying them with ϕ_j. For this case study, all the cyclone geometries followed the proportion of the Tengbergeb C cyclone separator for a diameter of 1.5 m, whereby the area conversion parameter changed in proportion to the diameter for the desired cyclone (j=7 for this case study), such as: $\phi_j=[a_jb_j/(a_{13}b_{13})]^{\frac{1}{2}}(1.5\text{m}/\phi_7)$, for instance, some selected sizes are in Table 7.6.

Fig. 7.9 contains the extended results for analyzing each cyclone separator with each pressure drop model. Based on the average pressure drop values, cyclone separators #14, #15, #7, #8, and #9 reached the highest pressure drop values, which relates to a high grade of efficiency, in contrast to cyclone separators #11, #12, and #13. The models do not present a pattern during calculation for the cyclone separators, for

Table 7.6 Resized values of selected cyclone separators for this case study.

#	Cyclone separator	Dc	De	S	H	h	a	b	B	Area
1	Muschelknautz E	3.022	0.755	1.382	4.150	0.77	0.77	0.258	1.013	0.198
5	Storch 2	1	0.482	1.07	4.892	2.07	0.838	0.236	0.375	0.198
7	Tengbergen C	1.5	0.5	0.645	4.139	0.832	0.445	0.445	0.5	0.198
13	Stairmand high flow	0.845	0.627	0.734	3.36	1.26	0.627	0.316	0.316	0.198
15	Vibco	1.274	0.5	0.552	3.21	1.015	0.5	0.4	0.3	0.198
16	Lapple GP	1.26	0.627	0.787	5.031	2.518	0.627	0.316	0.316	0.198

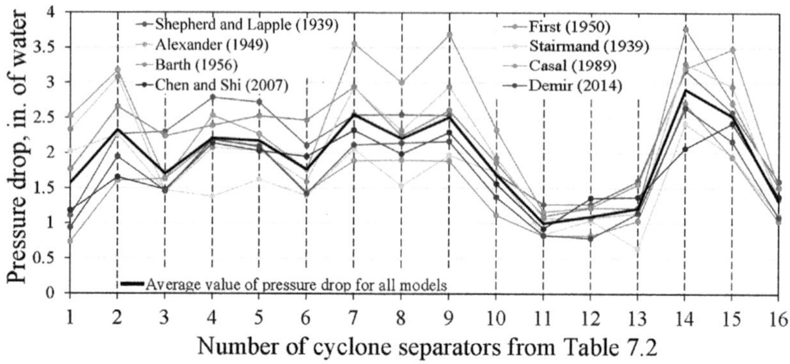

Figure 7.9 Pressure drop results for all the cyclone geometries.

instance, First's model reaches the high value of pressure drop in cyclone separator #2, but not the same for cyclone separator #6. This point demonstrates the significative effect of the geometrical sizes on the pressure drop calculation.

7.5.2 Collection efficiency

The initial studies of cyclone separators focused on determining the level at which the cyclones collect the dust at the bottom exit, and the losses correspond to the dust leaving through the vortex finder duct. Thus, the performance is quantified by a level of efficiency, in this case, the collection efficiency, η. The mass, density, and gravity force affect the heaviest dust, and the lightest dust complicates the level of collection efficiency. These elements consider the particle equilibrium principles: (a) gravitational, centrifugal, and drag forces, (b) gas density-particle density, and (c) radial velocity-angular velocity (Cortés & Gil, 2007). However, the collection efficiency concept applies per each dust size, which complicates the collecting level, whereby an approach adjusts the collection efficiency by the concepts of cut-size diameter or d_{50} and critical particle diameter or d_{100}. The first concept, d_{50}, describes the separated dust size at 50% of probabilities of remotion, and the last concept, d_{100}, applies to a particle with a collection efficiency of 100%. The cyclone separator theory has presented two principles to predict the collection efficiency by modeling (Hoffmann & Stein, 2008), that Table 7.7 presents.

The statistical treatment of the particle size yields a distribution of collection, from which Barth (1956) generated a universal curve of grade efficiency based on the d_{50}. The collection efficiency for an i-size might adjust to the function proposed by Barth (1956):

$$\eta_i = \frac{1}{1 + \left(\dfrac{d_{50}}{d_i}\right)^s} \tag{7.1}$$

Table 7.7 Models for compute the cut-size diameter.

Name and description	Model
Equilibrium-orbit model: a force balance is developed in a particle that rotates in a fictitious wall from the vortex finder (Barth's model)	$d_{50} = \sqrt{\frac{9\mu_G D_C v_r}{\rho_S v_{tx}^2}}$ Iozia and Leith (1990) have modified the model to: $d_{50} = \sqrt{\frac{9\mu_G F}{\pi \rho_S z_L v_{tx}^2}}$ Maximum tangential velocity: $v_{tx} = 6.1\mathrm{v}\left(\frac{ab}{\mathrm{D}^2}\right)^{0.61}\left(\frac{\mathrm{D}}{\mathrm{Dc}}\right)^{0.74}\left(\frac{\mathrm{D}}{\mathrm{H}}\right)^{0.33}$ Vortex core length: $d^*>B$: $z_L = \mathrm{H} - \mathrm{S}$ $d^*<B$: $z_L = \mathrm{H} - \mathrm{S} - \frac{(\mathrm{H}-\mathrm{h})(\mathrm{d}*-\mathrm{B})}{\mathrm{D}-\mathrm{B}}$ Core diameter: $\mathrm{d}* = 0.47\mathrm{D}\left(\frac{ab}{\mathrm{D}^2}\right)^{-1.26}\left(\frac{\mathrm{Dc}}{\mathrm{D}}\right)^{1.4}$
Time-of-flight model: a force balance applies in the particle motion from the injection to reach the cyclone wall or escape from the bottom (Lapple's model).	$d_{50} = \sqrt{\frac{9\mu_G b}{2\pi \rho_S \mathrm{v} \mathrm{N_T}}}$ Couper et al. (2012) presented a modified version of this model: $d_{50} = \sqrt{\frac{9\mu_G \mathrm{D}}{4\pi \mathrm{v} \mathrm{N_T}(\rho_S - \rho_G)}}$ Number of turns: Zenz (1982) $\mathrm{N_T} = \mathrm{v}(0.1079 - 7.7 \times 10^{-4}\mathrm{v}$ $+1.924 \times 10^{-6}\mathrm{v}^2)$ Hoffmann and Stein (2008) $\mathrm{N_T} = 6.1\left(1 - e^{-0.066\mathrm{v}}\right)$

where s is the curve slope. In addition, Zenz (1964) and Tawari and Zenz (1984) have presented a percent of remotion curve based on the ratio d_i/d_{50}, and herein Zenz's curve is adjusted to:

$$\eta_{R_i} = 9.7546 \ln(\zeta_i) + 121.288 \ln\left(\frac{\zeta_i}{\zeta_i + 1.0281} + 1\right), \zeta_i = \frac{d_i}{d_{50}}, R^2 = 0.9889 \quad (7.2)$$

where η_{Ri} is the percent of remotion of a i-size diameter.

The collection efficiency depends on the flow behavior inside the cyclone separator, which depends on the cyclone separator design. Table 7.8 contains the models proposed by Iozia and Leith (1990).

Table 7.8 Selected models to predict the collection efficiency.

Description	Model
Iozia and Leith model: estimates the η_i from expression (7.1) by including the slope from dimensionless geometric parameters. The d_{50} comes from the Equilibrium-orbit model (Table 7.7)	$\ln s = 0.62 - 0.87\ln(d_{50}\vert_{cm}) + 5.21\ln\left(\frac{ab}{\mathrm{D}^2}\right) + 1.05\left(\ln\left(\frac{ab}{\mathrm{D}^2}\right)\right)^2$
Leith and Licht model: computes η_i by its model, and the d_{50} comes from the Time-of-flight model (Table 7.7). This model includes the temperature effect, the geometric aspects, and the flowthrough, for instance: • Natural length, L* (distance below the vortex finder where the flow turns). • Volumetric flow rate, *F*. • Thermal correction parameter, *n* (see Alexander model from Table 7.3).	$\eta_i = 1 - \exp\left[-2\left(\frac{GF}{\mathrm{D}^3}(n+1)\tau_i\right)^{\frac{1}{2n+2}}\right]$ Relaxation time for the *i*-particle according to Stokes law: $\tau_i = \frac{\rho_S d_i^2}{18\mu_G}$ Cyclone configuration factor: $G = \left[2\pi\left(S - \frac{a}{2}\right)(\mathrm{D}^2 - \mathrm{Dc}^2) + 4\mathrm{V_V}\right]\frac{\mathrm{D}}{a^2b^2}$ Volume configuration: Lower zone: L*>(H−S) $\mathrm{V_V} =$ $\frac{\pi}{4}\left(\mathrm{Dc}^2(\mathrm{h}-\mathrm{S}) + \mathrm{D}^2\left[\left(\frac{\mathrm{H}-\mathrm{h}}{3}\right)\left(1 + \frac{\mathrm{B}}{\mathrm{D}} + \frac{\mathrm{B}^2}{\mathrm{D}^2}\right) - (\mathrm{H}-\mathrm{S})\right]\right)$ Upper zone: L*<(H−S) $\mathrm{V_V} =$ $\frac{\pi}{4}\left(\mathrm{Dc}^2(\mathrm{h}-\mathrm{S}-\mathrm{L}*) + \mathrm{D}^2\left[\left(\frac{\mathrm{L}*+\mathrm{S}-\mathrm{h}}{3}\right)\left(1 + \frac{\mathrm{dc}}{\mathrm{D}} + \frac{\mathrm{dc}^2}{\mathrm{D}^2}\right)\right]\right)$ Diameter of a central core in the flow turns: $\mathrm{dc} = \mathrm{D} - (\mathrm{D}-\mathrm{B})\left(\frac{\mathrm{S}+\mathrm{L}*-\mathrm{h}}{\mathrm{H}-\mathrm{h}}\right)$ Natural length: $\mathrm{L}* = 2.3\mathrm{Dc}\left(\frac{\mathrm{D}^2}{ab}\right)^{\frac{1}{3}}$

7.5.2.1 Case study of geometric design of cyclone separators

Reconsider the Tengbergen C cyclone separator from the case study 7.5.1.1 with 1.5 m in diameter to process an air stream (2 m^3s^{-1}) at 90°C and 1 atm (for the air: ρ_G=0.9718 kgm^{-3} and μ=2.135 × 10^{-5} $kg\cdot m^{-1}s^{-1}$ from Çengel and Cimbala [2017]). The cyclone separator receives a particle loading with density of 500 kgm^{-3}. For this case study, consider the size particles (mm): 0.025, 0.0365, 0.054, 0.082, and 0.115 to elaborate the efficiency curve and the percent of remotion curve by applying the Leith and Licht model, and determine the cyclone diameter to reach the lowest collection efficiency of 0.9.

Solution:

This problem uses the Leith and Licht model from Table 7.8, the expressions (7.1), (7.2), and some from Table 7.7. The cyclone separator dimensions correspond to Table 7.4 from case study 7.5.1.1.

The initial step computes the geometrical parameters from Table 7.8:

Natural length:

$$L* = 2.3Dc\left(\frac{D^2}{ab}\right)^{\frac{1}{3}} = 2.3(0.5m)\left(\frac{(1.5m)^2}{(0.445m)(0.445m)}\right)^{\frac{1}{3}} = 2.585m$$

Diameter of a central core in the flow turns:

$$\begin{aligned} dc &= D - (D - B)\left(\frac{S + L* - h}{H - h}\right) \\ &= 1.5m - (1.5m - 0.5m)\left(\frac{0.645m + 2.585m - 0.832m}{4.14m - 0.832m}\right) \\ dc &= 1.5m - (1m)(0.725) = 0.775m \end{aligned}$$

Conditional element: L*[>,<](H−S), 2.585m<3.495m, so:

Volume configuration: upper zone:

$$V_V = \frac{\pi}{4}\left(Dc^2(h - S - L*) + D^2\left[\left(\frac{L* + S - h}{3}\right)\left(1 + \frac{dc}{D} + \frac{dc^2}{D^2}\right)\right]\right)$$

$$V_V = \frac{\pi}{4}\left(\begin{array}{l} (0.5m)^2(0.832m - 0.645m - 2.585m) \\ +(1.5m)^2\left[\left(\frac{2.585m + 0.645m - 0.832m}{3}\right)\right. \\ \left.\left(1 + \frac{0.775m}{1.5m} + \left(\frac{0.775m}{1.5m}\right)^2\right)\right] \end{array}\right)$$

$$V_V = \frac{\pi}{4}\left(-0.6m^3 + (1.5m)^2[(0.8m)(1.7836)]\right) = 2.05m^3$$

Cyclone configuration factor:

$$G = \left[2\pi\left(S - \frac{a}{2}\right)\left(D^2 - Dc^2\right) + 4V_V\right]\frac{D}{a^2b^2}$$
$$G = \left[2\pi\left(0.645\text{m} - \frac{0.445\text{m}}{2}\right)\left((1.5\text{m})^2 - (0.5\text{m})^2\right) + 4\left(2.05\text{m}^3\right)\right] \times \frac{1.5\text{m}}{(0.445\text{m})^2(0.445\text{m})^2}$$
$$G = \left[(2.655\text{m})\left(2\text{m}^2\right) + 4\left(2.05\text{m}^3\right)\right]\left(38.2517\text{m}^{-3}\right) = 516.785$$

Secondly, the flow-related parameters are determined:
Inlet velocity:

$$v = \frac{F}{ab} = \frac{2\frac{\text{m}^3}{\text{s}}}{(0.445\text{m})(0.445\text{m})} = 10.1\frac{\text{m}}{\text{s}}$$

Number of turns (from Table 7.7):

$$N_T = v\left(0.1079 - 7.7 \times 10^{-4}v + 1.924 \times 10^{-6}v^2\right)$$
$$N_T = \left(10.1\frac{\text{m}}{\text{s}}\right)\left(0.1079 - 7.7 \times 10^{-4}\left(10.1\frac{\text{m}}{\text{s}}\right) + 1.924 \times 10^{-6}\left(10.1\frac{\text{m}}{\text{s}}\right)^2\right)$$
$$N_T = \left(10.1\frac{\text{m}}{\text{s}}\right)(0.1) = 1.01$$

Thermal correction parameter (Alexander model from Table 7.3):

$$n = 1 - \left(\frac{T|_K}{283K}\right)^{0.3}\left(1 - 0.67\left(D|_{meters}\right)^{0.14}\right)$$
$$= 1 - \left(\frac{(90 + 273)K}{283K}\right)^{0.3}\left(1 - 0.67(1.5\text{m})^{0.14}\right)$$
$$n = 1 - (1.2827)^{0.3}(0.2909) = 0.6866$$

Calculation of d_{50} (from Table 7.7):

$$d_{50} = \sqrt{\frac{9\mu_G D}{4\pi v N_T(\rho_S - \rho_G)}} = \sqrt{\frac{9\left(2.139 \times 10^{-5}\frac{\text{kg}}{\text{ms}}\right)(1.5\text{m})}{4\pi\left(10.1\frac{\text{m}}{\text{s}}\right)(1.01)(500 - 0.9718)\frac{\text{kg}}{\text{m}^3}}}$$
$$= 6.7186 \times 10^{-5}\text{m}$$

Finally, the parameters per particle size, where the size of 0.025mm (i=1) illustrates the calculation:

Relaxation time:

$$\tau_1 = \frac{\rho_S d_1^2}{18\mu_G} = \frac{\left(500\frac{\text{kg}}{\text{m}^3}\right)\left(2.5 \times 10^{-5}\text{m}\right)^2}{18\left(2.139 \times 10^{-5}\frac{\text{kg}}{\text{ms}}\right)} = 8.116 \times 10^{-4}\text{s}$$

Table 7.9 Results for the particle sizes by using the Leith and Licht model for case study 7.5.2.1 with D = 1.5 m.

Particle size, d_i (mm)	Relaxation time, τ_i (ms)	Collection efficiency, η_i	d_i/d_{50}	Percent of remotion, η_{R_i} (%)
0.025	0.8116	0.787	0.3721	0.1898
0.0365	1.85	0.861	0.5624	0.3095
0.054	4.05	0.917	0.832	0.4289
0.082	9.339	0.9587	1.2635	0.554
0.115	18.368	0.9796	1.772	0.6492

Efficiency:

$$\eta_1 = 1 - \exp\left[-2\left(\frac{GF}{D^3}(n+1)\tau_i\right)^{\frac{1}{2n+2}}\right]$$

$$\eta_1 = 1 - \exp\left[-2\left(\frac{(516.785)\left(2\frac{m^3}{s}\right)}{(1.5m)^3}(0.6866+1)\left(8.116\times10^{-4}s\right)\right)^{\frac{1}{2(0.6866)+2}}\right]$$

$$\eta_1 = 1 - \exp\left[-2(0.4192)^{0.2965}\right] = 1 - 0.2132 = 0.7868$$

Removed fraction (expression (7.2)):

$$\zeta_i = \frac{d_i}{d_{50}} = \frac{2.5\times10^{-5}\text{m}}{6.7186\times10^{-5}\text{m}} = 0.3721$$

$$\eta_{R_1} = 9.7546\ln(0.3721) + 121.288\ln\left(\frac{0.3721}{0.3721+1.0281}+1\right)$$

$$\eta_{R_1} = -9.6433 + 121.288\ln(1.2657) = 18.94\%$$

The reading from Zenz's graph is 17.8%.

Table 7.9 summarizes the remaining particle sizes, and Fig. 7.10 shows both graphs. The results correspond to the typical dust behavior in a cyclone separator.

The solution proposes the value of the cyclone separator diameter until constraint of η_i>0.9 is observed. The shortest diameter size reaches the lowest η_i in this case d_i=0.025mm. In principle, obtaining the value of D is complicated by a numerical method for any non-linear model from Table 7.7, whereby the reader might intend a trial-and-error search in spreadsheets. Herein, the Microsoft Excel package solves the problem by declaring the already described equations. Hence, the solution is D=1 m with d_{50}=0.023 mm, and Table 7.10 contains the solution parameters per particle size. These results remark how the reduction of size enhances the particle efficiency that corresponds to the observations by Azadi et al. (2010); nevertheless, the excessive size reduction might produce a risk of plugging and blockage.

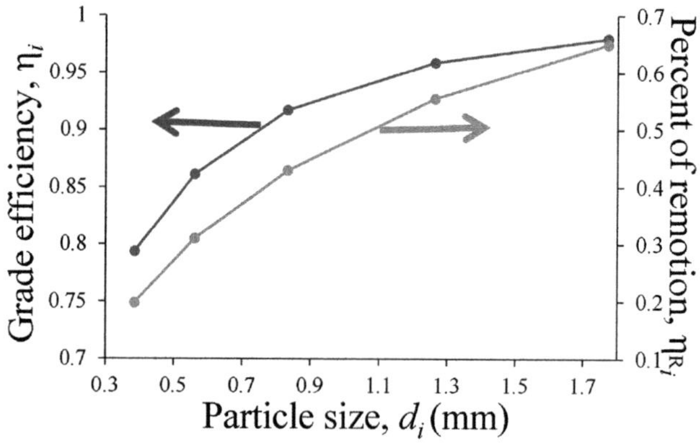

Figure 7.10 Grade efficiency and percent of remotion curves for the case study 7.5.2.1.

Table 7.10 Results for the particle sizes by using the Leith and Licht model for case study 7.5.2.1 and the solution of D = 1m.

Particle size, d_i (mm)	**Relaxation time, τ_i (ms)**	**Collection efficiency, η_i**	**d_i/d_{50}**	**Percent of remotion, ηR_i (%)**
0.025	0.868	0.9011	1.0876	0.51
0.0365	1.8503	0.9456	1.588	0.6191
0.054	4.05	0.975	2.3492	0.7226
0.082	9.339	0.9915	3.5674	0.82
0.115	18.368	0.9971	5.003	0.889

Altmeyer et al. (2004) have analyzed some models of performance. Similarly, herein, a complementary comparison applies to all the cyclone geometries under the Iozia and Leith model and the Leith and Licht model. Thus, the analysis considers the case studio such as baseline case (D=1.5 m, F=2 m^3s^{-1}, air at 90°C, ρ_S=500 kgm^{-3} and size particles (mm): 0.025, 0.0365, 0.054, 0.082 and 0.115). Fig. 7.11 contains the comparison for the d_{50} and the maxima of percent of remotion, ηR_i. The results are a remarkable difference between both models, and the Leith and Licht model did not evaluate the cyclone Storch 4 (#3) due to the large cylinder height. The results from the Leith and Licht model are similar across cyclone separators; in contrast, the Iozia and Leith model shows differences.

7.5.3 Particle tracking

The particle motion obeys the Newton's second law, where the mass acceleration equivalates to the sum of all the involved forces, such as: $m_S a_C = \Sigma\text{Force}$. By definition,

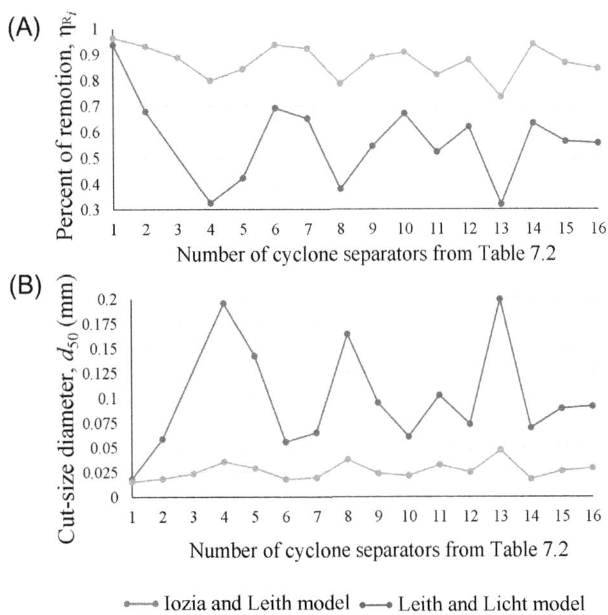

Figure 7.11 Comparison of Iozia and Leith model and Leith and Licht model for all the cyclone separators based on the case study 7.5.2.1: A) percent of remotion graph and B) cut-size diameter graph.

the acceleration corresponds to:

$$a_C = \frac{dv}{dt} = \frac{d}{dt}\left(\frac{dx}{dt}\right) = \frac{d^2x}{dt^2}$$

where x is the advance of position. Michaelides (2006), Yeoh and Tu (2010), and Michaelides et al. (2017) have indicated the involved forces in the particle tracking, for instance:

- Body force indicates the gravity force on the body.
- Fluid drag force indicates the effect of the particle opposition to the fluid flow (contrasting the kinetic energy) based on a relative velocity, $\mathbf{v}_R = |\mathbf{v} - \mathbf{v}_{S,i}|$, $\mathbf{v}_{S,i}$ is the particle velocity.
- Buoyancy force is the response of fluid to keep the particle in equilibrium by the fluid displacement.
- Basset force indicates the historicity of a particle changing of motion depending on the change rate.
- Saffman lift force corresponds to the rotation of a particle that yields a transversal component during flow (very low Re_G).

For solid particles in cyclone separators, the body, fluid drag, and buoyance forces have demonstrated the highest influence on the particle motion (Clift et al., 1978; Hoffmann & Stein, 2008; Lapple & Shepherd, 1940; Yeoh & Tu, 2010). Respectively, the representative model is:

$$\rho_S\left(\frac{\pi d_i^3}{6}\right)\frac{d\mathbf{v}_{S,i}}{dt} = \rho_S\left(\frac{\pi d_i^3}{6}\right)g - \rho_G\left(\frac{\pi d_i^3}{6}\right)g + C_{D,i}\left(\frac{\pi d_i^2}{4}\right)\frac{\rho_G\left(\mathbf{v}-\mathbf{v}_{S,i}\right)\left|\mathbf{v}-\mathbf{v}_{S,i}\right|}{2}$$

reduced to:

$$\frac{d\mathbf{v}_{S,i}}{dt} = \frac{g(\rho_S-\rho_G)}{\rho_S} + C_{D,i}\frac{3}{4}\frac{\rho_G\left(\mathbf{v}-\mathbf{v}_{S,i}\right)\left|\mathbf{v}-\mathbf{v}_{S,i}\right|}{d_i\rho_S} \quad (7.3)$$

by coupling in short terms:

$$\frac{d\mathbf{v}_{S,i}}{dt} = \frac{\mathbf{g}(\rho_S-\rho_G)}{\rho_S} + \frac{\left(\mathbf{v}-\mathbf{v}_{S,i}\right)}{\tau_i} \quad (7.4)$$

where τ_i is the relaxation time:

$$\tau_i = \frac{4}{3}\frac{\rho_S d_i^2}{\mu_G C_{D,i}\mathrm{Re}_{S,i}} \quad (7.5)$$

The drag coefficient, $C_{D,i}$, is approached to a Stokes flow such as: $24/\mathrm{Re}_{S,i}$, where $\mathrm{Re}_{S,i}=|\mathbf{v}-\mathbf{v}_{S,i}|\rho_G d_i(\mu_G)^{-1}$, whereby τ_i for a Stokes flow corresponds to expression in Table 7.8. Nevertheless, the drag coefficient, $C_{D,i}$, is obtained by diverse approaches. Clift et al. (1978) and Michaelides et al. (2017) have indicated some remarkably equations similar, for instance the Schiller and Naumann model:

$$C_{D,i} = \begin{cases} \dfrac{24}{\mathrm{Re}_{S,i}}, \mathrm{Re}_{S,i} < 0.5 \\ \dfrac{24}{\mathrm{Re}_{S,i}}\left(1+0.15\left(\mathrm{Re}_{S,i}\right)^{0.687}\right), 0.5 < \mathrm{Re}_{S,i} < 1000 \\ 0.44, \mathrm{Re}_{S,i} > 1000 \end{cases} \quad (7.6)$$

For a cyclone separator, the 3D motion of an i-sized particle is modeled by considering the centrifugal force, such as in Cartesian coordinates and gravity in z-direction (Azadi et al., 2010; Lapple & Shepherd, 1940; Wang et al., 2006):

$$\left.\begin{array}{l} \dfrac{dv_{Sx,i}}{dt} = \dfrac{\left(v_x - v_{Sx,i}\right)}{\tau_{x,i}} + \dfrac{\left(v_{Sx,i}\right)^2}{r_0} \\ \dfrac{dv_{Sy,i}}{dt} = \dfrac{\left(v_y - v_{Sy,i}\right)}{\tau_{y,i}} - \dfrac{v_{Sx,i}v_{Sy,i}}{r_0} \\ \dfrac{dv_{Sz,i}}{dt} = \dfrac{\left(v_z - v_{Sz,i}\right)}{\tau_{z,i}} + \dfrac{g(\rho_S-\rho_G)}{\rho_S} \\ r_0: \text{ cylone radius} \end{array}\right\} \quad (7.7)$$

The analysis of the particle tracking flow through a cyclone separator solves the expression (7.7) by considering the drag coefficient from expression (7.6).

7.5.3.1 *Case study of particle tracking*

Simulate during 10 s the 2D particle tracking flow through a cyclone separator of D=1.5 m. The particles are instant coffee with diameters of (mm):0.0965, 0.16, and 0.25, with a density of 470 kgm^{-3} (Onwulata, 2006). The air is v_x=0.1 ms^{-1} with density of 1.027 kgm^{-3} and viscosity of 2.14 × 10^{-5} kgm^{-1}s^{-1}. The Schiller and Nauman model is recommended for this case.

Solution:

The simulation is possible by solving the equation system from the expression (7.7). The description indicates only one velocity component of the fluid (v_x); and the particle suffers a descendent flow (z-direction for gravity effect) through advance with the fluid flow, whereby it corresponds to a 2D case. The expression (7.7) considers the centrifugal force with r_0=D/2=0.75 m, and reduces to:

$$\begin{cases} \dfrac{dv_{Sx,i}}{dt} = \dfrac{\left(0.1\frac{\text{m}}{\text{s}} - v_{Sx,i}\right)}{\tau_{x,i}} + \dfrac{\left(v_{Sx,i}\right)^2}{0.75\text{m}} \\ \dfrac{dv_{Sz,i}}{dt} = -\dfrac{v_{Sz,i}}{\tau_{z,i}} + \dfrac{\left(9.81\frac{\text{m}}{\text{s}^2}\right)\left(470\frac{\text{kg}}{\text{m}^3} - 1.027\frac{\text{kg}}{\text{m}^3}\right)}{470\frac{\text{kg}}{\text{m}^3}} = -\dfrac{v_{Sz,i}}{\tau_{z,i}} + 9.7886\frac{\text{m}}{\text{s}^2} \end{cases}$$
$$i = \{0.0965\text{mm}, 0.16\text{mm}, 0.25\text{mm}\} \tag{7.8a}$$

The particle tracking determines the position during the motion, which requires the calculation of the positions from the velocity concept: $v_x=dx/dt$. Thus, the model includes the expressions:

$$\begin{cases} \dfrac{dx}{dt} = v_{Sx,i} \\ \dfrac{dz}{dt} = v_{Sz,i} \end{cases} \tag{7.8b}$$

The relaxation times and Reynolds numbers are respective of each flow direction, from the expression (7.5):

$$\begin{cases} \tau_{x,i} = \dfrac{4}{3}\dfrac{\left(470\frac{\text{kg}}{\text{m}^3}\right)d_i^2}{\left(2.14x10^{-5}\frac{\text{kg}}{\text{ms}}\right)C_{D,x,i}\text{Re}_{Sx,i}} = \dfrac{29283.49\frac{s}{\text{m}^2}d_i^2}{C_{D,x,i}\text{Re}_{Sx,i}} \\ \tau_{z,i} = \dfrac{29283.49\frac{s}{\text{m}^2}d_i^2}{C_{D,z,i}\text{Re}_{Sz,i}} \end{cases}$$
$$i = \{0.0965\text{mm}, 0.16\text{mm}, 0.25\text{mm}\} \tag{7.8c}$$

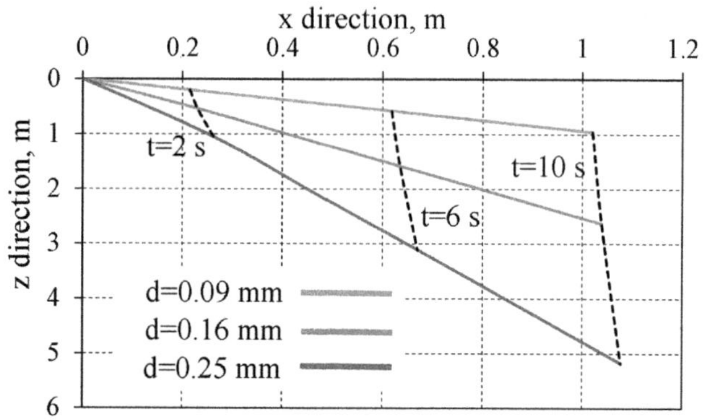

Figure 7.12 Particle tracking for case study 7.5.3.1.

with

$$\begin{cases} Re_{Sx,i} = \dfrac{\left|0.1\frac{m}{s} - v_{Sx,i}\right|\left(1.027\frac{kg}{m^3}\right)d_i}{2.14x10^{-5}\frac{kg}{ms}} = \left(47990.65\frac{s}{m^2}\right)\left|0.1\frac{m}{s} - v_{Sx,i}\right|d_i \\ Re_{Sz,i} = \dfrac{\left|-v_{Sz,i}\right|\left(1.027\frac{kg}{m^3}\right)d_i}{2.14x10^{-5}\frac{kg}{ms}} = \left(47990.65\frac{s}{m^2}\right)\left|v_{Sz,i}\right|d_i \end{cases}$$
$$i = \{0.0965\text{mm}, 0.16\text{mm}, 0.25\text{mm}\} \tag{7.8d}$$

and the drag coefficient is coming from the Schiller and Naumann model (expression 7.6).

The numerical solution is suitable by the Runge-Kutta method; however, the number of ordinary differential equations is four times the number of particles, herein there are 12 ordinary differential equations. To save time in programming, a swift strategy is to create an array to allocate the different values of the parameters from the Runge-Kutta method, as Table 7.11 and Table 7.12 indicates.

The solution was possible with a short time step (0.001 sec), and initial values: (a) starting from the origin: x=0 m, and z=0 m, and (b) without motion: $v_{x,i}$=0 ms^{-1}, and $v_{z,i}$=0.0001 ms^{-1}, this short value avoids $Re_{Sz,i}$=0 that yields an undefined error for computing $C_{Dz,i}$. Fig. 7.12 contains the results of this particle tracking simulation, and denotes the position of each particle at the same time. This is similar to the description from Lapple and Shepherd (1940), where the heaviest particle reaches a far position in z-coordinate.

7.6 CFD simulation for cyclone separators

The literature demonstrates suitable studies on cyclone separators by using computational fluid dynamics (CFD). CFD avoids developing experimental tests to understand

Table 7.11 Code in C++ of the generalized Runge-Kutta for case study 7.5.3.1.

Description	**Mathematical description**
Runge-Kutta method **y** is the dependent variables in vector form **f** is the functions of ordinary differential equations in vector form $\mathbf{k}_1$, $\mathbf{k}_2$, $\mathbf{k}_3$ and $\mathbf{k}_4$ are parameters in vector form Δt is the time step	$\mathbf{y}_{j+1} = \mathbf{y}_j + (\mathbf{k}_1 + 2 \cdot \mathbf{k}_2 + 2 \cdot \mathbf{k}_3 + \mathbf{k}_4) \cdot \frac{\Delta t}{6}$ $\mathbf{k}_1 = \mathbf{f}(t, \mathbf{y}_j)$ $\mathbf{k}_2 = \mathbf{f}\left(t + \frac{\Delta t}{2}, \mathbf{y}_j + \frac{\Delta t}{2} \cdot \mathbf{k}_1\right)$ $\mathbf{k}_3 = \mathbf{f}\left(t + \frac{\Delta t}{2}, \mathbf{y}_j + \frac{\Delta t}{2} \cdot \mathbf{k}_2\right)$ $\mathbf{k}_4 = \mathbf{f}(t + \Delta t, \mathbf{y}_j + \Delta t \cdot \mathbf{k}_3)$

Programming of Runge-Kutta method for *i* sized particles.
Code in C++.
	Generalized Runge-Kutta method (array form)
	for (j=1; j≤itera; j++)
	{
	for (i=1; i≤particles; i++)
	{
	for (k=1; k≤4; k++)//Loop 1 for k1's
	yy[1][k]=y1[k];
	for (k=1; k≤4; k++)//Loop 2 for k1's
	kk[1][k]=FF(1,k,t);
	for (i=1; k≤4; k++)//Loop 1 for k2's
	yy[2][k]=y1[k]+dt*kk[1][k]/2;
	for (k=1; k≤4; k++)//Loop 2 for k2's
	kk[2][k]=FF(2,k,t+dt/2);
	for (k=1; k≤4; k++)//Loop 1 for k3's
	yy[3][k]=y1[k]+dt*kk[2][k]/2;
	for (k=1; k≤4; k++)//Loop 2 for k3's
	kk[3][k]=FF(3,k,t+dt/2);
	for (k=1; k≤4; k++)//Loop 1 for k4's
	yy[4][k]=y1[k]+dt*kk[3][k];
	for (k=1; k≤4; k++)//Loop 2 for k4's
	kk[4][k]=FF(4,k,t+dt);
	for (k=1; k≤4; k++)//Loop for yj+1
	y2[k]=y1[k]+dt*(kk[1][k]+2*kk[2][k]
	+2*kk[3][k]+kk[4][k])/6;
	//****WRITTING NEW VALUES****
	for (k=1; k≤4; k++)//Loop for updating y1
	y1[k]=y2[k];
	t=t+dt;
	}
	}

the internal flow and the particle tracking, which might be expensive to probe the influence of geometrical elements. Nowadays, researchers make possible acceptable analysis of cyclone separators by applying CFD. This section describes this technique with a case study.

Table 7.12 Code in C++ of the numerical solution for case study 7.5.3.1.

Description	**Code for differential equations**

Arrays for dependent variables:
x=yy[j][1]
z=yy[j][2]
vx=yy[j][3]
vz=yy[j][4]
Arrays for relaxation time:
$\tau_{x,i}$=TD[1][i]
$\tau_{z,i}$=TD[2][i]
	double ClassF::FF(double j, double k, double t)
	{
	double FF_;
	switch(k){
	case 1:{//Expression 7.8b for x
	FF_= yy[j][3];
	break;}
	case 2:{//Expression 7.8b for z
	FF_= yy[j][4];
	break;}
	case 3:{//Expression 7.8a for vx
	FF_=(0.1-yy[j][3])/TD[1][i]-yy[j][3]*yy[j][3]/0.75;
	break;}
	case 4:{//Expression 7.8a for vz
	FF_=-yy[j][4]/TD[2][i]+9.7886;
	break;}
	return FF_;
	}

Arrays for *d*, Re and C_D:
d_i=D_[i]
$Re_{Sx,i}$=Re[1][i]
$Re_{Sz,i}$=Re[2][i]
$C_{D,x,i}$=CD[1][i]
$C_{D,z,i}$=CD[2][i]
	Code for additional variables
	//Expression 7.8c for x
	TD[1][i]=29283.49*D_[i]*D_[i]/(CD[1][i]*Re[1][i]);
	//Expression 7.8c for z
	TD[2][i]=29283.49*D_[i]*D_[i]/(CD[2][i]*Re[2][i]);
	//Expression 7.8d for x
	Re[1][i]=47990.65*abs(0.1-yy[j][3])*D_[i]
	//Expression 7.8d for z
	Re[2][i]=47990.65*abs(0.1-yy[j][4])*D_[i]
	//Expression 7.6 for x
	CD[1][i]=24*(1+0.15*pow(Re[1][i],0.687))/Re[1][i]
	//Expression 7.6 for z
	CD[2][i]=24*(1+0.15*pow(Re[2][i],0.687))/Re[2][i]

7.6.1 Elements of the CFD technique

CFD is a suitable technique to analyze the motion of fluids, demands knowledge of turbulence and numerical methods, and applies to most engineering topics and some relevant issues attend biological systems. Through the years, CFD has optimized existing processes, corrected the performance of processes, and aimed to develop new food processes (Scott & Richardson, 1997). Particularly, food engineering has received grants from CFD applications, such as Anandharamakrishnan (2013) and Sun (2007, 2019) have discussed. But, the CFD tool has succinctly treated the turbulence

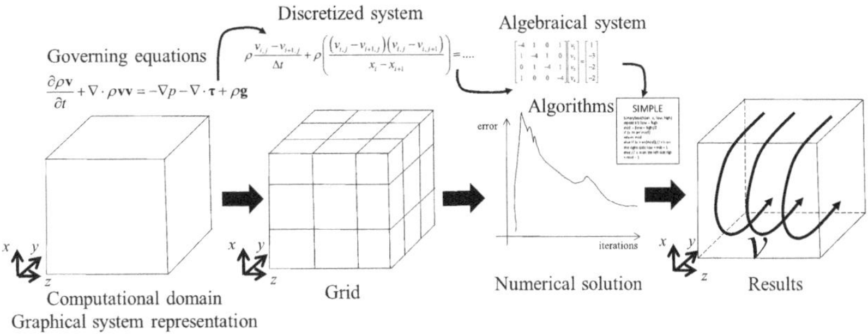

Figure 7.13 Graphical representation of the CFD procedure.
(From Díaz-Ovalle and Jafari (2023)).

phenomenon that is denoted by the literature, for instance, Blazek (2006), Chung (2010), Yeoh and Tu (2010), Jayanti (2018), Rodriguez (2019), etc.

The CFD technique solves the conservative equations present in fluid flow (Tu et al., 2013) which are partial differential equations, mainly represented by mass, momentum, and energy equations. Chung (2010) has provided solving procedures for these equations that demand a high mathematical and programming level, for instance, finite difference method (FDM), and finite volume method (FVM). These methods compute the variables $\mathbf{v}$, T, and p into spatial elements of the system; it means that the unknown variables will receive values at particular points of the space. This part is based on a meshing procedure used to convert the computational domain (drawn system) into volume control elements. Also, the CFD tool involves elements, such as boundary conditions, initial values, fluid properties, turbulence settings, radiative heat transfer, combustion, chemical reactions, magnetohydrodynamics, multiphase flow (granular species, bubbles, and drops), as well as non-Newtonian fluids phenomena (Díaz-Ovalle, 2024). Fig. 7.13 represents the conceptual ideal of the solving procedure of CFD.

The turbulence phenomenon rules the fluid behavior inside a cyclone separator and represents the unpredictable, irregular, and complex fluid motion under the effect of either surfaces or neighboring streams at a wide range of scales (Wilcox, 2006). Tennekes and Lumley (1972), Hinze (1987), Cebeci (2003, 2004), Wilcox (2006), Davidson (2015), and Rodriguez (2019) have explained the turbulence principles. Osborne Reynolds denoted the random behavior by time-averaging the velocity, $\langle\mathbf{v}\rangle$, and the instantaneous velocity, $\mathbf{v}$, is formed by $\langle\mathbf{v}\rangle$ and the instantaneous velocity fluctuation, $\mathbf{v}'$, such as: $\mathbf{v}=\langle\mathbf{v}\rangle+\mathbf{v}'$. Nevertheless, the Navier-Stokes equations formulate the nonlinear convective terms (product $\mathbf{vv}$) into $\mathbf{v}'\mathbf{v}'$ which is averaged to $\langle\mathbf{v}'\mathbf{v}'\rangle$ named Reynolds stress tensor, $\mathbf{R}$. Theoretical studies have determined $\mathbf{R}$ such as the Boussinesq hypothesis, $-\langle\mathbf{v}'\mathbf{v}'\rangle=\mathbf{R}=2\kappa$, where κ is turbulence kinetic energy. Hence, CFD solves the turbulence phenomena by adding additional mathematical elements to calculate κ (closure problem), for instance: (a) algebraic terms

Table 7.13 Governing equations for CFD.

Conservation of:	Equation
Mass (continuity)	$\frac{\partial \rho}{\partial t} + \nabla \cdot \rho \mathbf{v} = 0$
Momentum	$\frac{\partial \rho \mathbf{v}}{\partial t} + \nabla \cdot \rho \mathbf{v}\mathbf{v} = -\nabla p - \nabla \cdot \tau + \rho \mathbf{g} + \mathbf{S}_P$
Energy	$\frac{\partial \rho CpT}{\partial t} + \nabla \cdot \mathbf{v}\rho CpT = -\nabla \cdot \mathbf{v}p - \nabla \cdot \mathbf{q} + \nabla \cdot (\tau \cdot \mathbf{v})$
Turbulence kinetic energy	$\frac{\partial \rho \kappa}{\partial t} + \nabla \cdot \mathbf{v}\rho\kappa =$ $2\mu_t(\mathbf{S} : \mathbf{S}) - \rho\varepsilon + \nabla \cdot \left[\left(\mu + \frac{\mu_t}{\sigma_\kappa}\right)\nabla\kappa\right] + S_{p,\kappa}$
Reynolds stress tensor	$\frac{\partial \mathbf{R}}{\partial t} + \mathbf{v} \cdot \nabla \mathbf{R} =$ $-\mathbf{R} \cdot \nabla \mathbf{v} - \mathbf{R} \cdot (\nabla \mathbf{v})^{\mathrm{T}} - \varepsilon + \mathbf{\Pi} + \nabla \cdot \left(\frac{\mu}{\rho}\nabla \mathbf{R} + \mathbf{C}\right)$
Nomenclature	$\boldsymbol{\tau}$: shear stress tensor; Cp: heat capacity of the fluid; **q**: conductive heat flux vector; κ: turbulence kinetic energy; **S**: mean strain rate; $\boldsymbol{\varepsilon}$: dissipation rate tensor; **R**: Reynolds stress tensor; **C**: turbulence transport tensor; $\mathbf{\Pi}$: pressure strain tensor; $\mathbf{S}_P$: source term vector for the two-way coupling; $S_{p,\kappa}$: source term vector for turbulence

From Díaz-Ovalle, (2024).

instead of partial differential equations (zero equation models), (b) either one or two additional partial differential equations with additional parameters, such as the turbulence eddy dissipation, ε, rate dissipation of κ, ω, and eddy viscosity μ_t to relate them by the expression: $\mu_t \sim \rho \cdot \kappa^2/\varepsilon$, such as the κ-ε models, κ-ω model, κ-κ_L-ω model, and (c) either differential or constitutive expressions for the elements in the Reynolds stress transport model. Table 7.13 contains the governing equations for CFD.

Reynolds-stress model (RSM) is mathematically complex due to conserving the original formulation to treat the Reynolds-stress tensor (**R**=−<**v**'**v**'>). RSM does not follow the isotropic turbulence behavior and takes into account the streamline curvature; and, is appropriate for rapid changes in velocity and rotational flows. The mathematical structure is in Table 7.13, where the terms $\boldsymbol{\varepsilon}$, $\mathbf{\Pi}$, and **C** correspond to dissipation rate, pressure strain, and turbulence transport tensors. The $\boldsymbol{\varepsilon}$ and **C** tensors reduce to short mathematical expressions by applying anisotropy and invariant rotationality, respectively (Wilcox, 2006). The pressure strain tensor is treated in a particular way to represent an unmeasurable correlation that impacts the turbulence. Thus, some approaches have been developed to describe this term, such as:

- Linear pressure-strain model (Gibson & Launder, 1978).
- Quadratic pressure-strain model (Speziale et al., 1991).
- Low-Re stress-omega model (Wilcox, 2006).

Elghobashi (2006) has classified the particle motion based on the volume loading: (a) one-way coupling flow ($V_P<10^{-6}$, very dilute suspension), (b) two-way coupling flow ($10^{-6}<V_P<10^{-3}$, dilute suspension), (c) four-way coupling flow ($10^{-3}<V_P<1$, dense suspension), and d) granular flow ($V_P>1$): this flow characterizes the low presence of fluid motion. The two-way coupling flow characterizes the particle motion and follows the Lagrangian scheme of expression (7.3), by the expression (Yeoh & Tu, 2010):

$$\mathbf{S}_P = \mathrm{N_V} \sum_i -m_{S,i} \left(\frac{\mathbf{g}(\rho_S - \rho_G)}{\rho_S} + \frac{(\mathbf{v} - \mathbf{v}_{S,i})}{\tau_i} - \mathbf{g} \right) \tag{7.8}$$

where, N_V is the mean number of particles per unit volume.

The literature presents some computational packages to apply the CFD tool in cyclone separators, where ANSYS ® FLUENT is suitable to solve the discrete phase in cyclone separators. This package solves the continuity and momentum conservation equations in either segregated or coupled ways. Thus, the numerical solution of the algebraic equations, from the discretization of partial differential equations, follows particular algorithms, for instance semi-implicit methods for pressure-linked equations (SIMPLE) (Patankar & Spalding, 1972). Below, a case study demonstrates the use of CFD to predict the hydrodynamic behavior in a cyclone separation by including particle motion.

7.6.3 Case study

The Tengbergen C cyclone separator collects milk powder after the spray drying unit (Gamiño-Tovar et al., 2018). Table 7.14 concentrates on the information of this case study. The recommended CFD package is ANSYS ® FLUENT 2024 R1. In addition, the steady-state governing equations include the gravity effect, the Reynolds-Stress model (Linear pressure-strain approach), the two-way coupling particle motion, the drag coefficient based on the Morsi and Alexander model, and Rossin-Rammler particle size distribution, $x_i=\exp(-d_i/d_{ave})^n$. Díaz-Ovalle et al. (2023) have applied this case study to determine the influence of the loading concentration on the pressure drop, herein it is taken into account to understand the particle motion.

Solution:

The Design Modeler tool from ANSYS ® draws 3D computational domains from 2D profiles, such as the case of cyclone separators. The procedure corresponded to

Table 7.14 Information of the case study 7.6.3.

Parameter	**Value**	**Units**
Fluid data		
Cyclone separator diameter	1.2	m
Mass flow rate of air	1	kgs^{-1}
Dynamic viscosity of air	2.14×10^{-5}	$kgm^{-1}s^{-1}$
Particle data		
Milk powder density	500	kgm^{-3}
Particle inlet velocity	10	ms^{-1}
Particle mass flow rate	100	gs^{-1}
Number of particle sizes	10	-
Average particle diameter, d_{ave}	5	μm
Limit of particle size	$1 \leq d_i \leq 10$	μm
Parameter of the Rossin-Rammler distribution, n	3.5	-
Meshing data		
Number of tetrahedron elements	90,742	-
Average cell size	5.7	cm
Average skewness	0.226	-
Solution data		
Particle time step	0.5	s
Particle time steps per 20 fluid iterations	2	-
Total iterations	3000	-

a revolve instruction for a profile with the half-longitudinal shape of the cyclone separator. The 2D profile is sized 1.2 times the values from Table 7.2 (cyclone #7). An extrude instruction yielded the tangential inlet. Thus, the building of the cyclone separator Tengbergen C followed the similarity to the shape from Fig. 7.8.

The boundary conditions relate to the main element surfaces: inlet, outlet, and dust collector. The inlet boundary is a mass flow rate of air (1 kgs^{-1}), and the particles get into the cyclone at 100 gs^{-1} in a homogeneous surface distribution. The outlet boundary is a mass conservation exit to guarantee the same mass flow rate as the inlet boundary. The dust collector is a wall boundary to trap the milk powder. The remaining boundaries are no-slip walls.

In the meshing step, tetrahedron cells adapted to the cyclone separator with an inflation of five layers on the cyclone walls. Fig. 7.14 represents a cut view of the meshing, which indicates a homogeneous size inside the cyclone. The skewness parameter quantifies the mesh quality dimensionless and describes the quality of the cell shape in comparison to an optimal tetrahedron cell in a circumsphere. The values close to zero, indicate a suitable cell. In this case, the wall cells present flat shapes that demerit the mesh quality with high skewness values, but this mesh treatment is valid to emphasize the numerical approach for turbulence on the walls. In general, the mesh quality observed a suitable average value of 0.226. Table 7.14 contains the mesh information for this case study.

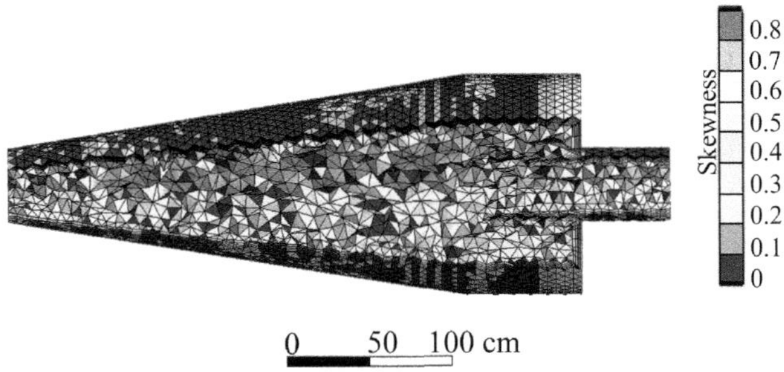

Figure 7.14 Skewness distribution into the cyclone separator for case study 7.6.3.

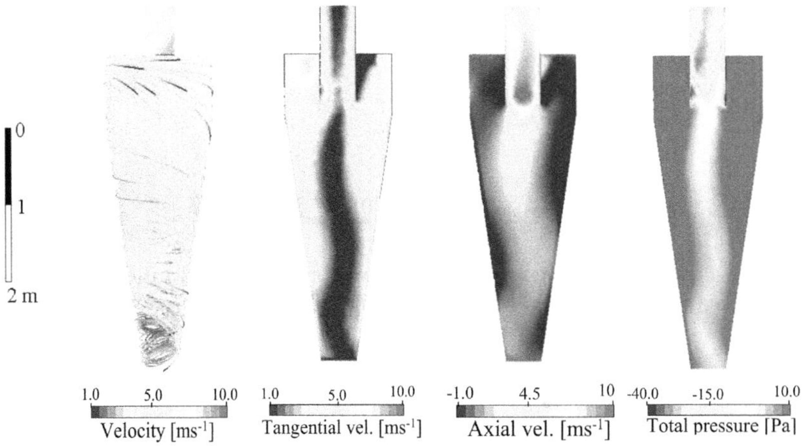

Figure 7.15 Hydrodynamic behavior in the cyclone separator with 100 gs^{-1} of milk powder for the case study 7.6.3.

The solution is possible with the high-order trapezoid scheme, the SIMPLE algorithm, and the second-order upwind interpolated cell values. The initialization procedure took the values from the inlet. For the steady-state scenario, the simulations required an average of 40 min to solve 3,000 iterations in a computer with 24GB RAM and Intel ® Corel^{TM} i7-3770 3.4 GHz processor.

Fig. 7.15 depicts hydrodynamic behavior for this case study, which is similar to Fig. 7.7. The velocity streamlines describe the expected behavior for a cyclone: tangential inlet flow, descendent flow by reduction (conical shape), and ascendent flow through the vortex finder. The streamlines represent a high rotational flow inside the cyclone separator. The tangential velocity produces the Rankine vortex, which indicates the typical distribution of the kinetic energy inside a cyclone separator. The axial velocity presents the descendent direction of the flow (negative values), close to the walls, and the ascendent direction (positive values) at the cyclone core. Finally, the pressure profile reaches low scale values (-40 Pa$<p<$10 Pa) with a narrow suction

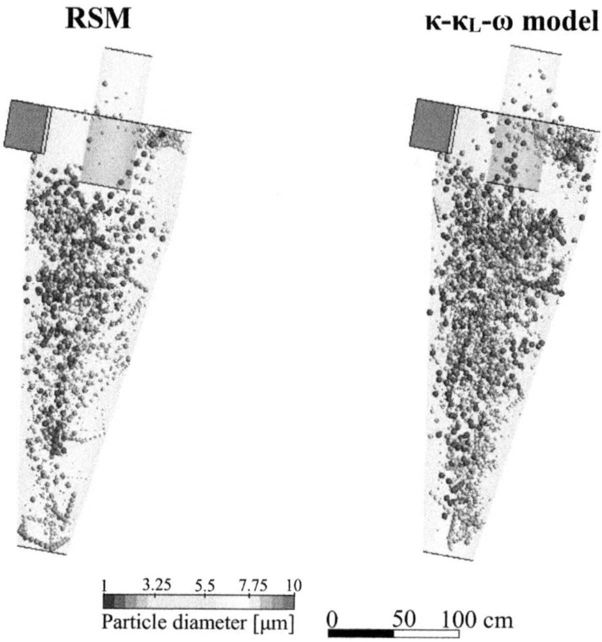

Figure 7.16 Particle distribution in the cyclone separator of milk powder (RSM for the case study 7.6.3 and κ-κ_L-ω model. From Díaz-Ovalle et al. (2023)).

region in the core. Through the vortex finder, the pressure experiences a vacuum effect in response to the acceleration of the axial velocity. Moreover, Fig. 7.16 a) shows the particle motion results in a particle diameter distribution for RSM. The milk powder suffers a low particle descent. However, Díaz-Ovalle et al. (2023) used the κ-κ_L-ω model for this case and observed a uniform descent between the particle diameters. Fig. 7.16 b) indicates the difference between the using RSM and the κ-κ_L-ω model.

7.7 Devices and configurations of cyclone separators

The underflow configuration is considered essential for high cyclone performance due to the capability to seal the conditions between the cyclone and the surroundings. For this, the pressure is the main parameter to keep under control, and without a sealing system, the cyclone separator might suction the surrounding air. Hoffmann and Stein (2008) have reported some sealing devices. However, the food industry demands high attention upon food safety, whereby the rotary valve and the flapper-type valve with closed protection are recommended.

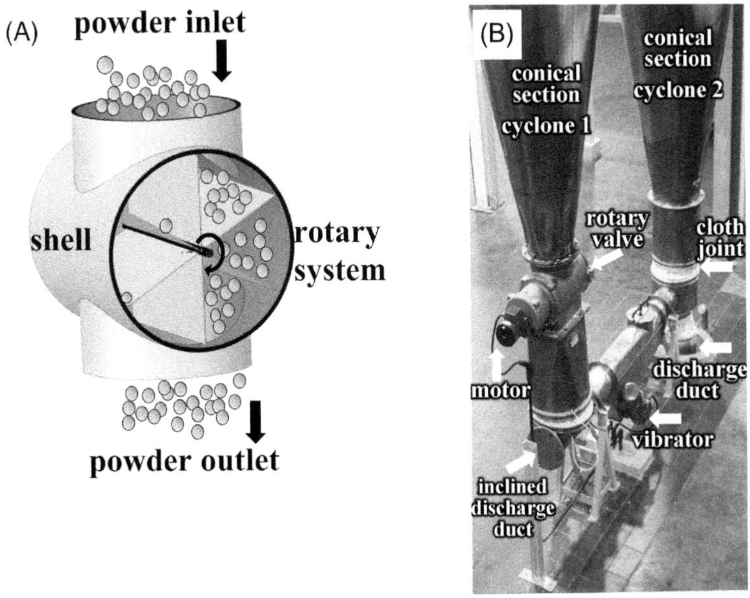

Figure 7.17 Rotary valve: A) schematic representation, and B) discharge system for a milk powder process.

- The rotary valve reclaims a motor and continuous maintenance attention. The chamber design lodges the inlet powder and unloads it in the half loop. This device contains rotary and stationary elements. The rotary element is a rotor with reels from chambers and receives the motion from a chain inserted in a gearing system and from the motor. The stationary element is characterized by a covering shape over the rotary element and is connected to the inlet and outlet ducts by air-lock seals. Fig. 7.3 identifies a rotary valve in a cyclone separator, and Fig. 7.17 depicts a rotary valve and a discharge system of cyclone separators for the milk powder process.
- The flapper-type valve with closed protection prevents gas leakage and passes the collected dust by pushing a hanging flapper plate. This element is recommended for the low difference of pressure between the discharge and the surroundings thus, the collected dust can force the flapper to open. The vacuum formed into the cyclone closes it after discharge. The closed protection is recommended to guarantee food safety and access immediately to the discharge for blockings.

The practical applications of cyclone separators have demonstrated that the shortest particles are not all collected, which demands additional elements to prevent losses of powder products. The filter cyclones add a bag recovery system after the cyclone separator outlet. The system receives the exhaust air with small-sized particles that cover the bag to increase the fluid flow resistance. To keep the filtering, the system cleans up by using pressurized air into the bags, this air gets into and injected through nozzles. A control system regulates the airflow in response to a desirable pressure

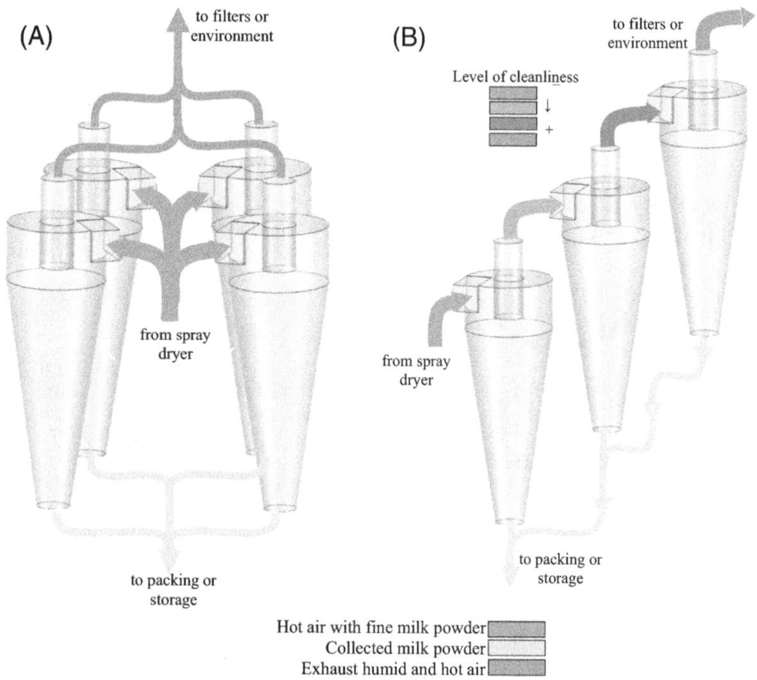

Figure 7.18 Representation of multiple cyclone separators for the milk powder process: A) parallel arrangement, and B) serial arrangement.

difference. Fig. 7.3 b) contains a cyclone separator in adjunction to the filter bags system (top section).

Multiple cyclone separator system is a common strategy to reduce losses of dust. This alternative conceptualizes the parallel and series arrangements. The parallel design is usual for considerable flow rates that compromise the separation efficiency. This cyclone separator arrangement has received attention due to its variability in the allocation design between the units. However, the nature of this arrangement complicates the prediction of the flow behavior and the separation efficiency. This design for a milk powder process is in Fig. 7.4. The series design is a successive interconnection from one cyclone separator to the next, where the outlet duct feeds the tangential inlet duct. For this case, the literature enounces the concept of overall separation efficiency that relates to the particular efficiency per each separator. Mainly, the dust loading concentration determines the preliminary designs and modifies the grade efficiency results. Fig. 7.18 contains both cyclone arrangements.

7.8 Elements of process safety in cyclone separators

A dust explosion is a sudden and swift combustion of confined particles resulting in an overpressure that yields deflagration and flame propagation through the dust cloud

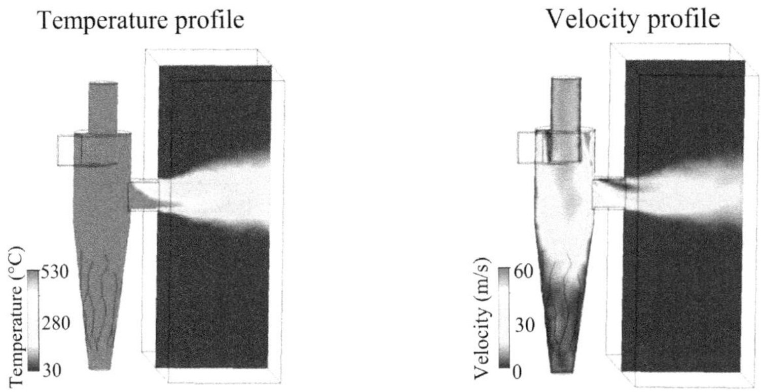

Figure 7.19 Demonstration of a venting system in cyclone separator by CFD simulation.

(Abbasi & Abbasi, 2007). Eckhoff (2003, 2009) indicated that the food processes have presented outstanding dust explosions in history, where the grain storage plants foreground. But generally, most of the food powders are well-known sources of dust explosions (Bhandari et al., 2013). The dust explosion occurrence is possible due to five elements that form the dust explosion pentagon (Wheat & Leali, 2022), where the elimination of one of these elements might prevent dust explosions (Jones, 2022):

1. Fuel is the combustible dust, such as grains, sugar, milk powder, and other agricultural raw materials.
2. Heat arises from the ignition source which might be machinery friction, faulty electrical equipment, etc.
3. Oxidant is the oxygen in the air.
4. Dispersion is the dust particles condition to maintain sufficient concentration during conveying and handling.
5. Confinement is the equipment condition to keep a dust dispersion inside without exterior disturbances. This concept characterizes the powder and dust processing equipment, which involves cyclone separators.

A cyclone separator, as a dust collector, retains for a short time a dust cloud, which under particular conditions could pose a dust explosion risk. In addition, the connectivity of cyclone separators to main process equipment (spray dryers, millers, sieves, etc.) comprises process safety. This demands special attention during the design, where the occurrence of dust explosion is not fully understood. To explain this phenomenon in real scenarios, Amyotte (2014) published important opinions that elucidate the myths and realities of this risk in the industry, where the inherently safe design is a valid option for food processes. Particularly, explosion and flameless venting methods are recommended for cyclone separators, Fig. 7.3 b) presents the venting system for a cyclone separator. In order to describe the venting system application, the CFD simulation of a fictitious case is presented in Fig. 7.19, where the conditions are overpressure and high temperature in the cyclone separator, particularly from the inlet duct. Fig. 7.19 depicts the system boundary conditions and the results in temperature and velocity profiles. This case demonstrates the benefit of applying a venting system

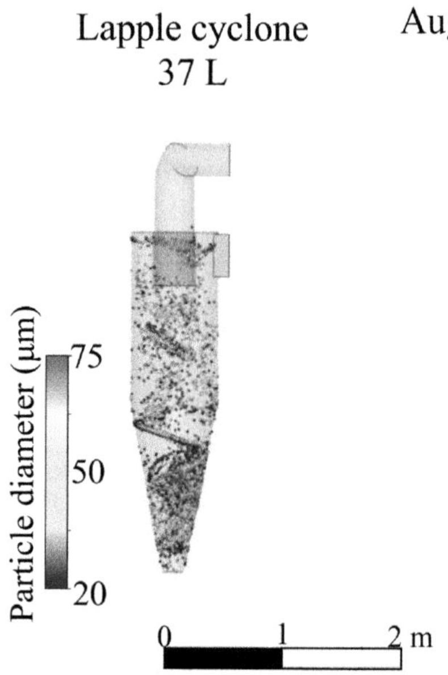

Figure 7.20 Comparison of cyclones based on the volume value of milk powder (>60 g/cm^3). (From Castrejón-González et al. (2019)).

through the fluid escapes at high velocity and temperature, which avoids an increment of pressure inside the cyclone separator.

Holbrow (2013) explored the requirements of this method in some food powders and carried out tests in cyclone separators. He concluded that the most severe test occurred in the ignition of the dust from the top outlet of the cyclone, the vortex finder might act as a vent duct, and the inhomogeneous concentration inside the cyclone and cone shape of the lower part promoted the flame during the test. Nevertheless, the cyclone separators must be protected to reduce the explosion severity due to the ignition sensitivity of the finest and driest dust inside (Bhandari et al., 2013). Castrejón-González et al. (2019) attempted to analyze cyclone separators inherently safe design by changing the geometrical elements, where the largest cyclone body might produce a high risk of dust explosion. Fig. 7.20 presents the dust cloud for two cyclone designs applied in the food industry, where the volume value of the cloud (>60 g/cm^3 as Eckhoff [2003] indicated) relates to the high level of risk.

7.9 Conclusions

Cyclone separators are well-known elements in the food processing industry due to the existence of powder products, whose recovery is necessary during transport, processing, or handling. The high demand for the spray drying process justifies the requirement of cyclone separators. However, the successful performance of the cyclone separators depends on a suitable design, which depends on its geometrical shape and size, and impacts the separation efficiency. Also, the performance requires enough understanding of the peculiar hydrodynamic behavior inside the cyclone separator. Thus, the conceptual elements of design and operation relate to the pressure drop, separation efficiency, and particle motion. The literature presents a vast list of studies concerning the prediction of these elements, but the complexity of the particle motion conducts particular research where the CFD has been the appropriate tool. This chapter contains the essential elements to understand the cyclone separators in the food processing industry, which can be scaled to solve real problems.

Nomenclature

A	total area of the cyclone walls, pressure drop models
A_C	dimensionless area of contact surface, pressure drop models
a	height of the inlet duct
a_C	particle acceleration
B	diameter of the tip diameter in the conical shape
b	width of the inlet duct
$\mathbf{C}$	turbulence transport tensor
Cp	heat capacity of the fluid
D	cyclone diameter
Dc	vortex finder diameter
dc	diameter of a central core in the flow turns, collection efficiency models
d*	core diameter
d_i	diameter of a particle of *i* size
d_{50}	cut-size diameter
d_{100}	critical particle diameter
Eu	Euler number
F	cyclone volumetric flow
$\mathbf{f}$	functions of ordinary differential equations in vector form
f	friction parameter, pressure drop models
H	cyclone total height
h	height of the cylindrical shape
h_H	hydrostatic height
hc	height of the conical shape
$\mathbf{k_1}, \mathbf{k_2}, \mathbf{k_3},$	parameters of the Runge-Kutta method in vector form
k $\mathbf{k_4}$	swirl exponent, pressure drop models
L	natural length, collection efficiency models
M	grouping parameter, pressure drop models
m_S	mass of a solid particle
N_H	number of heads for pressure drop
N_T	number of turns in flowthrough
N_V	mean number of particles per unit volume
n	thermal correction parameter, pressure drop models
p	static pressure
Q	grouping parameter, pressure drop models
$\mathbf{q}$	conductive heat flux vector
$\mathbf{R}$	Reynolds stress tensor
Re	Reynolds number

r	radial position
S	height of the immersed vortex finder
$\mathbf{S}$	mean strain rate
$\mathbf{S}_\mathrm{P}$	source term vector for two-way coupling
$S_{p,\kappa}$	source term vector for turbulence
s	curve slope of the collection efficiency
T	temperature
V	volume of the particle circumscription
V_E	volume of external pores between particles
V_I	volume of internal pores into a particle
$\mathrm{V_P}$	volume loading of particles
V_S	volume of a particle solid without pores
V_T	volume of a particle sample
$\mathrm{V_V}$	volume configuration
v	total velocity
$\mathbf{v}$	vector of velocity of gas
$\mathbf{v}_\mathrm{R}$	vector of relative velocity, $\lvert\mathbf{v}-\mathbf{v}_{\mathrm{S},i}\rvert$
$\mathbf{v}_{\mathrm{S},i}$	velocity of the solid of i size
v_inlet	velocity at the cyclone inlet
v_r	radial velocity component
v_t	tangential velocity
v_{tx}	maximum tangential velocity, collection efficiency models
v_x	x velocity component
v_y	y velocity component
v_z	axial velocity component
$v_{\mathrm{S}x,i}$	x velocity component of i size particle
$v_{\mathrm{S}y,i}$	y velocity component of i size particle
$v_{\mathrm{S}z,i}$	axial velocity component of i size particle
v_θ	angular velocity component
X	position in the coordinate system
$\mathbf{y}$	vector of dynamic variables
Z	grouping parameter, pressure drop models
z_L	vortex core length

Greek letters

α	groping parameter, pressure drop models
$\Delta\mathrm{P}$	pressure drop
ε	dissipation rate
$\boldsymbol{\varepsilon}$	dissipation rate tensor
ε_P	porosity in a particle
ζ_i	ratio of d_i/d_{50}
η_i	collection efficiency for the particle of i-size
θ	grouping parameter, pressure drop models
κ	turbulence kinetic energy
μ_G	viscosity of the gas
ν_t	eddy viscosity turbulence
ξ	grouping parameter, pressure drop models
ξ_B	pressure losses at the cyclone body, pressure drop models
ξ_{VF}	pressure losses at the vortex finder, pressure drop models

$\boldsymbol{\Pi}$	pressure strain tensor
ρ_A	apparent particle density
ρ_B	bulk particle density
ρ_G	gas density
ρ_S	true particle density
$\boldsymbol{\tau}$	shear stress tensor
τ_i	relaxation time of an i-size particle
ϕ_j	area conversion parameter for an j-cyclone
ω	angular velocity

Superscripts

m	exponent for a particle size distribution
n	exponent for the exponential profile
s	exponent for a particle size distribution

Subscripts

i	set of particle size
j	set of cyclone separator
G	gas
S	particle

Acronyms

CFD	Computational Fluid Dynamics
RNG	Re-Normalization Group
RSM	Reynolds-stress model
SST	Shear-stress transport

References

Abbasi, T., & Abbasi, S. A. (2007). Dust explosions—Cases, causes, consequences, and control. *Journal of Hazardous Materials, 140*(1–2), 7–44.

Alexander, R. M (1949). Fundamentals of cyclone design and operation. *Proceedings of Australian Institute of Mining and Metallurgy, 152*, 203–228.

Altmeyer, S., Mathieu, V., Jullemier, S., Contal, P., Midoux, N., Rode, S., & Leclerc, J. P. (2004). Comparison of different models of cyclone prediction performance for various operating conditions using a general software. *Chemical Engineering and Processing: Process Intensification, 43*(4), 511–522.

Amyotte, P. R. (2014). Some myths and realities about dust explosions. *Process Safety and Environmental Protection, 92*(4), 292–299.

Anandharamakrishnan, C. (2013). *Computational fluid dynamics applications in food processing*. Springer.

Azadi, M., Azadi, M., & Mohebbi, A. (2010). A CFD study of the effect of cyclone size on its performance parameters. *Journal of Hazardous Materials, 182*(1-3), 835–841.

Barbosa-Cánovas, G. V., & Juliano, P. (2005). Physical and chemical properties of food powders. In C. Onuwata (Ed.), *Encapsulated and powdered foods*. Taylor & Francis Group.

Barth, W. (1956). Design and layout of the cyclone separator on the basis of new investigations *Brennst.-Warme-Kraft, 8*, 1–9.

Bhandari, B., Bansal, N., Zhang, M., & Schuck, P. (2013). *Handbook of food powders: Processes and properties*. Woodhead Publishing.

Bhandari, B. R., Patel, K. C., & Chen, X. D. (2008). Spray drying of food materials-process and product characteristics. In X. D. Chen, & A. S. Mujumdar (Eds.), *Drying technologies in food processing* (pp. 113–159). Blackwell Publishing.

Blazek, J. (2006). *Computational fluid dynamics: Principles and applications*. Elsevier.

Casal, J., & Martinez-Benet, J. M. (1983). A better way to calculate cyclone pressure drop. *Chemical Engineering, 99*, 100–111.

Castrejón-González, E. O., Díaz-Ovalle, C. O., Jung, S., Cano-Gómez, J. J., Castillo-Borja, F., & Joaquín-Ramos, A. D. J (2019). Estudio computacional del efecto geométrico de separadores ciclones en nubes de polvo para prevenir explosiones [Computational study of the geometric effect of cyclone separators in dust clouds to prevent dust explosion]. *Avances en Ciencias e Ingeniería, 10*(2), 33–49.

Cebeci, T. (2003). *Turbulence models and their applications*. Horizons Publishing Inc.-Springer Verlag.

Cebeci, T. (2004). *Analysis of turbulent flows*. Elsevier.

Çengel, Y. A., & Cimbala, J. (2017). *Fluid mechanics: Fundamentals and applications*. WBC McGraw-Hill.

Chen, J., & Shi, M. (2007). A universal model to calculate cyclone pressure drop. *Powder Technology, 171*(3), 184–191.

Chung, T. J. (2010). *Computational fluid dynamics* (2nd ed.). Cambridge University Press.

Clift, R., Grace, J. R., & Weber, M. E. (1978). *Bubbles, drops, and particles*. Dover Publications, Inc.

Cortés, C., & Gil, A. (2007). Modeling the gas and particle flow inside cyclone separators. *Progress in Energy and Combustion Science, 33*(5), 409–452.

Couper, J. R., Penney, W. R., Fair, J. R., & Walas, S. M. (2012). *Chemical process equipment. Selection and design*. Butterworth-Heinemann.

Cuq, B., Rondet, E., & Abecassis, J. (2011). Food powders engineering, between knowhow and science: Constraints, stakes and opportunities. *Powder Technology, 208*(2), 244–251.

Davidson, P. A. (2015). *Turbulence: An introduction for scientists and engineers*. Oxford University Press.

Demir, S. (2014). A practical model for estimating pressure drop in cyclone separators: An experimental study. *Powder Technology, 268*, 329–338.

Díaz-Ovalle, C. O. (2024). Design, model, and simulation of evaporators for the food industry. In S. M. Jafari, E. Çapanoglu, & A. C. Karaca (Eds.), *Evaporation technology in food processing* (pp. 269–313). Woodhead Publishing.

Díaz-Ovalle, C. O., Ramírez-Rivera, M. J., Castillo-Borja, F., & Castrejón-González, E. O. (2023). CFD analysis of the dust loading in a cyclone separator for milk powder *Avances en Ciencias e Ingeniería, 14*, 109–117.

Eckhoff, R. (2003). *Dust explosions in the process industries, third edition: Identification, assessment and control of dust hazards*. Elsevier Science.

Eckhoff, R. K. (2009). Understanding dust explosions. The role of powder science and technology. *Journal of Loss Prevention in the Process Industries, 22*(1), 105–116.

Elghobashi, S. (2006). *An updated classification map of particle-laden turbulent flows*. Dordrecht.

Fayed, M. E., & Otten, L. (Eds.). (1997). *Handbook of powder science & technology*. Chapman & Hall.

First, M. W. (1950). *Fundamental factors in the design of cyclone dust collectors*. PhD diss. Harvard University.

Gamiño-Tovar, M. S., Castillo-Borja, F., Vázquez-Román, R., Díaz-Ovalle, C. O., Guzmán-Zazueta, A., & Herrera-Enciso, F. (2018). Análisis del efecto geométrico de ciclones en el secado por aspersión de leche usando CFD [Analysis of the geometric effect of cyclone separators over spray-dryer process of milk using CFD]. *Avances en Ciencias e Ingeniería, 9*(1), 11–23.

Gibson, M. M., & Launder, B. E. (1978). Ground effects on pressure fluctuations in the atmospheric boundary layer. *Journal of Fluid Mechanics, 86*(3), 491–511.

Guo, M., Yang, L., Son, H., Le, D. K., Manickam, S., Sun, X., et al. (2024). An overview of novel geometrical modifications and optimizations of gas-particle cyclone separators. *Separation and Purification Technology, 329*, 125136.

Hinze, J. (1987). *Turbulence*. McGraw-Hill.

Hoffmann, A. C., & Stein, L. E. (2008). *Gas cyclones and swirl tubes. Principles, design and operation*. Springer-Verlag.

Holbrow, P. (2013). Dust explosion venting of small vessels and flameless venting. *Process Safety and Environmental Protection, 91*(3), 183–190.

Iozia, D. L., & Leith, D. (1990). The logistic function and cyclone fractional efficiency. *Aerosol Science and Technology, 12*(3), 598–606.

Jafari, S. M., Arpagaus, C., Cerqueira, M. A., & Samborska, K. (2021). Nano spray drying of food ingredients: Materials, processing and applications. *Trends in Food Science & Technology, 109*, 632–646.

Díaz-Ovalle, C. O., & Jafari, S. M. (2023). Multiple-effect evaporators in the food industry: Fundamentals, design, simulation, control, and applications. *Food Engineering Reviews, 15*(4), 691–717. https://doi.org/10.1007/s12393-023-09350-6.

Jayanti, S. (2018). *Computational fluid dynamics for engineers and scientists*. Springer Verlag.

Jones, C. L. (2022). Dust control and explosion prevention. In K. A. Rosentrater (Ed.), *Storage of cereal grains and their products* (5th ed.). Woodhead Publishing.

Lanzerstorfer, C. (2020). Apparent density of compressible food powders under storage conditions. *Journal of Food Engineering, 276*, 109897.

Lapple, C. E., & Shepherd, C. B. (1940). Calculation of particle trajectories. *Industrial and Engineering Chemistry, 32*(5), 605–617.

Leith, D., & Jones, D. L. (1997). Cyclones. In M. E. Fayed, & L. Otten (Eds.), *Handbook of powder science & technology*. Chapman & Hall.

Michaelides, E. E. (2006). *Particles, bubbles and drops*. World Scientific Publishing Co.

Michaelides, E. E., Crowe, C. T., & Schwarzkopf, J. D. (Eds.). (2017). *Multiphase flow handbook*. CRC Press.

Ogawa, A. (1997). Mechanical separation process and flow patterns of cyclone dust collectors. *Applied Mechanics Reviews, 50*(3), 97–130.

Onwulata, C. (Ed.). (2006). *Encapsulated and powdered foods*. Taylor & Francis Group.

Patankar, S. V., & Spalding, D. B. (1972). A calculation procedure for heat, mass and momentum transfer in three-dimensional parabolic flows. *International Journal of Heat Mass Transfer, 15*, 1787–1806.

Perry, M. J. (2008). Agricultural health and safety. In H. K. Heggenhougen (Ed.), *International encyclopedia of public health* (pp. 69–76). Academic Press.

Pugliese, A., Cabassi, G., Chiavaro, E., Paciulli, M., Carini, E., & Mucchetti, G. (2017). Physical characterization of whole and skim dried milk powders. *Journal of Food Science and Technology, 54*(11), 3433–3442.

Rodriguez, S. (2019). *Applied computational fluid dynamics and turbulence modeling*. Springer-Verlag.

Rosentrater, K. A., & Bucklin, R. (2022). Chapter 6: Structural, physical, and engineering properties of cereal grains and grain products. In K. A. Rosentrater (Ed.), *Storage of cereal grains and their products* (5th ed.) (pp. 135–178). Woodhead Publishing.

Scott, G., & Richardson, P. (1997). The application of computational fluid dynamics in the food industry. *Trends in Food Science & Technology, 8*(4), 119–124.

Shenoy, P., Viau, M., Tammel, K., Innings, F., Fitzpatrick, J., & Ahrné, L. (2015). Effect of powder densities, particle size and shape on mixture quality of binary food powder mixtures. *Powder Technology, 272*, 165–172.

Shepherd, C. B., & Lapple, C. E. (1939). Flow pattern and pressure drop in cyclone dust collectors. *Industrial & Engineering Chemistry, 31*(8), 972–984.

Speziale, C. G., Sarkar, S., & Gatski, T. B. (1991). Modelling the pressure–strain correlation of turbulence: An invariant dynamical systems approach. *Journal of Fluid Mechanics, 227*, 245–272.

Stairmand, C. J. (1949). Pressure drop in cyclone separators *Engineering, 168 (4369)*, 409.

Sun, D.-W. (Ed.). (2007). *Computational fluid dynamics in food processing*. Taylor & Francis Group.

Sun, D.-W. (Ed.). (2019). *Computational fluid dynamics in food processing* (2nd ed.). CRC Press.

Tamime, A. Y. (Ed.). (2009). *Dairy powders and concentrated products*. Wiley-Blackwell.

Tawari, T. D., & Zenz, F. A. (1984). Evaluating cyclone efficiencies from stream compositions. *Chemical Engineering, 91*(9), 69–73.

Tennekes, H., & Lumley, J. L. (1972). *A first course in turbulence*. The MIT Press.

Tu, J., Yeoh, G.-H., & Liu, C. (2013). *Computational fluid dynamics: A practical approach* (2nd ed.). Butterworth-Heinemann, Elsevier.

Wang, B., Xu, D. L., Chu, K. W., & Yu, A. B. (2006). Numerical study of gas–solid flow in a cyclone separator. *Applied Mathematical Modelling, 30*(11), 1326–1342.

Wang, C., Ma, Y., & Sui, W. (2023). The secondary flows in a cyclone separator: A review. *Processes, 11*(10), 2935.

Wang, H., Zhang, Y., Wang, J., & Liu, H. (2012). Cyclonic separation technology: Researches and developments. *Chinese Journal of Chemical Engineering, 20*(2), 212–219.

Westergaard, V. (1994). *Milk powder technology: Evaporation and spray drying*. Niro A/S.

Wheat, J., & Leali, A. (2022). Chapter 13: Hazard monitoring equipment selection, installation, and maintenance. In K. A. Rosentrater (Ed.), *Storage of cereal grains and their products* (5th ed.) (pp. 371–405). Woodhead Publishing.

Wilcox, D. C. (2006). *Turbulence modeling for CFD*. DCW Industries Inc.

Williams, G. D., & Rosentrater, K. A. (2022). Design and engineering considerations for grain storage, handling, and processing facilities. In K. A. Rosentrater (Ed.), *Storage of cereal grains and their products* (5th ed.). Woodhead Publishing.

Woo, M. W., & Bhandari, B. R. (2013). Spray drying for food powder production. In B. R. Bhandari, N. Bansal, M. Zhang, & P. Schuck (Eds.), *Handbook of food powders: Processes and properties*. Woodhead Publishing.

Yamasaki, H., Kizilkan, Ö., Yamaguchi, H., Kamimura, T., Hattori, K., & Nekså, P. (2020). Experimental investigation of dry ice cyclone separator for ultra-low temperature energy storage using carbon dioxide. *Energy Storage, 2*(4), e149.

Yeoh, G.-H., & Tu, J. (2010). *Computational techniques for multi-phase flows*. Butterworth-Heinemann.

Zenz, F. A. (1964). Conveyability of materials of mixed particle size. *Industrial & Engineering Chemistry Fundamentals, 3*(1), 65–75.

Zenz, F. A. (1982). Cyclones. In J. J. McKetta (Ed.), *Encyclopedia of Chemical Processing and Design* (14, pp. 82–97). Taylor & Francis Group.

Deaeration

8

Nikita S. Bhatkar[a], *Vimal Challana*[a], *Geetanjali Pathania*[b] *and Shivanand S. Shirkole*[a,c]

[a] Department of Food Engineering and Technology, Institute of Chemical Technology Mumbai, Bhubaneswar, India, [b] Department of Food Science and Technology, Punjab Agricultural University, Ludhiana, Punjab, India, [c] Deprtament of Food Technology, School of Engineering and Technology, D. Y. Patil Agriculture and Technical University, Talsande, Kolhapur, India

8.1 Introduction

Deaeration is the process of removing air or oxygen from an environment. The phenomenon is hugely important to food processing and preservation as the method is ubiquitous in food processing, from production to packaging, storage, and food distribution. One of the most ancient yet natural methods where the deaeration process occurs in food processing is the release of carbon dioxide during fermentation, which leads to the removal of oxygen and air, creating an anaerobic environment (Tamime, 2002). Thus, deaeration is an essential step in the food processing industry. Deaeration is usually associated with removing oxygen from food products during processing or packaging. Oxygen is an inescapable element for the survival of organisms. On the other side, it can be detrimental to the food as oxygen can lead to the growth of unwanted organisms, leading to food spoilage. The percentage of annual food loss estimated by the Food and Agriculture Organization (FAO) is to be 13.23% for the year 2021 (FAO, 2021). The causes of food spoilage include microbial spoilage, enzymatic activity, oxidation, physical damage, temperature abuse, contamination, and infestation by insects and rodents. Among the eight listed reasons for spoilage, microbial spoilage, enzymatic activity, and oxidation can be easily accelerated by oxygen. Apart from these reasons, other reasons for spoilage can be controlled by appropriate storage and handling practices.

The removal of oxygen is mandatory to preserve the food from oxidative damage. The deaeration process is mandatory in the fruit juice processing industry. The process is carried out before the pasteurization of juices or juice concentrates. Fruit juices usually exhibit an acidic pH range, and it is well-established that the heat-labile compounds are stable in high-temperature environments in acidic conditions. However, the browning reactions and loss of nutrients, such as ascorbic acid, are inevitable due to oxygen and require a deaeration process (Ramesh, 2020). Like the juices, the deaeration of Milk before pasteurization is a common practice to preserve its quality. Other dairy products, such as cheese, ice cream, butter, and yogurt, require a deaeration step. Apart from maintaining the nutritional quality of the products by preservation of the sensitive nutrients, it is equally important to have the desired textural properties and aesthetic appearance; deaeration is instrumental in the confectionery and bakery industry to achieve the desired textural properties and appearance. Section 8.4 provides

Mechanical Separation Processes in the Food Industry: Unit Operations and Processing Equipment in the Food Industry.
DOI: https://doi.org/10.1016/B978-0-12-819523-9.00014-7

the information on the same. The deaeration process is also crucial for other operations such as canning, packaging, storage, and aging. The scope of deaeration in food is not only limited to the preservation, improving quality, and shelf life extension, but it is also essential for preventing corrosion in equipment. The presence of dissolved oxygen in the food product can also lead to the corrosion of the equipment and affect the longer life of the instruments. Scaling and precipitation of minerals from the food product in the equipment and food contact materials, as well as deaeration, also reduces the probability of scaling and precipitation in the equipment.

The method of deaeration in the food industry has evolved from boiling food products in an open container to remove the dissolved oxygen to introducing vacuum technology and, in recent times, to advanced methods such as membrane-based deaeration, ultrasound deaeration, and various chemical deaeration processes. This chapter discusses the importance of the deaeration process in food processing, the methods employed, and the application of this process in different industries.

8.2 Importance of deaeration

Deaeration, or degassing, is a process that is relatively underrepresented in the literature within the context of the food industry. There is vast literature available on other processing steps, from washing to packaging and storage of food. However, deaeration, though quite an essential step, is scarce in the literature, and there are limited articles where explicit mention of the process is observed. Nevertheless, it is an omnipresent step in many food processing industries. The section highlights the significance of deaeration in the food industry. Fig. 8.1 provides a visual representation of the process.

8.2.1 Prevention of oxidation

Oxidation of food is the major reason for food spoilage. Oxidation is inevitable in food such as fruit, vegetables, meat, and juices, and the most common one among these is lipids. Every year, huge losses are recorded due to oil deterioration and oxidation, which causes rancidity. When exposed to air or oxygen, unsaturated fatty acids in fats undergo oxidation, often accompanied by a free radical chain reaction. Lipid oxidation not only degrades the oil in terms of shelf life, color loss, and development of off-flavor, but it also deteriorates the nutritional value and antioxidants present in the oil (Geng et al., 2023). The deaeration of the oil to remove the dissolved oxygen and curb these reactions is carried out. Some studies suggest that removing dissolved oxygen from the oil enhanced its stability and quality in terms of peroxide value, K232, and K270 values (Guerrini et al., 2018). At an industrial scale, degassing operations are carried out before the deodorization process by applying a vacuum. Other food categories, such as fruit and vegetables, can develop off flavors and undergo browning on oxidation of phenols present in the fruits upon reaction with the oxygen in the environment accelerated by inherent enzymes. Also, with these enzymatic reactions, the oxidation of the bioactive compounds in the fruit or vegetables is of concern, such as the oxidation of ascorbic acid to dehydroascorbic acid. The oxidation of fruit juices

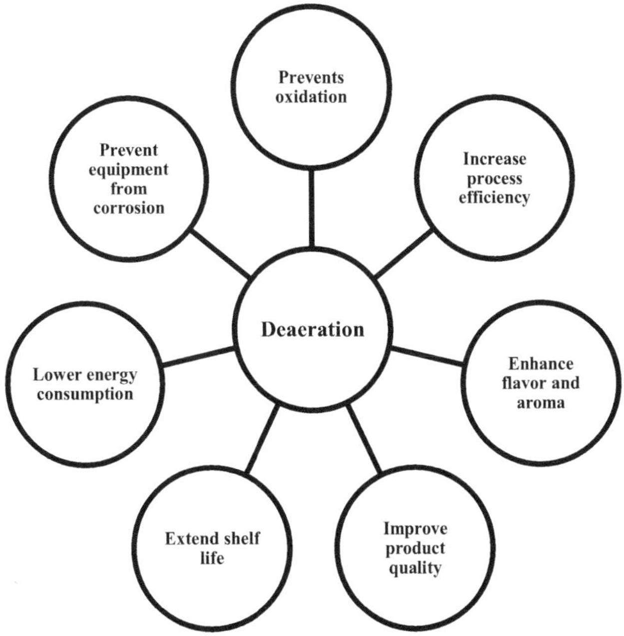

Figure 8.1 Importance of deaeration in the food industry.

or vegetable products is also an inescapable reaction. The deaeration of fruit juices before pasteurization is essential to prevent oxidative damage to the fruit juices or concentrates.

Oxidation is quite common in the case of meats, too. The lipid inside the meat contributes to both desirable and undesirable meat's aroma and flavor and is also essential for the tenderness and juiciness of meat (Amaral et al., 2018). The presence of oxygen reacts with polyunsaturated fatty acids, which can lead to oxidation. The oxidation reaction diminishes sensory and nutritional quality by the depletion of essential fatty acids and vitamins. Consequently, the changes in color, texture, and the presence of rancid odor and flavor contribute to consumer rejection. It is reported that hydroperoxides are the first product of the oxidation reaction. It is worth noting that hydroperoxides are odorless and do not contribute to aroma. Hydroperoxides, an unstable product, break into hydrocarbons, aldehydes, ketones, alcohols, esters, and acids, contributing to off-flavors and off-odors. These aldehydes are the major contributors to the volatile flavors in meat. Aldehydes also react with the protein in meat and cause modifications, degrading its nutritional and sensory value. Oshibanjo Tateo et al. reported that the vacuum aging of foal meat lowers the production of oxidative products and volatile compounds produced by lipid oxidation (Tateo et al., 2020).

8.2.2 Enhance flavor and aroma

Flavor and aroma are characteristic features of most food products, such as wine. Atmospheric oxygen causes deteriorative changes in food processing. Premature oxidation occurs (mostly in young dry white wines) when a wine is expected to age well, but it also leads to oxidation and is unfit for consumption. The oxidation leads to the production of off-flavors and loss of aroma. The primary contributors to premature oxidation are the presence of oxygen in the bottle's headspace, the amount of oxygen dissolved in the wine, and the permeation of oxygen through the bottle cap. Temperature and light also play significant roles in this process (Gabrielli et al., 2021). In the fruit and vegetable industry, oxygen leads to various oxidation reactions. These reactions lead to browning and a reduction in nutritional value, such as the degradation of essential fatty acids, vitamins, and flavonoids during processing. Oxygen also leads to textural changes and alterations in flavor or odor (Akyildiz & Ağçam, 2014). Additionally, the products formed by the oxidation, such as furfural juice content during storage, can harm health. It is seen that the deaeration process can reduce the formation of these products (Kanner et al., 1981). High oxygen is desirable in red meat to preserve its bright red color. However, high oxygen levels can significantly affect meat quality, as flavor production, reduced tenderness, and juiciness of meat can be attributed to the oxidation of lipids and proteins. To avoid this, modified atmospheric packaging can be a potential solution (Bao & Ertbjerg, 2015). In the dairy industry, oxygen is crucial, especially in producing products like yogurt, where microbial strains or cultures are used. However, oxygen can initiate the production of harmful metabolites, which leads to oxidative stress and can also result in cell death. The production of yogurt includes agitation and mixing, as well as oxygen diffusion during storage and diffusion of oxygen through packaging material, which increases oxygen levels within the product. The increase in oxygen concentration is directly associated with the toxicity of certain cells' release of peroxides and free radicals as the result of oxidation, which can lead to lipid peroxidation and damage the probiotic bacteria DNA. To mitigate these effects and to maintain probiotic bacteria, vacuum fermentation and oxygen-tolerable organisms can be employed (Meybodi et al., 2020).

8.2.3 Shelf life extension

The shelf life of food is often compromised by microbial spoilage. Microorganisms like obligate aerobes and microaerophiles require oxygen for their growth (Microorganisms, 2015). These organisms can produce undesirable compounds, leading to off-flavors. It was observed that the meat stored under vacuum packages showed 10 days more shelf life than the meat stored in high-oxygen packages (Jääskeläinen et al., 2016). It was also observed that the honeydew melon packed in a modified atmosphere shows better microbial stability against the food packed in the ambient air (Zhang et al., 2013). The consumer expects the product to have a specific color, flavor, aroma, appearance, and nutritional value. Consumers may only accept the product if these attributes are up to the mark. The oxygen leads to the oxidation of products, which can be visible in the color loss, production of off-flavor, loss of aroma, decreased nutritional value, and structure or integrity loss of certain food products. The shelf life of the food product is extended with oxygen scavengers, which aids in removing the residual oxygen present within the product (García-Torres et al., 2009). Kosiński et al. observed that apple juice's bioactive compound's stability increased when the dissolved oxygen concentration was minimized with the help of deaeration (Kosiński et al., 2024).

8.2.4 Improve process efficiency and reduce energy consumption

Liquid foam, commonly seen in food products during food processing, is a colloid where gas is either surrounded by liquid or entrapped within the liquid. The thin liquid film separates gas bubbles from each other. Liquid vaporization, release of gases during chemical and microbiological reactions, aeration, and agitation are the potential sources for producing foams during processing. Foaming can lead to issues like process control and equipment handling in fermentation and mixing. In addition, this foaming can result in product loss, environmental contamination, barriers to oxygen transfer, and a hindrance to the filling process, as well as promote microbial growth. Eliminating carbon dioxide can enhance fermentation, while the removal of oxygen can improve the quality and preservation of food products. It was observed that the application of moderate-intensity ultrasound reduces the fermentation time by 36-50% in the case of wine, beer, and sake. An ultrasonic defoaming system controls the foaming while filling bottles, carbonated beverages, reactors, and fermentation vessels (Villamiel et al., 2021).

The formation of air bubbles during processing reduces heat transfer efficiency, impacting product quality. When the air bubbles block the contact of fluid with the surface of heat exchangers, the burning of the product occurs, especially when the bubbles act as nuclei for protein fouling (Datta & Tomasula, 2015). Carbonation consistency is enhanced by lowering the dissolved oxygen value in water to less than 500 ppb. It reduces foam formation during bottling and increases line speed (Harbold & Park, 2008). Moreover, it is worth noting that all these increases in process efficiency eventually increase the rate of processing and reduce energy consumption. It was observed that the phase transition and freezing rates increased by 29.8% and 28.8%, respectively, when the degassing of radish was followed by ultrasound-assisted freezing. It also enhances the quality of the product (Taha et al., 2022). Additionally, the reduction in freezing time will decrease energy consumption. It is also observed that degassing the feed before the process also reduces the dead volume inside the equipment (Bellebna et al., 2014).

8.2.5 Prevention of equipment disruption

Foams in the food industry are one of the major reasons for equipment disruption and process interruption, leading to decreased efficiency. It was observed that the overfoaming results in decreased sugar yield and capacities of pumps, reaction tanks, and filters. Additionally, foaming negatively impacts the filtration process during corn starch extraction, resulting in decreased yield recovery of starch and protein (Leuner et al., 2020). Moreover, foaming can affect the capacity of vessels, blockage ducts, and exhaust valves, moisten outlet air filters, and control instrument malfunctions

(Bates & Patist, 2010). Additionally, foaming can reduce mass and heat transfer rates, alter circulation rates, block gas filter inlets and outlets, and disrupt electrode readings. These are some problems associated with foaming in bioprocesses (Vardar-Sukan, 1998). The presence of foams can also potentially corrode the equipment due to oxidation and mineral deposits from the food products. The usage of antifoam agents, equipment redesigning, usage of enzymes (to hydrolyze the surfactant), and usage of defoaming instruments with the application of antifoam agents are some of the potential solutions that can help to tackle the problem related to foams (Leuner et al., 2020).

8.3 Techniques or methods used for deaeration

8.3.1 Mechanical deaeration

Mechanical deaeration or degassing of food can be accomplished by different methods. Methods such as agitation, boiling, or tapping are usually employed in this method of deaeration. The frequency and amplitude of the agitation or shaking are critical, as above a certain frequency and amplitude, the process can lead to the incorporation of air into the food system and also potentially damage the food structure.

8.3.2 Vacuum deaeration

The most commonly practiced deaeration method in the food industry is the vacuum deaeration method. The method is suitable for liquid and semi-solid food as it mostly removes dissolved gasses, primarily oxygen, in most cases. The process is carried out by placing the liquid or semi-solid slurry of the food product inside a chamber. The pressure inside the chamber is reduced to a desired level with the help of a vacuum system. The method is based on Henry's law's principle that the gas's solubility within a liquid is proportional to the partial pressure of gases around the liquid. Thus, the dissolved oxygen is released when a vacuum is applied to the chamber. The components of a vacuum deaerator are a vacuum chamber, a vacuum pump, an inlet feeding system at the top, an outlet for the discharge of the liquid and air, and volatile gases at the bottom and top, respectively. Fig. 8.2 gives the schematic of the chamber for the vacuum deaerator. The process relies on the applied pressure within the vacuum chamber, which is influenced by the specific characteristics of the liquid undergoing deaeration. It is a usual practice to set the pressure such that the inlet temperature of the liquid is 2°**C** to 5°**C** above the boiling point of the liquid at that pressure (García-Torres et al., 2009). Historically, the batch mode has been prevalent in food operations. However, despite the convenient batch operation and process control, problems such as low throughput, higher energy consumption, and inconsistency in product quality have made the processors shift to centrifugal deaerators and film deaerators, which operate in continuous modes. The method is commonly used in beverage processing before the pasteurization of beverages, the processing of carbonated beverages, and the processing of milk and dairy products.

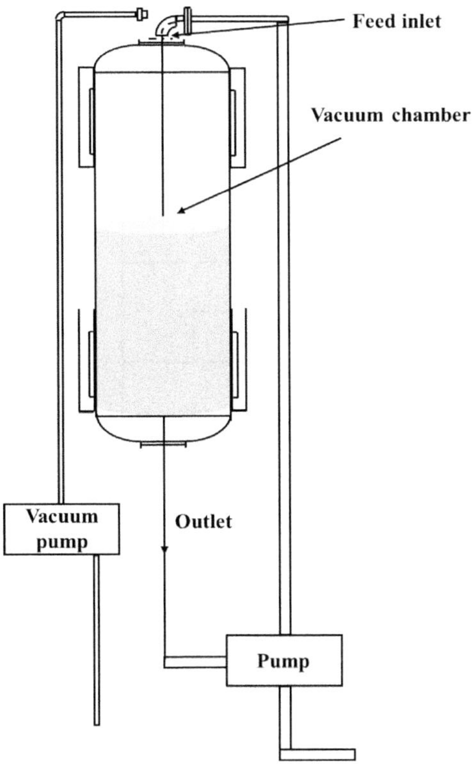

Figure 8.2 Schematic for vacuum deaeration tank.

8.3.3 Gas stripping

In the case of vacuum deaeration, where the pressure is within the vacuum zone, the gas stripping method operates under high pressure. Unlike vacuum deaeration with the absolute absence of air, in this case, inert air such as nitrogen or noble gases or steam is used as a medium to remove the dissolved oxygen from the food. Like the vacuum deaeration method, this method is mostly carried out for liquid food such as beverages, wine, beer, and dairy products. The liquid is exposed to air at high pressure. The underlying principle for the method is based on the concentration gradient of the gases in the food product, the gases used in the process, and the solubility of the gases at a given temperature. The mass transfer of the gases from the food is facilitated by the gas concentration gradient in the liquid and the medium applied with pressure. The temperature of the stripping gases is crucial as the solubility of the gases in liquid reduces at a higher temperature. The components of the stripping gas method include the stripping column, which is the core of the process, where the exchange of gases between the liquid and the stripping gas is carried out. Other components are a gas injection system, usually at the bottom of the column, inlet, and outlet for the feed at the top and bottom. There is also an outlet stream for stripping gas at the top. Fig. 8.3

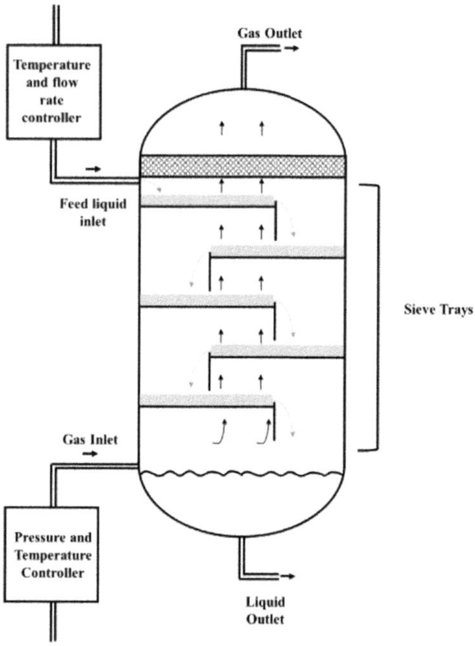

Figure 8.3 Gas stripping method for deaeration.

gives the pictorial representation of the stripping column. The gases used for stripping in the food industry are nitrogen or steam. When steam is used as a stripping gas, the method is called stream stripping; it is mostly employed for the restoration of aroma and flavor compounds of the juices or, in some cases, for the extraction of compounds. The solubility of the gases depends on the temperature and the pressure of the feed and the stripping gas used. A higher temperature of the liquid feed ensures lower gas solubility, while a higher temperature may be detrimental to the heat-labile components of the food. Thus, the feed liquid's optimum temperature and flow rate are desirable. Likewise, the stripping gas's pressure and flow rate dictate the contact surface area for the liquid and gases. Trays, sieves, and different types of packing are prevalent to increase the contact time inside the column. An attempt to increase the dissolved oxygen from the liquid using nitrogen stripping coupled with vacuum degassing is also done with appreciable results in removing the dissolved gases (Zhao et al., 2021).

8.3.4 Membrane techniques

The main purpose of deaeration is to remove oxygen to preserve the bioactive contents from oxidation. However, the degassing process, such as gas stripping and vacuum deaeration, may involve high-temperature exposure to the bioactive material, causing thermal degradation. In products where the thermal losses of the bioactives are to be

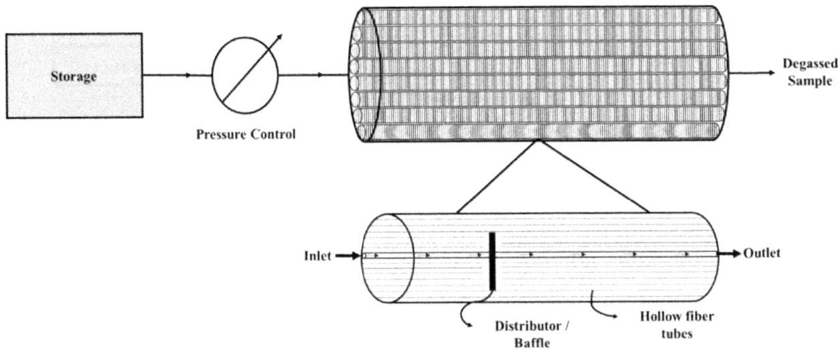

Figure 8.4 Hollow fiber membrane separation for deaeration.

reduced, membrane-based deaeration can be used. Along with ambient temperature operations, membrane degassing offers other advantages, such as lower operational cost, better preservation of the quality, higher efficiency, and lower energy consumption (Chen et al., 2023; Van Pelt & Churman, 2015). The process involves passing the food products (liquid) through the hollow fiber tubes, semi-permeable under applied pressure to separate the gases from the liquid food. The components of the membrane deaeration system include the storage tank for food, a chamber with distributed hollow fiber tubes as the membrane for separation, baffles to facilitate even distribution of the liquid food in the bundle of hollow tubes, pressure controlling devices, and an outlet to collect the degassed food. Fig. 8.4 gives the schematic of the hollow fiber membrane deaerator. The principle of separating gases from the food is based on the permeability of the gases through the hollow tube and the pressure gradient applied along either side of the hollow fiber. The applied pressure for food and beverage products is usually less than 0.41 MPa, with flow rates ranging from 1 to 48 m^3/h and temperatures between 5 to 50°C (Chen et al., 2023).

8.3.5 Ultrasonication

Ultrasonic techniques in the food industry are increasing significantly due to their promising results. The use of ultrasonication for degassing as a method can be traced back to 1954 by Kapustin (1954). Since then, ultrasonication has been used to degas liquid food products. Sometimes, the gases are trapped in the food system as foam. The formation of foam causes the entrapment of gases within the food matrix. In such cases, ultrasound defoaming is used; these are quite prevalent in fermentation tanks, canning, bottling lines, distillers, etc. The foam formation not only traps the gases within the food but also hinders the escape or transfer of gases (Gallego-Juárez et al., 2015). The ultrasonic defoaming and degassing method is based on the principle of cavitation of bubbles or foam to release the entrapped gases. The cavitation of the gases is ensured by applying high-intensity sound waves to the liquid food. The ultrasound degassing of food on an industrial scale is performed by ultrasonic baths or a sonicator probe

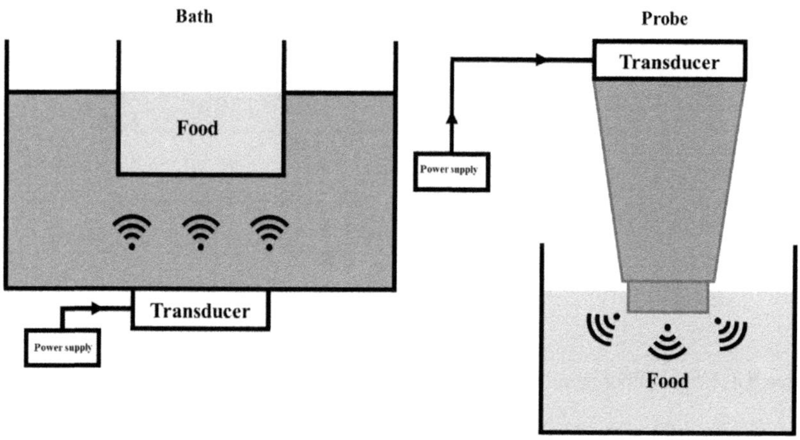

Figure 8.5 Bath and probe sonication for degassing of food.

directly into the bottles (Villamiel et al., 2021). The components of an ultrasound bath at an industrial scale include a tank, transducer, and temperature controller, if required. Fig. 8.5 shows a pictorial diagram of the bath and probe sonicator used in the industry. Though the method is quite promising for liquids, it has a limited utilization for liquids with high viscosity, such as molten chocolate.

8.3.6 Vacuum packaging

Vacuum packaging is a food preservation method that enhances shelf life by removing air from the headspace of the package. It is done by placing food inside the packaging material. The packaging material is partially sealed. The air inside the food package is either removed with the help of suction or displaced with inert gases. The pouch is sealed once the desired vacuum level is achieved inside the package. Unlike other deaeration methods, this method can also be applied to deair solids (Narasimha Rao & Sachindra, 2002).

8.3.7 Enzymatic deaeration

Enzyme deaeration is the process where the enzymatic reactions cause the release of oxygen from the food system. One commonly used combination of enzymes is the glucose peroxidase and catalase system. Glucose oxidase causes the oxidation of glucose to gluconolactone and hydrogen peroxide. The catalase enzyme then acts on hydrogen peroxide to release the oxygen molecule from the food matrix. One mole of oxygen is released for every 2 moles of glucose oxidizing (Meyer & Isaksen, 1995). The dissolved oxygen can be reduced to as low as 1 ppm on enzymatic treatment in juices (Sagi & Mannheim, 1990). Similarly, laccase enzyme is utilized to stabilize beer by removing oxygen.

8.3.8 Centrifugal degassing

Centrifugal degassing is commonly applied in metallurgy, semiconductor manufacturing, and other processing industries. This finds fewer applications in the food industry as the food product process will be exposed to high shear and mechanical forces, which can damage the food matrix. The method could be more efficient and consistent in removing the oxygen to a desired level. The principle of centrifugal degassing is based on differences in the density of the gases dissolved and the liquid to be degassed. The separation occurs under the applied centrifugal forces.

8.3.9 Miscellaneous methods

This section covers the processes implemented to achieve certain goals, and the deaeration of the food system occurs as a secondary objective. Processes such as carbonation of the beverages, heating, boiling, and so on are among them. The carbonation of the beverage is carried out in a storage tank, and carbon dioxide gas is pressurized to dissolve it. During this process, dissolved oxygen can be released, which may be attributed to the changes in the pH level of the beverage after the addition.

8.4 Applications

In the dynamic world of fresh food preservation, degassing and deaeration play pivotal roles in ensuring that your perishable goods maintain their quality and taste for an extended period. This section will discuss specific industries where the deaeration of food is required to accomplish specific objectives.

8.4.1 Liquid food

8.4.1.1 Juices and concentrates

As discussed earlier, the degassing operation is most commonly practiced for fruit juices and concentrates to overcome the oxidative deterioration of the juices and concentrates. The conventional method of deaerating liquid food, such as juices and concentrates, uses vacuum deaeration in hermetically sealed tanks with 200 mbar or less pressure. The upper limit for pressure is suggested to be 500 mbar (Kosiński et al., 2024). Other methods are employed, such as membrane-based degassing, gas sparging, and gas stripping. The process of deaeration or degassing ensures a protective effect on bioactives such as ascorbic acid and total phenolic and flavonoid content during further processing. It also enables the production of fruit juices and concentrates with superior quality in color and flavor. Table 8.1 gives the different conditions used for the deaeration of juices and concentrates and the effect of degassing on the final product.

Table 8.1 Method of deaeration for liquid foods.

Fruit juice	Method of deaeration	Conditions	Effect of deaeration	References
Mango juice	Ultrasonication	40 kHz US bath 130 W, 25°C, 15-60 minutes	Reduction in ascorbic acid by 13 to 28 %	Aguilar et al. (2017)
Apple juice	Ultrasonication	25 kHz US bath 0.06 WmL^{-1} 20°C for 90 min	Increase in conc. of Ascorbic acid by 34 %	Aguilar et al. (2017)
Orange juice	Effect of Temperature while Degassing	10, 20, and 30°C and DO	Temperature and DO are inversely proportional to the quality of juice	Cai et al. (2019)
Orange juice	Vacuum pump	Vacuum maintained at 25 mm Hg	Loss of flavor and removal of air bubbles to avoid arcing during treatment	Jia et al. (1999)
Blood Orange juice	Replacement of Oxygen with Argon	Bubbling the argon gas for 30 minutes or more until the concentration of DO is reduced to less than 1 mg/L	Prevents degradation of ascorbic acid and color intensity	Remini et al. (2015)
Apple juice	Spiral-filter press	Pressure 150 mbar	Degradation of the Patulin	Assaf et al. (2020)
Apple juice	Vacuum tank with a rotating spray nozzle	25°C temperature and 150 rpm rotation speed	Reduction of oxygen by 80.4% in comparison with the traditional degassing process, bioactive components stability enhanced	Kosiński et al. (2024)
Orange juice	Ultrasonic-Vacuum Combined	900 W and 15.6 minute	Prevents degradation of ascorbic acid	Chen et al. (2023)
Dairy Beverage	Electroreduction	Anodic Part - 0.15 N H_2SO_4 and Cathodic Part - dairy beverage (280 mL each). Electrolysis was conducted for 60 minutes with a potentiostatic controller	Reduced lipid oxidation and non-enzymatic browning reactions	Giroux et al. (2008)

(*continued on next page*)

Table 8.1 Method of deaeration for liquid foods—cont'd

Fruit juice	Method of deaeration	Conditions	Effect of deaeration	References
Dairy Beverage	Vacuum pump	559 mm Hg for 60 minutes	Reduced lipid oxidation and non-enzymatic browning reactions	Giroux et al. (2008)
Carrot Juice	Vacuum pump	0.1 Mpa Pressure	Color lesser affected	Liao et al. (2007)
Extra Virgin Olive Oil	Nitrogen Stripping Technique	The flow rate of N_2 was 17 L/min, and the Olive oil flow rate and the temperature were 6 L/min and 21°C. The stripping pipe was 10 m long and had a 5 cm internal diameter.	Peroxide value improved, bio phenol content, K232 parameter, tocopherol, prevent depletion of Secoiridoids and reduce oxidation	Guerrini et al. (2018)
Cherry Concentrate	Ultrasound	380-Watt Power, 1.14 Watt/cm^2 intensity, 37, 80 kHz frequency, 20±2°C Temperature	Degassing of the feed	Shahraki et al. (2017)
Water in a rotor-stator reactor	Coupled vacuum-N_2 technique	Optimum conditions: Vacuum degree of 0.06 MPa, gas and liquid volumetric flow rate of 3 m^3/h and 0.4 m^3/h with rotational speed of 600 rpm, Composition gas used 99.5% N_2 and 0.5% O_2	Feasible approach for removing DO from water	Zhao et al. (2021)
Blood Orange juice	Replacement of Oxygen with Argon	Bubbling the argon gas for 30 min or more until the conc. of DO reduced to less than 1 mg/L	Prevents degradation of ascorbic acid and color intensity	Remini et al. (2015)
Water	Hollow Fiber membrane	The water and nitrogen flow rate is 20 m^3/h, and 18 m^3/h, inlet temperature of the water is 12°C, 150 mm Hg vacuum level	–	Martić et al. (2015)

8.4.1.2 Carbonated soft drinks

Carbonated beverages are highly valued by consumers for their fizziness. pH, titratable acidity, and sugar are the characteristic parameters that determine the quality of carbonated beverages (Gomes, 2011). The presence of dissolved oxygen causes obstacles in the process of carbonation. Deaeration is crucial for carbonated soft drink bottling. It is not used for the final product. Instead, the water used to produce carbonated beverages is degassed. Process water deaerated to low dissolved oxygen levels can improve carbonation consistency without needing constant operator adjustments and line stoppage. This leads to better product quality, faster line speeds, and higher yields on syrup consumption (Harbold & Park, 2008). Having less than 500 ppb of dissolved oxygen in the water used for carbonated beverages is recommended. Table 8.1 gives the conditions for the deaeration of water.

8.4.1.3 Alcoholic beverages

Wine and beer are the alcoholic beverages used for deaeration. The production of wine involves different stages, such as harvesting, crushing, and fermentation. The degassing is a mandatory step followed by the fermentation of the wine. The fermentation renders a lot of carbon dioxide in the wine; the presence of carbon dioxide affects the mouthfeel and the taste of the wine. Thus, deaerating the wine after the fermentation step is common.

In some cases, secondary malic to lactic acid fermentation is carried out; even here, the fermentation process follows a deaeration step. The carbon dioxide produced after fermentation is about 2000 mg/L (Pambianchi, 2012). Different wines contain different carbon dioxide levels in the final product; still, wines contain low levels of carbon dioxide, around 0.5 g/L of carbon dioxide (AtlasScientific, 2023). The degassing process is done using vacuum deaeration or simple agitation. Unlike still wine, it is desirable to dissolve carbon dioxide in beer, as it provides the characteristic effervescence to the beer.

Nevertheless, deaeration removes oxygen to avoid the oxidation of flavor compounds. A higher amount of dissolved oxygen in beer can affect the flavor and aroma of the beer. It is suggested that less than 50 ppb of dissolved oxygen be added to the final product for lager beer to have minimal impact on the quality attributes (Barnette & Shellhammer, 2019).

8.4.1.4 Milk and dairy products

Besides microbial spoilage, dairy products are most susceptible to oxidative damage, especially lipid oxidation. The milk and milk products undergo a deaeration process to overcome this oxidative damage. Similar to the beverages, this process is conducted before pasteurization. The method of physical deaeration of milk is more applicable than other methods as this poses less risk of indirect or intentional addition of an additive to the milk (Giroux et al., 2008). Another common issue in milk and other products is the presence of foam, which potentially reduces the product's yield and expedites oxidative damage. In such circumstances, ultrasonication of the milk and related products is a promising method (Ashokkumar et al., 2010). However, it should be noted that only the low frequency is effective in such cases (Villamiel et al., 2000). Similarly, a vacuum applied for a shorter duration (5 min) helped to get a smaller bubble size and increased the viscosity of the ice cream. However, a long vacuum application adversely affected the quality of ice creams, such as lower overrun (Lomolino et al., 2023).

8.4.1.5 Vegetable juices and puree

Similar to fruit juices, vegetable juices, and puree are loaded with substrate and inherent enzymes, which undergo oxidative changes and deteriorate. Deaeration of the juices and puree presents a prospectus option to achieve a clean label and quality product without adding additives and thermal treatments. The deaeration process is followed by the homogenization of the juices or puree before the sterilization or pasteurization (Liao et al., 2007). For instance, pumpkin purée was stabilized using a deaeration process. The process helps to reduce the loss of color (Contador et al., 2014).

8.4.1.6 Edible oil

The addition of antioxidant compounds, particularly natural herbs and spices, to edible oil has been practiced for ages; the physical degassing of oil is a relatively new method to prevent oxidative damage. The method can be done by simply heating the oil to release the gases, or advanced methods such as gas stripping and vacuum degassing can be practiced to eliminate oxygen. As previously discussed, the primary objective of the deaeration of edible oil is to extend the shelf life by depleting oxygen (Guerrini et al., 2018).

8.4.2 Solid foods

8.4.2.1 Confectionary

Conching is one of the steps involved in the manufacturing of chocolate. It is accomplished by continuous agitation or shaking of the chocolate mass to develop the chocolate's characteristic flavor, texture, and aroma as the primary object. During agitation, unwanted gases are removed, causing deaeration of the chocolate mass (Perego et al., 2004). The undesirable gases released mostly constitute oxygen, which prevents the oxidation of flavor and aroma compounds. On the other hand, confectionery products like Jimmy Candies or jellies are deaerated by heating to remove the bubbles to get a clear surface and texture (DeMars & Ziegler, 2001). Flashing water is also used during the extrusion of such candies (DeMars & Ziegler, 2001; Hartel et al., 2018). Table 8.2 summarizes the deaeration method employed for the solid food products, their conditions, and the effect of deaeration.

Table 8.2 Method of deaeration for solid food.

Food product	Method of deaeration	Conditions	Effect of deaeration	References
Cocoa-based confectionery	Vacuum Pump	25°C, 0.3 atmospheres absolute for 15 minutes	Remove bubbles	Meza et al. (2018)
Candy	Coil (Static) Cooker [Flashing off]	–	Sugar mass deaerated	Hartel et al. (2018)
Jelly-Candies	Steam Injection Cooker	Releasing the pressure leads to flashing off the water	Bubble-free candy syrup	Hartel et al. (2018)
Cocoa-based confectionery	Conching	75°C with the residence time 8 hour	Improves aroma	Perego et al. (2004)
Baking Industry	Manual Kneading or mechanical chewing up	–	Overall quality increases, scaling accuracy improves, and run time also increases.	Haegens (2014)
Bread	Squeezing and Vacuum Pump	Manually squeeze dough after fermentation, packing 20 g dough, and then vacuum treat for 1 minute.	The moisture content of degassed crumbs and the value of Young modulus improves.	Le-Bail et al. (2009)
Bread	Vacuum Pump	–	Resting time, porosity fraction, and specific volume of bread increased.	Jha et al. (2017)

8.4.2.2 Bakery

In contrast to other products, the deaeration or degassing of the product is not carried out to improve oxidative stability. Rather, degassing in the bakery industry is intended to create a desirable texture for the final product and to achieve scaling accuracy. The deaeration of the dough ensures a uniform density, consistent product, and improved handling. Another crucial advantage of deaeration is the higher run time for a batch, which ensures minimal scaling inaccuracies and better handling of the product without interruption or discontinuous production (Haegens, 2014).

8.4.3 Specific processing method

8.4.3.1 Canning

Deaeration is an important step in the thermal sterilization of canned foods. It helps to prevent the formation of air pockets in the can, which can lead to uneven heating and potential spoilage. Deaeration is part of the retort process, which involves heating food containers in pressurized retorts at defined temperatures for pre-established time intervals followed by sudden cooling (Gedraite et al., 2005). The deaeration while canning is done by tapping, replacing the gas in the headspace with an inert gas, or vacuum sealing the containers.

4.3.2 Packaging

Food, along with the packaging material, is a dynamic system where continuous exchange of gases or moisture occurs between the headspace, food material to and from the packaging materials, and the external environment. For example, as seen in the case of coffee, which releases carbon dioxide during storage, degassing or appropriate packaging is critical in this case as the released carbon dioxide can even lead to the bursting of coffee pouches (Shimoni & Labuza, 2000). To avoid such situations, the coffee is packed with gases to equilibrate the carbon dioxide in the headspace. This facilitates the headspace's degassing and a modified packaging atmosphere with gases. Similarly, fried food products are flushed with inert gases such as nitrogen to prevent oxidation and loss of textural properties. The flushing of gases causes the deaeration of gases in the head space. Another method of degassing while packaging is to vacuum packaging the food product, which is quite popular among dried dairy products. Milk is often spray-dried to prevent microbial spoilage; however, lipid oxidation is an inevitable challenge for the dried milk product industry. To overcome this effect, vacuum packaging or modified atmospheric packaging of these dried products can be used.

8.5 Challenges and future scope

The degassing process has evolved, and more advanced methods are now being applied to eliminate oxygen and entrapped air inside the food matrix. These methods work well for liquid food; their applications are limited to low-viscosity fluids. On the other hand, very few options are available for the deaeration of viscous fluids. Nevertheless, a novel approach was presented by Zhan et al. (2017), where low-frequency vertical vibrations were used for degassing the viscous fluid. Another consideration while degassing is that excess movement or displacement of the fluid food can lead to the entrapment of the air from the environment into the food, potentially countering the deaeration process (Zhan et al., 2017). This phenomenon is based on the amplitude and vibration applied during degassing. Thus, estimating a region where the frequency and amplitude of the applied method cause food deaeration is essential. Another aspect to consider with degassing, typically in vacuum-assisted degassing operations, is that in certain conditions, the food constituents may be separated from the food matrix under the applied vacuum conditions (Blekhman et al., 2013). Along similar lines, it is quite evident that during the degassing operations, along with the unwanted gasses, the volatile aromatic constituents can also be lost during the process. The loss of bioactive compounds when deaeration is carried out by thermal deaeration methods, such as vacuum or gas stripping, is another challenge to address. Furthermore, the process is inevitable to eliminate the gases; the deterioration of the textural properties amidst the deaeration process can be a concern for many food products (Dobiáš et al., 2006).

8.6 Concluding remarks

Deaeration of food as an intermediate step in food processing is critical for a longer shelf life and product quality. The process also ensures the processing line's smooth and continuous operation, which is important for maintaining the equipment. The method of deaeration and the operational conditions must be selected based on the nature of the product. Simple methods such as agitation or tapping to sophisticated methods such as ultrasound-assisted deaeration can be employed based on the degree of degassing required. The selection of the deaeration method should also be based on the economic consideration of the entire process.

References

Aguilar, K., Garvín, A., Ibarz, A., & Augusto, P. E. (2017). Ascorbic acid stability in fruit juices during thermosonication. *Ultrasonics Sonochemistry, 37*, 375–381. https://doi.org/10.1016/j.ultsonch.2017.01.029.

Akyildiz, A., & Ağçam, E. (2014). Citrus juices technology. In *Food processing: Strategies for quality assessment* (pp. 37–103). New York, NY: Springer.

Amaral, A. B., Silva, M. V. D., & Lannes, S. C. D. S (2018). Lipid oxidation in meat: Mechanisms and protective factors: A review. *Food Science and Technology, 38*, 1–15. https://doi.org/10.1590/fst.32518.

Ashokkumar, M., Bhaskaracharya, R., Kentish, S., Lee, J., Palmer, M., & Zisu, B. (2010). The ultrasonic processing of dairy products: An overview. *Dairy Science & Technology, 90*(2-3), 147–168. https://doi.org/10.1051/dst/2009044.

Assaf, C. E. H., De Clercq, N., De Paepe, E., Vlaemynck, G., Van Coillie, E., & Van Pamel, E. (2020). Effect of ascorbic acid, oxygen and storage duration on patulin in cloudy apple juice produced on a semi-industrial scale. *World Mycotoxin Journal, 13*(4), 487–497. https://doi.org/10.3920/WMJ2019.2528.

AtlasScientific. (2023). The role of CO_2 in wine production. https://atlas-scientific.com/blog/co2-in-wine/.

Bao, Y., & Ertbjerg, P. (2015). Relationship between oxygen concentration, shear force and protein oxidation in modified atmosphere packaged pork. *Meat Science, 110*, 174–179. https://doi.org/10.1016/j.meatsci.2015.07.022.

Barnette, B. M., & Shellhammer, T. H. (2019). Evaluating the impact of dissolved oxygen and aging on dry-hopped aroma stability in beer. *Journal of the American Society of Brewing Chemists, 77*(3), 179–187.

Bates, D., & Patist, A. (2010). Industrial applications of high power ultrasonics in the food, beverage, and wine industry. In *Case Studies in Novel Food Processing Technologies* (pp. 119–138). Woodhead Publishing.

Bellebna, Y., Ouiddir, R., Dey, Z., Miloudi, M., & Tilmatine, A. (2014). Robustness testing of the juice extraction process using a pulsed electrical field. *International Journal of Environmental Studies, 71*(3), 360–371. https://doi.org/10.1080/00207233.2014.916495.

Blekhman, I., Blekhman, L., Sorokin, V., Vaisberg, L., Vasilkov, V., & Yakimova, K. (2013). Motion of gas bubbles and rigid particles in vibrating fluid-filled volumes. *Procedia Iutam, 8*, 43–50. https://doi.org/10.1016/j.piutam.2013.04.007.

Cai, D., Tan, A., Sun, Z., Huang, L., & Dou, H. (2019). Effect of dissolved oxygen content on the quality of NFC orange juice degassed at different temperatures. *Food and Fermentation Industries, 45*(20), 169–175. https://doi.org/10.13995/j.cnki.11-1802/ts.021183.

Chen, H., Lu, Z., Cheng, Y., Drioli, E., Wang, Z., Zhang, F., & Cui, Z. (2023). Development and emerging application of membrane degassing technology. *Advanced Membranes, 3*, 100076. doi:10.1016/j.advmem.2023.100076.

Chen, Y., Zhang, M., Mujumdar, A. S., & Liu, Y. (2023). Study on the regulation of browning of concentrated orange juice by ultrasonic-vacuum combined with high-pressure CO_2 during storage. *Food and Bioprocess Technology*, 1–14. https://doi.org/10.1007/s11947-023-03276-x.

Contador, R., González-Cebrino, F., García-Parra, J., Lozano, M., & Ramírez, R. (2014). Effect of hydrostatic high pressure and thermal treatments on two types of pumpkin purée and changes during refrigerated storage. *Journal of food processing and preservation, 38*(2), 704–712. https://doi.org/10.1111/jfpp.12021.

Datta, N., & Tomasula, P. M. (2015). *Emerging dairy processing technologies: Opportunities for the dairy industry*. John Wiley & Sons.

DeMars, L. L., & Ziegler, G. R. (2001). Texture and structure of gelatin/pectin-based gummy confections. *Food Hydrocolloids, 15*(4-6), 643–653. https://doi.org/10.1016/S0268-005X(01)00044-3.

Dobiáš, J., Voldřich, M., & Čurda, D. (2006). Heating of canned fruits and vegetables: Deaeration and texture changes. *Journal of Food Engineering, 77*(3), 421–425. https://doi.org/10.1016/j.jfoodeng.2005.07.008.

FAO. (2021). SDG indicators. https://www.fao.org/faostat/en/#data/SDGB. Retrieved on August 20, 2024.

Gabrielli, M., Fracassetti, D., Romanini, E., Colangelo, D., Tirelli, A., & Lambri, M. (2021). Oxygen-induced faults in bottled white wine: A review of technological and chemical characteristics. *Food Chemistry, 348*, 128922. https://doi.org/10.1016/j.foodchem.2020.128922.

Gallego-Juárez, J., Rodríguez, G., Riera, E., & Cardoni, A. (2015). Ultrasonic defoaming and debubbling in food processing and other applications. In J. A. Gallego-Juárez, & K. F. Graff (Eds.), *Power ultrasonics* (pp. 793–814). Elsevier.

García-Torres, R., Ponagandla, N., Rouseff, R., Goodrich-Schneider, R., & Reyes-De-Corcuera, J. (2009). Effects of dissolved oxygen in fruit juices and methods of removal. *Comprehensive Reviews in Food Science and Food Safety, 8*(4), 409–423.

Gedraite, R., Ferreira, N. L., de Faria, M. A. S., & Gallego, A. G. (2005). The use of supervisory software as a optimization tool for can sterilization processes. Proceedings of COBEM 2005, 18th International Congress of Mechanical Engineering November 6–11, 2005, Ouro Preto, MG.

Geng, L., Liu, K., & Zhang, H. (2023). Lipid oxidation in foods and its implications on proteins. *Frontiers in Nutrition, 10*, 1192199. https://doi.org/10.3389/fnut.2023.1192199.

Giroux, H. J., St-Amant, J. B., Fustier, P., Chapuzet, J.-M., & Britten, M. (2008). Effect of electroreduction and heat treatments on oxidative degradation of a dairy beverage enriched with polyunsaturated fatty acids. *Food Research International, 41*(2), 145–153.

Gomes, C. É. (2011). Development of a robust degassing process for carbonated beverage using gas stripping. *The Boolean: Snapshots of Doctoral Research at University College Cork*, 71–73.

Guerrini, L., Masella, P., Angeloni, G., & Parenti, A. (2018). Stripping of dissolved oxygen from extra virgin olive oil: Effects on oxidation and biophenols. *Journal of Food Processing and Preservation, 42*(12), e13832.

Haegens, N. (2014). Mixing, dough making, and dough make-up. *Bakery Products Science and Technology*, 307–324. https://doi.org/10.1002/9781118792001.ch17.

Harbold, G., & Park, J. (2008). Using the GasTran deaeration system to achieve low dissolved oxygen levels for superior line speed and product quality: A case study in carbonated soft drink bottling. In *Paper presented at the Proceedings of the 7th International Conference on Process Intensification.*

Hartel, R. W., von Elbe, J. H., Hofberger, R., Hartel, R. W., von Elbe, J. H., & Hofberger, R. (2018). Jellies, gummies and licorices. *Confectionery Science and Technology*, 329–359.

Jääskeläinen, E., Hultman, J., Parshintsev, J., Riekkola, M.-L., & Björkroth, J. (2016). Development of spoilage bacterial community and volatile compounds in chilled beef under vacuum or high oxygen atmospheres. *International Journal of Food Microbiology, 223*, 25–32. https://doi.org/10.1016/j.ijfoodmicro.2016.01.022.

Jha, P. K., Chevallier, S., Cheio, J., Rawson, A., & Le-Bail, A. (2017). Impact of resting time between mixing and shaping on the dough porosity and final cell distribution in sandwich bread. *Journal of Food Engineering, 194*, 15–23. https://doi.org/10.1016/j.jfoodeng.2016.07.016.

Jia, M., Zhang, Q. H., & Min, D. B. (1999). Pulsed electric field processing effects on flavor compounds and microorganisms of orange juice. *Food Chemistry, 65*(4), 445–451. https://doi.org/10.1016/S0308-8146(98)00186-1.

Kanner, J., Harel, S., Fishbein, Y., & Shalom, P. (1981). Furfural accumulation in stored orange juice concentrates. *Journal of Agricultural and Food Chemistry, 29*(5), 948–949. https://doi.org/10.1021/jf00107a015.

Kapustin, A. (1954). The degassing of a liquid in an ultrasonic field. *Zu. Tehn. Fiz., 24*, 1008.

Kosiński, J., Cywińska-Antonik, M., Szczepańska-Stolarczyk, J., Jasińska, U. T., Woźniak, Ł., Kaniewska, B., & Marszałek, K. (2024). Application of an electromagnetic field for extending the shelf-life of not from concentrate (NFC) apple juice. *Applied Sciences, 14*(2), 662.

Le-Bail, A., Boumali, K., Jury, V., Ben-Aissa, F., & Zuniga, R. (2009). Impact of the baking kinetics on staling rate and mechanical properties of bread crumb and degassed bread crumb. *Journal of Cereal Science, 50*(2), 235–240. https://doi.org/10.1016/j.jcs.2009.05.008.

Leuner, H., Gerstenberg, C., Lechner, K., McHardy, C., Rauh, C., & Repke, J.-U. (2020). Overcoming unwanted foam in industrial processes of the chemical and food industry: An ongoing survey. *Chemical Engineering Research and Design, 163*, 281–294.

Liao, H., Sun, Y., Ni, Y., Liao, X., Hu, X., Wu, J., & Chen, F. (2007). The effect of enzymatic mash treatment, pressing, centrifugation, homogenization, deaeration, sterilization and storage on carrot juice. *Journal of Food Process Engineering, 30*(4), 421–435.

Lomolino, G., De Iseppi, A., Bravo, S., Vegro, M., Marangon, M., Crapisi, A., & Curioni, A. (2023). Effect of vacuum application on the physical, rheological and sensory characteristics of an artisanal ice cream. *International Journal of Dairy Technology, 76*(2), 418–428. https://doi.org/10.1111/1471-0307.12928.

Martić, I., Maslarević, A., Mladenović, S., Lukić, U., & Budimir, S. (2015). Water deoxygenation using hollow fiber membrane module with nitrogen as inert gas. *Desalination and Water Treatment, 54*(6), 1563–1567. https://doi.org/10.1080/19443994.2014.888677.

Meybodi, N. M., Mortazavian, A. M., Arab, M., & Nematollahi, A. (2020). Probiotic viability in yoghurt: A review of influential factors. *International Dairy Journal, 109*, 104793. https://doi.org/10.1016/j.idairyj.2020.104793.

Meyer, A. S., & Isaksen, A. (1995). Application of enzymes as food antioxidants. *Trends in Food Science & Technology, 6*(9), 300–304. https://doi.org/10.1016/S0924-2244(00)89140-2.

Meza, B. E., Carboni, A. D., & Peralta, J. M. (2018). Water adsorption and rheological properties of full-fat and low-fat cocoa-based confectionery coatings. *Food and Bioproducts Processing, 110*, 16–25. https://doi.org/10.1016/j.fbp.2018.04.005.

Microorganisms, F. S. (2015). food-spoilage-microorganisms-and-their-prevention. *Asian Journal of Plant Science and Research, 5*(4), 47–56.

Narasimha Rao, D., & Sachindra, N (2002). Modified atmosphere and vacuum packaging of meat and poultry products. *Food Reviews International, 18*(4), 263–293. https://doi.org/10.1081/FRI-120016206.

Pambianchi, D. (2012). Characterization of degassing equipment and its impact on wine chemistry.

Perego, P., Fabiano, B., Cavicchioli, M., & Del Borghi, M. (2004). Cocoa quality and processing: A study by solid-phase microextraction and gas chromatography analysis of methylpyrazines. *Food and Bioproducts Processing, 82*(4), 291–297.

Ramesh, M. (2020). Pasteurization and food preservation. In M. S. Rahman (Ed.), *Handbook of food preservation* (pp. 599–608). CRC Press.

Remini, H., Mertz, C., Belbahi, A., Achir, N., Dornier, M., & Madani, K. (2015). Degradation kinetic modelling of ascorbic acid and colour intensity in pasteurised blood orange juice during storage. *Food Chemistry, 173*, 665–673.

Sagi, I., & Mannheim, C. (1990). The effect of enzymatic oxygen removal on quality of unpasteurized and pasteurized orange juice. *Journal of Food Processing and Preservation, 14*(4), 253–266. https://doi.org/10.1111/j.1745-4549.1990.tb00132.x.

Shahraki, M. H., Maskooki, A., & Faezian, A. (2017). Ultrafiltration of cherry concentrate under ultrasound with carbonated feed as a new fouling control method. *Journal of Food Processing and Preservation, 41*(2), e12795. https://doi.org/10.1111/jfpp.12795.

Shimoni, E., & Labuza, T. P. (2000). Degassing kinetics and sorption equilibrium of carbon dioxide in fresh roasted and ground coffee. *Journal of Food Process Engineering, 23*(6), 419–436. https://doi.org/10.1111/j.1745-4530.2000.tb00524.x.

Taha, A., Mehany, T., Pandiselvam, R., Siddiqui, Anusha, S. , Mir, N. , A., Malik, M. A., Sujayasree, O. J., Alamuru, K. C., Khanashyam, A. C., Casanova, F., Xu, X., Pan, S., & Hu, H. (2022). Sonoprocessing: Mechanisms and recent applications of power ultrasound in food. *Critical Reviews in Food Science and Nutrition, 64*(17), 6016–6054. https://doi.org/10.1080/10408398.2022.2161464.

Tamime, A. Y. (2002). Fermented milks: A historical food with modern applications: A review. *European Journal of Clinical Nutrition, 56*(4), S2–S15. https://doi.org/10.1038/sj.ejcn.1601657.

Tateo, A., Maggiolino, A., Domínguez, R., Lorenzo, J. M., Dinardo, F. R., Ceci, E., Marino, R., Malva, A. D., Bragaglio, A., & Palo, P. (2020). Volatile organic compounds, oxidative and sensory patterns of vacuum aged foal meat. *Animals, 10*(9), 1495. https://doi.org/10.3390/ani10091495.

Van Pelt, S., & Churman, H. (2015). Offshore membrane deaeration as a replacement for vacuum tower deaeration: A comparative study. In *Paper presented at the SPE Annual Technical Conference and Exhibition,*.

Vardar-Sukan, F. (1998). Foaming: Consequences, prevention and destruction. *Biotechnology Advances, 16*(5-6), 913–948. https://doi.org/10.1016/S0734-9750(98)00010-X.

Villamiel, M., Verdurmen, R., & Jong, P. d. (2000). Degassing of milk by high-intensity ultrasound.

Villamiel, M., Riera, E., & García-Pérez, J. V. (2021). The use of ultrasound for drying, degassing and defoaming of foods. *Innovative Food Processing Technologies*. Elsevier http://doi.org/10.1016/b978-0-08-100596-5.22957-0.

Zhan, X., Shen, B., He, Y., Sun, Z., & Li, X. (2017). Application of low frequency vibration to degassing of highly viscous shear-thinning fluids. *Vibroengineering Procedia, 16*, 1–6. https://doi.org/10.21595/vp.2017.19007.

Zhang, B.-Y., Samapundo, S., Pothakos, V., Sürengil, G., & Devlieghere, F. (2013). Effect of high oxygen and high carbon dioxide atmosphere packaging on the microbial spoilage and shelf-life of fresh-cut honeydew melon. *International Journal of Food Microbiology, 166*(3), 378–390. https://doi.org/10.1016/j.ijfoodmicro.2013.08.002.

Zhao, Z., Liu, Z., Xiang, Y., Arowo, M., & Shao, L. (2021). Removal of dissolved oxygen from water by nitrogen stripping coupled with vacuum degassing in a rotor–stator reactor. *Processes, 9*(8), 1354.

Section II

Application of mechanical separation operations in the food industry

Application of sedimentation in the dairy industry

9

Celale Kirkin and Nisa Akkuzu
Department of Food Engineering, Faculty of Chemical and Metallurgical Engineering, Istanbul Technical University, Maslak, Istanbul, Türkiye

9.1 Introduction

Sedimentation is a unit operation that can be applied for the purpose of mechanical separation in the food industry. It utilizes gravitational forces to separate particles including dust and dirt from a fluid generally according to Stoke's law (Earle, 1983). Sedimentation can be used in the separation of milk components, quality evaluation, and treatment of wastewater in the dairy industry. On the other hand, sedimentation naturally occurs in milk as a layer of mostly protein accumulated at the bottom of the package (Gaur et al., 2018). The occurrence of sedimentation in dairy products depends on several parameters, such as the method and temperature of the heat treatment, time and temperature of storage, type of milk, milk proteins, pH, calcium concentration, etc (Anema, 2019). Adjusting the pH and concentration of calcium, and addition of polysaccharides are some of the methods that can be used to control sedimentation in dairy products (Achaw & Danso-Boateng, 2021; Anema, 2019; Barone et al., 2021; Crowley et al., 2014). The factors that affect sedimentation in milk and dairy products and prevention methods are further explained in the following sections. The significance of the sedimentation process and its uses in the dairy industry are summarized in Fig. 9.1.

This chapter provides a comprehensive review of the applications of sedimentation in the dairy industry mainly focusing on the studies of the last decade. The use of sedimentation in the dairy processing industry, the occurrence of sedimentation in dairy products as a quality defect, methods used for the prevention of sedimentation, the utilization of the sedimentation process as a quality control tool. Additionally, the chapter explores the treatment of dairy industry wastewater by sedimentation.

9.2 Sedimentation

Sedimentation is the separation of solid or liquid particles from a fluid employing gravitational forces (Earle, 1983). It follows Stoke's law as shown in Eq. 9.1.

$$v_m = D^2 g(\rho_p - \rho_f)/18\mu \tag{9.1}$$

where v_m is the terminal velocity, D is the particle diameter, g is the gravitational acceleration, ρ_p is the particle density, ρ_f is the fluid density, and μ is the fluid viscosity

Mechanical Separation Processes in the Food Industry: Unit Operations and Processing Equipment in the Food Industry.
DOI: https://doi.org/10.1016/B978-0-12-819523-9.00013-5

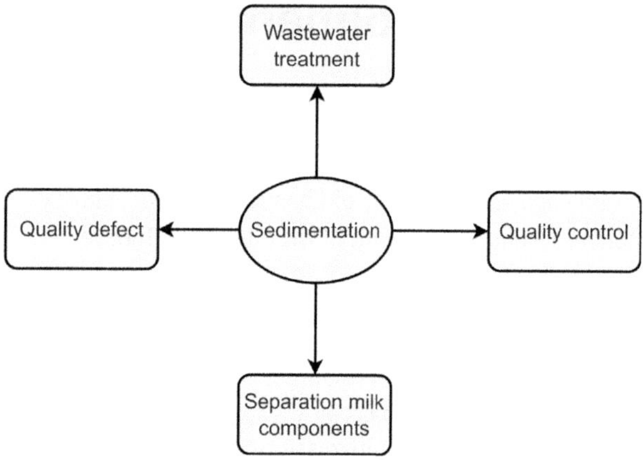

Figure 9.1 Sedimentation process and its uses in the dairy industry.

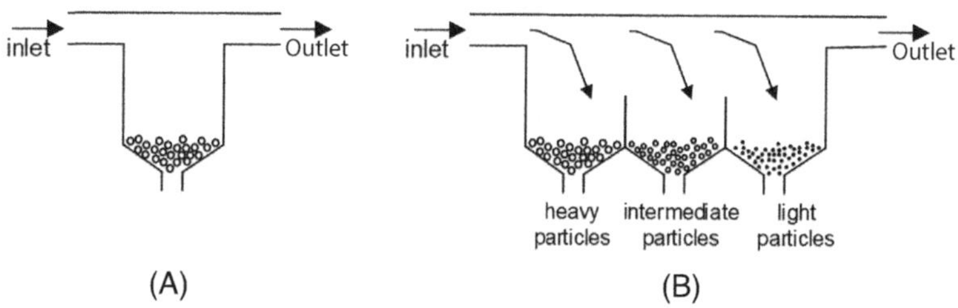

Figure 9.2 Schematic of (A) simple sedimentation tank and (B) gravity settling tank.

(Earle, 1983). Accordingly, it can be concluded that low viscosity, high particle size, and high density differences promote sedimentation.

In the food industry, sedimentation is primarily used to remove dirt, debris, dust, and particles from fluid streams or separate crystals from the mother liquor (Earle, 1983). Solids in a liquid can be separated according to Stoke's law using a sedimentation tank, where the solid particles are settled and collected from the bottom (Sayyad et al., 2022). Besides, the separation of dried material from the hot air at the end of a spray drying process can be given as an example of the sedimentation of particles in a gas (Sayyad et al., 2022). Gravity settling tanks, Spitzkasten gravity settling chambers, simple classifiers, and clarifier-thickener units can be used for sedimentation (Bürger et al., 2005; Sayyad et al., 2022). Cyclones, in which gravitational force is combined with centrifugal force, can increase the rate of sedimentation (Earle, 1983). A schematic of a sedimentation tank and a gravity settling tank is demonstrated in Fig. 9.2.

9.3 Sedimentation in dairy products

Sedimentation occurs as a quality defect upon heat treatment and storage of milk (Akkerman et al., 2021). It can be described as a layer observed at the bottom of the bottle during storage (Akkerman et al., 2021). Sedimentation starts after production and is increased by increasing temperature and prolonged storage (Akkerman et al., 2021; Anema, 2019). Sedimentation is considered a quality loss when it reaches high levels (Anema, 2019). In addition, it compromises the sensory quality of dairy products (Jansson et al., 2019).

Native casein micelles can sediment in UHT milk during storage, but generally, it is not regarded as a problem, as it is slow and can be dispersed again upon shaking (Anema, 2019). Factors that affect sedimentation in ultra-high temperature (UHT) milk include storage time, type of UHT processing, heating surface, processing time, temperature, level of protein, homogenization pressure, pH, amount of soluble calcium, whey protein to casein ratio, milk quality, proteolytic enzymes, microbial contamination, etc (Anema, 2019; Deeth, 2020; de Souza et al., 2023). During UHT treatment of milk, proteins are denatured and protein-protein crosslinks are formed causing sedimentation (Rutigliano et al., 2020). Sedimentation in milk is also associated with the content of κ-casein (Anema, 2019). Heat treatment can cause κ-casein to separate from the surface of the micelles lowering the stability (Anema, 2019). Sediments from UHT milk are generally composed of κ-casein-depleted casein and denatured whey proteins at lower levels (Anema, 2019; Gaur et al., 2018). Even though the sediment of UHT milk mainly consists of κ-casein depleted micelles; α-lactalbumin and β-lactoglobulin are also present in lower levels (Eisner, 2021). It was observed that α-lactalbumin and β-lactoglobulin form complexes at room temperature and aggregate after UHT treatment (Zhang et al., 2023). These aggregates have high molecular weight and particle size causing a decrease in zeta potential and increase in sedimentation (Zhang et al., 2023). The presence of large natural casein micelles was associated with a higher amount of sediment compared to smaller micelles (Aguilera-Toro et al., 2022). Also, the rate of sedimentation increases with the increase in the molecular weight of micelles (Anema, 2022). The incorporation of fibers into milk can also favor sedimentation due to the increase in the amount of large particles that are prone to aggregation (Rezvani & Goli, 2024). On the other hand, it was suggested that buffalo milk had higher fat content and bigger fat globules than bovine milk, and the fat globules around the casein micelles can prevent aggregation but force them to rise to the top (Mejares et al., 2022).

In addition, it was claimed that sedimentation in goat milk was high, as it had a high ionic calcium content and was low in α-s1-casein (Wu et al., 2020). Furthermore, the heat stability of donkey milk was lower and sedimentation was higher than that of bovine milk (Luo et al., 2019). The authors also indicated that the κ-casein content in the donkey milk casein micelles was lower than the detection limit, and the micelles were mainly made up of α_S- and β-casein. The lower the amount of κ-casein in micelles the higher the susceptibility to aggregation (Luo et al., 2019).

The sedimentation of whey protein was higher in the sediment from indirect UHT milk compared to direct UHT (Gaur et al., 2018). Also, direct UHT milk was reported to be more susceptible to sedimentation than indirect UHT milk (Gaur et al., 2018). In addition, sedimentation was higher after steam injection than steam infusion (Malmgren et al., 2017). Substitution of milk with whey in the production of ayran caused higher sedimentation and lower viscosity and pH (Rezazadeh-Bari et al., 2019).

On the other hand, (Karlsson et al., 2019) observed that the sedimentation was higher in UHT milk stored at 4°C than at 37°C. The authors claimed that the sedimentation was related to β-casein dissociation at 4°C.

Sedimentation develops slowly and naturally in milk products by the settling of the colloidal particles according to Stoke's law (Anema, 2019). It has been claimed that at low pH and/or high ionic calcium levels the particles in milk aggregate and increase the particle size accelerating the sedimentation (Lewis et al., 2011). The unstable casein micelles can aggregate and sediment when the pH is low or the level of calcium ions is high (Anema, 2019). Furthermore, it has been reported that sedimentation in sheep milk during heating was related to the pH (Pan et al., 2023).

Proteolysis can also contribute to sedimentation in UHT milk (Akkerman et al., 2021). If plasmin is not inactivated during milk processing, it can promote proteolysis and sedimentation (Fan et al., 2024; Graf et al., 2021; Yun & Imm, 2021). It was reported that remaining plasmin activity can cause protein hydrolysis and subsequently, sedimentation in heat-treated milk when stored at a relatively high temperature (37°C) (Fan et al., 2024). Additionally, sedimentation increased with the increase in plasmin concentration during storage at 23°C, but extensive proteolysis at 37°C resulted in a negative correlation between the plasmin concentration and sediment amount (Yun & Imm, 2021). Moreover, it was claimed that sedimentation is induced by the production of agglomerates that are composed of casein particles and micro-fat globules in addition to proteolysis (Malmgren et al., 2017).

Apart from being a quality problem, sedimentation can be valorized in the separation of milk components. For example, casein can be extracted from milk by isoelectric sedimentation (Yukalo et al., 2019). In addition, it can be utilized in the separation of whey proteins (Haller, 2023). For the production of α-lactalbumin, whey is subjected to heating for denaturation and precipitation. The resulting precipitate is then recovered by settling and decanting (or centrifugation) followed by washing, drying, grinding, blending, and bagging (Pearce, 1992).

9.4 Controlling sedimentation in dairy products

A summary of studies evaluating sedimentation in dairy products upon several treatments conducted in the last decade is demonstrated in Table 9.1.

Sedimentation in UHT milk can be controlled by using calcium chelators or by increasing the pH (Anema, 2019). Trisodium citrate was found effective in decreasing ionic calcium content and increasing pH (Gaur et al., 2018). The addition of polysaccharides can also contribute to the control of sedimentation in products, such as heat-treated dairy products and acidified milk drinks (de Souza et al., 2023; Kazemalilou & Alizadeh, 2017; Ntazinda et al., 2014). Carboxymethyl cellulose was found effective in improving the stability of milk drinks in some studies (Wu et al., 2013). It has been reported that carboxymethyl cellulose can be used to cover the casein micelles and prevent aggregation (Wu et al., 2013).

Preventing the contamination of heat-treated milk with microorganisms with high thermal resistance is crucial not only for food safety but also for quality. For example,

Table 9.1 Summary of recent studies evaluating the effects of different treatments on sedimentation in dairy products.

Sample	Treatment	Major findings	References
Acidified milk drink	Carboxymethyl cellulose (CMC) addition	Higher degrees of substitution and molecular weights of CMC led to a decrease in the sedimentation fractions. Unadsorbed CMC increases the viscosity delaying sedimentation.	Cai et al. (2018)
Acidified milk drink	Addition of soybean soluble polysaccharide (SSPS)	As the degree of esterification of SSPS coating the casein micelles decreased, the stability of acidified drinks increased by improving the electrostatic repulsion.	Cai et al. (2020)
Acidified milk drink	Addition of pea soluble polysaccharide (PSPS) and SSPS	Higher molecular weight PSPS exhibited stability in acidified milk drinks comparable to that of SSPS, whereas lower molecular weight PSPS required a higher concentration to achieve the same level of stability.	Cheng et al. (2018)
Acidified milk drink	Addition of potato pulp pectic polysaccharide (PPP), low methoxyl pectin (LMP), and high methoxyl pectin (HMP)	The stability of beverages containing PPP and HMP was higher compared to those with LMP. PPP adsorbs onto casein micelles creating steric hindrance. Also, the non-adsorbed PPP increases the viscosity of the serum.	Sun et al. (2020)
Acidified milk drink	Addition of methyl alginate	The use of methyl alginate in combination with other polysaccharide stabilizers showed better sedimentation prevention than methyl alginate alone. Increased degree of esterification improved stability.	Xia et al. (2019)
Acidified milk drink	Addition of agar oligosaccharides	The combination of agar oligosaccharides with other polysaccharides created a structured network with casein micelles and decreased the sedimentation fractions; however, it was not enough for stabilizing casein micelles when used alone.	Zhao et al. (2018)

(continued on next page)

Table 9.1 Summary of recent studies evaluating the effects of different treatments on sedimentation in dairy products—cont'd

Sample	Treatment	Major findings	References
Cocoa milk	Addition of flaxseed mucilage powder and stevia followed by ultrasonication	Increasing the level of flaxseed mucilage and stevia increased sedimentation. Sedimentation also increased with prolonged exposure to ultrasonication.	Azarpazhooh et al. (2021)
Chocolate flavoured milk	Addition of microcrystalline cellulose	The addition of microcrystalline cellulose at a concentration of 0.10%(w/w) was recommended to decrease sedimentation.	Sumonsiri et al. (2018)
Enteral high protein dairy beverage	Modification of calcium amount of milk protein concentrate and pH adjustment using sodium hexametaphosphate	Sedimentation was prevented in samples with 20% Ca reduction and 0% sodium hexametaphosphate.	Pandalaneni et al. (2019)
Fermented milk beverage	Addition of modified high methoxyl pectin	Concentrations above 0.4% provided better stability.	Li et al. (2018)
Goat milk	pH adjustment	Sedimentation was significant in full fat goat milk when the pH was adjusted from 6.7 to 5.2.	Wu et al. (2020)
High protein dairy beverage	Preheating in combination with homogenization Addition of sugar or corn soluble fibre	Sedimentation was reduced after the pre-treatment. The addition of sugar or corn fibre did not affect sedimentation.	de Souza et al. (2023)
Low sugar flavoured milk drink	Addition of liquorice flavour with varying cacao powder to liquorice extract powder ratios.	The optimum formulation included 0.65:0.35 cacao powder: liquorice extract powder and 5% sugar with increased sedimentation stability.	Azami et al. (2018)
Milk-based smoothie	Addition of horse gram extract	Sedimentation increased with the level of the horse gram extract.	Panda et al. (2023)
Probiotic yogurt drink	Addition of bitter almond gum	The addition of bitter almond gum increased the stability. Absorption of the gum onto the casein micelles helped prevent casein micelle aggregation.	Hashemi and Hosseini (2021)
Stirred type yogurt and yogurt drink	Addition of apple pomace	The addition of apple pomace improved stability.	Wang et al. (2020)
UHT milk	Microfiltration and double homogenization	Microfiltration delayed sedimentation.	D'Incecco et al. (2018)

Bacillus sporothermodurans can contribute to proteolysis during storage, and the proteolysis of casein can lead to sedimentation (Tabit, 2018). Moreover, the bacterial protease AprX from *Pseudomonas fluorescens* increased sedimentation in milk (Zhang et al., 2018, 2020). This can be explained by the fact that plasmin and bacterial proteases remain in milk after the UHT treatment can cause proteolysis of proteins during storage (Matéos et al., 2015; Rauh et al., 2014; Zhang et al., 2018, 2020). While plasmin causes the hydrolysis of α- and β-casein on the fat globules leading to creaming, AprX causes the hydrolysis of κ-casein leading to the flocculation of casein micelles and fat globules (Zhang et al., 2020). Therefore, heat treatment of dairy products to inactivate microorganisms as well as enzymes is crucial for the shelf life and sensory quality.

Homogenization at low or high pressure can also contribute to the controlling of sedimentation (Tribst et al., 2020). It is considered essential in dairy processing to prevent sedimentation and creaming (Wu et al., 2014). The combination of high-temperature short time and high pressure in the treatment of donkey milk reduced microbial contamination, slowed pH change, and decreased sedimentation rate (Gong et al., 2024).

Sedimentation is also a challenge for dairy products fortified with calcium, as the addition of calcium increases the amount of ionic calcium, decreasing the zeta potential and stability (Barone et al., 2021; Crowley et al., 2014). In this context, the selection of the calcium source becomes important. Using calcium hydroxide as the fortificant in reconstituted skim milk powder resulted in an increased sedimentation rate (Crowley et al., 2014). Furthermore, decreasing the particle size of calcium salts can also decrease the sedimentation rate (Barone et al., 2021).

In addition, ingredients such as green tea catechins can also help reduce sedimentation in milk products (Jansson et al., 2019). The use of stabilizers, such as carboxymethyl cellulose, pectin, and starch is recommended to control sedimentation in yogurt drinks (Achaw & Danso-Boateng, 2021).

Sedimentation is high in high-protein dairy drinks (Abdallah et al., 2022; Pandalaneni et al., 2019). Additionally, during the rehydration of milk protein powders, it is very possible that some of the powder cannot be dissolved. Pathania et al. (2018) suggested that hydrodynamic cavitation treatment can improve the stability of rehydrated milk protein concentrate solutions.

Pretreatment of UHT high protein drink containing casein and whey, using a two-step preheating combined with homogenization, effectively prevented further denaturation of the whey protein and aggregation (de Souza et al., 2024). A similar finding was also reported in another study (de Souza et al., 2023).

Preheated whey protein isolate and pectin complexes could decrease the precipitation of proteins with polyphenols (Thongkaew et al., 2014). Polysaccharides adsorbed on proteins increase electrostatic repulsion (Wagoner et al., 2016). Moreover, crosslinking of whey protein isolate microgels with tannic acid reduced sedimentation improving the physical stability and potential to be used in pickering emulsion formulation (do Prado Silva et al., 2021). Furthermore, dos Santos et al. (2024) suggested that adding nisin and *Lacticaseibacillus casei* could improve the stability of goat and sheep cheese whey after acidification, and these products can be used in developing new food formulations. Moreover, the size reduction of whey protein concentrate can improve stability by preventing stability (Kelly, 2019).

9.5 Sedimentation test for quality control of dairy products

Sedimentation has been employed as a tool to assess the quality parameters of dairy products. It is one of the methods that can be used to evaluate the heat stability of milk products (Singh, 2004).

Hu et al. (2020) developed a method utilizing sedimentation, which combined recombinase polymerase amplification with polymer flocculation sedimentation to detect *Staphylococcus aureus* in food products, including milk. Sedimentation has also been used to evaluate the physical stability of emulsions containing dairy ingredients (do Prado Silva et al., 2021). Phase separation is an important quality defect in stirred yogurt, sedimentation tests have been used to observe the physical stability of these products (Wijesekara et al., 2022). Furthermore, sedimentation of milk was also related to gastric digestion (Mulet-Cabero et al., 2017, 2019). It is linked with an early emptying profile of nutrients (Mulet-Cabero et al., 2017; Zaeim et al., 2022). Thus, sedimentation tests can provide information on the digestion of dairy products. Additionally, sedimentability tests can also be utilized to assess the contribution of different methods in the treatment of wastewater from the dairy industry. For example, Rivas et al. (2010) evaluated the settling of sludge from cheese production after the addition of ferrous sulfate and aluminum sulfate by a sedimentability test.

9.6 Treatment of wastewater from dairy processes

It was claimed that 0.2-20 L of wastewater is produced per L of milk processed (Balannec et al., 2005). Dairy industry wastewater also has high chemical and biological oxygen demand (Balannec et al., 2005; Formentini-Schmitt et al., 2013; Tchamango et al., 2010). This can cause great environmental concern if not treated. Sedimentation is one of the methods that has been used for the treatment of dairy wastewater.

A sedimentation tank is utilized to remove microorganisms and small particles (Garg et al., 2020). The settling of the sediment is improved by the longevity of the undisturbed effluent (water) (Garg et al., 2020). Aluminum sulphate, ferric sulphate, poly aluminum chloride, or natural coagulants can be used to accelerate sedimentation (Garg et al., 2020).

Moringa oleifera has been suggested as a potential coagulant for the treatment of dairy wastewater during the coagulation/flocculation/sedimentation process (Mateus et al., 2017). It was stated that when the coagulation/flocculation/sedimentation process using *M. oleifera* as a coagulant was applied in combination, the quality of the treated water was improved (Formentini-Schmitt et al., 2013). Also, it was successfully used in removing chemical oxygen demand, turbidity, color, and calcium when combined with nanofiltration (Formentini-Schmitt et al., 2023). A similar finding was also reported by Mateus et al. (2017). In addition, Bialeski et al. (2024) recommended the treatment of coagulation/flocculation/sedimentation using *M. oleifera* functionalized with magnetic nanoparticles as the coagulant for the treatment of dairy wastewater.

9.7 Conclusions

Sedimentation technology has a wide range of applications in the dairy industry, such as quality assessment, separation of components, and wastewater treatment. As sedimentation naturally occurs due to gravitational forces, it can be observed in stored dairy products as a precipitate at the bottom of the container. However, this is reversible and not regarded as a quality defect unless at extreme levels. Sedimentation can generally be prevented by controlling the pH and concentration of ionic calcium, addition of stabilizers, inactivating microorganisms and proteolytic enzymes, etc. Future studies can focus on further explaining the occurrence and control mechanisms of sedimentation in dairy products, mainly high-protein, and acidified products.

References

Abdallah, M., Azevedo-Scudeller, L., Hiolle, M., Lesur, C., Baniel, A., & Delaplace, G. (2022). Review on mechanisms leading to fouling and stability issues related to heat treatment of casein-based RTD beverages. *Food Bioprod. Process., 136*, 67–83. https://doi.org/10.1016/j.fbp.2022.09.009.

Achaw, O.-W., & Danso-Boateng, E. (2021). *Chemical and process industries*. Springer International Publishing https://doi.org/10.1007/978-3-030-79139-1_11.

Aguilera-Toro, M., Poulsen, N. A., Akkerman, M., Rauh, V., Larsen, L. B., & Nielsen, S. D.-H. (2022). Development in Maillard reaction and dehydroalanine pathway markers during storage of UHT milk representing differences in casein micelle size and sedimentation. *Foods, 11*(10), 1525. https://doi.org/10.3390/foods11101525.

Akkerman, M., Johansen, L. B., Rauh, V., Poulsen, N. A., & Larsen, L. B. (2021). Contribution of casein micelle size and proteolysis on protein distribution and sediment formation in UHT milk during storage. *Int. Dairy J., 117*, 104980. https://doi.org/10.1016/j.idairyj.2021.104980.

Anema, S. G. (2019). Age gelation, sedimentation, and creaming in UHT milk: A review. *Compr. Rev. Food Sci. Food Saf., 18*(1), 140–166. https://doi.org/10.1111/1541-4337.12407.

Anema, S. G. (2022). Layering of proteins in stored UHT milk samples. *Int. Dairy J., 129*, 105345. https://doi.org/10.1016/j.idairyj.2022.105345.

Azami, T., Niakousari, M., Hashemi, S. M. B., & Torri, L. (2018). A three-step sensory-based approach to maximize consumer acceptability for new low-sugar licorice-chocolate-flavored milk drink. *LWT, 91*, 375–381. https://doi.org/10.1016/j.lwt.2018.01.064.

Azarpazhooh, E., Rashidi, H., Sharayei, P., Behmadi, H., & Ramaswamy, H. S. (2021). Effect of flaxseed-mucilage and Stevia on physico-chemical, antioxidant and sensorial properties of formulated cocoa milk. *Food Hydrocolloids for Health, 1*, 100017. https://doi.org/10.1016/j.fhfh.2021.100017.

Balannec, B., Vourch, M., Rabiller-Baudry, M., & Chaufer, B. (2005). Comparative study of different nanofiltration and reverse osmosis membranes for dairy effluent treatment by dead-end filtration. *Sep. Purif. Technol., 42*(2), 195–200. https://doi.org/10.1016/j.seppur.2004.07.013.

Barone, G., O'Regan, J., Kelly, A. L., & O'Mahony, J. A (2021). Physicochemical and bulk handling properties of micronised calcium salts and their application in calcium fortification of whey protein-based solutions. *J. Food Eng., 292*, 110213. https://doi.org/10.1016/j.jfoodeng.2020.110213.

Bialeski, D. D., Triques, C. C., Carvalho, J. K., Rodrigues, M. L. F., Dall'Oglio, I. C., Silva-Lucca, R. A., & Fagundes-Klen, M. R. (2024). Natural coagulant functionalized with magnetic nanoparticles better treated dairy wastewater compared to bioremediation. *Waste Biomass Valorization*. https://doi.org/10.1007/s12649-024-02585-2.

Bürger, R., Karlsen, K. H., & Towers, J. D. (2005). A model of continuous sedimentation of flocculated suspensions in clarifier-thickener units. *SIAM J. Appl. Math., 65*(3), 882–940. https://doi.org/10.1137/04060620X.

Cai, Z., Wei, Y., Guo, Y., Ma, A., & Zhang, H. (2020). Influence of the degree of esterification of soluble soybean polysaccharide on the stability of acidified milk drinks. *Food Hydrocolloids, 108*, 106052. https://doi.org/10.1016/j.foodhyd.2020.106052.

Cai, Z., Wu, J., Du, B., & Zhang, H. (2018). Impact of distribution of carboxymethyl substituents in the stabilizer of carboxymethyl cellulose on the stability of acidified milk drinks. *Food Hydrocolloids, 76*, 150–157. https://doi.org/10.1016/j.foodhyd.2016.12.034.

Cheng, M., Qi, J.-R., Feng, J.-L., Cao, J., Wang, J.-M., & Yang, X.-Q. (2018). Pea soluble polysaccharides obtained from two enzyme-assisted extraction methods and their application as acidified milk drinks stabilizers. *Food Res. Int., 109*, 544–551. https://doi.org/10.1016/j.foodres.2018.04.056.

Crowley, S. V, Kelly, A. L., & O'Mahony, J. A (2014). Fortification of reconstituted skim milk powder with different calcium salts: Impact of physicochemical changes on stability to processing. *Int. J. Dairy Technol., 67*(4), 474–482. https://doi.org/10.1111/1471-0307.12150.

Deeth, H. C. (2020). The effect of UHT processing and storage on milk proteins. In M. Boland, & H. Singh (Eds.), *Milk proteins* (3rd ed.) (pp. 385–421). Elsevier https://doi.org/10.1016/B978-0-12-815251-5.00010-4.

de Souza, A. B., Stephani, R., & Tavares, G. M. (2023). Stability of milk proteins subjected to UHT treatments: Challenges and future perspectives. *Crit. Rev. Food Sci. Nutr.*, 1–11. https://doi.org/10.1080/10408398.2023.2250865.

de Souza, A. B., Xavier, A. A. O., Stephani, R., & Tavares, G. M. (2023). Sedimentation in UHT high-protein dairy beverages: Influence of sequential preheating coupled with homogenisation or supplementation with carbohydrates. *Int. Dairy J., 137*, 105505. https://doi.org/10.1016/j.idairyj.2022.105505.

de Souza, A. B., Xavier, A. A. O., Stephani, R., & Tavares, G. M. (2024). Prior denaturation and aggregation of whey proteins: Is this a useful strategy for increasing the content of these proteins in UHT high-protein dairy beverages? *Food Bioprod. Process., 144*, 43–49. https://doi.org/10.1016/j.fbp.2023.12.008.

D'Incecco, P., Rosi, V., Cabassi, G., Hogenboom, J. A., & Pellegrino, L. (2018). Microfiltration and ultra-high-pressure homogenization for extending the shelf-storage stability of UHT milk. *Food Res. Int., 107*, 477–485. https://doi.org/10.1016/j.foodres.2018.02.068.

do Prado Silva, J. T., Benetti, J. V. M., Alexandrino, T. T. de B., Assis, O. B. G., de Ruiter, J., Schroën, K., & Nicoletti, V. R. (2021). Whey protein isolate microgel properties tuned by crosslinking with organic acids to achieve stabilization of Pickering emulsions. *Foods, 10*(6), 1296. https://doi.org/10.3390/foods10061296.

dos Santos, F. R., Martins Filho, C. M., de Cerqueira, R. F. L., Yada, R. Y., Augusto, P. E. D., Leite Junior, B. R. de C., & Tribst, A. A. L. (2024). Strategies to extend the shelf life of sheep and goat cheese whey under refrigeration: Nisin, bioprotective culture, and acidification. *Food Biosci., 57*, 103495. https://doi.org/10.1016/j.fbio.2023.103495.

Earle, R. L. (1983). *Unit operations in food processing* (2nd ed.). Pergamon Press https://doi.org/10.1016/B978-0-08-025536-1.50003-5.

Eisner, M. D. (2021). Direct and indirect heating of milk: A technological perspective beyond time–temperature profiles. *Int. Dairy J., 122*, 105145. https://doi.org/10.1016/j.idairyj.2021.105145.

Fan, K., Wu, P., Guo, M., Wang, Y., Cao, Y., Wang, P., Ren, F., & Luo, J. (2024). Destabilization of ultra-instantaneous ultra-high-temperature sterilized milk stored at different temperatures. *J. Dairy Sci., 107*(8), 5460–5472. https://doi.org/10.3168/jds.2024-24705.

Formentini-Schmitt, D. M., Alves, Á. C. D., Veit, M. T., Bergamasco, R., Vieira, A. M. S., & Fagundes-Klen, M. R. (2013). Ultrafiltration combined with coagulation/flocculation/sedimentation using *Moringa oleifera* as coagulant to treat dairy industry wastewater. *Water Air Soil Pollut., 224*(9), 1682. https://doi.org/10.1007/s11270-013-1682-2.

Formentini-Schmitt, D. M., Veit, M. T., Trigueros, D. E. G., Bergamasco, R., Triques, C. C., Mateus, G. A. P., & Fagundes-Klen, M. R. (2023). Primary treatment with a natural coagulant improved the nanofiltration stage to recover dairy wastewater. *J. Chem. Technol. Biotechnol., 98*(7), 1769–1780. https://doi.org/10.1002/jctb.7402.

Garg, S., Rumjit, N. P., Thomas, P., Sikander, Lai, C. W., & George, P. J (2020). Green technologies for the treatment and utilisation of dairy product wastes. In Inamuddin, & A. Asiri (Eds.), *Sustainable green chemical processes and their allied applications. Nanotechnology in the life sciences* (pp. 311–338). Springer https://doi.org/10.1007/978-3-030-42284-4_11.

Gaur, V., Schalk, J., & Anema, S. G. (2018). Sedimentation in UHT milk. *Int. Dairy J., 78*, 92–102. https://doi.org/10.1016/j.idairyj.2017.11.003.

Gong, H., Wu, X., Du, J., & Mao, X. (2024). High hydrostatic pressure combined with high-temperature short time pasteurization improves the quality characteristics and extends the shelf life of donkey milk. *LWT, 196*, 115789. https://doi.org/10.1016/j.lwt.2024.115789.

Graf, B., Kohler, E., Rosenberger, M., Schäfer, J., & Hinrichs, J. (2021). Shelf-stable milk produced by microfiltration and microwave heating: Effects of processing and storage. *J. Food Eng., 311*, 110734. https://doi.org/10.1016/j.jfoodeng.2021.110734.

Haller, N. (2023). *Separation of the major whey proteins α-Lactalbumin and β-Lactoglobulin by means of selective precipitation and continuous centrifugal separation* [PhD Thesis]. Technische Universität München.

Hashemi, K., & Hosseini, E. (2021). The stabilizing and prebiotic potential of water-soluble phase of bitter almond gum exudate in probiotic yogurt drink. *Carbohydr. Polym., 255*, 117395. https://doi.org/10.1016/j.carbpol.2020.117395.

Hu, J., Wang, Y., Ding, H., Jiang, C., Geng, Y., Sun, X., Jing, J., Gao, H., Wang, Z., & Dong, C. (2020). Recombinase polymerase amplification with polymer flocculation sedimentation for rapid detection of *Staphylococcus aureus* in food samples. *Int. J. Food Microbiol., 331*, 108691. https://doi.org/10.1016/j.ijfoodmicro.2020.108691.

Jansson, T., Waehrens, S. S., Rauh, V., Danielsen, B. P., Sørensen, J., Bredie, W. L. P., Petersen, M. A., Ray, C. A., & Lund, M. N. (2019). Effect of green tea catechins on physical stability and sensory quality of lactose-reduced UHT milk during storage for one year. *Int. Dairy J., 95*, 25–34. https://doi.org/10.1016/j.idairyj.2019.03.007.

Karlsson, M. A., Lundh, Å., Innings, F., Höjer, A., Wikström, M., & Langton, M. (2019). The effect of calcium, citrate, and urea on the stability of ultra-high temperature treated milk: A full factorial designed study. *Foods, 8*(9), 418. https://doi.org/10.3390/foods8090418.

Kazemalilou, S., & Alizadeh, A. (2017). Optimization of sugar replacement with date syrup in prebiotic chocolate milk using response surface methodology. *Korean Journal for Food Science of Animal Resources, 37*(3), 449–455. https://doi.org/10.5851/kosfa.2017.37.3.449.

Kelly, P. (2019). Manufacture of whey protein products. In H. C. Deeth, & N. Bansal (Eds.), *Whey proteins* (pp. 97–122). Elsevier https://doi.org/10.1016/B978-0-12-812124-5.00003-5.

Lewis, M., Grandison, A., Lin, M.-J., & Tsioulpas, A. (2011). Ionic calcium and pH as predictors of stability of milk to UHT processing. *Milchwissenschaft, 66*(2), 197–200.

Li, Y., Wang, X., Meng, Y., Zhang, F., Shao, Z., & Hu, L. (2018). Effect of the modified high methoxyl pectin on the stability of the fermented milk beverage. *Int. J. Food Prop., 21*(1), 2075–2086. https://doi.org/10.1080/10942912.2018.1485029.

Luo, J., Jian, S., Wang, P., Ren, F., Wang, F., Chen, S., & Guo, H. (2019). Thermal instability and characteristics of donkey casein micelles. *Food Res. Int., 119*, 436–443. https://doi.org/10.1016/j.foodres.2019.02.023.

Malmgren, B., Ardö, Y., Langton, M., Altskär, A., Bremer, M. G. E. G., Dejmek, P., & Paulsson, M. (2017). Changes in proteins, physical stability and structure in directly heated UHT milk during storage at different temperatures. *Int. Dairy J., 71*, 60–75. https://doi.org/10.1016/j.idairyj.2017.03.002.

Matéos, A., Guyard-Nicodème, M., Baglinière, F., Jardin, J., Gaucheron, F., Dary, A., Humbert, G., & Gaillard, J.-L. (2015). Proteolysis of milk proteins by AprX, an extracellular protease identified in *Pseudomonas* LBSA1 isolated from bulk raw milk, and implications for the stability of UHT milk. *Int. Dairy J., 49*, 78–88. https://doi.org/10.1016/j.idairyj.2015.04.008.

Mateus, G. A. P., Formentini-Schmitt, D. M., Nishi, L., Fagundes-Klen, M. R., Gomes, R. G., & Bergamasco, R. (2017). Coagulation/flocculation with *Moringa oleifera* and membrane filtration for dairy wastewater treatment. *Water Air Soil Pollut., 228*(9), 342. https://doi.org/10.1007/s11270-017-3509-z.

Mateus, G. A. P., Pinto, L. A. de M., Baptista, A. T. A., Nishi, L., Klen, M. R. F., Gomes, R. G., Araújo, A. A., & Bergamasco, R. (2017). Evaluation of natural coagulant *Moringa oleifera* Lam. in the treatment of dairy wastewater at different pH. *Acta Hortic., 1158*, 357–364. https://doi.org/10.17660/ActaHortic.2017.1158.40.

Mejares, C. T., Huppertz, T., & Chandrapala, J. (2022). Thermal processing of buffalo milk: A review. *Int. Dairy J., 129*, 105311. https://doi.org/10.1016/j.idairyj.2021.105311.

Mulet-Cabero, A.-I., Mackie, A. R., Wilde, P. J., Fenelon, M. A., & Brodkorb, A. (2019). Structural mechanism and kinetics of in vitro gastric digestion are affected by process-induced changes in bovine milk. *Food Hydrocolloids, 86*, 172–183. https://doi.org/10.1016/j.foodhyd.2018.03.035.

Mulet-Cabero, A.-I., Rigby, N. M., Brodkorb, A., & Mackie, A. R. (2017). Dairy food structures influence the rates of nutrient digestion through different in vitro gastric behaviour. *Food Hydrocolloids, 67*, 63–73. https://doi.org/10.1016/j.foodhyd.2016.12.039.

Ntazinda, A., Cheserek, M. J., Sheng, L.-X., Meng, J., & Lu, R.-R. (2014). Combination effect of sodium carboxymethyl cellulose and soybean soluble polysaccharides on stability of acidified skimmed milk drinks. *Dairy Science & Technology, 94*(3), 283–295. https://doi.org/10.1007/s13594-014-0162-9.

Pandalaneni, K., Bhanduriya, K., Amamcharla, J. K., Marella, C., & Metzger, L. E. (2019). Influence of milk protein concentrates with modified calcium content on enteral dairy beverage formulations: Storage stability. *J. Dairy Sci., 102*(1), 155–163. https://doi.org/10.3168/jds.2018-15239.

Panda, P., Meena, S., Meena, K., Rai, D. C., Bunkar, S., & Gautam, P. B. (2023). Development of functional milk-based smoothie by incorporating horse gram extract. *Current Research in Nutrition and Food Science Journal, 11*(3), 1218–1228. https://doi.org/10.12944/CRNFSJ.11.3.26.

Pan, Z., Ye, A., Dave, A., Fraser, K., & Singh, H. (2023). pH-dependent sedimentation and protein interactions in ultra-high-temperature-treated sheep skim milk. *J. Dairy Sci., 106*(3), 1626–1637. https://doi.org/10.3168/jds.2022-22637.

Pathania, S., Ho, Q. T., Hogan, S. A., McCarthy, N., & Tobin, J. T. (2018). Applications of hydrodynamic cavitation for instant rehydration of high protein milk powders. *J. Food Eng., 225*, 18–25. https://doi.org/10.1016/j.jfoodeng.2018.01.005.

Pearce, R. J. (1992). Whey protein recovery and whey protein fractionation. In J. G. Zadow (Ed.), *Whey and lactose processing* (pp. 271–316). Elsevier.

Rauh, V. M., Sundgren, A., Bakman, M., Ipsen, R., Paulsson, M., Larsen, L. B., & Hammershøj, M. (2014). Plasmin activity as a possible cause for age gelation in UHT milk produced by direct steam infusion. *Int. Dairy J., 38*(2), 199–207. https://doi.org/10.1016/j.idairyj.2013.12.007.

Rezazadeh-Bari, M., Najafi-Darmian, Y., Alizadeh, M., & Amiri, S. (2019). Numerical optimization of probiotic Ayran production based on whey containing transglutaminase and *Aloe vera* gel. *J. Food Sci. Technol., 56*(7), 3502–3512. https://doi.org/10.1007/s13197-019-03841-3.

Rezvani, Z., & Goli, S. A. H. (2024). Production of milk-based drink enriched by dietary fiber using carrot pomace: Physicochemical and organoleptic properties during storage. *Food Hydrocolloids, 151*, 109834. https://doi.org/10.1016/j.foodhyd.2024.109834.

Rivas, J., Prazeres, A. R., Carvalho, F., & Beltrán, F. (2010). Treatment of cheese whey wastewater: Combined coagulation–flocculation and aerobic biodegradation. *J. Agric. Food Chem., 58*(13), 7871–7877. https://doi.org/10.1021/jf100602j.

Rutigliano, M., Rusco, G., Picariello, G., Bulgari, O., Spadaccino, G., Gagliardi, R., Di Luccia, A., Addeo, F., & la Gatta, B. (2020). Protein aggregation mechanism in UHT milk: Supramolecular evidences. *Eur. Food Res. Technol., 246*(5), 1081–1094. https://doi.org/10.1007/s00217-020-03474-3.

Sayyad, F. G., Sharma, H. K., & Kumar, N. (2022). Cleaning and separation. In H. K. Sharma, & N. Kumar (Eds.), *Agro-processing and food engineering* (pp. 307–352). Springer https://doi.org/10.1007/978-981-16-7289-7_8.

Singh, H. (2004). Heat stability of milk. *Int. J. Dairy Technol., 57*(2–3), 111–119. https://doi.org/10.1111/j.1471-0307.2004.00143.x.

Sumonsiri, N., Kundacha, N., & Pom-Iam, N. (2018). Effect of microcrystalline cellulose on physical characteristics and sensory acceptance of chocolate flavored milk. *Curr. Res. Nutr. Food Sci, 6*, 781–788.

Sun, W., Yang, W., Zheng, Y., Zhang, H., Fang, H., Liu, D., Kong, X., Chen, S., Ye, X., & Tian, J. (2020). Effect of potato pulp pectic polysaccharide on the stability of acidified milk drinks. *Molecules, 25*(23), 5632. https://doi.org/10.3390/molecules25235632.

Tabit, F. T. (2018). Prevalence and growth characteristics of *Bacillus sporothermodurans* in UHT milk. *British Food Journal, 120*(10), 2250–2260. https://doi.org/10.1108/BFJ-02-2018-0126.

Tchamango, S., Nanseu-Njiki, C. P., Ngameni, E., Hadjiev, D., & Darchen, A. (2010). Treatment of dairy effluents by electrocoagulation using aluminium electrodes. *Sci. Total Environ., 408*(4), 947–952. https://doi.org/10.1016/j.scitotenv.2009.10.026.

Thongkaew, C., Gibis, M., Hinrichs, J., & Weiss, J. (2014). Polyphenol interactions with whey protein isolate and whey protein isolate–pectin coacervates. *Food Hydrocolloids, 41*, 103–112. https://doi.org/10.1016/j.foodhyd.2014.02.006.

Tribst, A. A. L., Falcade, L. T. P., Carvalho, N. S., Cristianini, M., Leite Júnior, B. R. de C., & Oliveira, M. M. de (2020). Using physical processes to improve physicochemical and structural characteristics of fresh and frozen/thawed sheep milk. *Innovative Food Sci. Emerg. Technol., 59*, 102247. https://doi.org/10.1016/j.ifset.2019.102247.

Wagoner, T., Vardhanabhuti, B., & Foegeding, E. A. (2016). Designing whey protein–polysaccharide particles for colloidal stability. *Annual Review of Food Science and Technology, 7*(1), 93–116. https://doi.org/10.1146/annurev-food-041715-033315.

Wang, X., Kristo, E., & LaPointe, G. (2020). Adding apple pomace as a functional ingredient in stirred-type yogurt and yogurt drinks. *Food Hydrocolloids, 100*, 105453. https://doi.org/10.1016/j.foodhyd.2019.105453.

Wijesekara, A., Weerasingha, V., Jayarathna, S., & Priyashantha, H. (2022). Quality parameters of natural phenolics and its impact on physicochemical, microbiological, and sensory quality attributes of probiotic stirred yogurt during the storage. *Food Chemistry: X, 14*, 100332. https://doi.org/10.1016/j.fochx.2022.100332.

Wu, C.-S., Guo, J.-H., & Lin, M.-J. (2020). Stability evaluation of pH-adjusted goat milk for developing ricotta cheese with a mixture of cow cheese whey and goat milk. *Foods, 9*(3), 366. https://doi.org/10.3390/foods9030366.

Wu, J., Du, B., Li, J., & Zhang, H. (2014). Influence of homogenisation and the degradation of stabilizer on the stability of acidified milk drinks stabilized by carboxymethylcellulose. *LWT - Food Sci. Technol., 56*(2), 370–376. https://doi.org/10.1016/j.lwt.2013.12.029.

Wu, J., Liu, J., Dai, Q., & Zhang, H. (2013). The stabilisation of acidified whole milk drinks by carboxymethylcellulose. *Int. Dairy J., 28*(1), 40–42. https://doi.org/10.1016/j.idairyj.2012.05.005.

Xia, K., Zong, P., Liu, X., Zhao, J., & Zhang, X. (2019). Preparation of methyl alginate and its application in acidified milk drinks. *Int. J. Biol. Macromol., 132*, 651–657. https://doi.org/10.1016/j.ijbiomac.2019.03.243.

Yukalo, V., Krupa, O., & Storozh, L. (2019). Characteristics of proteolytic processes during the isolation of natural casein phosphopeptides. *Ukrainian Food Journal, 8*(1), 61–69. https://doi.org/10.24263/2304-974X-2019-8-1-7.

Yun, S.-Y., & Imm, J.-Y. (2021). Changes in particle pize, sedimentation, and protein microstructure of ultra-high-temperature skim milk considering plasmin concentration and storage temperature. *Molecules, 26*(8), 2339. https://doi.org/10.3390/molecules26082339.

Zaeim, D., Mulet-Cabero, A.-I., Read, S. A., Liu, W., Han, J., & Wilde, P. J. (2022). Effect of oil droplet size on the gastric digestion of milk protein emulsions using a semi-dynamic gastric model. *Food Hydrocolloids, 124*, 107278. https://doi.org/10.1016/j.foodhyd.2021.107278.

Zhang, C., Bijl, E., & Hettinga, K. (2018). Destabilization of UHT milk by protease AprX from *Pseudomonas fluorescens* and plasmin. *Food Chem., 263*, 127–134. https://doi.org/10.1016/j.foodchem.2018.04.128.

Zhang, C., Bijl, E., Muis, K. E., & Hettinga, K. (2020). Stability of fat globules in UHT milk during proteolysis by the AprX protease from *Pseudomonas fluorescens* and by plasmin. *J. Dairy Sci., 103*(1), 179–190. https://doi.org/10.3168/jds.2019-17150.

Zhang, T., Liu, Y., Wang, P., Li, Y., Ren, F., & Yi, H. (2023). Exploration of interaction between α-lactalbumin and β-lactoglobulin under dUHT treatment and storage: Experimental and molecular dynamics study. *Food Hydrocolloids, 138*, 108469. https://doi.org/10.1016/j.foodhyd.2023.108469.

Zhao, J., Chen, Z., Zou, R., & Zhang, X. (2018). The application of agar oligosaccharides in directly acidified milk drinks. *Food Hydrocolloids, 77*, 421–426. https://doi.org/10.1016/j.foodhyd.2017.10.018.

Separation of cream from milk by centrifugation

10

Maria Turtoi
Life Sciences Department, Cross-Border Faculty, "Dunarea de Jos" University of Galati, Galati, Romania

10.1 Introduction

Archaeological discoveries, such as traces of milk lipids in pottery residues (Evershed et al., 2008) or human dental calculus (Warinner et al., 2014), show humans have consumed milk from ancient times. The pieces of pottery with tiny holes from the Linear Pottery Culture, a European Neolithic civilization distributed from Ukraine to France and Hungary to the Baltic Sea, are evidence of milk processing over 5000 years BC (Bogucki, 1984; Curry, 2013; Salque et al., 2012).

Beyond doubt, human ancestors observed that cream tends to separate from milk because raw milk if poured into various vessels and left undisturbed, slowly but surely separates into a cream layer and fat-depleted skimmed milk (Meng et al., 2022; Salque et al., 2012; Subbaraman, 2012). Thus, for thousands of years, the cream was separated from milk by rising on the surface due to the influence of gravity, which allowed it to be creamed off (Moore, 2008).

Up to around 150 years ago, the technique used for separating one substance from another was sedimentation by gravity, a natural process influenced by the difference in density of the substances involved in the process (Tetra Pak, 2003, p. 100). Then, during the last half of the 19th century, centrifugal separation was developed for milk skimming, and afterward, it was introduced to other technologies (Heymann, 2011; Tetra Pak, 2003; Vogel-Prandtl, 2004, 2014). Therefore, the centrifugal separator can be considered a recent invention.

Milk centrifugal separators enabled a fast (instantaneous) separation of cream from milk and a better microbiological quality of cream and skimmed milk than gravity separation. The first centrifugal separators were manually operated using a crank handle. They were used mainly on farms and revolutionized milk processing. In their evolution, milk separators have gone through numerous stages, such as spiral channels, conical discs, electric motors, self-discharging bowls, automatic cleaning, semiopen and hermetic bowls, different driving systems, warm and cold separation, etc., discussed in this chapter. A bit of history of centrifugal separation of milk fat is presented below.

10.1.1 *Early history of centrifugal separators*

A centrifugal separator or centrifuge is "a machine using centrifugal force for separating substances of different densities, for removing moisture, or for simulating

Mechanical Separation Processes in the Food Industry: Unit Operations and Processing Equipment in the Food Industry.
DOI: https://doi.org/10.1016/B978-0-12-819523-9.00002-0

gravitational effects" (Merriam-Webster Online, 2024) or "a machine that turns a container around very quickly, causing the solids and liquids inside it to separate by centrifugal action" (Cambridge English Dictionary, 2024). The word centrifuge was borrowed from the French *centrifuge,* derived from the New Latin *centrifugālis* or Modern Latin *centrifugus*. In turn, the *centrifugus* is made up of the Latin *centri-*, an alternative combining form of the *centrum* ("center" in English) + *fugere* ("to flee" in English) (Etymonline, 2024).

The term "centrifugal force" was first attested in the work *De vi centrifuga* (1659) of Christiaan Huygens, a Dutch mathematician, and physicist. Huygens derived geometrically the formula of centrifugal force in classical mechanics 5 years before Isaac Newton (Andriesse, 2005; Yoder, 1991).

The idea of a centrifuge to separate cream from milk emerged around two centuries later, during the nineteenth century. It sprouted in many minds and has been practiced in various ways, from swiftly rotating a barrel of milk (a German brewmaster) or two pails of milk hung to a rotating crosspiece (Jensen, a Danish horse doctor) to the machine of Lefeldt, a German engineer. The machine of Lefeldt was a large tube filled with milk and spun by a steam engine at 800 rotations per minute (RPM). The separation of cream from milk took 15 to 20 minutes. The machine was available in four sizes, handling 110, 220, 440, or 660 pounds (50, 100, 200, and 300 L) of milk per hour. Neilsen & Peterson was another prominent maker of separators at that time, whose Danish–Weston machines were widely exported to the United States. These large-capacity machines at that time were used primarily in large creameries and were expensive and slow (Moore, 2008).

Another idea of improving and speeding up the separation of cream from milk arose in 1964 and belongs to Antonin Prandtl. According to Vogel-Prandtl (2004, p. 10), in the biography of her father, Ludwig Prandtl, the design of the hand-operated centrifuge of Antonin Prandtl was improved by his brother, Alexander Prandtl. After 5 years of work in the laboratory of the experimental dairy station at Weihenstephan, Alexander Prandtl exhibited a machine for butterfat extraction at the World's Fair in Frankfurt am Main, Germany, in 1875. The machine was considered, at that time, the first continuously hand-operating milk centrifuge in the world and was further exhibited in the *Deutsches Museum* (German Museum) in the Department of Dairy Farming (Vogel-Prandtl, 2014, p. 4).

As a professor and scientist, Alexander Prandtl also published scientific papers, one of them being "On the theoretically expected effect of creaming caused by centrifugal forces" in the *Weihenstephan Milk Newspaper* (*Weihenstephaner Milchzeitung*), 1877 (Vogel-Prandtl, 2014, p. 4). The article described a newly invented appliance for separating cream from milk, "a drum which is made to rotate and which, after turning for a time, leaves the cream floating on the surface so that it can be skimmed off in the usual fashion." The article was probably the one Carl Gustaf Patrick de Laval, a young Swedish engineer, read and decided to show that "centrifugal force will act in Sweden as well as in Germany" (Tetra Pak, 2003, p. 99).

De Laval started developing a centrifugal separator of cream from milk and, in less than 2 years, patented the first continuous separator (ERIH European Route of Industrial Heritage, 2024). According to the news reported on January 15, 1879, in

Figure 10.1 The Alfa A1 separator, manufactured from 1882 (Tetra Pak, 2003, p. 99, Fig. 6.2.2).

the daily newspaper *Stockholms Dagblad* (*Stockholm Daily Newspaper*), the separator looked like a drum driven around by a belt and pulley. The cream, which is lighter than the milk, is driven by centrifugal force to the surface of the milk and flows off through a channel toward a collecting vessel. The milk is forced out toward the periphery of the drum and flows through another channel toward a separate collecting vessel (Tetra Pak, 2003, p. 99).

The separator revolutionized milk processing in Europe and led to the establishment of other companies, such as Bergdorfer Eisenwerk of Hamburg, Boeke & Huidekooper of Groningen, Peterson & Co. of Copenhagen, and Pilter of Paris, which began the production of separators under license in the same period (ERIH European Route of Industrial Heritage, 2024).

In 1888, the German Clemens von Bechtolsheim obtained a patent on conical discs named "Alfa discs," specially designed for centrifugal separators. The Swedish company AB Separator acquired the patent in 1889, and after 1890, the separators built by Carl Gustaf Patrick de Laval, part-owner of the company, were equipped with conical discs (ERIH European Route of Industrial Heritage, 2024; Tetra Pak, 2003, p. 99). Fig. 10.1 presents the Alfa A1, one of the first milk separators provided with conical discs manufactured by AB Separator (Tetra Pak, 2003, p. 99).

A few years later, on September 1, 1893, the brothers-in-law Franz Ramesohl and Franz Schmidt opened a workshop under the name of Ramesohl & Schmidt oHG in Oelde. With only two turners and three locksmiths, they started the production of milk separators (Fig. 10.2A) in four sizes: 75 L/h (1), 100 L/h (2), 125 L/h (3), and 150 L/h (4) at prices between 200 and 320 marks (Kersten, 1993, pp. 8–9). The company developed and eventually became the most powerful concurrent of AB Separator. In 1897, it reached 60 employees and sold more than 2000 centrifuges. The company was transformed into an Aktiengesellschaft (*Stock corporation*) in 1899, based on a

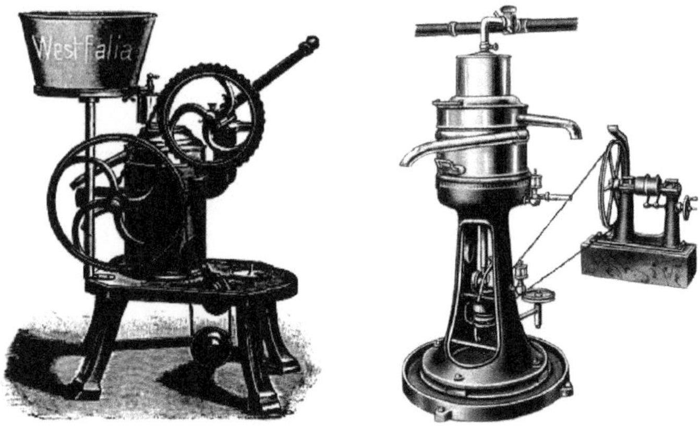

(A) (B)

Figure 10.2 The first milk centrifuges from Ramesohl and Schmidt oHG: (A) The first hand-operated milk separator from 1893 (Kersten, 1993, p. 9); and (B) the milk separator from 1900 (Kersten, 1993, p. 16).

shareholders' agreement. The following year, around 10,000 milk centrifuges from Ramesohl and Schmidt (Fig. 10.2B) were in service (Kersten, 1993, p. 16). By 1913, Ramesohl and Schmidt were joined by 68 companies in Germany making centrifuges, around 10 of which were located near Oelde. However, Ramesohl and Schmidt continued growing in the face of such stiff competition (Kersten, 1993, pp. 32–33).

Ramesohl and Schmidt was named "Westfalia Separator" in 1941 and was incorporated into the GEA Group in 1994 (GEA, 2023c). In 1963, the AB Separator Company changed its name to Alfa Laval AB, in which "Alfa" derives from the Alpha discs and "Laval" from the founder of the company, de Laval (Laval, 2015).

10.1.2 Evolution of centrifugal separators

Farmers were happy with the hand-operated milk separators at the beginning of the manufacture of centrifugal separators. However, the manufacturers keep a close eye on the competition.

Thus, as stated before, the AB Separator company introduced the conical discs inside the bowl in 1890 (Laval, 2015; Tetra Pak, 2003, p. 99).

Ten years later, Ramesohl and Schmidt equipped the bowl of the milk separators with a spiral insert (Fig. 10.2B) and increased the throughput to 600 L/h (A type separator), 1000 L/h (B type), 1500 L/h (C type), and 2000 L/h (D type). The operation principle of the spiral insert is similar to that of the modern disc set (Kersten, 1993, p. 16). After 1910, Ramesohl and Schmidt replaced the original spiral channels with conical discs to increase the efficiency of cream separation from milk (Kersten, 1993, p. 28).

The development of industrial dairies demanded higher capacities and the machines were equipped with electric motors. Other important steps forward were the following:

- The replacement of sliding bearings, which required a relative amount of oil for lubrication with ball bearings (Kersten, 1993, p. 41).
- The first self-cleaning centrifuge with the self-initiated discharge of the sludge (GB366873A, 1932; Kersten, 1993, p. 58).
- The introduction of the first hermetic separator by Alfa Laval at an exhibition in Berlin in 1933 (Laval, 2015).
- The introduction of the foamless separator for cream separation, model MD in 1933, and models MDD 1000 to MDD 5000, by Ramesohl and Schmidt (Kersten, 1993, p. 62).
- The beginning of sales of the "self-cleaning" centrifugal separator by Alfa Laval in 1951 (Laval, 2015).
- The introduction in dairies of the self-discharging bowls by Westfalia Separator in 1968 enabled automatic cleaning or cleaning in place (CIP) (GEA, 2023b; Kersten, 1993, p. 117).
- The introduction of milk separators with the "SoftStream" system in 1978 (Kersten, 1993, p. 140).

Since the first milk centrifugal separators were built, thousands of patents have been granted to inventors and assigned to companies manufacturing separators. Therefore, milk separators have been and are in a continuous evolution.

10.2 Milk and cream

10.2.1 Definitions and composition of milk

Milk is the secretion of the mammary glands of mammals and feeds young mammals in their first period of life before they can digest other food (Tetra Pak, 2003, p. 2; Walstra et al., 2006, p. 3).

The milk of some animals, especially cows, buffaloes, sheep, and goats (Walstra et al., 2006, p. 3), but also horses (Doreau & Martin-Rosset, 2022), donkeys (Salimei & Fantuz, 2022), reindeer (Holand et al., 2022), yak (Wiener, 2022), and camels (Alhadrami & Faye, 2022) in some regions, is processed and used for human consumption as drinking milk and a large variety of dairy products (Walstra et al., 2006, p. 3). The cow is the most widespread milking animal in the world, and it is found on all continents and in almost all countries (Tetra Pak, 2003, p. 2).

According to Codex Alimentarius standards, milk is "the normal mammary secretion of milking animals obtained from one or more milkings without either addition to it or extraction from it, intended for consumption as liquid milk or for further processing" (Codex, 1999).

The composition of milk includes lactose, protein, fat, mineral substances (Table 10.1), vitamins, and other miscellaneous compounds in trace amounts.

Lactose or milk sugar is a reducing disaccharide composed of glucose and galactose. Milk *protein* is composed of about 80% casein (α_{s1}–, α_{s2}–, β–, γ–, and κ–casein), milk-serum proteins (α–lactalbumin, β–lactoglobulin), and other proteins in tiny amounts, such as immunoglobulins, membrane proteins, and enzymes (peroxidase, catalase,

Table 10.1 Composition of milk from various mammals (medium values).

Mammal	Water (%)	Dry matter (%)	Carbo-hydrate (%)	Total protein (%)	Fat (%)	Ash (%)	References
Buffalo	83.0	17.0	4.8	4.0	7.5	0.7	Tetra Pak (2003, p. 2, Table 1.1)
Camel	87.7	12.3	4.5	3.3	3.7	0.8	Alhadrami and Faye (2022, p. 57)
Cow	87.3	12.7	4.8	3.5	3.6	0.8	Tetra Pak (2003, p. 2, Table 1.1) Walstra et al. (2006, pp. 3–4)
Donkey	91.1	9.9	6.2	1.6	1.7	0.4	Salimei and Fantuz (2022, pp. 67–69)
Goat	86.8	13.2	4.7	3.6	4.1	0.8	Tetra Pak (2003, p. 2, Table 1.1)
Horse	89.2	10.8	6.2	2.2	1.7	0.5	Doreau and Martin-Rosset, (2022, p. 100)
Human	87.3	12.7	7.0	1.0	4.5	0.2	Tetra Pak (2003, p. 2, Table 1.1)
Sheep	82.6	17.4	4.8	4.6	7.2	0.8	Tetra Pak (2003, p. 2, Table 1.1)

phosphatase, lipase, lysozyme, plasmin, and proteinase). The *fat* comprises a complex mixture of triglycerides containing fatty acids with chain lengths varying from 2 to 20 carbon atoms and 0 to 4 double bonds. Other milk lipids are monoglycerides, diglycerides, cholesterol, free fatty acids, and phospholipids. The *mineral substances*, primarily Ca, K, Na, and Mg, occur as phosphates, chlorides, citrates, and caseinates. Potassium and calcium salts are the most abundant in regular milk. *Organic acids* are present as ions or salts, the principal being citrate, with trace amounts of lactic and pyruvic acid in milk serum (Tetra Pak, 2003, pp. 22–36; Walstra et al., 2006, pp. 3–4; Vaclavik & Christian, 2014, pp. 201–229).

All known *vitamins* are present in milk, which is a good source of most vitamins, especially vitamins A and B (B_1, B_2, B_3, B_5, and B_6). Vitamins C, D, E, and K_2 are also found in milk in tiny amounts but have specific roles, for example, the yellow color of milk fat ensured by carotenes (precursors of vitamin A), antioxidant activity of tocopherols (precursors of vitamin E), or redox reactions in which ascorbic acid (vitamin A) is involved (Walstra et al., 2006, pp. 93–94).

Other constituents of milk are somatic cells—white blood corpuscles or leucocytes—whose content is low in milk from a healthy udder (less than 200,000 cells/mL) and gases, 5% to 6% by volume in fresh milk and as high as 10% by volume on arrival at the dairy. The gases consist primarily of carbon dioxide, nitrogen, and oxygen and exist dissolved in the milk, dispersed in the milk, and bound and nonseparable from the milk. Dissolved and dispersed gases are a severe problem in the processing of milk and should be eliminated by degassing (Tetra Pak, 2003, pp. 35–36).

Table 10.2 Quantitative composition of cow milk (Tetra Pak, 2003, pp. 2, 22; Walstra et al., 2006).

Main constituent	Water	Total solids	Fat	Proteins	Lactose	Minerals
Limits of variation (%)	85.5–89.5	10.5–14.9	2.4–5.0	2.9–5.0	3.6–5.5	0.6–0.9
Mean value (%)	87.3	12.7	3.6	3.5	4.8	0.8

The chemical composition of milk determines its nutritional and energetic values, the extent to which the microorganisms can grow in milk, the flavor of milk, and the chemical reactions occurring in milk. The latter include reactions that produce off-flavors in milk (Walstra et al., 2006, p. 4).

Also, cow milk is defined as a white liquid of about 87% water and 13% dry matter suspended or dissolved in the water (Tetra Pak, 2003, p. 19). The quantities of the main constituents of milk vary considerably between individual cows of the same breed and between cows of different breeds. Besides genetic factors, milk composition is highly influenced by many other factors, such as the stage of lactation, the healthy state of the cow, the amount and the quality of the feed, the period of the year, the contamination of the milk after milking, etc. (Walstra et al., 2006, pp. 12–13). The limits and mean values of the main constituents of cow milk are presented in Table 10.2.

The total content of all milk components except water is called the content of dry matter or total solids. Another difference is between solid nonfat (SNF) and fat content in dry matter (Walstra et al., 2006, p. 4). According to Table 10.2, the mean SNF of cow milk is the difference between total solids or dry matter content and fat content: 12.7–3.6 = 9.1% (Tetra Pak, 2003, p. 22).

10.2.2 Milk fat and cream

Milk is a complex system in which lactose and ash form a true solution, proteins are found in a colloidal solution/suspension, and fat is distributed in an oil-in-water or *fat-in-water* emulsion (Tetra Pak, 2003, p. 19).

The milk fat exists as small globules dispersed in the milk serum (Fig. 10.3) with diameters ranging from 0.1 to 20 μm and an average of 3 to 4 μm. There are about 15 billion globules per mL of milk. The fat globule is surrounded by a skinny membrane, only 5 to 10 nm thick, that stabilizes the emulsion (Tetra Pak, 2003, p. 22). Fat globules are more complicated than emulsion droplets because they have an intricate structure consisting of many substances (Walstra et al., 2006, pp. 5–6).

The distribution of fat globules in milk is the following:

- Small globules with less than 1 μm, about 75% to 80% of the number of globules but only about 2% of the globular fat.
- Medium globules, about 95 % of the fat.
- Large globules, about 2% of the fat (Walstra, 1983, p. 120).

Milk fat is around 98% a mixture of triacylglycerols, generally called triglycerides. Besides triglycerides, milk fat contains small quantities of di- and monoglycerides,

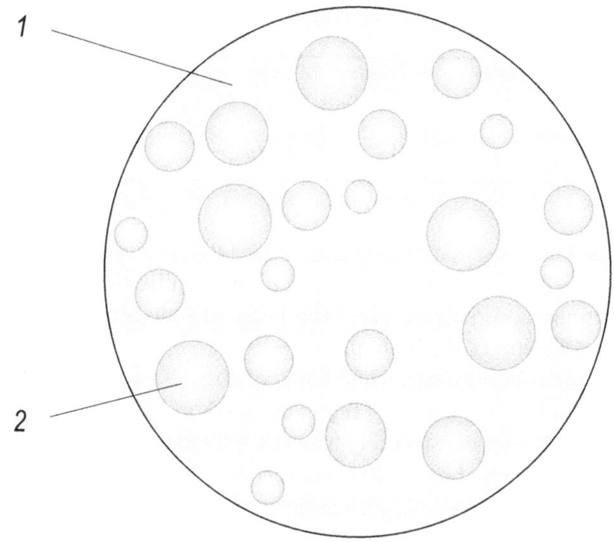

Figure 10.3 Skim milk and fat globules: 1—skimmed milk; 2—fat globule (author's drawing adapted from Tetra Pak, 2003, p. 22, Fig. 2.15).

fatty acids, sterols, carotenoids (responsible for the yellow color of the fat), and liposoluble vitamins (A, D, E, and K) (Tetra Pak, 2003, p. 22). The structure of milk fat globules is outlined in Fig. 10.4.

The membrane of fat globules contains phospholipids, lipoproteins, glycerides, cerebrosides, proteins, nucleic acids, enzymes, trace metals, and bound water. Its composition and thickness are not constant because components are continuously exchanged with the surrounding milk plasma. The internal layers consist of crystalized fat with various melting points and contain triglycerides, diglycerides, fatty acids, sterols, carotenoids, and liposoluble vitamins (A, D, E, and K). The core liquid fat consists of relatively low-melting triglycerides, which contain fatty acids that are liquid at room temperature, for example, saturated with short-chain (butyric, caproic, and caprylic acids) and unsaturated (oleic, linoleic, linolenic, and arachidonic acids) (Tetra Pak, 2003, p. 22).

The fat globules are the largest particles in the milk and also the lightest, with a density of 0.93 g/cm^3 (930 kg/m^3) at 15.5°C (Tetra Pak, 2003, p. 22). When the milk is left to stand in a vessel for a while, the fat globules tend to rise and concentrate at the surface of the milk because of the difference in density between fat globules and milk plasma. The milk fat separated on the surface of the milk is called cream, a thick, yellowish liquid with a smooth and velvety texture. The property is essential because it causes the separation of milk into cream and skim milk during storage, a process known as creaming (Walstra et al., 2006, p. 136).

Figure 10.4 The structure of fat globules (author's drawing adapted from Tetra Pak, 2003, p. 22, Fig. 2.16): 1—membrane; 2—internal solid layers; 3—liquid fat.

10.2.3 Creaming

Creaming is a slow process due to the small size of the fat globules (Huppertz et al., 2020; Tetra Pak, 2003, p. 22). In cow milk, it is faster at low temperatures (Fox et al., 2015a, b), being enhanced if the fat globules have been aggregated into floccules or clusters under the influence of a complex of cryoglobulins, predominantly immunoglobulin M (ImM), called *agglutinin* (Fox et al., 2015b; Walstra et al., 2006, pp. 13, 136). The aggregates rise much faster than individual fat globules but are easily broken up by heating or mechanical treatment. A time-temperature combination such as 65°C/10 min, 75°C/2 min, or equivalent denatures the agglutinin, making the aggregation impossible (Tetra Pak, 2003, p. 22).

Milk fat is the least affected by heat from the main components of milk. Thus, no agglutination occurs at 37°C. The creaming is quicker if the milk temperature decreases faster and is lower. The creaming is seriously impaired if cold raw milk is agitated for a while. Also, warming cold raw milk restores the ability to flocculate due to agglutinin dissociation (Walstra et al., 2006, pp. 137–138).

However, heat induces changes in milk fat and influences the creaming of the fat globules. Thus, cold agglutination in raw or low-pasteurized milk determines a thick layer of cream containing 20% to 25% fat and many bacteria and somatic cells from milk. In high-heated milk, a thin layer of cream with a high content of fat, 40% to 50%, is formed (Walstra et al., 2006, p. 138).

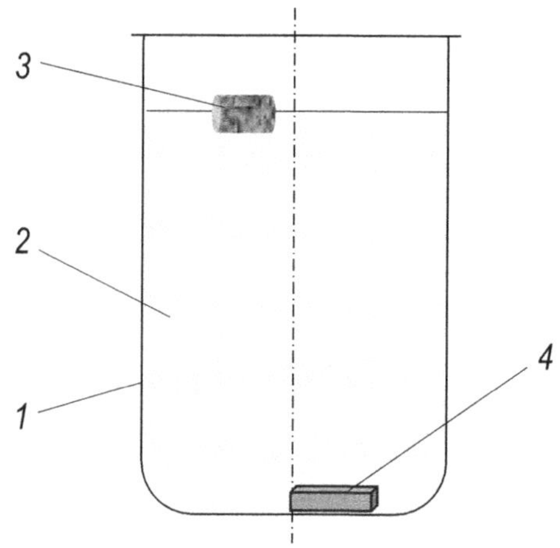

Figure 10.5 Drawing representing the sedimentation and flotation (author's drawing adapted from Tetra Pak, 2003, p. 100, Fig. 6.2.4): 1—Vessel; 2—liquid (water); 3—cork; 4—metal piece.

After natural creaming, skimmed milk rarely has a fat content lower than 0.5% and still contains medium and large fat globules (Walstra et al., 2006, p. 138).

A cream plug formation in bottled milk was observed and studied almost a century ago to find its causes and suggest prevention (Sommer & Royer, 1927). Later, Hoyle and Burton (1984) studied the cream plug formation in bottled cream. Various theories have been discussed over time, and it is considered that the free fat released from fat globules in plasma contributes to the agglutination of fat globules when they collide. The cream plug formation is avoided when milk and cream are homogenized (Huppertz et al., 2020; Tetra Pak, 2003, p. 37).

Since almost all the milk fat is in fat globules, it is separated by different methods, that is, gravity creaming and centrifugation, to obtain products rich in fat, such as cream, butter, and standardized milk with various fat content (whole milk, semiskimmed milk, skimmed milk) used either as drinking milk or to obtain dairy products with a standardized fat content.

10.3 Centrifugation theory

Centrifugation is a mechanical process of separating particles from biphasic systems, for example, solid–liquid systems composed of solid particles dispersed in a liquid medium or liquid–liquid systems composed of liquid droplets emulsified in another liquid medium. While the liquid phase is continuous, the individual particles or droplets form a discontinuous phase.

Fig. 10.5 illustrates what can happen in a vessel (1) containing a liquid (2) with a certain density, that is, water, in which objects with different densities are introduced, for example, a cork (3), and a metal piece (4). The metal piece has a higher density than the liquid medium, so it will sink and settle on the bottom of the vessel. The process is called sedimentation. Conversely, the cork has a lower density than the liquid medium and will rise and reach the surface of the liquid. The process is called flotation (Tetra Pak, 2003, p. 100).

The separation process is described by the law of Stokes, from which the sedimentation or flotation velocity in the gravitational field is calculated (Coulson et al., 1978, p. 174; Steinour, 1944):

$$v_g = \left[d^2 \cdot \left(\rho_p - -\rho_l\right) \cdot g\right]/18 \cdot \eta \tag{10.1}$$

Where, v_g is the velocity of sedimentation or flotation of the particle or droplet in m/s; d the diameter of the particle or droplet in m; ρ_p the density of the particle or droplet in kg/m^3; ρ_l the density of the liquid medium in kg/m^3; g gravitational acceleration in m/s^2; and η the dynamic viscosity of the continuous phase (the liquid medium) in Pa•s.

Eq. (10.1) shows that the velocity of sedimentation or flotation of a particle or droplet increases as the square of the particle or droplet diameter, with increasing differential density between the phases and with reducing viscosity of the continuous phase.

In a vessel (1) containing fresh milk (2), the fat globules (3) will rise slowly toward the milk surface because of their lower density than milk serum or plasma density, generating the creaming process (Fig. 10.6).

Fig. 10.6 shows schematically how milk fat globules of different sizes move through the milk serum during time and under the influence of gravity. The fat globules are at the bottom of the vessel at the beginning of the experiment, considered zero time. During time, the fat globules move upwards with different velocities as a function of their size, that is, the diameter, the largest fat globules rising faster. Thus, when the largest fat globules reach the surface, the medium-sized fat globules are at a point halfway to the surface, and the smallest fat globules are only at a quarter of the distance (Tetra Pak, 2003, p. 101).

Fig. 10.7 illustrates what happens in a vessel (1) containing a biphasic system (2) if spun. Particles or droplets with a higher density (5) than that of the liquid medium will move outwards, forming a heavy phase (6), and particles with a lower density (4) will move inwards, creating a light phase (3).

The movement of particles is due to generating a field of centrifugal force when the vessel filled with liquid is spun (Fig. 10.8). The action produces a centrifugal acceleration a, which is not constant but increases with the distance between the particles suspended in the liquid and the axis of rotation called radius (r) and the speed of rotation, expressed as angular velocity ω.

$$a = r \cdot \omega^2 \tag{10.2}$$

The expression of the law of Stokes for a centrifugal field is obtained by replacing the gravitational acceleration (g) in Eq. (10.1) with the centrifugal acceleration (a) and is used to calculate the velocity of sedimentation or flotation (v_c) of a particle in the

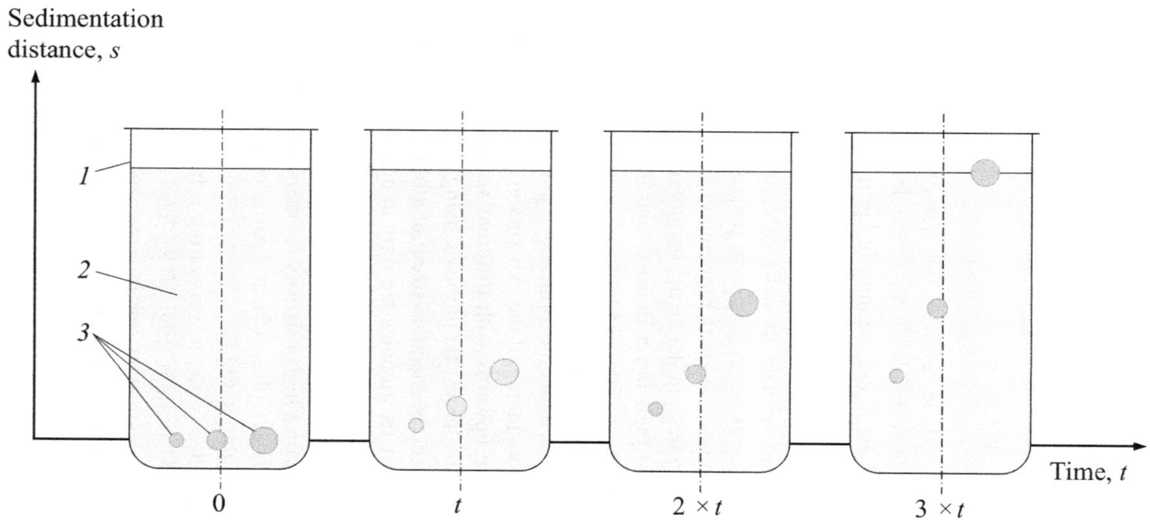

Figure 10.6 Drawing representing the flotation of fat globules with different sizes (author's drawing adapted from Tetra Pak, 2003, p. 102, Fig. 6.2.6): 1—vessel; 2—milk; 3—fat globules with different sizes.

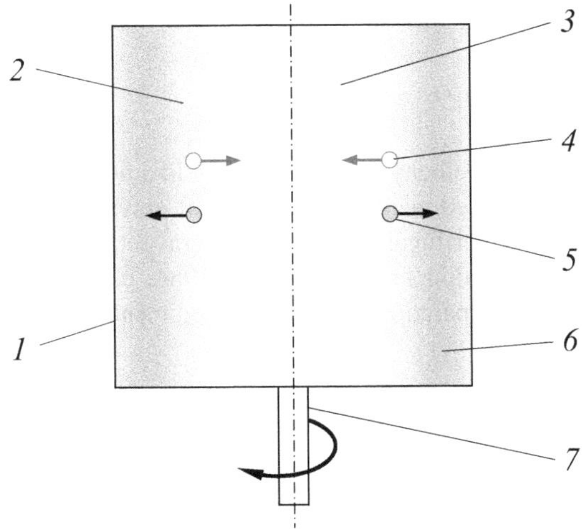

Figure 10.7 Drawing representing the centrifugation (author's drawing): 1—vessel; 2—biphasic system; 3—layer containing particles with low density; 4—a particle with low density; 5—a particle with high density; 6—layer containing particles with high density; 7—shaft.

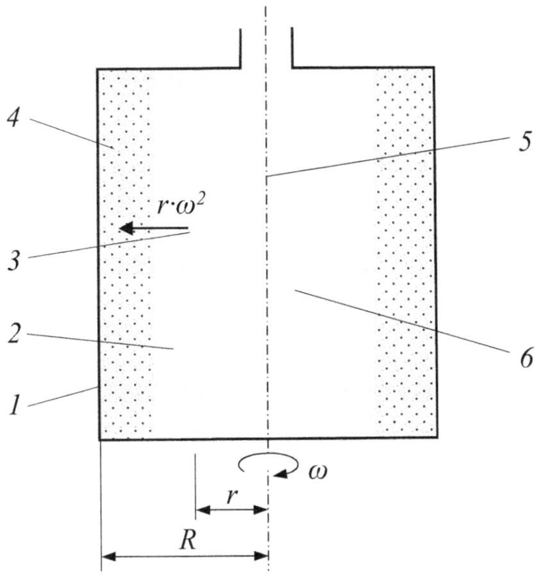

Figure 10.8 Principle of centrifugation (author's drawing adapted from Tetra Pak, 2003, p. 104, Fig. 6.2.14): 1—vessel; 2—suspension; 3—layer containing heavier particles; 4—particle suspended in liquid; 5—axis; 6—central tube without suspension formed during spinning. R: radius of the vessel; r: distance between a suspended particle and the axis of rotation; w: angular velocity; $r \bullet \omega^2$: centrifugal acceleration.

centrifugal field (Tetra Pak, 2003, p. 104):

$$v_c = \left[d^2 \cdot \left(\rho_p - -\rho_l\right) \cdot r \cdot \omega^2\right]/18 \cdot \eta \quad (10.3)$$

The ratio between the velocity of sedimentation in a field of centrifugal force (Eq. 10.3) and the velocity of sedimentation in a field of gravitational force (Eq. 10.1), is reduced to the ratio of centrifugal acceleration to gravitational acceleration, z, named centrifugation or acceleration factor (Hemfort, 1984, p. 10), which represents the efficiency of centrifugal separation, compared to sedimentation by gravity:

$$z = v_c/v_g = r \cdot \omega^2/g \quad (10.4)$$

The angular velocity is equal to the revolutions per minute n (rpm) multiplied by one revolution (2π) and divided by 60 (to obtain radians per second):

$$\omega = 2 \cdot \pi \cdot n/60 \quad (10.5)$$

The centrifugation factor z varies between 5000 and 13,000 for disc-type centrifuges (Hemfort, 1984, p. 10). Usually, centrifugal separators work at a spinning of at least 5000 rpm. Therefore, the velocity of sedimentation is thousands of times higher in a field of centrifugal force compared to the gravitational field. Consequently, the centrifugal separation is thousands of times faster than separation by gravity.

10.4 Backgrounds of cream separation by centrifugation

The centrifugal separation of cream from milk is a liquid–liquid process applied in the industrial manufacture of almost all dairy products. It serves to obtain cream and skim milk and allows milk standardization and milk products with a desired fat content production.

Compared to natural creaming, the separation of cream from milk by centrifugation is faster, more hygienic, and far more complete because the process is continuous, the fat globules move much faster under the action of centrifugal acceleration, and the distance that fat globules have to move is limited by dividing the space in which separation occurs into very thin compartments with the help of conical discs (Walstra et al., 2006, p. 273). Fig. 10.9 shows a section through a conical disc (1) provided with distribution holes (2) and spacers (3), which allow the formation of compartments between the discs in the stack.

The discs are arranged one on top of the other (Fig. 10.10) so that the distribution holes (2) are vertically aligned, creating distribution channels through which the feed flow of the milk (7) is distributed. In the rising channels, the milk flow (7) is split into parallel flows between the conical discs (1) of the disc stack. The caulks welded onto the discs ensured the thin space between the discs. Under centrifugal force, the light phase (4), consisting of fat globules (6) and some skim milk necessary to transport

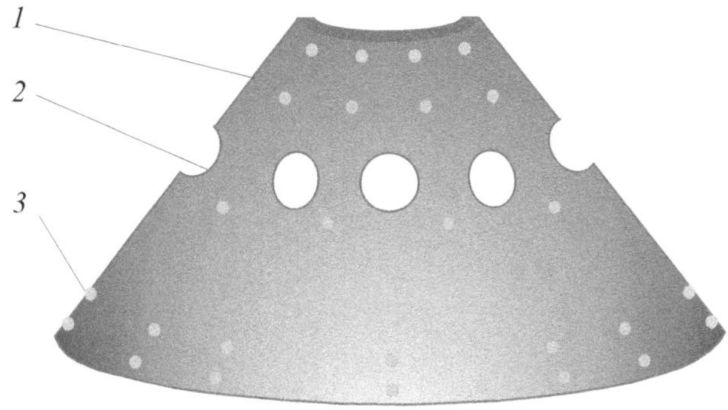

Figure 10.9 Section through a disc with distribution holes and caulks (author's drawing adapted from Tetra Pak, 2003, p. 106, Fig. 6.2.21): 1—conical disc; 2—distribution hole; 3—caulk (spacer).

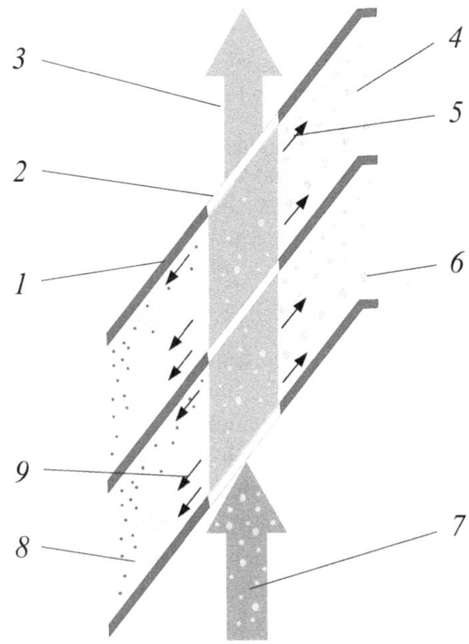

Figure 10.10 Drawing representing a sectional view of a disc stack in a centrifugal separator (author's drawing adapted from Tetra Pak, 2003, p. 106, Figure 6.2.20) 1—conical disc; 2—distribution hole; 3—biphasic system (milk) flow direction; 4—light phase (cream); 5—light phase (cream) flow direction inwards; 6—light phase droplet; 7—biphasic system (milk) flow entering the distribution channel; 8—heavy phase (skim milk); 9—heavy phase (skim milk) direction outwards.

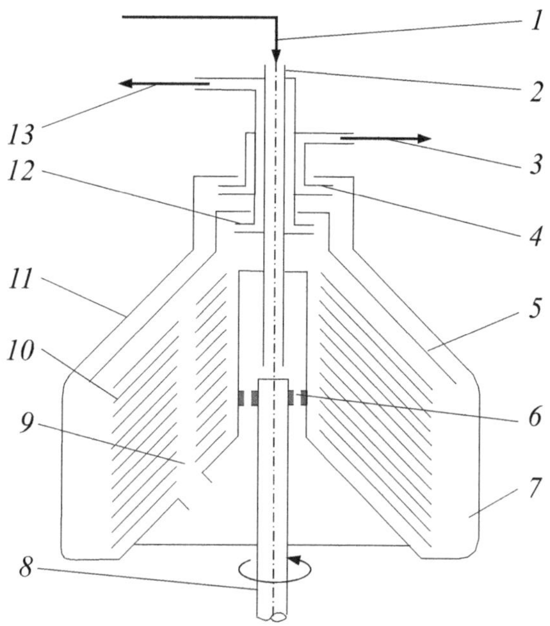

Figure 10.11 The basic principle of cream separation from milk in a centrifugal separator (author's drawing): 1—milk inlet; 2—central feeding tube; 3—skimmed milk discharge; 4—skimmed milk centripetal pump; 5—cream disc; 6—holes; 7—sludge holding space; 8—shaft; 9—distribution channels; 10—conical discs; 11—revolving bowl; 12—cream centripetal pump; 13—cream discharge.

them (cream), moves inwards (5) in the centrifugal separator, while the heavy phase (8), the skim milk, moves outwards (9).

Fig. 10.11 illustrates the principle of cream separation from milk by centrifugation. In this machine, milk (1) is fed under the hydrostatic pressure of the liquid column in the supply system. Thus, the milk flows into the separator through the central feeding tube (2) and reaches the revolving bowl (11) through several holes (6). Then, the milk enters and flows up through the distribution channels (9) of the stack of conical discs (10) obtained by the vertical alignment of the distribution holes of conical discs. The flow is further divided through the slim spaces between the conical discs (5), where the cream is separated. The centrifugal force drives the fat globules inwards, then upwards in the form of cream, toward the cream centripetal pump (12). The skimmed milk, consisting of milk plasma and some small fat globules that have escaped from separation, is driven outwards and then flows upwards between the cream disc (5) and the bowl (11) toward the skimmed milk centripetal pump (4). The separated phases, cream, and skimmed milk, have a high kinetic energy due to the fast spinning of the bowl (usually at least 5000 rpm) and are forced into the centripetal pumps. Also called pairing discs, the centripetal pumps are fixed devices that use the high energy of the cream and skimmed milk to produce pressure and determine the flow through the specific discharge pipe in closed discharge systems. In this way, the revolving and the static parts of the machine are separate.

The system avoids foaming and air intake. Also, it allows the integration of the milk separator in a closed plant comprising a balance tank, pumps, pipelines, heat exchanges, etc. (Buecker & Hoecker, 2009; Tetra Pak, 2003, p. 107).

10.5 Centrifugal separators

The centrifugal separator used to separate cream from milk by centrifugation, also called a milk separator or a skimming separator, is a vertical machine with a stack of conical discs arranged inside a bowl in such a way that allows the rising flow of milk through the distribution holes and then between the discs, as shown before, in Section 10.4.

Because milk always contains a certain amount of solids, such as small impurities, microbial cells and spores, somatic cells, leukocytes, and some coagulated proteins, in skimming separators, a third phase of heavy solids called sediment or sludge is obtained. Therefore, the skimming separators are three-phase centrifuges: cream, skimmed milk, and sediment (Heymann, 2011, p. 168; Tetra Pak, 2003, p. 107).

The sediment in milk varies around 1 kg/10,000 L and is collected in the space of the bowl farthest from the axis. The volume of the sediment space depends on the size of the separator, usually 10 to 20 L (Tetra Pak, 2003, p. 107).

10.5.1 Types of milk separators

The first hand-driven centrifugal separators were open, meaning milk was poured into a container above the separating chamber, and the cream and skimmed milk freely flowed into separate containers (Tetra Pak, 2003, p. 99).

Then, the open-type milk separators were improved to allow higher working capacity. Thus, the milk was fed into the rotating bowl through a fixed feed pipe. The skimmed milk and the cream were discharged tangentially from the bowl into separate but open collectors in the hood of the separator, from where they were evacuated through closed or open discharge lines (Hemfort, 1984, p. 23).

The disadvantages of an open centrifugal separator are the generation of foam and the need for collecting vessels for the discharged cream and skimmed milk. Nowadays, only very small centrifugal separators feature open design.

Subsequently, semiopen (half-hermetic) and hermetic separators were built. Also, milk separators are built using two types of construction: a solid-wall bowl and a self-cleaning bowl.

The separators with a solid-wall bowl are used for discontinuous processing, which means that the solids separated from milk accumulate in the solids holding space on the inner wall of the bowl. The bowl must be dismantled after each production phase and manually cleaned. The duration of a production phase depends on the amount of sediment, that is, the solids content in the product feed. The production must be

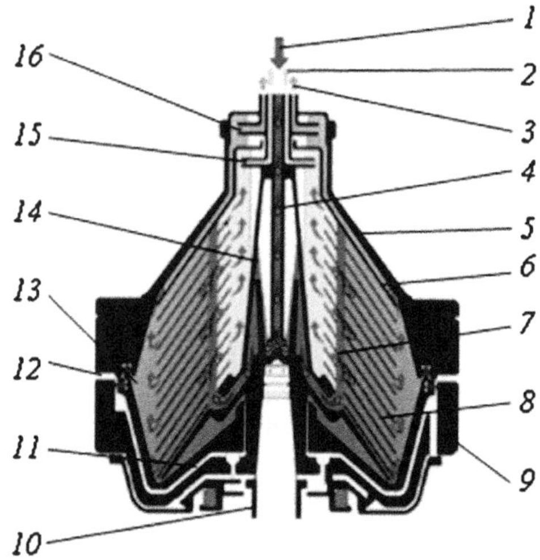

Figure 10.12 Section through the bowl of an Alfa Laval semiopen (paring disc) self-cleaning separator (Tetra Pak, 2003, p. 107, Fig. 6.2.23): 1—milk inlet; 2—cream outlet; 3—skimmed milk outlet; 4—central feeding tube; 5—revolving bowl upper part; 6—cream disc; 7—distribution channels; 8—conical discs; 9—revolving bowl lower part; 10—shaft; 11 —operating water channels; 12—holes for sediment ejection; 13—sediment collecting space; 14—distributor; 15—cream paring disc; 16—skimmed milk paring disc.

interrupted to clean the bowl when the sediment reaches the edge of the conical disc stack (GEA WS Food Tec, 2009, p. 28; Lehmann & Zettier, 1994, p. 27).

The separators with a self-cleaning bowl are built for continuous processing. The sediment is separated in the solids holding space, which incorporates ejection ports that can be hydraulically opened and closed by acting the sliding piston. Thus, during production, the sediment is instantaneously evacuated at preset intervals by lowering the sliding piston using an operating water system. The sliding piston is immediately closed without affecting the continuity of the separation process. At the end of the production phase, the separator is automatically subjected to a CIP procedure. The operation can be fully automated by installing suitable control units (GEA WS Food Tec, 2009, p. 29; Lehmann & Zettier, 1994, p. 27).

Modern milk separators are semiopen and hermetic and are usually provided with self-cleaning bowls.

10.5.1.1 Semiopen milk separators

Semiopen milk separators are equipped with paring discs at the outlet (Fig. 10.12). The milk is supplied to the bowl of the separator from a top inlet through a stationary central tube (4) and a ribbed distributor (14), in which it is accelerated to the speed of rotation of the bowl. Until this point, the milk is in contact with the air at atmospheric pressure. Then, the milk enters the distribution channels (7) in the conical discs (8) stack and

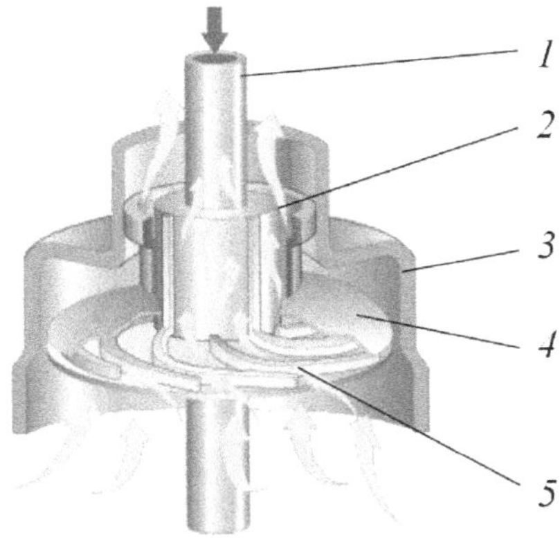

Figure 10.13 A paring disc outlet (Tetra Pak, 2003, p. 107, Fig. 6.2.24): 1—axial inlet tube; 2—evacuation space; 3—paring disc hood; 4—stationary paring disc; 5—rims of the paring disc.

its pressure progressively increases with the distance from the axis of rotation to the periphery of the bowl (5). The cream is separated from the milk alongside the thin spaces between conical discs (8), as shown in Fig. 10.10. The cream moves inwards and reaches the cream-paring disc (15), from where it is evacuated under pressure. The skimmed milk moves outwards and flows between the cream disc (6) and the bowl hood to the skimmed milk paring disc (16). The solid particles heavier than milk settle outwards in the sediment collecting space (13). Modern separator bowls are equipped with operating water systems (11) for automatic ejection of the sediment at preset intervals; therefore, they are called self-cleaning separators. The water operating system opens the holes (12) through which the sediment is ejected in tenths of a second (Tetra Pak, 2003, p. 107).

Fig. 10.13 illustrates a paring disc outlet at the top of the semiopen milk centrifugal separator. The device gives its name to semiopen separators equipped with it, "paring disc separators." The stationary paring disc has rims (5), continuously paring out a certain amount of product. The kinetic energy of the rotating liquid from the bowl is converted into pressure, which always equals the pressure drop in the evacuation line (Tetra Pak, 2003, p. 107).

10.5.1.2 Hermetic milk separators

In a hermetic separator, the milk inlet pipe and the discharge pipes are connected to the revolving parts by mechanical seals (hermetic sealing) that contain a flexible packing ring able to withstand the pressure of the liquid (Hemfort, 1984, p. 27; Walstra et al., 2006, p. 273). The mechanical seals are fragile components prone to wear and tear. Also, because the discharge pressure is created mainly by the feed pump, the machine drive can be smaller, but a bigger feed pump is required (Heymann, 2011, p. 168).

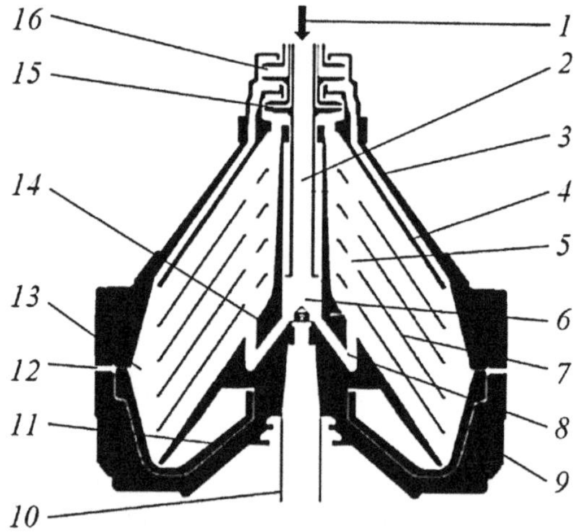

Figure 10.14 Section through the bowl of a Westfalia Separator hermetic milk self-cleaning centrifuge (GEA WS Food Tec, 2009, p. 30, Fig. 32): 1—milk feeding; 2—stationary central feeding tube; 3—upper part of the bowl; 4—separating disc; 5—rising channels; 6 —central bore; 7—conical disc stack; 8—radial bores; 9—lower part of the bowl (bowl bottom); 10—vertical spindle; 11—operating water channels; 12—holes for sediment ejection; 13—sediment collecting space; 14—distributor; 15—cream centripetal pump; and 16—skimmed milk centripetal pump.

In the hermetic milk self-cleaning separators from Westfalia Separator (Fig. 10.14), the milk is fed to the bowl from the top through a stationary central feeding tube (2), which allows the product to flow in the central bore (5) of the rotating distributor (14). The central bore has no ribs to avoid the shear effect on the milk. Further, the milk gently flows through the radial bores (8) into the inner rising channels (5) of the disc stack, where it is stratified into many fine layers between the conical discs (7) and separated into skimmed milk and cream. The central bore and radial bores ensure a "SoftStream" inlet of the milk inside the bowl. After separation, the cream travels toward the center of the bowl and is discharged under pressure by the lower centripetal pump (15). The skimmed milk flows outwards over the separating disc (4) and is discharged under pressure by the upper centripetal pump (16). The bowl is fixed to a vertical spindle (10) driven by an electric motor via a clutch and a worm wheel drive (GEA WS Food Tec, 2009, pp. 29–30; Hemfort, 1984, p. 27).

The cast-iron frame of separators is clad with stainless steel. Also, the three-phase alternative current (AC) motor is additionally clad with stainless steel. All parts of separators coming into contact with the products (milk, cream, and skimmed milk) are made of austenitic Cr–Ni–Mo steel or, in the case of mechanically highly stressed bowl parts, of soft martensitic Cr–Ni–Mo steel or duplex alloy. The gaskets are made of rubber-elastic materials (elastomers) (GEA WS, 2000).

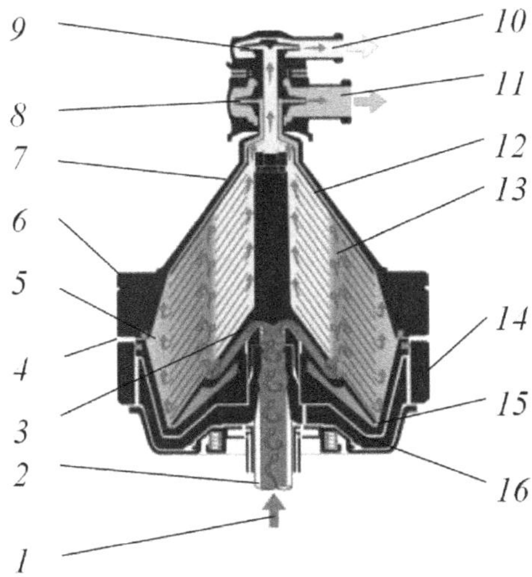

Figure 10.15 Section through the bowl of an Alfa Laval hermetic milk self-cleaning separator (Tetra Pak, 2003, p. 108, Fig. 6.2.25): 1—milk feeding; 2—hollow bowl spindle; 3—distributor; 4—holes for sediment ejection; 5—sediment collecting space; 6—lock ring; 7—bowl hood; 8—outlet pump for the skimmed milk; 9—outlet pump for the cream; 10—cream discharge; 11—skimmed milk discharge; 12—cream disc; 13—distribution channels; 14—revolving bowl lower part; 15—sliding bowl bottom; and 16—operating water system.

Because mechanical seals require additional space, in some centrifugal separators, for example, Alfa Laval hermetic milk separators (Fig. 10.15), the milk is fed to the bowl from below through a hollow spindle (2). (Tetra Pak, 2003, p. 108).

The milk reaches the distributor (3), is accelerated to the same rotation speed as the bowl, and then flows through the distribution channels (13) in the disc stack. The separation occurs as described before, with the skimmed milk being evacuated through the pump (8) and the cream through the pump (9). The sediment is collected in the collecting space (5), from where it is evacuated through the holes (4) in a self-discharge procedure with the help of the operating water system (16).

The bowl of a hermetic separator is filled with milk during operation, which can be considered part of a closed piping system. The pressure the milk feed pump produces is sufficient to overcome the flow resistance through the separator until the skimmed milk and cream outlet pumps. The diameter of the pump impellers can be engineered to suit the pressure requirements at the skimmed milk and cream outlet (Tetra Pak, 2003, p. 109).

10.5.1.3 Driving systems of milk separators

Apart from the hand-driven centrifuges, milk separators are provided with drive systems that evolved from gear drive (Fig. 10.16A) via flat belt drive (Fig. 10.16B) to direct

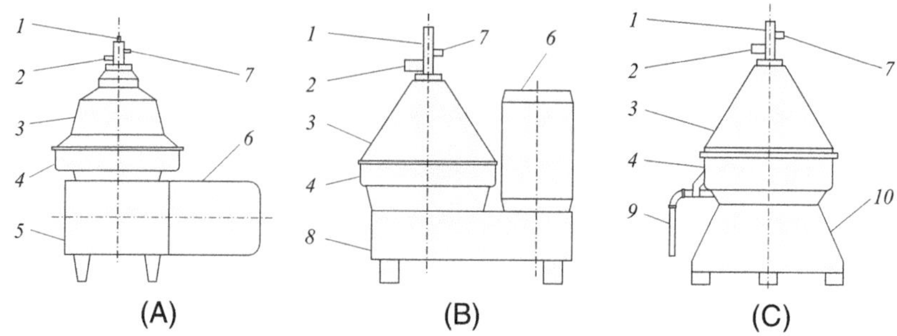

Figure 10.16 Drawings of centrifugal separators with different types of drives (author's drawings): (A) gear drive centrifugal separator; (B) belt drive centrifugal separator; (C) direct drive centrifugal separator; 1—milk inlet pipe; 2—skimmed milk discharge pipe; 3—upper part of the bowl; 4—lower part of the bowl; 5—gear drive; 6—motor; 7—cream discharge pipe; 8—belt drive; 9—operating water system; and 10—integrated direct drive.

Figure 10.17 Picture of a belt drive separator (GEA WS Food Tec, 2009, p. 28).

drive (Fig. 10.16C) (GEA WS, 2000). The changes in the direction of driving improved the mechanical efficiency of separation, reduced the noise level and decreased the number of components of the centrifuge (Hemfort, 1984; Heymann, 2011, p. 169).

Pictures of a belt drive and a direct drive separator are presented in Figs. 10.17 and 10.18, respectively, with the GEA Westfalia Separator acknowledgment.

Figure 10.18 Picture of a direct drive separator (MSI skimming separator installed at DMK Zeven milk powder plant).

10.5.1.4 Sediment discharge

The sediment collected in the sediment space is evacuated manually or automatically as a function of the separator construction type. Separators with solid-wall bowls are used for discontinuous processing and separators with self-cleaning bowls for continuous operation (Lehmann & Zettier, 1994).

The solid bowl wall forms the solids holding space in the discontinuously operating milk separators. The sediment separated from the milk accumulates on the inner wall of the bowl in the solids holding space and is evacuated manually at relatively frequent intervals after the bowl is manually dismantled, which involves immense manual labor. The duration of one production phase depends on the amount of sediment, that is, the solids content in the milk feed. Production must be halted and the bowl cleaned when the sediment reaches the edge of the disc stack at the latest (Lehmann & Zettier, 1994; Tetra Pak, 2003, p. 107).

Self-cleaning milk separators were manufactured to eliminate the manual cleaning of the bowl and allow prolonged operation. The sediment is separated in the holding space, which is of double conical form and incorporates ejection ports opened and closed by hydraulically lowering and raising a sliding piston of the lower bottom part. During production, the accumulated sediment is instantaneously ejected at preset intervals, usually at 30 to 60 minutes. At the end of the production phase, the separator is automatically cleaned without dismantling (CIP). The operation can be fully automated by installing suitable control units (Lehmann & Zettier, 1994; Tetra Pak 2003, p. 107).

10.5.2 *Warm and cold milk skimming*

The cream should be separated from the milk when the milk is fresh to avoid changes in the membrane of the fat globule due to prolonged storage, which increases the density of the fat globule, decreases the $\Delta\rho = \rho_p - \rho_l$ (Eq. 10.3), and, in turn, reduces the separating velocity of fat globule and thus the efficiency of skimming.

The milk centrifugal separator is joined with an in-line fat standardization system for milk and cream (Burke et al., 2018; GEA WS Food Tec, 2009; Tetra Pak, 2003, p. 113).

Usually, the cream is separated after heating the milk, but cold milk skimming is increasingly used in dairies, especially in the USA, Mexico, Australia, and New Zealand (GEA WS Food Tec, 2010, p. 3).

10.5.2.1 *Warm milk skimming*

Dairy legislation stipulates the pasteurization of raw milk worldwide (Regulation, 1995; EC Regulation 1441/2007, 2007; CFR, 2124. 21 CFR 1240.61). The centrifugal separator is an essential unit in the pasteurization plant and is integrated into the process mostly at temperatures of 50°C to 60°C (Heymann, 2011, p. 166).

Separators specially designed for warm milk separation (see Fig. 10.14) operate at a minimum product temperature between 20°C and 25°C because the viscosity of the cream highly increases, and its fluidity is reduced with falling temperature (GEA WS Food Tec, 2009, p. 19; Lehmann & Zettier, 1994, p. 20).

However, the optimum milk temperature is around 55°C because, at such a value, the dynamic viscosity of the milk is stable (GEA WS Food Tec, 2009, p. 18). A separation temperature above 55°C leads to negligible improvements in efficiency. Also, a temperature above 60°C determines the protein precipitation and the deposition of denatured whey proteins on the discs, disrupting the smooth two-way flow pattern in the spaces between the discs and negatively affecting the separability (GEA WS Food Tec, 2009, p. 18; Lehmann & Zettier, 1994, p. 20). Moreover, the tendency of fat globules to agglomerate decreases drastically above 60°C (GEA WS Food Tec, 2009, p. 19).

Centrifugal separators designed for warm separation of cream from milk have efficiencies of about 99% because the fat globules with a diameter of less than 1 μm cannot be removed from the milk due to their very low separating velocity. As a result, skimmed milk has a fat content of 0.05% to 0.1% w/w (Hupperts, 2022, p. 390; McCarthy, 2011, p. 176). The efficiency of separators designed for warm separation does not drop much when used for cold separation at temperatures above 20°C to 25°C if the fat content required for the cream is lower than 15% (GEA WS Food Tec, 2009, p. 19; Lehmann & Zettier, 1994, p. 20).

10.5.2.2 *Cold milk skimming*

Skimming at cold temperatures of 4°C to 15°C is desired in certain situations to avoid thermal changes in the cream when no heating energy is required, for example, before

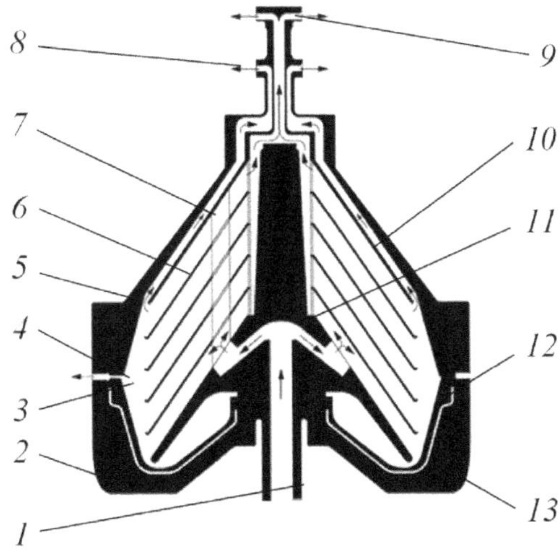

Figure 10.19 Cold milk separator: 1—milk feeding; 2—lower bowl part; 3—sediment collecting space; 4—holes for sediment ejection; 5—upper bowl part; 6—disc stack; 7—rising channels; 8—skimmed milk discharge; 9—cream discharge; 10—cream disc; 11—distributor; 12—sliding piston; and 13—operating water system (GEA WS Food Tec, 2009, p. 31, Fig. 33).

the ultra-high treatment of milk to prevent thermal overtreatment (GEA, 2024a). The separation of the cream is challenging, with the fat content of the milk after separation being about 2% w/w (McCarthy, 2011, p. 176).

Cold milk skimming is less efficient than warm separation. Thus, the skimming efficiency decreases with temperature to about 50% at 5°C if the cold milk separator is fed with its total nominal capacity because the milk fat is more crystalline than the warm status, and the thickness of the flow path and viscosities of the cream and skimmed milk are much higher. The efficiency increases to about 94% by gently increasing the temperature to around 20°C or reducing the flow rate by half to enhance the residence time of the milk in the bowl (Huppertz, 2022, p. 390).

Because cold milk skimming is helpful in specific situations, it is increasingly used, offering several advantages, such as better control of the microbiological quality of milk and improved quality of the cream, for instance, higher whippability. Also, it offers higher flexibility in processing chilled milk received from farms and enhanced quality of unpasteurized milk intended for cheese manufacture (McCarthy, 2011, pp. 176–177).

Fig. 10.19 shows a cold milk separator (GEA WS Food Tec, 2009, p. 31) in which the milk is fed to the bowl from below through a feeding pipe (1). The bowl of the cold milk skimming separator is designed to handle the cream with a much higher viscosity, avoid the churning of the cream inside the separator, and maximize the skimming efficiency (McCarthy, 2011, p. 176). Thus, the milk reaches the distributor (11) and is redirected to the rising channels (7) of the disc stack (6), where it is separated into cream and

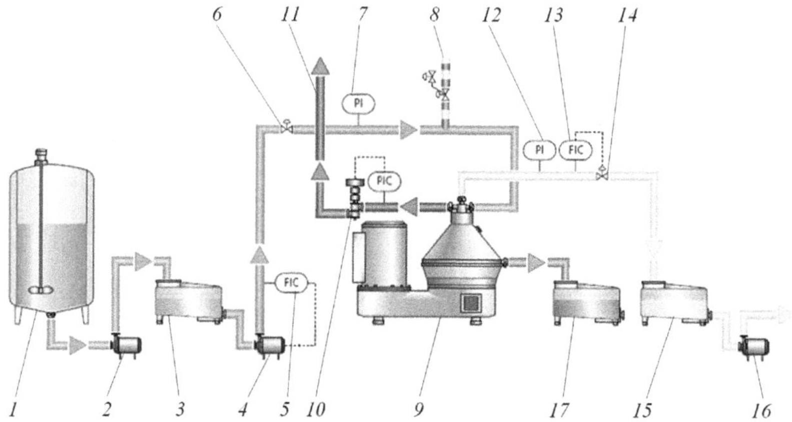

Figure 10.20 Installation diagram of a cold milk separation: 1—storage tank for whole raw milk; 2—pump; 3—balance tank; 4—feed pump; 5—inductive flowmeter on the whole milk pipe; 6—automatic valve on the whole milk pipe; 7—pressure gauge on the whole milk pipe; 8—rinse line; 9—cold milk separator; 10—constant pressure valve on the skimmed milk pipe; 11—skimmed milk outlet; 12—pressure gauge on the cream pipe; 13—inductive flowmeter on the cream pipe; 14—control valve on the cream pipe; 15—cream storage tank; 16—cream pump; 17—sludge tank. FIC: flowmeter inductive controller; PI: pressure indicator (pressure gauge); and PIC: pressure inductive controller (GEA WS Food Tec, 2009, p. 25, Fig. 30.)

skimmed milk. The cream flows to the center of the bowl and the skimmed milk to the opposite direction. Both products are discharged foam-free and under pressure through specific outlets (8) and (9) using an enclosed pipeline system. The operating pressure is set using a control valve in the discharge line of skimmed milk. Another control valve adjusts the discharging cream.

The solids separated from milk are deposited in the sediment collecting space (3), from where they are instantly ejected at preset intervals through holes (4) using the sliding piston (12), hydraulically operated by the water system (13) and a piston valve. The ejection cycles are initiated via the control unit.

Fig. 10.20 presents the diagram of the installation of a cold milk separator (GEA WS Food Tec, 2009, p. 25, Fig. 30). The cold whole milk stored in the tank (1) is transported to the balance tank (3) by the pump (2). The balance tank has to be adequately dimensioned to contain enough milk and prevent air from being drawn into the adjustable positive displacement pump (4). The pump (4) feeds the separator (9) with milk at a constant flow rate as a result of the control carried out by the flowmeter inductive controller (5). The pressure of the separator feed, indicated by the pressure gauge (7), is 3 to 4 bar and depends on the throughput regulated by the automatic valve (6).

After separation, the concentrated cream is discharged toward the cream storage tank (15) and the skimmed milk flows through the skimmed milk pipe (11). The fat content of the cream is adjusted by the skimmed milk pressure set at 2 to 5 bar by the constant pressure valve (10) and regulated by the pressure inductive controller (PIC) according to the milk feeding throughput. Fine adjustment is realized by slightly throttling the cream discharge with the control valve (14) controlled by the inductive flowmeter (13). The pressure gauge (12) indicates the pressure value in the cream discharge line, which must not exceed 2 bars.

The signals from the inductive flowmeters situated on the feed line (5) and the cream line (13) are compared in a control unit, allowing automatic adjustment of the fat content in the cream. The condition for a correct function of this automation is a constant fat content in the feed line. The cream collected in the tank (15) is further transported with the positive displacement pump (16). The sediment ejected from the separator (9) is collected in the sludge container (17).

10.5.3 Efficiency of centrifugal separators

10.5.3.1 Skimming efficiency

The separating efficiency is the percentage of the fat in the whole milk separated into the cream. It is usually called skimming efficiency and can be determined from a partial mass balance in fat content across the separator after a simplification related to the fat concentration of skimmed milk much lower than that of whole milk as follows (McCarthy, 2011, p. 176):

$$E = (1 - -f_s/f_w) \cdot 100, \% \tag{10.6}$$

Where, E is the skimming efficiency in %; f_S—the fat content of skimmed milk in %, w/w; and f_w—the fat content of whole milk in %, w/w. The skimming efficiency varies with how milk is skimmed, that is, hot or cold milk skimming (see Section 10.5.2).

The separating efficiency is not of significant concern if the separator is used for standardization. In such a case, a higher capacity and a longer operation time of the separator are more important than the fat content in skimmed milk, which could vary a little and will be standardized anyway (Walstra et al., 2006, p. 276).

10.5.3.2 Factors that affect skimming efficiency

The efficiency of skimming is affected by temperature, as described above (see Section 10.5.2), and several other factors, such as the size of fat globules, the content of the air in the milk, the impurities from milk, and the rotational speed of the bowl (McCarthy, 2011, p. 177; Walstra et al., 2006, pp. 274–276).

The velocity of separation (Eq. 10.3) shows a direct proportion to the square of fat globule diameter. Therefore, the pipelines are designed to avoid damage to the fat globules and the formation of fat globule agglomerations. Also, the flow rates are controlled by variable speed drives on pumps rather than through pump throttling (Walstra et al., 2006, p. 274).

Milk should be deaerated before cream separation to avoid damage to fat globules and loss of efficiency caused by the presence of air in the separator, which can obstruct the flow of cream toward the inner edge of the disc stack, increase cream viscosity and result in substantial fat loss to the skimmed milk (McCarthy, 2011, p. 177).

Modern centrifugal separators designed for cream separation from milk are self-cleaning, meaning they can separate a third phase consisting of impurities, leukocytes, microbial cells and spores, blood, dirt particles, somatic cells, and some milk proteins. The sludge is ejected at preset intervals with the help of an operating water system, which descends the lower part of the bowl and opens the sludge holes for fractions of a second (GEA WS Food Tec, 2009, p. 29; Hupperts, 2022, p. 391; McCarthy, 2011, p. 177).

Rotational speed, n, greatly influences the velocity of separation (Eq. 10.3). Therefore, separators should be run at the bowl speed recommended by the manufacturer, thus avoiding the reduction of skimming efficiency.

10.5.4 Control of the fat content of cream

The separation of cream from milk should be controlled to ensure the desired fat content and optimum productivity of milk separators. Proper control of the fat content of cream is achieved by manual or automatic adjustment of the pressure on the cream outlet pipe, while the pressure in the skimmed milk outlet pipe is usually kept constant. The fat content of cream can be adjusted to any desired value up to around 50% without compromising the separation efficiency (GEA WS Food Tec, 2009, p. 24; Huppertz, 2022, p. 390; Lehmann & Zettier, 1994, pp. 24–25; McCarthy, 2011, p. 176).

In a semiopen separator (Fig. 10.21), a throttling valve (5) and a cream flowmeter (6) exist in the cream outlet pipe. Adjusting the throttling valve increases or decreases the volume of cream discharged from the cream centripetal pump (7) with corresponding fat content. Thus, if the valve is opened more, the volume of cream discharged will be higher and the fat content lower. The flowmeter controls the volume of cream discharged and indicates the size of the throttling valve aperture as a function of calculations based on mass balance and the desired cream fat content. Also, a pressure-regulating valve (3) exists on the skimmed milk outlet pipe (2), set at a specific value according to the throughput and the separator (Tetra Pak, 2003, p. 109).

In a hermetic separator (Fig. 10.22), an automatic constant pressure unit (6) consisting of a diaphragm valve mounted on the skimmed milk outlet pipe (7) controls the skimmed milk pressure. The pressure is adjusted with compressed air above the diagram. During separation, the pressure of compressed air is kept constant. The pressure difference between compressed air and skimmed milk affects the diaphragm. Thus, the diaphragm descends if the pressure in the skimmed milk drops. The valve plug fixed to the diaphragm moves downwards and reduces the flow section (passage), increasing the pressure in the skimmed milk outlet. On the contrary, the diaphragm rises if the pressure in the skimmed milk increases and the valve plug moves upwards, restoring the preset pressure (Tetra Pak, 2003, p. 109)

Fig. 10.23 presents control and measuring devices mounted on the feed and discharge lines of a centrifugal separator (GEA WS Food Tec, 2009, pp. 23–24; Lehmann & Zettier, 1994, pp. 24–25), for example, a flowmeter (2) on the cream discharge line, pressure indicators on all the lines, and a constant pressure valve (10) on the skimmed milk discharge line.

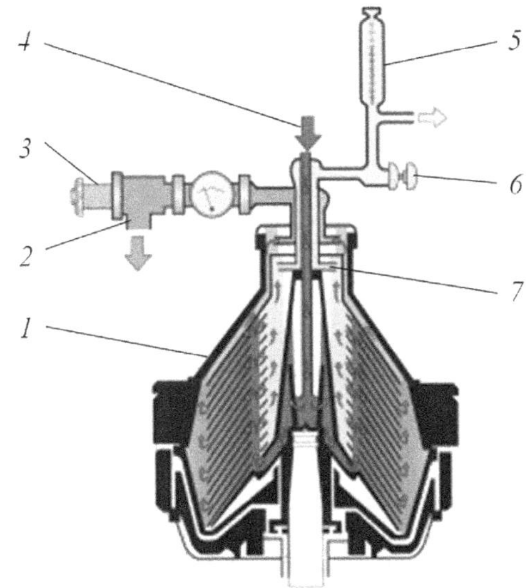

Figure 10.21 Paring-disc separator with manual control devices in the outlets (Tetra Pak, 2003, p. 109, Fig. 6.2.27):1—paring-disc separator; 2—skimmed milk outlet; 3—pressure-regulating valve; 4—milk inlet; 5—cream flowmeter; 6—cream throttling valve; and 7—cream centripetal pump.

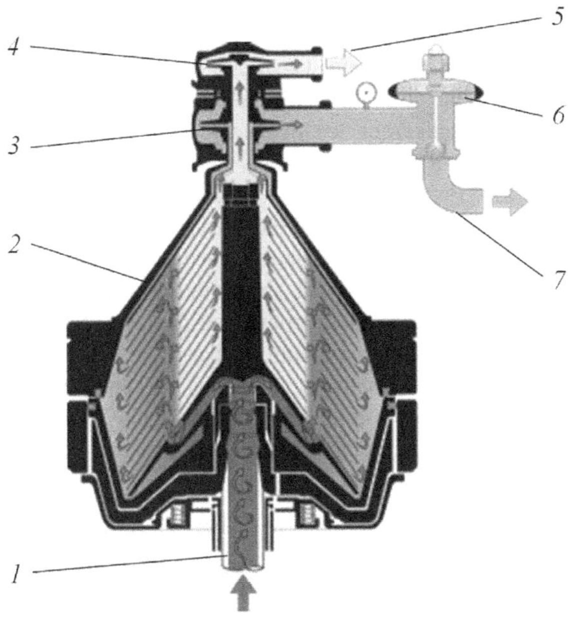

Figure 10.22 Hermetic separator with an automatic constant pressure unit on the skimmed milk outlet (Tetra Pak, 2003, p. 109, Fig. 6.2.28):1—milk inlet; 2—hermetic separator; 3—skimmed milk centripetal pump; 4—cream centripetal pump; 5—cream outlet; 6—automatic constant pressure unit; and 7—skimmed milk outlet.

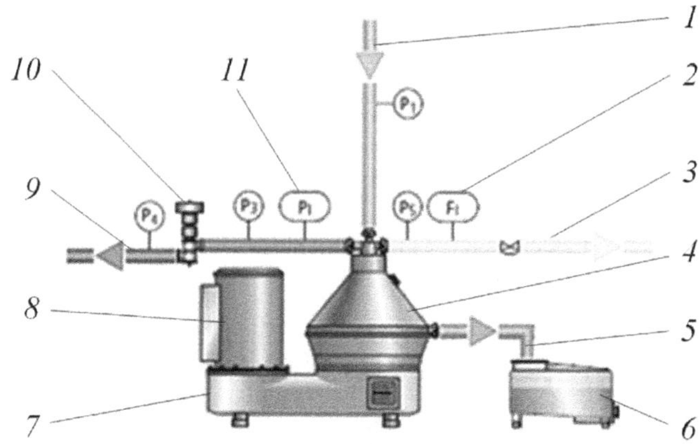

Figure 10.23 Control and measuring devices for a separator feed and discharge (GEA WS Food Tec, 2009, p. 24, Fig. 29): 1—milk feeding; 2—flowmeter; 3—cream discharge; 4—centrifugal separator; 5—sediment discharge; 6—sediment tank; 7—belt drive; 8—motor; 9—skimmed milk discharge; 10—constant pressure valve; and 11—pressure gauge. Measuring devices: P_1—feed pressure (approx. 1.0 bar); P_3—discharge pressure of skimmed milk (5 bar); P_4—pressure of downstream apparatus (max. 4.5 bar); and P_5—discharge pressure of cream (5.2 bar).

The purpose of using control and measuring devices is to correctly adjust the pressure in the separator to achieve a low residual fat content in the skimmed milk and a constant fat content in the cream. The pressures P_3 and P_4 are responsible for the separation efficiency on the skim milk side. Also, P_3 determines the separation zone in the disc stack, so its fluctuations negatively affect the cream separation. Therefore, pressure P_3 has to be set to a level that prevents an overflow on the skimmed milk side. In practice, the pressure P_3 is set to a value of 0.5 bar less than the overflow pressure. The constant pressure valve (10) installed on the skimmed milk line prevents pressure P3 from being affected by pressure variations in the equipment downstream of the separator. Thus, the pressure $P_4 = 4.5$ bar, shown in Fig. 10.23, must not be exceeded. The maximum pressure of cream, P_5, depends on pressure P_3. In the example from Fig. 10.23, P3 is 5 bar and P_5 is a maximum 5.2 bar.

The above are the simplest methods to control the fat content in cream. Other instruments can be used, for example, automatic in-line systems, to measure the fat content of cream in combination with control systems, which keep the fat content at a constant value.

10.5.5 Technical data and specific features of milk separators

Because milk skimming is the core task for any dairy company, the manufacturers of milk separators offer a large variety of machines in terms of productivity, separation efficiency, flexibility, and many other advantages and benefits.

The capacity of the modern milk separator can be up to 60,000 L/h (temperature up to 50°C–60°C) or 75,000 L/h (temperature up to 45°C) for warm separation and up to 50,000 L/h for cold separation (temperature up to 10°C) (GEA WS Food Tec, 2009, p. 33; GEA WS Food Tec, 2010, p. 5; Huppertz, 2022, p. 390; McCarthy, 2011, p. 176). Alfa Laval separators have up to 80,000 L/h for hot milk separation (Alfa Laval H885) and 55,000 L/h for cold milk separation (Alfa Laval C855). Skimming efficiency is, in terms of fat content in skimmed milk, 0.08% to 1.00% when product temperature exceeds 4°C (cold separation) (GEA WS Food Tec, 2009, p. 5), 0.04% to 0.07% (Tetra Pak, 2003, p. 98; Walstra et al., 2006, p. 275), 0.04% to 0.05% at 45°C (warm separation) (Walstra et al., 2006, p. 275).

The fat content of cream can be 40% in cold milk separation up to 50% if the fat content in the whole milk changes to 4.5% or up to 60% in warm milk separation (Alfa Laval H885; GEA WS Food Tec, 2009, pp. 20–21).

The size of the milk separator varies with capacity. Thus, the disc stack consists of up to 120 discs stacked together at an angle of 45 to 60 degrees and separated by a 0.4 to 2.0 mm gap or separation channel (Goff et al., 2023; Hemfort, 1984, p. 17). The overall dimensions could be up to around 1.8 m × 1.3 m × 2.8 m (L × l × H) (Alfa Laval H885).

The tensile strength of the bowl material limits the rotary speed, which varies from 4800 rpm for bowl diameters higher than 800 mm to 11,000 rpm for bowl diameters lower than 300 mm (Buecker & Hoecker, 2009).

The total amount of sediment in milk varies but can be about 0.01% (1 kg/10,000 L). The sediment space volume varies depending on the size of the separator, typically between 10 and 20 L (GEA WS Food Tec, 2010; Tetra Pak, 2003, p. 107).

The installed power of the electric motor varies from 7.5 kWh to more than 45 kWh as a function of capacity (milk inlet flow) (Alfa Laval C855, H885; GEA WS Food Tec, 2010).

Milk separators should have a hygienic design compliant with the European Hygienic Engineering and Design Group practical guidance on the hygienic engineering aspects of manufacturing safe and healthy food and with the European legislation related to hygienic handling, preparation, processing, and packaging of food (EC Directive 2006/42/EC, 2006).

Also, the manufacturers of milk separators are focused on reducing water consumption for the self-cleaning feature and energy consumption (e.g., integrated direct drive option), monitoring relevant parameters, predicting maintenance costs, and ensuring low operating costs (GEA, 2023a, p. 14). Thus, the innovation and improvements of milk separators are continuously evolving, as shown in the following subsection.

10.6 Recent advances in separation of cream from milk by centrifugation

The separation of cream from milk by centrifugation and the construction of milk separators have been continuously changing since the production of the first centrifugal separators in the second half of the nineteenth century. Most of the innovations and improvements belong to the leading manufacturers of centrifugal separators, Alfa Laval Corporate AB and Westfalia Separator, which is part of GEA Group Aktiengesellschaft

(AG). However, studies have begun to be published in the literature presenting various research that aim to improve the process of separating cream from milk by centrifugation.

This subsection presents the advances in the separation of cream from milk by centrifugation from the last decades.

10.6.1 Recent advances from the leading manufacturers of centrifugal separators

10.6.1.1 Recent advances from westfalia separator

In the last decades, Westfalia Separator brought to light several innovations, such as HydroSoft feed, Hyvol Proplus, Procool, and DirectDrive systems.

The HydroSoft feed system was meant to treat the product gently by combining the advantages of the SoftStream and Hydrohermetic feed systems.

The SoftStream inlet system was created to provide low pressure in the feeding line of the separator through an adequate flow cross-section and high flexibility for treating the raw milk softly (GEA WS Food Tec, 2009, p. 5).

The Hydrohermetic feed system was designed to eliminate the ribs in the feed chamber for gentle product treatment since the absence of ribs means no shear force. Also, the separator is provided with a hydraulic seal through a hydrohermetic disc at the end of the feed pipe, so no air is trapped in the product (GEA, 2024a, b).

In the HydroSoft feed system, the product flowing through the stationary feed pipe reaches the centrally located bore of the rotating distributor. Without a hydrohermetic disc at the end of the feed pipe, the milk is subjected to high shear forces which destroy the fat globules, that is, they become smaller. The centrifuge separates the smaller globules more difficult and fat losses in the skimmed milk occur. Combining SoftStream and Hydrohermetic feed systems allows very gentle acceleration of the product at low pressure and prevents air contact with the product due to the existence of the hydrohermetic disc, eliminating the friction between the ambient air and the liquid. Thus, the fat globules are not destroyed and can be adequately separated from the milk (GEA WS Food Tec, 2009, p. 30; GEA, 2024a).

The Hyvol Proplus system combines the performances of the Westfalia Separator Hyvol with the Westfalia Separator Proplus (GEA, 2020).

The Proplus system was designed to achieve higher product yield, minimize water consumption, lower maintenance and wear costs, and longer machine service life. It extends the ejection intervals of milk separators from 20 to 30 minutes to up to 90 minutes, reducing up to three times the shots and the sludge volume, saving up to 28% water and reducing product waste. Thus, it ensures added value and a significant increase in protein yield from the milk subjected to separation (GEA, 2024a).

Hyvol is the name GEA gave to its separators, equipped with an ejection system via piston valves. The adverse effects of shearing forces are minimized with a hydrohermetic feed system, resulting in perfect separation and product quality. The hydrohermetic seal prevents product contamination by ambient air and allows perfect CIP capability. Another feature of the hyvol separators is the extremely quick-acting ejection valves, which reduce the product losses during separator ejections to a minimum (GEA, 2024a).

With the Hyvol Proplus system, the benefits are substantially increased protein yield and significant cost savings based on reduced water consumption and reduction of the solids volume. The features of milk separators also include high throughputs, high efficiency, flexible processing, high product quality, gentle feed system and product treatment, easy operation, uniform solids discharge, low maintenance and operating costs, and low energy requirement (GEA, 2024a; GEA WS Food Tec 2009, p. 5)

The Procool process is an innovative process of Westfalia Separator that advances dairies in cold milk skimming, complying with the quality requirements for products such as milk powder and drinking milk. The cold separation of cream from milk at temperatures between 4°C and 25°C maintains optimum product quality and provides high microbiological safety (GEA, 2021, p. 21; GEA, 2024a).

Westfalia Separator direct drive is a separator equipped with an integrated direct drive that uses the power transmission with almost no loss of efficiency. Thus, it ensures lower energy and infrastructure costs, leaner processes, and reduced space requirements. (GEA, 2021, pp. 14, 21).

10.6.1.2 Recent advances from alfa laval

Alfa Laval innovations from the last decades in milk separation are the Hermetic Design, eDrive, and UniDisc.

Hermetic Design is the trademark of the Alfa Laval air-tight separator, which provides gentle product handling through the gentle acceleration of the bowl and the reduction of shear forces, smooth flow, improved product quality through the prevention of oxygen pickup, and reduced power consumption. The design of the separator optimizes the CIP cycle, allowing the contact of CIP fluids with all product-wetted parts (Alfa Laval, 2015–2024).

The eDrive is the Alfa Laval direct drive system designed to simplify the separator drive system and reduce energy costs. It offers mechanical simplicity as a system with fewer rotating parts than conventional gear-drive systems and a built-in control system of the lubricating oil quality, ensuring longer intervals between services, better energy efficiency, and lower maintenance costs (Alfa Laval, 2015–2024).

The UniDisc is an innovation of Alfa Laval that uses a weld-free technology featuring micro caulks molded simultaneously with the disc from a single metal sheet to form a conical disc with spacers. These spacers are smaller and shorter than the welded ones, enabling more discs to fit in the disc stack. The results are increased separation surface area and flow rate, boosted separation performance, and ensured superior hygiene. Also, the spacers are uniformly distributed across the disc surface, providing even distances between the discs and resulting in a rigid and robust disc stack (Alfa Laval, 2015–2024).

10.6.2 Recent advances from academia

Research performed in academia and research institutes worldwide has focused on various topics of milk processing and dairy products related to cream separation from milk. Thus, recently, many researchers have been interested in fractionation methods based on the size and density of milk fat globules (MFG) and investigated different separation techniques, such as conventional gravity separation (Geer & Barbano, 2014; Ma & Barbano, 2000), two-stage gravity separation (O'Mahony et al., 2005), combined gravity and centrifugal separation (Edén et al., 2016), compared gravity and centrifugal separation (Meng et al., 2022), centrifugal separation (Dickow et al., 2011; Holzmüller et al., 2016), centrifugal separation with four discs in the disc stack (Edén et al., 2016), two-stage centrifugal separation (Dhungana et al., 2017, 2022), ultrasonic separation (Juliano et al., 2011; Leong et al., 2014a, b; Mahmoud et al., 2022), multistage ultrasonication separation (Leong et al., 2016), one-step separation of fat globules using crossflow microfiltration (Kapoor et al., 2024), microfiltration through ceramic silicon carbide membranes (Dons et al., 2023, 2024), microfiltration (Jukkola et al., 2016), centrifugation compared with microfiltration (Jukkola et al., 2019), and impact of pulsed electric field on milk fat separation (Walkling-Ribeiro et al., 2024).

Among the thousands of published studies in the last decades, few stand out regarding the separation of cream from milk by centrifugation and are discussed below.

A modified bowl disk centrifuge with four disks, each having three caulks evenly distributed, was used to fractionate commercial cream and cream separated from milk by creaming to investigate the effect of native MFG size on whipping properties. The combination of gravity and centrifugal separation showed that the whipping time to maximum firmness varied inversely proportional to the MFG mean size (Edén et al., 2016).

In another study, Meng et al. (2022) compared the gravitational and centrifugal separation to determine the influence of heating, addition of calcium-chelating agents, sonication, and freezing on the creaming of bovine milk. The creaming capacity of bovine milk slightly decreased after heating and keeping for 10 minutes at 65°C to 85°C but did not change significantly ($P > .05$). Adding trisodium citrate enhanced the creaming rate, while milk sonication decreased it significantly ($P < .05$). Freezing the raw milk and thawing at 4°C increased the creaming rate through gravitational separation for 24 hours but did not influence it in centrifugal separation. The work indicates that choosing specific techniques to study milk creaming can have certain advantages.

Holzmüller et al. (2016) used a cream separator to investigate the effect of cream washing on MFG stability and the loss of proteins from the milk fat globule membrane (MFGM). They also considered the pumping step using a pump and a heat exchanger to determine which process step impacted the MFG destabilization more. The research suggested that the milk pumping and the shear forces in the cream separator cause damage to MFGM, which are released as fragments.

Milk from two breeds, cooled to 4°C or kept at the milking temperature of 31°C, was stored for up to 24 hours and then separated into cream and skimmed milk by centrifugation to investigate the influence of storage time and temperature on lipoprotein lipase (LPL) activity, the free fatty acid (FFA) levels, and the effect of cooling on the composition of proteins associated with MFGM (Dickow et al., 2011). The authors found that the LPL activity increased in the cream phase after cooling for up to 24 hours and decreased in the cream phase and skimmed milk phase after storage at 31°C, whereas the FFA levels depend on the LPL activity. They also reported different behavior of the proteins after cooling. Thus, the proteomic analysis showed that after storage at 4°C, adipophilin, lactadherin, and lactoferrin were quantitatively more associated with the MFGM, whereas other proteins, such as β-casein, heat shock protein, and fatty acid-binding protein migrate toward the skimmed milk. This major change in the protein distribution between the cream and skimmed milk phases contributes to the stability of MFGM alongside the changes in the LPL association with the MFGM after cooling.

Dhungana et al. (2017) developed a two-stage centrifugal separation process to fractionate MFG based on size. The first stage consisted of a modified separator with the separating cones removed, whereas the second stage used a conventional cream separator. Two phases of milk, one rich in larger fat globules and the other rich in smaller fat globules, were obtained in the modified separator. Each phase was subjected to separation in the conventional separator. The fractionation was conducted at various separation temperatures and milk feed rates to evaluate the influence of these factors on the size and integrity of MFG. Thus, the size range of MFG achieved without coalescence and without affecting the globule integrity was 1.35 to 4.28 μm. The optimum temperature was 35°C for all feed rates. Larger MFGs were more damaged at lower temperatures.

The method allows the obtaining of fractionated milk and cream with various nutritional values and physical functionalities. It can provide the basis for a new industrial process with further developments regarding throughput, separation efficiency, and use of conventional processing equipment. Having these in mind, the authors continued the research using the two-stage centrifugal separation to define the parameters that could increase the difference in globule size in the fractionated creams (Dhungana et al., 2022). Focused on a controlled decrease in the efficiency of the cream centrifugal separation, the authors increased the average size of MFG in the cream containing larger globules to 5.2 μm. Thus, they obtained fractionated cream with the smallest (1.32 μm) and largest (5.2 μm) average MFG sizes with the fat content of the cream as 33.6% (w/w) and 32.75% (w/w). The fractionating performance was reduced for increased feed rate and decreased rotational speed of the modified cream separator. The authors concluded that their results could help scale up the fractionation of MFG based on size and design a new type of milk centrifugal separator with enhanced performance in MFG fractionation (Dhungana et al., 2022).

10.7 Concluding remarks and future perspectives

The separation of cream from milk by centrifugation is essential for processing raw milk to obtain a large variety of dairy products with reduced and standardized fat content and specific products with higher fat content than milk, such as cream, butter, or anhydrous milk fat. Although a recent invention, the milk separator is vital for any milk processing plant and highly contributes to the quality of life and well-being of consumers. The manufacturers of centrifugal separators have done impressive research concerning cream separation from milk by centrifugation and milk separators.

They are keen to continue improving milk separators through specific innovations and destinations from small-scale to large-capacity dairy processing units. Thus, some manufacturers found specific solutions to obtain special features of the separators they sell, such as avoiding mechanical seals for hermetic design and air-tight bowls, maximizing protein yield in separated milk, minimizing water consumption, using an integrated direct-drive system with fewer rotating parts that reduce the maintenance and energy costs or a radical new disc design that decreases cleaning needs while increasing the flow rate.

References

Alfa Laval (C855). Disc stack separator for the dairy industry. Alfa Laval Corporate AB, Retrieved December 2024 from, www.alfalaval.com.

Alfa Laval (H885). Disc stack separator for the dairy industry. Alfa Laval Corporate AB, Retrieved December 2024 from, www.alfalaval.com.

Alfa Laval (2015). History of Alfa Laval. Retrieved June 28, 2024, from https://www.alfalaval.com/en-AU/about-us/our-company/history-of-alfa-laval/.

Alfa Laval (2015-2024). Innovative separation solutions for dairy processing. Separators for dairy processing. Centrifugal separators, Retrieved December 2024 from https://www.alfalaval.com/products/separation/centrifugal-separators/separators/dairy/dairy-separation-solutions/.

Alhadrami, G. A., & Faye, B. (2022). Camel. In P. L. H. McSweeney, & J. P. McNamara (Eds.), *Encyclopedia of dairy sciences* ((3rd ed.)) (1, pp. 48–64). Elsevier.

Andriesse, C. D. (2005). *Huygens: The man behind the principle. translated in English by S. Miedema.* Cambridge University Press.

Bogucki, P. (1984). Ceramic sieves of the linear pottery culture and their economic implications. *Oxford Journal of Archaeology, 3*(1), 15–30. doi:10.1111/j.1462-0092.1984.tb00113.x.

Buecker, F. J., Hoecker, J., et al. (2009). *Separators from GEA Westfalia separator for the dairy industry.* Oelde, Germany: GEA Westfalia Separator.

Burke, N., Zacharski, K. A., Southern, M., Hogan, P., Ryan, M. P., & Adley, C. C. (2018). The dairy industry: Process, monitoring, standards, and quality. . In A. V. Díaz, & R. M. García-Gimeno (Eds.), *Descriptive food science* (pp. 1–23). IntechOpen. doi:10.5772/intechopen.80398.

Cambridge English Dictionary Online (2024). Centrifuge definition. Retrieved November 1, 2024, from https://dictionary.cambridge.org/us/dictionary/english/centrifuge.

CFR (2024). 21 CFR 1240.61 Mandatory pasteurization for all milk and milk products in final package form intended for direct human consumption. Retrieved August 11, 2024, from https://www.ecfr.gov/current/title-21/section-1240.61.

Coulson, J. M., Richardson, J. F., Backhurst, J. R., & Harker, J. H. (1978). Sedimentation. *Chemical engineering, vol. 2 unit operations* ((3rd ed.), pp. 172–229). Pergamon Press.

Codex, S. 206-1999. (1999). Codex Alimentarius Commission. Codex general standard for the use of dairy terms. Retrieved January 17, 2024, from www.fao.org.

Curry, A. (2013). Archaeology: The milk revolution. *Nature, 500*(7460), 20–22. doi:10.1038/500020a.

Dhungana, P., Truong, T., Bansal, N., & Bhandari, B. (2022). A novel continuous method for size-based fractionation of natural milk fat globules by modifying the cream separator. *International Dairy Journal, 125*, 105209. doi:10.1016/j.idairyj.2021.105209.

Dhungana, P., Truong, T., Palmer, M., Bansal, N., & Bhandari, B. (2017). Size-based fractionation of native milk fat globules by two-stage centrifugal separation. *Innovative Food Science & Emerging Technologies, 41*, 235–243. doi:10.1016/j.ifset.2017.03.011.

Dickow, J. A., Larsen, L. B., Hammershøj, M., & Wiking, L. (2011). Cooling causes changes in the distribution of lipoprotein lipase and milk fat globule membrane proteins between the skim milk and cream phase. *Journal of Dairy Science, 94*(2), 646–656. doi:10.3168/jds.2010-3549.

Dons, T., Candelario, V., Andersen, U., & Ahrné, L. (2024). Separation of milk fat using silicon carbide support ceramic membranes with different pore sizes. *Innovative Food Science & Emerging Technologies, 94*, 103671. doi:10.1016/j.ifset.2024.103671.

Dons, T., Candelario, V., Andersen, U., & Ahrné, L. M. (2023). Gentle milk fat separation using silicon carbide ceramic membranes. *Innovative Food Science & Emerging Technologies, 84*, 103299. doi:10.1016/j.ifset.2023.103299.

Doreau, M., & Martin-Rosset, W. (2022). Horse. In P. L. H. McSweeney, & J. P. McNamara (Eds.), *Encyclopedia of dairy sciences* ((3rd ed.)) (1, pp. 123–164). Elsevier.

EC Directive 2006/42/EC. (2006). Directive 2006/42/EC of the European Parliament and of the Council of 17 May 2006 on machinery, and amending Directive 95/16/EC (recast). *Official Journal of the European Union, L 157*, 24–86. https://eur-lex.europa.eu/legal-content/EN/TXT/PDF/?uri=CELEX:32006L0042.

EC Regulation 1441/2007. (2007). Commission Regulation (EC) No 1441/2007 amending Regulation (EC) No 2073/2005 on microbiological criteria for foodstuffs with regards to milk and dairy products. *Official Journal of the European Union, L 322*, 12–29.

Edén, J., Dejmek, P., Löfgren, R., Paulsson, M., & Glantz, M. (2016). Native milk fat globule size and its influence on whipping properties. *International Dairy Journal, 61*, 176–181. doi:10.1016/j.idairyj.2016.06.004.

ERIH (European Route of Industrial Heritage) (2024). Carl Gustaf Patrick de Laval (1845–1913). Retrieved June 24, 2024, from https://www.erih.net/how-it-started/stories-about-people-biographies/biography/laval.

Etymonline (2024). Centrifuge. Etymology of centrifuge by Etymonline. Retrieved November 1, 2024, from https://www.etymonline.com/word/centrifuge.

Evershed, R. P., Payne, S., Sherratt, A. G., Copley, M. S., Coolidge, J., Urem-Kotsu, D., Kotsakis, K., Ozdoğan, M., Ozdoğan, A. E., Nieuwenhuyse, O., Akkermans, P. M. M. G., Bailey, D., Andeescu, R.-R., Stuart Campbell, S., Farid, S., Hodder, I., Yalman, N., Ozbaşaran, M., Biçakci, E., … Burton, M. M (2008). Earliest date for milk use in the Near East and southeastern Europe linked to cattle herding. *Nature, 455*(7212), 528–531. doi:10.1038/nature07180.

Fox, P. F., Uniacke-Lowe, T., McSweeney, P. L. H., & O'Mahony, J. A. (2015b). Milk lipids. *Dairy chemistry and biochemistry*. Springer. doi:10.1007/978-3-319-14892-2_3.

Fox, P. F., Uniacke-Lowe, T., McSweeney, P. L. H., & O'Mahony, J. A. (2015a). Heat-induced changes in milk. *Dairy chemistry and biochemistry*. Springer. doi:10.1007/978-3-319-14892-2_9.

GB366873A. (1932). *Improvements relating to centrifugal separating machines*. Ramesohl und Schmidt A.G.

GEA. (2020).Mechanical separation solutions for the dairy industry. Process Stages, systems, and equipment from GEA. GEA Westfalia Separator Group GmbH. Retrieved July 2024 from, https://www.gea.com/en/assets/252179/.

GEA. (2021). *Mechanical separation solutions from GEA Westfalia Separator*. GEA Westfalia Separator.

GEA. (2023a). *Better world. Better products*. GEA Westfalia Separator Group GmbH.

GEA (2023b). Getting old is easy. It's getting better that matters. Retrieved June 2024 from, https://www.gea.com/en/stories/125-years-gea-separation/.

GEA (2023c). Separate from the rest (GEA separation technology celebrates 130 years.). Retrieved June 2024 from, https://www.gea.com/en/stories/separate-from-the-rest/.

GEA (2024a). Products & services. Dairy products. Centrifuges & separation equipment. Centrifugal separator. Retrieved December 2024 from, https://www.gea.com/en/products/centrifuges-separation/.

GEA. (2024b). *Separators for milk and whey skimming, cold milk separator*. GEA Group Aktiengesellschaft Retrieved December 2024 from, https://www.gea.com/en/products/centrifuges-separation/centrifugal-separator/separator/separators-milk-whey-skimming/.

GEA WS. (2000). *Take the best—Separate the rest. Skimming separators for milk and whey*. GEA Westfalia Separator Food Tec GmbH.

GEA WS Food Tec. (2009). *Separators for the dairy industry*. GEA Westfalia Separator Food Tec GmbH.

GEA WS Food Tec. (2010). Cold milk separation with Westfalia Separator PROCOOL technology. *Leadership solutions*. Westfalia Separator Food Tec GmbH.

Geer, S. R., & Barbano, D. M. (2014). The effect of immunoglobulins and somatic cells on the gravity separation of fat, bacteria, and spores in pasteurized whole milk. *Journal of Dairy Science, 97*(4), 2027–2038. doi:10.3168/jds.2013-7629.

Goff, H. D., Hill, A., & Ferrer, M. A. (2023). *Dairy science and technology ebook* Retrieved July 2024 from, https://books.lib.uoguelph.ca/dairyscienceandtechnologyebook/.

Hemfort, H. (1984). *Separators. Centrifuges for clarification, separation and extraction processes. Technical scientific documentation No. 1*. GEA Westfalia Separator AG.

Heymann, B. (2011). Centrifuges and separators: Types and design. In J. W. Fuquay, P. F. Fox, & P. L. H. McSweeney (Eds.), *Encyclopedia of dairy sciences* (4, pp. 166–175). Elsevier/Academic Press.

Holand, Ø., Gjøstein, H., & Nieminen, M. (2022). Reindeer. In P. L. H. McSweeney, & J. P. McNamara (Eds.), *Encyclopedia of dairy sciences* ((3rd ed.)) (1, pp. 123–164). Elsevier.

Holzmüller, W., Müller, M., Himbert, D., & Kulozik, U. (2016). Impact of cream washing on fat globules and milk fat globule membrane proteins. *International Dairy Journal, 59*, 52–61. doi:10.1016/j.idairyj.2016.03.003.

Hoyle, J. B., & Burton, H. (1984). The problem of cream plug in bottled unhomogenized pasteurized milk. *International Journal of Dairy Technology, 37*(2), 42–45. doi:10.1111/j.1471-0307.1984.tb01236.x.

Huppertz, T. (2022). Separators and centrifuges. In P. L. H. McSweeney, & J. P. McNamara (Eds.), *Encyclopedia of dairy science* (3rd ed) (5, pp. 389–395). Elsevier/Academic Press. doi:10.1016/B978-0-12-818766-1.00132-X.

Huppertz, T., Uniacke-Lowe, T., & Kelly, A. L. (2020). Physical chemistry of milk fat globules.. In P. L. H. McSweeney, P. F. Fox, & J. A. O'Mahony (Eds.), *Advanced dairy chemistry* ((4th ed.)) (2: Lipids, pp. 133–167). Springer Nature Switzerland AG.

Jukkola, A., Hokkanen, S., Kämäräinen, T., Partanen, R., Heino, A., & Rojas, O. J. (2019). Changes in milk fat globules and membrane lipids under the shear fields of microfiltration and centrifugation. *Journal of Membrane Science, 573*, 218–225. doi:10.1016/j.memsci.2018.12.007.

Jukkola, A., Partanen, R., Rojas, O. J., & Heino, A. (2016). Separation of milk fat globules via microfiltration: Effect of diafiltration media and opportunities for stream valorization. *Journal of Dairy Science, 99*(11), 8644–8654. doi:10.3168/jds.2016-11422.

Juliano, P., Kutter, A., Cheng, L. J., Swiergon, P., Mawson, R., & Augustin, M. A. (2011). Enhanced creaming of milk fat globules in milk emulsions by the application of ultrasound and detection by means of optical methods. *Ultrasonics Sonochemistry, 18*(5), 963–973. doi:10.1016/j.ultsonch.2011.03.003.

Kapoor, A., Datta, S., Gupta, G., Vishwakarma, A., Singh, A., Chattopadhyay, S., & Ambatipudi, K. (2024). One-step separation of fat globules based on size from bovine milk using a cross-flow microfiltration. *Separation and Purification Technology, 330*(Part A), 125251. doi:10.1016/j.seppur.2023.125251.

Kersten, M. (1993). *Innovation is our tradition. 100 years of Westfalia Separator AG, Oelde*. Westfalia Separator AG.

Lehmann, H. R., & Zettier, K. H. (1994). *Separators for the dairy industry. Technical scientific documentation no. 7* ((4th revised ed.)). GEA Westfalia Separator AG.

Leong, T., Johansson, L., Juliano, P., Mawson, R., McArthur, S., & Manasseh, R. (2014a). Design parameters for the separation of fat from natural whole milk in an ultrasonic liter-scale vessel. *Ultrasonics Sonochemistry, 21*(4), 1289–1298. doi:10.1016/j.ultsonch.2014.01.007.

Leong, T., Johansson, L., Mawson, R., McArthur, S.L., Manasseh, R., & Juliano,P. (2016). Ultrasonically enhanced fractionation of milk fat in a litre-scale prototype vessel. Ultrasonics Sonochemistry, 28, 118–129. doi:10.1016/j.ultsonch.2015.06.023.

Leong, T., Juliano, P., Johansson, L., Mawson, R., McArthur, S. L., & Manasseh, R. (2014b). Temperature effects on the ultrasonic separation of fat from natural whole milk. *Ultrasonics Sonochemistry, 21*(6), 2092–2098. doi:10.1016/j.ultsonch.2014.02.003.

Ma, Y., & Barbano, D M (2000). Gravity separation of raw bovine milk: Fat globule size distribution and fat content of milk fractions. *Journal of Dairy Science, 83*(8), 1719–1727. doi:10.3168/jds.S0022-0302(00)75041-7.

Mahmoud, M. Z., Davidson, R., Abdelbasse, W. K., & Fagiry, M. A. (2022). The new achievements in ultrasonic processing of milk and dairy products. *Journal of Radiation Research and Applied Sciences, 15*, 199–205. doi:10.1016/j.jrras.2022.03.005.

McCarthy, O. J. (2011). Centrifuges and separators: Applications in the dairy industry. In J. W. Fuquay, P. F. Fox, & P. L. H. McSweeney (Eds.), *Encyclopedia of dairy sciences* (4, pp. 175–183). Elsevier/Academic Press.

Meng, F., Uniacke-Lowe, T., & Kelly, A. L. (2022). Factors affecting the creaming of raw bovine milk: A comparison of natural and accelerated methods. *LWT Food Science and Technology, 161*, 113288. doi:10.1016/j.lwt.2022.113288.

Merriam-Webster Online Dictionary (2024). Centrifuge definition and meaning. Retrieved November 1, 2024, from https://www.merriam-webster.com/dictionary/centrifuge.

Moore, S. (2008). Early separators helped the cream rise to the top. Retrieved December 2, 2024, from https://www.farmcollector.com/farm-life/early-separators-helped-cream-rise/.

O'Mahony, J. A., Auty, M. A., & McSweeney, P. L. (2005). The manufacture of miniature Cheddar-type cheeses from milks with different fat globule size distributions. *Journal of Dairy Research, 72*, 338–348. doi:10.1017/S0022029905001044.

Salimei, E., & Fantuz, F. (2022). Donkey. In P. L. H. McSweeney, & J. P. McNamara (Eds.), *Encyclopedia of dairy sciences* ((3rd ed.)) (1, pp. 65–76). Elsevier.

Salque, M., Bogucki, P. I., Pyzel, J., Sobkowiak-Tabaka, I., Grygiel, R., Szmyt, M., & Evershed, R. P. (2012). Earliest evidence for cheese making in the sixth millennium BC in Northern Europe. *Nature, 493*(7433), 522–525. doi:10.1038/nature11698.

Somme, r H. H., & Roye, r K. M (1927). The cream plug: Its causes and prevention. *Journal of Dairy Science, 10*(5), 416–430. doi:10.3168/jds.S0022-0302(27)93859-4.

Steinour, H. H. (1944). Rate of sedimentation. Nonflocculated suspensions of uniform spheres. *Industrial & Engineering Chemistry, 36*(7), 618–624. doi:10.1021/ie50415a005.

Subbraman, N. (2012). Art of cheese-making is 7,500 years old. *Nature*. doi:10.1038/nature.2012.12020.

Tetra Pak. (2003). *Dairy processing handbook* ((2nd revised ed.)). Tetra Pak Processing Systems AB.

UK Regulation (1995). The Dairy Products (Hygiene) Regulations 1995, UK Statutory Instruments No. 1086/1995. Retrieved July 12, 2024, from https://www.legislation.gov.uk/uksi/1995/1086/made/data.xht?wrap=true.

Vaclavik, V. A., & Christian, E. W. (2014). Milk and milk products. *Essentials of food science*. Springer. doi:10.1007/978-1-4614-9138-5_11.

Vogel-Prandtl, J. (2004). *Ludwig Prandtl: A biographical sketch, remembrances, and documents. Translated in English by V. Vasanta Ram*. The International Centre for Theoretical Physics.

Vogel-Prandtl, J. (2014). *Ludwig Prandtl: A personal biography drawn from memories and correspondence. Translated in English by David A. Tigwell*. Universitätsverlag Göttingen.

Walkling-Ribeiro, M., Jacob, T., & Ahrné, L. (2024). Impact of pulsed electric field intensity on the cream separation efficiency from bovine milk and physico-chemical properties of the cream. *Food Research International, 180*, 114074. doi:10.1016/j.foodres.2024.114074.

Walstra, P. (1983). Physical chemistry of milk fat globules. In P. F. Fox (Ed.), Developments in dairy chemistry—2. Lipids (pp. 119–158). Springer. doi:10.1007/978-94-010-9231-9_4.

Walstra, P., Wouters, J. T. M., & Geurts, T. J. (2006). *Dairy science and technology* (2nd ed). Taylor & Francis Group.

Warinner, C., Hendy, J., Speller, C., Cappellini, E., Fischer, R., Trachsel, C., Arneborg, J., Lynnerup, N., Craig, O. E., Swallow, D. M., Fotakis, A., Christensen, R. J., Olsen, J. V., Liebert, A., Montalva, N., Fiddyment, S., Charlton, S., Mackie, M., Canci, A., … Collins, M. J. (2014). Direct evidence of milk consumption from ancient human dental calculus. *Scientific Reports, 4*, 7104 1–6. doi:10.1038/srep07104.

Wiener, G. (2022). Yak. In P. L. H. McSweeney, & J. P. McNamara (Eds.), *Encyclopedia of dairy sciences* (3rd ed) (1, pp. 157–164). Elsevier.

Yoder, J. (1991). Christiaan Huygens' great treasure. *Tractris, 3*, 1–13 Retrieved November 2024 from, https://www.gewina.nl/journals/tractrix/yoder91.pdf.

Application of filtration in the fruit juice processing

11

S.P. Schwaminger[a,b] and M. Eigenfeld[b]
[a]NanoLab Graz, Division of Medicinal Chemistry, Otto Loewi Reseaerch Center, Medical University of Graz, Neue Stiftingtalstraße, Graz, Austria, [b]BioTechMed Graz, Mozartgasse, Graz, Austria

11.1 Introduction

Filtration is a crucial step in the fruit juice production process, ensuring that the final product is not only clear and visually appealing but also free from unwanted particulate matter and microorganisms that can compromise its quality and shelf life (Hu & Wu, 2023; Sarbatly et al., 2023). The rich, tantalizing color of fruit juices and the bioactive compounds that we often take for granted are a result of meticulous processing (Mandha et al., 2023), with filtration playing a pivotal role.

The process of fruit juice processing begins with the raw juice, which, in its natural state, harbors a mixture of suspended solids like pulp, seeds, and skin fragments, along with potential enzymes and microorganisms. These constituents contribute to the juice's opacity and texture (Beveridge, 2002). While certain juices are celebrated for their natural cloudiness and reflect a rich, unfiltered essence, the production of clear juices demands a rigorous filtration process (Sarbatly et al., 2023). Achieving such clarity involves more than just removing visible particulates; it also entails a subtle balance of preserving the juice's intrinsic nutritional value and flavor profile while ensuring microbial stability.

As we progress, the discussion will shift towards a detailed examination of the various filtration techniques that have been tailored to suit different types of fruit juices and their specific characteristics. We will cover mechanical filtration methods, which rely on physical barriers to strain out larger particles, and advanced membrane-based filtration, which offers a finer level of purification capable of targeting microorganisms and tiny particulates. Additionally, the role of innovative filtration technologies, such as crossflow filtration and the use of adsorptive materials, will be explored for their ability to enhance product quality and process efficiency.

Understanding the selection and use of proper filtration media and equipment is essential. It influences the efficiency and cost of juice production and its environmental impact. The chapter will cover critical filtration parameters such as flow rates, pressure, and equipment maintenance to prevent clogging and extend its life. It will emphasize the role of filtration in maintaining microbial stability, enhancing safety, and extending the shelf life of fruit juices without compromising nutritional and sensory qualities through excessive heat. Quality assurance and regulatory compliance in juice production will also be discussed. Additionally, illustrative case studies will demonstrate

Mechanical Separation Processes in the Food Industry: Unit Operations and Processing Equipment in the Food Industry.
DOI: https://doi.org/10.1016/B978-0-128-19523-9.00007-X

real-world applications of filtration techniques, offering insights into practical challenges and solutions in juice processing.

11.2 Principles and techniques of filtration

Fruit juice processing, essential for converting fruit into tasty and durable beverages, increasingly aims to preserve the original flavor and nutritional value of the fruit. As consumer demand for nutrient-rich juices rises, especially during health crises like the COVID-19 pandemic, the industry faces pressure to innovate while retaining key vitamins, minerals, and antioxidants (Sarbatly et al., 2023). This innovation includes not only removing solid particles but also reducing turbidity to meet expectations for clear, high-quality juice. It is important to ensure that filtration processes do not strip away beneficial compounds (Bump, 1989) or nutrients sensitive to heat, such as anthocyanins (Kechinski et al., 2010), carotenoids (Provesi et al., 2011), vitamin C (Barros et al., 2011), and bioactive compounds (Bao et al., 2023).

Therefore, several filtration strategies, from initial coarse removal of particulates to advanced purification stages are developed, each designed to refine the juice's quality while ensuring microbial safety (Lu et al., 2021; Hu & Wu, 2023). These strategies range from primary filtration, where coarse sieves remove larger particles, to advanced stages using microfiltration (MF) and ultrafiltration (UF) for finer contaminants (Fig. 11.1). The removed products can be used for further processes, using side streams (Kandemir et al., 2022).

1. Primary Filtration: This is the preliminary stage where large particulate matter, like seeds and large pulp fragments, are removed. This can be achieved through traditional clarification methods like coarse sieves or screens (Bump, 1989; Urošević et al., 2017).

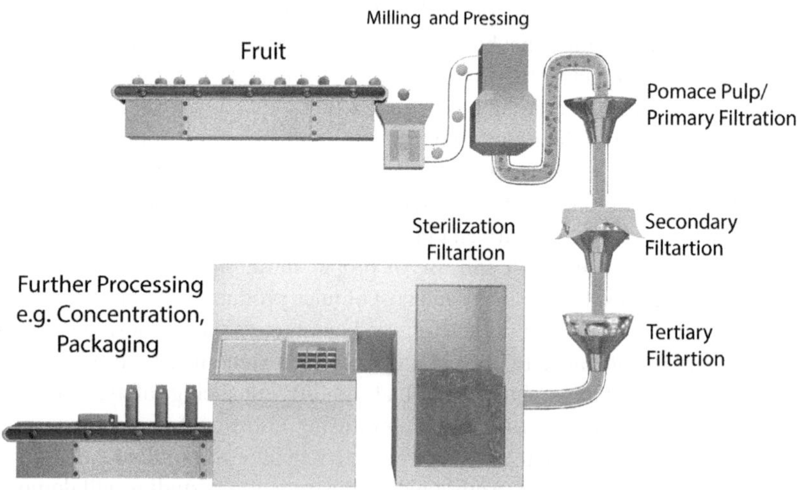

Figure 11.1 Diagram illustrating the fruit processing sequence and various filtration stages.

2. Secondary Filtration: Fine mesh screens or cloth filters are used for further clarification steps that include finer filtration methods to achieve clearer juice (Urošević et al., 2017). This step is more refined and ensures that the juice has a consistent texture.
3. Tertiary Filtration: Advanced techniques like MF come into play here. These processes can remove very fine particles, turbidity, and even some microorganisms, from the juice, resulting in a clear and clean product (Hu & Wu, 2023; Urošević et al., 2017).
4. Sterilization Filtration: Ensuring a juice is free from harmful microorganisms is critical for its safety and shelf life. Filtration methods that can effectively remove bacteria and other pathogens, like UF and some MF, are employed in this final stage (Hu & Wu, 2023) and are also known as biological filters. Alternatively, nowadays, pasteurization is used in this step (Mandha et al., 2023).

In examining these filtration methods, we delve into the scientific and technological principles underlying each process, emphasizing their roles in enhancing the quality and safety of fruit juice. The transformation from raw, turbid juice to a clear, refined product represents a complex interplay of physical and occasionally chemical processes, each vital for attaining desired outcomes. By comprehending these processes, producers can refine their filtration techniques to ensure that every glass of fruit juice meets the highest standards of quality and safety.

11.2.1 Key metrics in fruit juice filtration

11.2.1.1 Flow rate

The flow rate in a filtration system is one of the most critical parameters, especially according to fouling effects, as it directly influences the efficiency and effectiveness of the process (Lu et al., 2021). It is defined as the volume of fluid that passes through a filter per unit of time, typically measured in liters per minute (L/min). A higher flow rate can lead to a reduced contact time between the fluid and the filter media, potentially decreasing the removal efficiency of contaminants. Conversely, a lower flow rate may increase filtration effectiveness but can also lead to higher operational costs and reduced throughput. Optimal flow rates must be determined based on the specific requirements of the filtration process, the nature of the fluids and contaminants, and the type of filter media used.

11.2.1.2 Pressure drop

Pressure drop is a critical indicator of filter performance and durability, particularly noted in studies on juice filtration systems. It measures the loss of pressure as the fluid passes through the filter medium, encountering resistance, often quantified in pascals (Pa) or pounds per square inch (psi). This parameter significantly impacts the energy consumption and efficiency of filtration systems, as observed in nanofiltration (NF) processes for apple and pear juices where low pressures optimize output quality and reduce energy requirements (Warczok et al., 2004). A notable increase in pressure drop can indicate filter clogging or a need for maintenance, as also emphasized by the practical applications in juice processing, where maintaining low-pressure drops is crucial to prevent overburdening the pump and other mechanical components, ensuring the system operates within its designed specifications (Warczok et al., 2004).

11.2.1.3 Filtration efficiency

Filtration efficiency is the measure of how effectively a filtration system removes specific contaminants from a fluid, expressed as a percentage that represents the ratio of contaminants removed to those present in the influent stream. This efficiency depends on factors such as the size and nature of particles, filter media properties like pore size and surface area, and operational conditions including flow rate and pressure. Achieving high filtration efficiency is essential for industries requiring high purity levels, such as pharmaceutical manufacturing, drinking water treatment, and critical industrial processes.

In juice filtration, this efficiency is assessed through metrics such as exemplary the clarity of the juice, retention of alcohol insoluble solids, and the protein and polyphenol content (Severcan et al., 2020a; Sagu et al., 2014). These factors are critical in the design, operation, and maintenance of filtration systems and vary depending on the type of juice processed. By understanding and optimizing these parameters, engineers and operators can enhance system performance, extend equipment lifespan, and ensure high-quality outputs (Table 11.1).

11.2.2 Basic filtration concepts

Filtration is an essential procedure in juice production, utilizing filter media with complex structures to separate solid elements and fluids from a mixture. This process is critical for ensuring the purity and quality of the final product, allowing only the fluid to pass through the filter. Filtration can be categorized into physical, biological, or chemical processes, depending on the nature of the separation, like clarification, microbial removal or to ensure safety for consumption. Table 11.2 presents a summary of the various types of physical filtration methods and their applications in different juices. In summary, different types of fruit juices require specific filtration methods and operational parameters for optimal processing. UF is suitable for juices like apple, banana, dragon fruit, guava, and sour cherry, with effective pressures ranging from 200 to 540 kPa. NF is preferred for juices like apple and pear, operating at temperatures around 25°C and pressures between 8 and 50 bar. MF is appropriate for juices like cashew apple, citrus, carrot, mosambi, orange prickly pear, pomegranate, and tomato, with operating temperatures ranging from 20°C to 28°C and pressures typically ranging from 50 to 200 kPa.

11.2.2.1 Filtration stages in juice production

Given the significant concentration of organisms and particles in raw juice, even after bulk filtering, multiple filtering stages are vital. These stages can be broadly

Table 11.1 Indication of the operational parameter of juice filtration in comparison to the juice.

Juice	Filtration type	Flow rate	Pressure drop	Filtration efficiency	References
Apple	UF	26.5 to 44.6 L $m^{-2}h$	Not explicitly stated	- Evaluated through changes in color, turbidity, total soluble solids, total phenolic content, and antioxidant capacity	Severcan et al. (2020a)
Apple	UF	Crossflow velocities of 2 and 7 m s^{-1}	150 and 400 kPa	Achieved juice permeation at optimal conditions was 60 $m^3/(m^2 \times kWh)$, with minimal fouling at high feed velocity and low pressure. Excellent quality attributes meeting commercial specifications.	de Bruijn et al. (2002)
Apple	NF	Feed flow rate of 40 L/h	Operating pressure of 50 bar	Achieved sugar rejection of approximately 95% while retaining significant levels of bioactive compounds. #Total phenolic content rejection was 99.1% ± 0.2% and ash content rejection was 64.7% ± 2.9%.	Pruksasri et al. (2020)
Apple and Pear	NF	Not explicitly stated, but the operational pressures were between 8 and 12 bar.	Operational pressures used were low, between 8 and 12 bar.	Assessed the feasibility of using NF at low pressures for juice concentration	Warczok et al. (2004)
Banana	UF	10 to 20 L h^{-1}	35, 52, 69, and 104 kPa	- Increased clarity in the permeate - 100% retention of alcohol insoluble solids - Retention and reduction in protein and polyphenols	Sagu et al. (2014)

(continued on next page)

Table 11.1 Indication of the operational parameter of juice filtration in comparison to the juice—cont'd

Juice	Filtration type	Flow rate	Pressure drop	Filtration efficiency	References
Apple, orange, and watermelon, among others	Various	Not specified	As low as 20 kPa up to 500 kPa	Significant variations in permeate flux, fouling rates, and quality of filtered juice.	Lu et al. (2021)
Dragonfruit	UF	7.9 kg m^{-2} h	1–3 bar (TMP)	High permeate flux, low betacyanin and phenolic retention, excellent clarity	Le et al. (2021)
Pomegranate	MF	Adjusted to maintain constant TMP of 1.4 bar	1.4 bar (TMP)	- Improved color values; highest was 5,781 $\pm$ 4 Pt-Co - Drastic reduction in turbidity, e.g., 0.69 NTU with 0.05% Al2O3 MF membrane - Maintained Total Soluble Solids ($\approx$16.2° Brix) - Significant retention of phenolic compounds and antioxidants, e.g., 2,642.1 $\pm$ 46.4 mg GAE/L phenolics and 62.4 $\pm$ 0.2 TEAC/L antioxidant activity	Severcan et al. (2020b)
Pomegranate	MF	10 mL s^{-1}	0.5 bar	Increased permeate flux and reduced fouling with laser enhancement	Salehinia et al. (2021)

Table 11.2 Comparison of physical filtration methods for the clarification of fruit juices: Data adapted from (Sarbatly et al., 2023) and extended.

Fruit juice	Filtration type	Filtration mode	Membrane type	Operating parameters	References
Apple	UF	Crossflow	PSF/PEI (20/2 wt.%) with TiO_2 and Al_2O_3, flat sheet, 150 cm^2	Temperature: n.d. Pressure: 540 kPa	Severcan et al. (2020a)
Apple	UF	Dead-End	PES and PVP	Pressure: 1 bar	Borneman et al. (2001)
Apple	UF	Crossflow	Zirconium oxide membranes with nominal molecular weight cut-off (MWCO) of 15 and 50 kDa	Temperature: Regulated at 50–55°C Transmembrane pressures: 150 and 400 kPa	de Bruijn et al. (2002)
Apple	NF	Not explained*	Polyamide-thin film composite NF membrane, Desal-5-DK with a molecular weight cut-off of 150–300 Da	Temperature: 25°C Pressure: 50 bar	Pruksasri et al. (2020)
Apple and pear Juices	NF	Not explained*	AFC80 (pH range: 1.5–10.5, Maximum Temp: 70°C); MPT-34 (pH range: 0–14, Maximum Temp: 80°C); Desal-5DK (pH range: 1–11.5, Maximum Temp: 50°C)	Temperature: Ranges from 25 to 35°C for optimal juice concentration Pressure: Operated between 8 and 12 bar	Warczok et al. (2004)
Banana	UF	Dead-End	PES, flat sheet, 100 kDa, 32 cm^2	Temperature: nd Pressure: 200 kPa	Yee et al. (2021)
Cashew apple	MF	Crossflow	PSF, hollow fiber, 0.2 μm, 51.8 cm^2	T: 25 ± 2°C; Transmembrane pressure: 138 kPa; Q: 10 L h^{-1}	Abdullah et al. (2022)

(continued on next page)

Table 11.2 Comparison of physical filtration methods for the clarification of fruit juices: Data adapted from (Sarbatly et al., 2023) and extended—cont'd

Fruit juice	Filtration type	Filtration mode	Membrane type	Operating parameters	References
Citrus (orange and lemon) and carrot juices	UF	Crossflow	PVDF, 15 kDa MWCO	Temperature: 0–55°C Pressure: 0.8–5.5 bar.	Cassano et al. (2003)
Dragon fruit	UF	Dead-end	PES, flat sheet, 10 kDa, 6.15 cm^2	T: 25 °C; TMP: 300 kPa	Le et al. (2021)
Guava	UF	Dead-end	PES, flat sheet, 100 kDa, 32 cm^2	P: 200 kPa, 80% permeate recovery	Omar et al. (2020)
Mosambi	MF	Dead-end	PA, flat sheet, 2.5 μm, 28.3 cm^2	T: 28 ± 2°C; TMP: 69 kPa	Singh and Das (2021)
Orange prickly pear	MF	Crossflow	PSF, hollow fiber, 0.1 μm, 1000 kDa, 420 cm^2	T: 20 ± 1°C; TMP: 69 kPa	Mejia and Yáñez-Fernandez (2021)
Pomegranate	MF	Crossflow	PVDF, hollow fiber, 0.13 μm, 27.6 cm^2	T: 25°C; TMP: 60 kPa;	Morittu et al. (2020)
Pomegranate	MF	Dead-end	PSF/PEI (17/2 wt.%) with TiO_2 and Al_2O_3, flat sheet, 14.6 cm^2	T: 25 ± 3°C; P: 140 kPa	Severcan et al. (2020b)
Pomegranate	MF	Crossflow	PVDF, flat sheet, 0.22 μm, 78 cm^2	T: 25°C; P: 50 kPa	Salehinia et al. (2021)
Pomegranate	MF	Crossflow	Cellulose ester, flat sheet, 0.22 μm, 78 cm^2	T: 20°C; TMP: 50.66 kPa	Zarouk et al. (2022)
Sour cherry	UF	Not explained*	Paper filter	Not given	Toydemir et al. (2013)
Tomato	MF	Crossflow	Al_2O_3 ceramic, tubular, 0.2 μm, 392 cm^2	T: 26°C; P: 200 kPa	Alaei et al. (2021)

* The work does not explicitly mention if the filtration mode is crossflow or dead-end, but NF generally employs a crossflow configuration to manage fouling more effectively.

categorized into mechanical and membrane-based processes (Su, 2018; Wang et al., 2020; Men et al., 2023). Mechanical filtration employs physical barriers, such as sieves or filters, to capture larger particles (Ersahin et al., 2012). Typically these sieves indicate diameters in the smaller millimeter range of 0.8 mm (Ilame & Singh, 2015). In contrast, membrane-based filtration uses semi-permeable membranes capable of separating particles at the molecular level (Gandhi et al., 2022), thus providing a finer level of clarification and microbial reduction.

11.2.2.2 Clarification and prefiltration

After most of the fruit pulp has been removed by mechanical filtration, raw fruit juices undergo prompt clarification. This step eliminates particles too small for the main filtering process, preventing the introduction of undesirable tastes. Clarifying processes further reduce the quantity of smaller particles, enhancing the product's clarity and ensuring a cleaner taste and appearance (Echavarría et al., 2011).

11.2.2.3 Bacteria reduction/removal

The presence of organisms that can withstand heat treatments, such as pasteurization, poses a significant challenge. Thermo-acidophilic bacteria (TAB), especially species of *Alicyclobacillus*, are of particular concern due to their resilience against pasteurization and natural occurrence in fruits (Lee et al., 2007; Oke et al., 2012; Tranquillini et al., 2017). Classical preservation methods prove ineffective against spores, which can germinate in the finished drink, potentially spoiling it. The introduction of external substances can also bring other thermophilic microorganisms into the process, posing additional risks. The introduction of external substances can also bring other thermophilic microorganisms into the process, posing additional risks. *Alicyclobacillus* species have been notably problematic in a variety of fruit juices, especially fruit concentrates (Steyn et al., 2011), reflecting their widespread relevance in the beverage industry. Incidents of spoilage have been reported increasingly in recent years in juices such as orange concentrate (Eiroa et al., 1999), apple (Chen et al., 2006; Tayefe et al., 2014), mango (Gouws et al., 2005), passion fruit (McKnight et al., 2010), pear, peach and citrus (Durak et al., 2010; Groenewald et al., 2009). This indicates that the contamination risk from *Alicyclobacillus* spans across a broad spectrum of both tropical and temperate fruit juices, necessitating vigilant monitoring and targeted mitigation strategies in their production processes.

Addressing these primary contaminants requires higher temperatures (>70°C) (*Byssoschlamys nivea*: 94°C -*Talaromyces flavus*: 113°C), which can be costly and potentially damage the product's properties, such as vitamins and color. Contaminants, including molds and yeasts from the environment, can be introduced at various stages of the manufacturing process (Pereira & Sant'Ana, 2018). This necessitates stringent filtration methods to ensure the microbial safety of the fruit concentrates by targeting both heat-resistant organisms and external contaminants.

Organism	Heat tolerance/ time for inactivation	Effect of contamination	References
Acetobacter pasteurianus	heat sensitive	Acetic acid from ethanol	Lima Tribst et al. (2009)
Alicyclobacillus sp.	88–90°C		Kotzekidou (2014), Pettipher et al. (1997)
Byssochlamys fulva	70°C/30 min		Erkmen (2022)
B. nivea	94°C 88–99°C	Mycotoxins	Sant'Ana et al. (2009) Kotzekidou (2014)
Bougainvillea spectabilis	85°C/75 min		Kotzekidou (2014)
Gluconobacter		Turbidity Spoilage of grape juice	Marsellés-Fontanet et al. (2009)
Lactobacillus	D_{90} = 16–23 min	Off-Flavor (acetic acid and ethanol) Gummy slime	ICMSF (1998), Splittstoesser et al. (1994)
Leuconostoc		Off-Flavor (acetic acid and ethanol) Gummy slime	ICMSF (1998), Splittstoesser et al. (1994)
Neosartoria fischeri	70°C/30 min		Erkmen (2022)
Heat-Resistant Molds (HRM)	Ascospores can survive pasteurization and high-pressure processing (HPP); specific inactivation times are not provided but require extensive heat or combined treatments for inactivation.	Production of off-flavors and odors; potential mycotoxin production which can lead to health hazards; economic losses due to spoilage and product recalls.	Rico-Munoz (2017)

11.2.2.4 *Tank vent and process gas filtration*

To prevent the entry of airborne microorganisms, tank vent filters are employed during the emptying process of tanks. Similarly, process gas filters are crucial in maintaining the high quality of the packed juice by preventing the deposition of particles and germs during the filling process (Lawlor et al., 2009). These filters, typically rated with a pore size of 0.22 μm, also serve to protect the product in storage tanks by filtering the process gas used as a protective layer (Cassano & Basile, 2011). Additionally, the use of gas sparging, as detailed in the study of pineapple juice clarification, further exemplifies advanced filtration techniques (Laorko et al., 2011). In this process, gas (nitrogen) is injected into an MF system to enhance permeate flux and reduce fouling by promoting local mixing near the membrane surface (Laorko et al., 2011). This method effectively mitigates concentration polarization and fouling, ensuring a higher quality of juice by maintaining a clean and efficient filtration process (Laorko et al., 2011).

11.2.2.5 *Detailed exploration of filtration methods*

It is feasible to physically eliminate all the suspended solids and other particulate matter. However, the microorganisms that transmit disease to humans through drinking juice or water have varying sizes: protozoa (4–100 μm), bacteria (0.2–1.0 μm), and viruses (0.01–0.1 μm). This requires the use of different filtration methods.

11.2.3 Advanced filtration concepts

11.2.3.1 *Microfiltration*

MF utilizes membranes with pore sizes typically between 0.1 and 10 mm (Hu & Wu, 2023), effectively removing bacteria, yeast, and larger colloids from liquids. In juice processing, MF's primary goal is juice clarification, removing suspended solids (SS), fat, and high molecular weight proteins. It's also used in the dairy industry for applications such as whey clarification and milk fat reduction (Bhattacharjee et al., 2017). Notably, MF serves as a pre-treatment stage to reduce fouling in subsequent membrane processes, thereby extending their operational life (Charcosset, 2016).

11.2.3.2 *Ultrafiltration*

UF membranes, featuring smaller pores than those used in MF (0.001 to 0.1 mm) (Hu & Wu, 2023; McCarron et al., 2016), are crucial in various industries for their ability to concentrate juices by removing water and small solutes like sugars, salts, and proteins. This process not only enhances clarity and maintains pH levels but also reduces turbidity and improves color properties (Omar et al., 2020). Specifically, in the juice industry, UF focuses on the molecular weight cut-off (MWCO) to optimize the permeate for a purified product and ensure stability (Arriola et al., 2014). However, it's important to note that UF may lead to a reduction in total phenolic content by up to 40% (Borneman et al., 2001) and ascorbic acid (Omar et al., 2020), which are vital for nutritional quality.

In the dairy sector, UF is essential for enhancing cheese production efficiency by concentrating milk proteins and segregating other milk components, which increases cheese yields and improves control over the final curd composition (Gavazzi-April et al., 2018). Additionally, UF is extensively used to concentrate skimmed milk to produce high-protein milk protein concentrates (MPCs), containing over 80% protein on a dry weight basis, through a combined process with diafiltration (DF) (Gavazzi-April et al., 2018). This membrane technology is also adept at eliminating bacteria, viruses, and macromolecules, making it suitable for fractionation, fixation, and filtering purposes (Hankins & Singh, 2016).

Selecting membrane materials and configurations is essential to address the unique challenges and goals of exemplary apple juice UF. This includes optimizing clarity, preserving flavor, enhancing operational efficiencies, and managing the intrinsic fouling properties typical of fruit juice filtrations. Specifically, the use of PSF/PEI with TiO_2 and Al_2O_3 enhances fouling resistance (Severcan et al., 2020a), while polyethersulfone (PES) combined with polyvinylpyrrolidone (PVP) optimizes quality retention (Borneman et al., 2001). Additionally, Zirconium oxide membranes are selected for their robust performance across different operational pressures and temperatures (de Bruijn et al., 2002).

11.2.3.3 *Nanofiltration*

NF continues to be a highly effective method for concentrating and purifying fruit juices by utilizing its fine pore size range, typically indicated in molecular weight cut-off (<1,000 Da). It excels in the separation of small molecules such as organic acids, pigments, and ions, which facilitates the adjustment of juice taste profiles and clarity through selective separation based on both size and charge exclusion (Bhattacharjee et al., 2017). This precision allows for the effective concentration of bioactive compounds (Salehi, 2014), such as polyphenols, lycopene concentration in watermelon juice, and certain vitamins, without the high energy costs associated with reverse osmosis (RO).

Recent applications of NF in the food industry include the demineralization of dairy products, specifically for the removal of minerals from whey, enhancing its value for further processing (Bhattacharjee et al., 2017). In juice processing, NF is particularly valued for its ability to reduce sugar content while maintaining or even enhancing the concentration of valuable nutrients and antioxidants (Pruksasri et al., 2020). For example, a study demonstrated the use of NF in reducing sugar content in apple juice while retaining significant levels of polyphenols and antioxidant capacity, although some losses were inevitable (Pruksasri et al., 2020).

Additionally, in another study focused on apple and pear juices, NF was utilized at low pressures (8–12 bar) with temperatures maintained between 25–35°C to optimize juice concentration while minimizing damage from high processing pressures (Warczok et al., 2004). This process not only ensured the preservation of juice quality but also demonstrated a lower degree of irreversible fouling, making NF a more sustainable option for long-term industrial applications (Warczok et al., 2004).

By combining both mechanical fractionation and NF, innovative approaches have been developed to produce juices with reduced sugar levels yet enriched with bioactive compounds. These methods underline NF's versatility and its growing importance in producing healthier beverage options without compromising on taste or nutritional value.

Therefore, in following table summarizes the features of the three presented filtration types (Table 11.3).

11.2.3.4 *Dead-end filtration process*

Dead-end filtration, or surface filtration, separates particles from liquids by trapping them on the filter media's surface as the liquid passes through. This method contrasts

Table 11.3 Comparison of the parameters of NF, UF, and MF. Adopted from Ilame and Singh (2015) and extended.

Parameter	Nanofiltration (NF)	Ultrafiltration (UF)	Microfiltration (MF)
Membrane pore size (nm)	-	1–10,000	100–100,000
molecular weight cut-off (Dalton)	>1,000	ranging from 30,000 to 100,000	-
Permeate	Water	Water and small molecules	Water and dissolved solutes
Application	Concentrating divalent salts, bacteria, proteins, and other constituents that have a molecular weight > 1 kDa.	Clarification and chillprofing of clear fruit juices.	Clarification of apple juice.
Reference	Bhattacharjee et al. (2017)	Bhattacharjee et al. (2017), da Silva et al. (2008), McCarron et al. (2016)	Bhattacharjee et al. (2017), da Silva et al. (2008), Ismail and Goh (2015)

with cross-flow filtration by directing the liquid at a right angle to the filter surface (Zhu et al., 2015). Selecting the appropriate filter media, such as filter sheets or membrane filters, depends on particle size and juice content compatibility. The process involves preparing and disinfecting the filtration media and equipment, ensuring product integrity. Solid components like pulp and seeds are retained either on the surface or within the filter matrix, producing a clarified juice. Dead-end filtration is ideal for batch processing processes that need clarification or filtering of smaller amounts of juice (Arikrishnan & Keshav, 2023).

11.2.3.5 Crossflow filtration

Crossflow (tangential flow) filtration operates by allowing the liquid to flow parallel to the membrane surface, effectively mitigating particle buildup and fouling. This method separates the feed into permeate (filtered liquid) and retentate (concentrated solids) (Mai, 2017). System preparation includes sterilizing to maintain product purity, essential for producing clear juice products while optionally recovering concentrated solids, thus reducing environmental impact. In fruit juice processing, such as pineapple juice, gas sparging enhances crossflow filtration by introducing gas into the feed stream to disrupt the particle layer near the membrane surface, improving flux and reducing fouling (Laorko et al., 2011).

The use of CF is typically implemented in situations where it is necessary to manage high levels of membrane fouling or when processing large volumes or viscous feeds. In the context of fruit juice processing, as detailed in (Bhattacharjee et al., 2017), crossflow filtration is employed mainly due to its effectiveness in handling large amounts of feed and its ability to maintain a more consistent flow and separation efficiency over time compared to traditional dead-end filtration.

Crossflow filtration, including MF and UF, is particularly advantageous for clarifying juices and removing particulates, including bacteria and other spoilage organisms, which can considerably extend the shelf life and quality of the juice (Bhattacharjee et al., 2017). These systems are suitable not only for large-scale operations but also for juices with high viscosity or a significant quantity of suspended solids. The decision to use crossflow filtration is based on specific processing needs rather than the size of the operation, making it ideal for managing high fouling potentials and complex juice compositions.

11.3 Filtration media and materials

This chapter explores the critical role of filtration media and materials in the fruit juice processing industry, highlighting how they influence the efficacy, selectivity, and overall quality of the filtration process (Sadeghi et al., 2018). When selecting the appropriate media, factors such as pore size are crucial, as they determine the size of the contaminants or fractions that can be effectively filtered out (Jørgensen et al., 2016; Wakeman, 2007). Additionally, compatibility of the materials with the liquid phase is essential to avoid adverse interactions that could affect the juice's flavor, aroma, and nutritional value. The desired filtration efficiency, whether targeting the smallest particles or optimizing flow rates and pressure drop, guides the selection process (Chen et al., 2018).

11.3.1 Mechanical filter

Mechanical filters are pivotal in the manufacture of juice, fruit puree, and fruit concentrate used to remove enzymes or undesired solid particles such as debris, seeds, and pulp, leaving behind a clean and smooth finished product (Klopotek et al., 2005; Morris & Spayd, 1981). These filters are essential components of the fruit processing operations, ensuring that only the desired elements remain in the final juice or puree.

11.3.1.1 Paper filters

Clarification of fruit juices represents one of the main uses of filter sheets in the processing of fruit. Juice that has been extracted from fruits like apples, grapes, or berries frequently includes pulp, suspended particles, and other contaminants. These undesirable particles are eliminated using filter sheets with particular pore diameters, leaving behind clear and aesthetically pleasing fruit juices (Oke et al., 2012). Pectin, which can result in cloudiness or gelling, must be removed in several fruit processing activities, such as the creation of fruit concentrates or jams. To deliberately remove pectin and ensure the correct texture and purity of the finished product, filter sheets with small pore diameters are used (Toydemir et al., 2013). Cellulose fibers make up the majority of paper filters. The permeable nature of the cellulose substance makes it useful for trapping and eliminating contaminants from liquids. Infiltration systems, and paper filters are frequently used as pre-filters or polishing filters. By eliminating solid particles and other pollutants, they are essential for improving the quality of liquids. Filter sheets are employed for separating the desired extracts from the fruit matrix when certain compounds are extracted from fruits for flavorings, colorings, or additions. As a result, it is easier to obtain concentrated and pure extracts for use in a variety of applications related to food and beverages. To assess fruit-based goods for criteria such as sugar content, filter sheets are employed in quality assurance and testing methods (Rodriguez-Saona et al., 2001). In the production of sour cherry nectar, filtration techniques such as UF and the use of paper filters are specifically adapted to the process to ensure optimal removal of particulates and maintain the high quality of the final product (Toydemir et al., 2013).

11.3.1.1.1 Challenges

One of the primary challenges of using paper filters is efficiently extracting phenolic compounds, such as anthocyanins and procyanidins, from substances like sour cherry. If the filters are not designed to retain these compounds effectively, valuable phenolics may be lost during filtration. Paper filters might not always provide the required selectivity to maintain specific substances in suspension. These filters vary in their chemical retention capabilities; hence optimization of filter parameters is necessary for maximum retention. Additionally, there is a risk of clogging, which can impede the flow of liquid and necessitate maintenance or replacement of the filters (Toydemir et al., 2013).

11.3.1.2 Mesh filters

Mesh filters, which are frequently made of components like stainless steel or man-made fibers, are essential to filtration operations. Larger pores on these filters provide them the ability to collect and filter out particles from beverages like murky apple juice (Youn et al., 2004). Stainless steel or synthetic fibers are frequently used in the construction of mesh filters. Typical sieve sizes are designed to capture larger particles, usually with diameters in the smaller millimeter range, around 0.8 mm (Ersahin et al., 2012; Ilame & Singh, 2015). The durability and corrosion resistance of these materials make them appropriate for use in a variety of filtration applications. Mesh filters have bigger pores than other filtration techniques, which is one of their distinguishing characteristics. Because of their bigger pores, they can efficiently capture and hold particles of various sizes that are present in the liquid being.

11.3.1.2.1 Challenges

Mesh filters excel at removing larger particles but may struggle with finer particles, particularly those in the micron or submicron range. As the filter mesh number increases, the effectiveness of these filters can decrease, particularly when filtering beverages with smaller particles that can pass through the larger pore (Zhu et al., 2020). Mesh filters are generally made to separate liquids from solid particles. They are less successful at getting rid of dissolved substances like polyphenols, which are crucial for the flavor and nutrient content of the beverage. Instead, these are often separated by membrane technology (Kelly et al., 2019; Tylkowski et al., 2017).

11.3.1.3 Gelatin

Gelatin is a widely used filtration medium in the beverage industry, renowned for its effectiveness in clarifying and stabilizing fruit juices and wines. Extracted from animal collagen, gelatin, sometimes combined with bentonite, efficiently binds with colloidal particles responsible for haze and turbidity in beverages (Singh & Das, 2021). In the juice and wine industries, gelatin is primarily utilized as a fining agent, interacting with tannins, phenolic compounds, and other particles that cause turbidity. By forming larger aggregates, these particles are more easily removed in subsequent filtration steps, enhancing both the clarity and sensory qualities of the beverage (Singh & Das, 2021; Fenercioğlu & Vardin, 2003).

11.3.1.3.1 Challenges

Gelatin must comply with food safety and regulatory standards specific to each region. It is important for manufacturers to ensure that the gelatin used is of food-grade quality and sourced responsibly to meet consumer expectations and regulatory requirements.

Furthermore, the clarification by use of gelatin takes a long and is therefore in large scale ness practical due to sedimentation times of 6-18 h (Meyer et al., 2001).

11.3.2 Thermodynamic driven filters

This section explores advanced filtration methods that rely on thermodynamic principles to enhance the quality and safety of fruit-based products.

11.3.2.1 Activated carbon

Activated carbon is a highly versatile chemical filter known for its extensive surface area and porous structure, which make it ideal for adsorbing a wide range of contaminants. It is produced by activating carbon-rich materials such as wood, coal, coconut shells, or peat at high temperatures with gases like steam (Selvaraju & Bakar, 2017). This process enhances the carbon's adsorptive capacity significantly (Gan, 2021).

Activated carbon is extensively used in the fruit processing industry to remove unwanted substances such as colorants, off-tastes, and various chemical components, thereby enhancing the nutritional value of fruit-based products (Carabasa et al., 1998; Pap et al., 2017; Solís-Fuentes et al., 2019).

11.3.2.1.1 Challenges

The ability of activated carbon to adsorb substances is not very selective. Colorants, flavorings, organic impurities, and certain toxins are just a few of the substances that activated carbon can adsorb from the processing of fruit. The physical and chemical characteristics of these substances are the basis for the adsorption. Instead, some studies describe a selective separation of food colorants at optimized conditions, like the pH of the original product (Duygu Ozsoy & van Leeuwen, 2010).

11.3.2.2 Ion-exchange resins

Ion exchange is a pivotal technique in the deacidification of fruit-based beverages, mainly orange and citrus juice, as well as in enhancing other juices like pineapple mill juice, apple juice, pear juice, and grape juice, addressing specific processing needs such as mineral removal, decolorization, and the improvement of flavor and stability (Akyıldız et al., 2022; Norman, 1995), addressing the crucial aspect of acidity control in fruit processing. The control of acidity during fruit processing is important because too much acidity can degrade the flavor and quality of the finished product (Thompson et al., 2012). Ion-exchange resins have proven to be an efficient method for deacidifying fruit-based beverages, and recent research has focused heavily on this usage (Vera et al., 2003).

Several ion-exchange resins, including D301, D318, D314, and 330, have been tested in the search for the best fruit deacidification (Zhong et al., 2018). The 330-resin stood out among them for its exceptional effectiveness in eliminating citric and total acids from kiwifruit wine. This resin selection procedure emphasizes how crucial it is to pick the ideal resin to get the desired results (Kammerer et al., 2011; Zhong et al., 2018). Several quality metrics were found to be considerably impacted by deacidification utilizing ion-exchange resins. It successfully decreased the chromaticity of kiwifruit wine, which was caused by the adsorption of colorants. The procedure also increased pH, putting it closer to an acceptable range. Surprisingly, the deacidified wine's scent was mostly unaffected, proving the ability of ion-exchange resins to raise product caliber (Zhong et al., 2018).

11.3.2.2.1 Challenges

Choosing the appropriate ion-exchange resin is complex due to the variability in fruit types and desired product characteristics. This selection must consider the type and concentration of acids present, the desired flavor profile, and the quality of the final product (Akyıldız et al., 2022; Li et al., 2019). The effectiveness of deacidification and the quality of the finished product can be significantly affected by the resin selected. Additionally, optimizing process variables such as the resin-to-fruit ratio, flow rate, and contact duration is crucial to achieving the desired outcomes while preserving the sensory and nutritional properties of the fruit product (Lue & Chiang, 1989). These resins also impact other quality parameters such as color and nutritional content by altering levels of ascorbic acid, phenolics, and carotenoids, which are vital for consumer acceptance and health benefits (Akyıldız et al., 2022).

11.4 Quality and safety considerations

This chapter highlights the crucial importance of maintaining the quality and safety of fruit juices – fundamental to the food industry's commitment to consumer health and satisfaction. Ensuring that fruit juices retain their inherent sensory attributes and nutritional value, while also mitigating risks such as microbial contamination and oxidation, is essential for preserving the product's appeal and safety. Moreover, adherence to stringent food safety regulations and standards is a legal and moral imperative to protect consumers and enhance public trust.

11.4.1 Quality and safety considerations: filtration techniques overview

In the juice processing industry, microbial control is critical for ensuring product safety and extending shelf life. Therefore, the goal is to achieve zero tolerance towards heat-resistant microbes, which requires strict microbiological quality control. For example, the current standard method IFU4 (alternatively: MM4), published by the "International Fruit and Vegetable Juice Association" (IFU) (IFU method MM4, 1996), allows for the detection of heat-resistant spores of Patulin-producing fungi in juice or raw materials after a heating step like pasteurization. The analysis duration for this method ranges from 6 to 15 days (Ishara & Gunasena, 2021). Filtration techniques such as MF, UF, NF, and RO play therefore significant roles in this process by removing bacteria and limiting their growth, thereby preventing spoilage and maintaining product quality (Aguayo et al., 2006; Nath et al., 2018; Lawlor et al., 2009). These techniques utilize membranes with varying pore sizes to capture particles and microorganisms, effectively reducing microbial load and enhancing the safety of the juice for consumption.

MF employs membranes with larger pore sizes that physically trap particles and larger microorganisms like yeast, molds, and certain bacteria, leaving behind cleaner juice with a reduced microbial burden (Vieira et al., 2020). For instance, *E. coli*, which range from about 0.5–5 mm in length, and *Listeria monocytogenes*, typically 0.5 to 2.0 mm wide, are effectively removed by this process (Norrrahim et al., 2021).

UF, utilizing slightly smaller pore sizes, targets smaller bacteria, effectively separating molecules based on size. This method can eliminate bacteria such as *Campylobacter jejuni* and *Vibrio cholerae*, which are smaller in size compared to those removed by MF (Fukumoto et al., 1998; Norrrahim et al., 2021). By further refining the size exclusion, UF ensures a higher level of purity and safety in the juice.

NF and RO employ even narrower membranes, which are essential for removing the smallest of viruses and other microscopic pollutants. These advanced filtration techniques enhance the overall safety of the juice, significantly extend shelf life, and improve the storage stability of the finished product by removing nearly all microbial presence (Alvarez et al., 1998; Fukumoto et al., 1998; Salehi, 2014).

Each of these filtration methods contributes uniquely to the control of microorganisms in fruit juices, playing an integral role in maintaining the safety and quality of the products. By effectively reducing or eliminating hazardous microorganisms, these filtration technologies ensure that fruit juices not only meet the stringent safety standards required by regulatory bodies but also maintain the high quality expected by consumers.

11.4.2 Clarity, flavor, and nutritional preservation

Quality in the food and beverage industry encompasses sensory attributes, nutritional content, and overall safety of the product. An important aspect of juice processing is maintaining juice quality, particularly with various filtration techniques that improve clarity, remove particulate matter, and preserve flavor and nutritional components. For example, in guava juice processing, both plate and frame filtering and UF are employed to enhance clarity—a crucial characteristic of premium juice. Plate and frame filtration uses several filter pads of varying pore diameters to effectively remove particles, thereby enhancing juice clarity (Bhattacharjee et al., 2017). In contrast, UF utilizes selectively permeable membranes that can filter out even tiny particles and colloids, contributing to a clear finished product (Jawaheer et al., 2003).

The retention of soluble solids and ascorbic acid (vitamin C) in guava juice is significantly influenced by the chosen filtration technique. While UF provides excellent clarity, it may lead to a slight loss of soluble solids due to the removal of tiny molecular weight molecules (Sharoba & Ramadan, 2011), and oxidative processes involved in UF could also decrease the levels of ascorbic acid (Girard & Fukumoto, 2000). On the other hand, plate and frame filtering tends to retain more of these soluble solids and ascorbic acid, thereby preserving the juice's desirable sweetness, flavor character, and nutritional value (Jawaheer et al., 2003).

11.4.3 Oxidation management

Oxidation in fruit juices can lead to undesirable changes in flavor, color, and nutritional value. Filtration techniques help manage oxidation by removing oxygen and other oxidizing agents, thus increasing shelf life, and maintaining product quality. MF and UF can efficiently filter out larger particles and pathogens that contribute to oxidation and spoilage (Dushkova et al., 2022), while NF and RO can remove divalent ions, larger molecules, and certain color compounds, further reducing the impact of oxidation (Maksimovic et al., 2019).

11.4.4 Allergen removal

Allergen removal is critical due to the potential health risks allergenic substances pose to sensitive individuals. Filtration methods such as MF and UF are effective in mitigating allergen content in fruit products. Microfiltration removes larger allergenic particles like pollen grains, while UF can target larger allergenic proteins (Brenna et al., 2000; Daufin et al., 2000). NF's more restrictive process enables the targeting of even smaller allergenic proteins, aiding in allergen removal while preserving important elements of the fruit product (de Bruijn et al., 2002; Brenna et al., 2000; Echavarría et al., 2011).

This streamlined version focuses on each section's core aspects, clarifies the role of different filtration techniques, and avoids repetitive descriptions of how each filtration method works, thereby enhancing overall readability and coherence.

11.5 Packaging and storage stability

In the fruit juice production chain, filtration, packaging, and storage are interconnected stages, each profoundly influencing product quality, safety, and shelf life. This chapter provides an overview of the advancements and considerations in this holistic process, emphasizing the synergies between filtration methods and packaging technologies that enhance product shelf life and safety by maintaining sensory qualities (Bodbodak & Rafiee, 2016). The aspects of stability and shelf life of fruit juices need to be differentiated between two main types: Not from Concentrate (NFC) and reconstituted juices (Ashurst, 2016). NFC juices include unpasteurized versions, which require cold distribution and must be consumed within a few days, and lightly pasteurized varieties, which have a shelf life of several weeks (Ashurst, 2016). Reconstituted juices, made from concentrated juice, are aseptically packaged and can have a shelf life of up to 1 year (Ashurst, 2016).

11.5.1 Emerging trends and innovations in packaging

Packaging innovations in response to shifting consumer preferences have led to the development of active and intelligent packaging technologies aimed at maintaining the quality and safety of food products, particularly fruit juices. These advanced packaging concepts focus on enhancing the packaged food's overall integrity while extending its shelf life and enhancing sensory qualities.

Active packaging techniques such as oxygen scavengers significantly reduce oxygen levels within the package, thus preventing damage caused by oxidative processes including lipid oxidation and the breakdown of delicate substances such as vitamins. This method is especially effective in preserving the color and flavor of fruit juices (Han & Mehyar, 2011). Moisture regulators help maintain optimal humidity levels, preventing moisture-related issues that could impair the quality of packed juice (Day & Potter, 2011). Ethylene scavengers extend the shelf life by inhibiting ripening and senescence, and antimicrobial-releasing devices help maintain the microbiological quality by inhibiting the growth of spoilage microorganisms (Han et al., 2005).

Intelligent packaging transforms the fruit juice industry by incorporating sensors and indicators that monitor temperature variations and time-temperature changes from manufacturing to consumption. Features include sensors to detect packaging breaches, freshness indicators for volatile compounds that signal juice quality, and dynamic labels providing information on production dates, nutritional content, and storage guidelines (Kruijf et al., 2002). Intelligent packaging not only ensures product safety and extends shelf life but also reduces waste and enhances the overall consumer experience (Han et al., 2005; Kruijf et al., 2002).

Aseptic packaging employs various sterilization techniques such as steam sterilization, hydrogen peroxide vapor, gamma irradiation, and electron beam to sterilize packaging materials to eliminate bacteria before filling. Hydrogen peroxide sterilization followed by hot air has emerged as a preferred method for in-line sterilization due to its efficacy and minimal adverse reports (Alves et al., 2001; Ansari & Datta, 2003).

11.5.2 Storage stability enhancements

Essential for maintaining the quality and safety of filtrated perishable products like fruit juices, cold chain management involves keeping a consistent and optimal temperature throughout the supply chain. Cold storage slows microbial growth, inhibits enzymatic activity, and minimizes chemical reactions, thus preserving the sensory qualities, nutritional value, and overall quality of fruit juices (Soto-Silva et al., 2016).

High-pressure processing (HPP) is a nonthermal preservation method that uses high hydrostatic pressure to inactivate pathogens, enzymes, and harmful bacteria while preserving the sensory qualities and nutritional value of the products. HPP avoids high temperatures that can alter the food's flavor, color, and nutritional content, although it may cause a temperature increase that needs to be controlled (Sreedevi et al., 2020). Due to the focus on filtration, no further detail will be provided in this chapter.

11.5.3 The direct influence of filtration on packaging and storage

Filtration plays a pivotal role in the processing of fruit juice, it is primarily utilized to eliminate germs, particles, and contaminants that can lead to spoilage. By effectively removing these potential pollutants, filtration significantly extends the shelf life of juices, proving especially beneficial for juices without preservatives. This capability allows juices to remain fresh and of high quality for extended periods, regardless of storage conditions. Such an extension of shelf life aligns seamlessly with contemporary packaging strategies like Modified Atmosphere Packaging (MAP) or aseptic packaging, which rely on minimizing the presence of contaminants to ensure the durability and safety of the product (Pereira et al., 2013; Sharma et al., 2018).

11.5.4 Improved product integrity

Through meticulous filtration, the regularity and quality of fruit juices are maintained as they transition from processing to packaging. Filtration ensures that the juice reaching the packaging maintains an unwavering degree of purity by thoroughly removing any contaminants. This rigorous purification process not only guarantees consistent flavor, color, and texture but also lays the foundation for a harmonious consumer sensory experience. This consistency extends into the storage phase, reducing the likelihood of unexpected quality deviations or the need for product recalls, fostering a strong sense of reliability and customer trust, thus securing the juice's reputation as a dependable and preferred choice (Bhattacharjee et al., 2017; Echavarría et al., 2011).

11.5.5 Enhanced visual appeal

Meticulous filtration significantly impacts the aesthetic appeal of fruit juices, particularly when packaged in transparent or semi-transparent materials. The filtration process ensures that the juice is free of any contaminants or suspended particles, resulting in a product with remarkable clarity and translucence. This visual clarity allows consumers to observe the juice's purity and quality directly, enhancing perceptions of freshness and authenticity. The clear and unobstructed view of the juice inside the packaging not only demonstrates the product's commitment to excellence but also the stringent quality control processes it has undergone, influencing consumer choice and fostering brand loyalty (Gil et al., 2006).

11.5.6 Cost efficiency

Effective filtration makes the production and storage of fruit juice more cost-effective. By ensuring the juice is of the highest quality and stability at the onset of storage, filtration significantly reduces the need for costly storage solutions or complex packaging methods that might otherwise be necessary to address potential contamination issues. This allows for the use of more conventional and economical storage facilities, reducing energy consumption and operational costs. Furthermore, because the filtration process maintains the fundamental quality of the juice, there is less reliance on complex packaging designed to mask flaws or extend shelf life. This cost optimization carries over to downstream processes, simplifying marketing, distribution, and shelf management, allowing resources to be allocated towards enhancing customer value and expanding the business rather than rectifying quality issues or taking unnecessary precautions (Bhattacharjee et al., 2017; Koupaie et al., 2014).

In summary, the integration of advanced filtration techniques in the juice production process not only enhances product quality and safety but also optimizes packaging and storage solutions, offering significant benefits in terms of shelf-life extension, visual appeal, and cost efficiency.

11.6 Real-world applications illustrate the practical significance of filtration in fruit juice processing

Fruit and vegetable juices are rich in nutrients and packed with vitamins, minerals, and antioxidants beneficial for human health. However, during industrial processing, the quality of the fresh product can change dramatically due to factors like heat oxidation and chemical degradation, which weaken more delicate components, thereby reducing the end product's quality (Cassano et al., 2003). Filtration techniques have become crucial in enhancing product quality, prolonging shelf life, and maximizing processing efficiency in the dynamic sector of fruit juice processing (Echavarría et al., 2011).

11.6.1 Applications of membrane filtration

11.6.1.1 Citrus juice clarification

Citrus juice clarification can be challenging due to an abundance of suspended particles and bacteria, which can lead to erratic product appearance and flavor. MF has emerged as an effective solution for these issues, providing excellent product clarity. For instance, in the processing of bittersweet orange juice, MF has successfully removed large particles and reduced turbidity by over 98%, significantly enhancing product clarity. Interestingly, while the acidity and ascorbic acid levels are reduced due to their susceptibility to oxidation, the total polyphenol content and antioxidant properties are largely preserved, showcasing the capability of MF to protect these valuable nutritional elements (Mirsaeedghazi & Emam-Djomeh, 2017).

Additional studies with mandarin and sweet orange juice have shown that MF, whether used with or without enzyme treatment, effectively reduces alcohol-insoluble substances and viscosity by removing pectin and suspended particles. The clarity of the juices also improved markedly (Satyannarayana et al., 2023). The importance of MF in citrus juice clarification extends to preventing the burning of particles on the walls of evaporators during juice concentration, which can lead to off-flavors and undesirable taste profiles (Sidlagatta et al., 2023; Urošević et al., 2017).

11.6.1.2 UF in apple juice processing

In the food and beverage sector, UF is employed for various purposes, including the processing and concentration of fruit juices like apple juice. By carefully removing water from the juice, UF concentrates the natural sugars, flavors, and nutrients, resulting in a concentrated apple juice that retains the original flavors, and compounds in a smaller volume. This concentration also enhances the apple juice's shelf life by reducing the water content, which could otherwise promote microbial growth and enzymatic reactions (Cai et al., 2020; Dushkova et al., 2022).

11.6.1.3 NF for enhanced quality and reduced sugar

NF is a pressure-driven separation method influenced by steric (sieving) and charge (Donnan) effects, using less energy compared to RO. NF has been employed for recovering natural aromas in fruit juices, treating juice beverage manufacturing effluent, and controlling sugar content (Pruksasri et al., 2020). Its moderate processing conditions prevent the denaturation of heat-sensitive components like enzymes, vitamins, and polyphenols, thus maintaining the nutritional value and flavor of the fruit juices (Warczok et al., 2004). One of the major benefits of NF is recovering fruit juices' natural smells, treating juice beverage manufacturing effluent, and controlling sugar content and its ability to remove allergenic proteins that can cause allergic symptoms, reducing allergen levels in fruit juices to below detectable thresholds (García-Martín et al., 2010; Warczok et al., 2004).

11.6.2 Integrating filtration with other technologies

The integration of filtration techniques with other processing technologies can significantly enhance product quality and operational efficiency. For example, combining membrane filtration (MF and UF) has proven effective in removing particles and germs, enhancing juice clarity, and extending shelf life. These methods provide a continuous and efficient processing solution, often used in tangential or crossflow filtration settings. The combination of MF and UF not only maintains key components like antioxidants and vitamins but also minimizes the use of fining agents and reduces enzyme usage (Onsekizoglu et al., 2010; de Carvalho et al., 2008; Urošević et al., 2017).

11.6.3 Crossflow filtration applications

In the food sector, crossflow filtration, often referred to as tangential flow filtration, is a commonly used membrane separation method that is used for the clarification and concentration of fruit juices, among other things. This method allows for ongoing filtration while lowering clogging and preserving greater effectiveness of filtration by running the initial solution obliquely down the surface of a filter (Lipnizki, 2010).

Because of its high nutritional value and vitamin C content, cashew apples, a byproduct of the Vietnamese cashew nut industry, have promise for juice manufacturing. Cashew apple juice can be processed using crossflow UF, which allows for the separation of sediments and contaminants while preserving important nutrients. This method provides a creative way to use cashew apples that would otherwise be wasted. Crossflow filtration can be used to turn dragon fruit into juice. By extracting beneficial elements including fibers, vitamins, and minerals, this method can improve the juice product's clarity and stability. Crossflow filtration can be used to concentrate and cleanse the juice of gac fruit, a traditional Vietnamese fruit rich in carotenoids. Higher carotenoid concentrations in the finished product may result from it. Crossflow filtering can help pineapple juice become clearer. This method can successfully remove particles, suspended sediments, and contaminants, leaving behind a juice that is cleaner and more aesthetically pleasing (Mai, 2017; Vaillant et al., 2001).

The use of crossflow UF with hollow fiber membranes of 30 kDa has shown promise in extending the shelf life of sugarcane juice. The ultrafiltered juice had enhanced clarity, reduced total solids, turbidity, and color, as well as minimal polyphenol oxidase (PPO) activity. Additionally, a considerable decrease in the yeast and mold count (YMC) and total plate count (TPC) suggested increased juice stability. Without requiring heat treatment or chemical preservatives, sugarcane juice can be successfully preserved via crossflow filtering using hollow fiber UF modules. The juice's prolonged shelf life—up to nine weeks when refrigerated—holds a lot of promise for the food manufacturing sector. Consumer desires for minimally processed, high-quality, and long-lasting beverages are in line with this novel strategy (Panigrahi et al., 2020).

11.6.4 Applications of combination of microfiltration and ultrafiltration

Combining MF with UF procedures has proven to be a successful method for removing particles and germs from juice in the current day, improving product quality and shelf life. These methods have many advantages over traditional ones and help to provide a continuous and straightforward processing procedure. They are frequently used in tangential or crossflow filtration settings. A high flux and product quality must always be maintained for a clarifying process to be effective. These objectives are accomplished using MF and UF processes, which also offer better control over the properties of the finished juice (Urošević et al., 2017).

The use of MF and UF removes the requirement for fining agents and minimizes the consumption of enzymes, both of which are problems in conventional procedures (Onsekizoglu et al., 2010). By successfully eliminating spoilage bacteria, the combination of MF and UF improves the permeate juice's microbiological quality and creates a safer product (de Carvalho et al., 2008). MF and UF procedures help maintain important elements including antioxidants, vitamins, amino acids, and minerals, which are frequently found in the clarified juice, notwithstanding possible changes in the juice content throughout the processes.

Cactus pear juice was subjected to MF and UF processes, and it was discovered that both techniques produced a clear solution with little protein and high quantities of antioxidants, carbohydrates, amino acids, and minerals. The fruit's nutritive and practical qualities were preserved by these procedures (Cassano et al., 2010). The use of MF and UF on pomegranate juice revealed variances in the two methods' efficacy. Although UF processing resulted in lower quantities of important components like total anthocyanins and antioxidant activity, pre-treatment with MF minimized membrane fouling (Mirsaeedghazi et al., 2012). When utilizing a ceramic membrane for MF, West Indian Cherry Juice showed significant permeate flow and ascorbic acid retention. The juice's color and nutrients were retained by the addition of MF (Wang et al., 2005).

11.6.5 Applications of filtration techniques

To reduce water usage and make it easier to produce concentrated juices, the use of advanced filtration techniques, particularly membrane technology, in the field of juice processing has expanded. These ground-breaking methods optimize water use and boost the effectiveness of juice concentration procedures to provide sustainable solutions (Bhattacharjee et al., 2017).

Fruit juices are clarified and concentrated at the same time using MF and UF. The juice is successfully cleaned of suspended particles, bacteria, and particulate matter using MF, which is distinguished by its larger pore sizes. The juice is further clarified and concentrated using UF, which has smaller pores, by eliminating specific macromolecules like proteins and polysaccharides. This simultaneous reduction in the volume of liquid that must be treated and improvement in water efficiency (Mirsaeedghazi et al., 2012). A crucial technology for concentrating fruit juices while using the least amount of water is NF. Higher juice concentrations are achieved by separating water and other tiny solutes from the juice using membranes with selective permeability. Juice concentrates with improved flavor and nutritional value are produced using this procedure, which also conserves water resources (Nath et al., 2018; Salehi, 2014).

The most efficient membrane technology for getting optimum juice concentration is RO. It enables the removal of water molecules while retaining dissolved solutes by using membranes with incredibly small pores. The delivery of juice products that are highly concentrated and the reduction of water waste are two areas where this technology excels (Rastogi, 2018). The possibility of sequentially combining various membrane processes, including RO, MF, UF, and NF, has been acknowledged. In all phases of juice processing, from early clarifying to final concentration, this holistic approach optimizes water use, reducing water waste and increasing total process effectiveness (Echavarría et al., 2011). Juice manufacturers can significantly reduce their water usage by implementing cutting-edge membrane technologies. Lowering water-related expenses and promoting a more resource-efficient production process, not only supports sustainability goals but also has a positive economic impact (Bhattacharjee et al., 2017).

11.7 Conclusion and future directions

The integral role of filtration in fruit juice production cannot be overstated. This process ensures that the final product is not only visually appealing and free from unwanted particulate matter and microorganisms but also retains its inherent nutritional value and flavor profiles (Panigrahi et al., 2020; Hu and Wu, 2023; Mandha et al., 2023; Sarbatly et al., 2023). Various filtration techniques, ranging from mechanical methods to advanced membrane-based applications, have been pivotal in enhancing the clarity, quality, and safety of fruit juices. These methods effectively remove solids, manage turbidity, and preserve essential nutrients such as anthocyanins, carotenoids, and vitamins which are susceptible to degradation during processing (Barros et al., 2011; Provesi et al., 2011; Bao et al., 2023; Kechinski et al., 2010).

The future of filtration technology in the fruit juice industry looks promising with several innovative trends on the horizon. Emerging membrane technologies are enhancing both the efficiency and sustainability of filtration operations. Innovations such as adjustable pore-size membranes and improved aseptic packaging methods facilitated by advanced filtration promise to further enhance the safety and extend the shelf life of fruit juices. These advancements are instrumental in meeting the increasing consumer demand for high-quality, preservative-free juices.

For industry practitioners, particularly in small to medium-sized enterprises (SMEs), the adoption of these advanced filtration technologies involves a careful analysis of cost benefits and potential implementation barriers. The practical application of these filtration techniques can lead to significant improvements in product quality and production efficiency, making a compelling case for their broader adoption across the industry.

There remains scope for further research to explore the long-term effects of new filtration technologies on the retention of nutrients in fruit juices. Studies could also focus on the economic viability of these technologies across different scales of production, providing deeper insights into their potential for wider adoption.

Reflecting on the environmental impact of these technologies, there is a growing need to integrate sustainable practices within the filtration processes. This includes optimizing water use and reducing energy consumption, which not only supports environmental sustainability but also enhances the economic viability of juice production.

References

Abdullah, S., Karmakar, S., Pradhan, R. C., & Mishra, S. (2022). Pressure-driven cross-flow microfiltration coupled with centrifugation for tannin reduction and clarification of cashew apple juice: Modeling of permeate flux decline and optimization of process parameters. *Journal of Food Processing and Preservation, 46*(6), e16497. https://doi.org/10.1111/jfpp.16497.

Akyıldız, A., Önür, E., Ağçam, E., Dündar Kirit, B., & Uçan Türkmen, F. (2022). Changes in quality parameters of orange juice deacidified by ion exchange resins. *Food Chemistry, 375*, 131837. https://doi.org/10.1016/j.foodchem.2021.131837.

Alaei, B., Amiri Chayjan, R., & Ali Zolfigol, M. (2021). Tomato juice microfiltration process assisted with pressure-vacuum combination condition: A physicochemical investigation and optimization. *Biosystems Engineering, 212*, 62–76. https://doi.org/10.1016/j.biosystemseng.2021.10.005.

Alvarez, S., Alvarez, R., Riera, F. A., & Coca, J. (1998). Influence of depectinization on apple juice ultrafiltration. *Colloids and Surfaces A: Physicochemical and Engineering Aspects, 138*(2), 377–382. https://doi.org/10.1016/S0927-7757(98)00235-0.

Alves, R. M. V., Sarantópoulos, C. I. G. L., Saron, E. S., & Bordin, M. R. (2001). Stability of fruit juice drinks in aseptic packages. *Packaging Technology and Science, 14*(2), 79–86. https://doi.org/10.1002/pts.538.

Ansari, I. A., & Datta, A. K. (2003). An overview of sterilization methods for packaging materials used in aseptic packaging systems. *Food and Bioproducts Processing, 81*(1), 57–65. https://doi.org/10.1205/096030803765208670.

Arikrishnan, D., & Keshav, A. (2023). Pineapple juice clarification by continuous dead-end microfiltration using a low-cost ceramic membrane. *Journal of Food Measurement and Characterization, 17*(1), 863–881. https://doi.org/10.1007/s11694-022-01634-5.

Arriola, N. A., dos Santos, G. D., Prudêncio, E. S., Vitali, L., Petrus, J. C. C., & Castanho Amboni, R. D. M. (2014). Potential of nanofiltration for the concentration of bioactive compounds from watermelon juice. *International Journal of Food Science & Technology, 49*(9), 2052–2060. https://doi.org/10.1111/ijfs.12513.

Ashurst, P. (2016). 12: The stability and shelf life of fruit juices and soft drinks. In P. Subramaniam (Ed.), *The Stability and Shelf Life of Food* (2nd ed.) (pp. 347–374). Woodhead Publishing https://doi.org/10.1016/B978-0-08-100435-7.00012-5.

Bao, S., Yin, D., Zhao, Q., Zhou, Y., Hu, Y., Sun, X., Liu, X., & Ma, T. (2023). Comprehensive evaluation of the effect of five sterilization methods on the quality of black carrot juice based on PCA, TOPSIS and GRA models. *Food Chemistry: X, 17*, 100604. https://doi.org/10.1016/j.fochx.2023.100604.

Barros, A. I. R. N. A., Nunes, F. M., Gonçalves, B., Bennett, R. N., & Silva, A. P. (2011). Effect of cooking on total vitamin C contents and antioxidant activity of sweet chestnuts (Castanea sativa Mill.). *Food Chemistry, 128*(1), 165–172. https://doi.org/10.1016/j.foodchem.2011.03.013.

Beveridge, T. (2002). Opalescent and cloudy fruit juices: Formation and particle stability. *Critical Reviews in Food Science and Nutrition, 42*(4), 317–337. https://doi.org/10.1080/10408690290825556.

Bhattacharjee, C., Saxena, V. K., & Dutta, S. (2017). Fruit juice processing using membrane technology: A review. *Innovative Food Science & Emerging Technologies, 43*, 136–153. https://doi.org/10.1016/j.ifset.2017.08.002.

Bodbodak, S., & Rafiee, Z. (2016). 3: Recent trends in active packaging in fruits and vegetables. In M. W. Siddiqui (Ed.), *Eco-Friendly Technology for Postharvest Produce Quality* (pp. 77–125). Academic Press https://doi.org/10.1016/B978-0-12-804313-4.00003-7.

Borneman, Z., Gökmen, V., & Nijhuis, H. H. (2001). Selective removal of polyphenols and brown colour in apple juices using PES/PVP membranes in a single ultrafiltration process. *Separation and Purification Technology, 22-23*, 53–61. https://doi.org/10.1016/S1383-5866(00)00142-8.

Brenna, O., Pompei, C., Ortolani, C., Pravettoni, V., Pastorello, E. A., & Farioli, L. (2000). Technological processes to decrease the allergenicity of peach juice and nectar. *Journal of Agricultural and Food Chemistry, 48*(2), 493–497. https://doi.org/10.1021/jf9906681.

Bump, V. L. (1989). Filtration of apple juice. In D. L. Downing (Ed.), *Processed Apple Products* (pp. 121–136). Springer US https://doi.org/10.1007/978-1-4684-8225-6_6.

Cai, M., Xie, C., Lv, Y., Yang, K., & Sun, P. (2020). Changes in physicochemical profiles and quality of apple juice treated by ultrafiltration and during its storage. *Food Science & Nutrition, 8*(6), 2913–2919. https://doi.org/10.1002/fsn3.1593.

Carabasa, M., Ibarz, A., Garza, S., & Barbosa-Cánovas, G. V. (1998). Removal of dark compounds from clarified fruit juices by adsorption processes. *Journal of Food Engineering, 37*(1), 25–41. https://doi.org/10.1016/S0260-8774(98)00075-2.

Cassano, A., Drioli, E., Galaverna, G., Marchelli, R., Di Silvestro, G., & Cagnasso, P. (2003). Clarification and concentration of citrus and carrot juices by integrated membrane processes. *Journal of Food Engineering, 57*(2), 153–163. https://doi.org/10.1016/S0260-8774(02)00293-5.

Cassano, A., & Basile, A. (2011). 20: Membranes for industrial microfiltration and ultrafiltration. In A. Basile, & S. P. Nunes (Eds.), *Advanced Membrane Science and Technology for Sustainable Energy and Environmental Applications* (pp. 647–679). Woodhead Publishing https://doi.org/10.1533/9780857093790.5.647.

Cassano, A., Conidi, C., & Drioli, E. (2010). Physico-chemical parameters of cactus pear (*Opuntia ficus-indica*) juice clarified by microfiltration and ultrafiltration processes. *Desalination, 250*(3), 1101–1104. https://doi.org/10.1016/j.desal.2009.09.117.

Charcosset, C. (2016). Ultrafiltration, microfiltration, nanofiltration and reverse osmosis in integrated membrane processes. *Integrated Membrane Systems and Processes* (pp. 1–22) https://doi.org/10.1002/9781118739167.ch1.

Chen, F., Ji, Z., & Qi, Q. (2018). Effect of pore size and layers on filtration performance of coalescing filters with different wettabilities. *Separation and Purification Technology, 201*, 71–78. https://doi.org/10.1016/j.seppur.2018.03.004.

Chen, S., Tang, Q., Zhang, X., Zhao, G., Hu, X., Liao, X., Chen, F., Wu, J., & Xiang, H. (2006). Isolation and characterization of thermo-acidophilic endospore-forming bacteria from the concentrated apple juice-processing environment. *Food Microbiology, 23*(5), 439–445. https://doi.org/10.1016/j.fm.2005.07.005.

Daufin, G., Berot, S., Fillaudeau, L., Decloux, M., Carrere, H., & Escudier, J. (2000). Application of Membrane Processes in Food and Dairy Industry.

Day, B. P. F., & Potter, L. (2011). Active packaging. *Food and Beverage Packaging Technology* (pp. 251–262) https://doi.org/10.1002/9781444392180.ch9.

de Bruijn, J., Venegas, A., & Borquez, R. (2002). Influence of crossflow ultrafiltration on membrane fouling and apple juice quality. *Desalination, 148*(1), 131–136. https://doi.org/10.1016/S0011-9164(02)00666-5.

de Carvalho, L. M. J., de Castro, I. M., & da Silva, C. A. B. (2008). A study of retention of sugars in the process of clarification of pineapple juice (*Ananas comosus*, L. Merril) by micro- and ultra-filtration. *Journal of Food Engineering, 87*(4), 447–454. https://doi.org/10.1016/j.jfoodeng.2007.12.015.

Durak, M. Z., Churey, J. J., Danyluk, M. D., & Worobo, R. W. (2010). Identification and haplotype distribution of Alicyclobacillus spp. from different juices and beverages. *International Journal of Food Microbiology, 142*(3), 286–291. https://doi.org/10.1016/j.ijfoodmicro.2010.07.003.

Dushkova, M., Mihalev, K., Dinchev, A., Vasilev, K., Georgiev, D., & Terziyska, M. (2022). Concentration of polyphenolic antioxidants in apple juice and extract using ultrafiltration. *Membranes, 12*(11), 1032. https://www.mdpi.com/2077-0375/12/11/1032.

Duygu Ozsoy, H., & van Leeuwen, J. (2010). Removal of color from fruit candy waste by activated carbon adsorption. *Journal of Food Engineering, 101*(1), 106–112. https://doi.org/10.1016/j.jfoodeng.2010.06.018.

Echavarría, A. P., Torras, C., Pagán, J., & Ibarz, A. (2011). Fruit juice processing and membrane technology application. *Food Engineering Reviews, 3*(3), 136–158. https://doi.org/10.1007/s12393-011-9042-8.

Eiroa, M. N. U., Junqueira, V. C. A., & Schmidt, F. L. (1999). Alicyclobacillus in Orange Juice: Occurrence and Heat Resistance of Spores. *Journal of Food Protection, 62*(8), 883–886. https://doi.org/10.4315/0362-028X-62.8.883.

Erkmen, O. (2022). Practice 12- Counting of thermoduric microorganisms. In O. Erkmen (Ed.), *Microbiological Analysis of Foods and Food Processing Environments* (pp. 99–102). Academic Press https://doi.org/10.1016/B978-0-323-91651-6.00032-X.

Ersahin, M. E., Ozgun, H., Dereli, R. K., Ozturk, I., Roest, K., & van Lier, J. B. (2012). A review on dynamic membrane filtration: Materials, applications and future perspectives. *Bioresource Technology, 122*, 196–206. https://doi.org/10.1016/j.biortech.2012.03.086.

Fukumoto, L. R., Delaquis, P., & Girard, B. (1998). Microfiltration and Ultrafiltration Ceramic Membranes for Apple Juice Clarification. *Journal of Food Science, 63*(5), 845–850. https://doi.org/10.1111/j.1365-2621.1998.tb17912.x.

Gan, Y. X. (2021). Activated Carbon from Biomass Sustainable Sources. *C, 7*(2), 39. https://www.mdpi.com/2311-5629/7/2/39.

Gandhi, K., Sharma, N., Gautam, P. B., Sharma, R., Mann, B., & Pandey, V. (2022). Membrane Processes. *Advanced Analytical Techniques in Dairy Chemistry* (pp. 131–145). Springer US https://doi.org/10.1007/978-1-0716-1940-7_6.

García-Martín, N., Perez-Magariño, S., Ortega-Heras, M., González-Huerta, C., Mihnea, M., González-Sanjosé, M. L., Palacio, L., Prádanos, P., & Hernández, A. (2010). Sugar reduction in musts with nanofiltration membranes to obtain low alcohol-content wines. *Separation and Purification Technology, 76*(2), 158–170. https://doi.org/10.1016/j.seppur.2010.10.002.

Gavazzi-April, C., Benoit, S., Doyen, A., Britten, M., & Pouliot, Y. (2018). Preparation of milk protein concentrates by ultrafiltration and continuous diafiltration: Effect of process design on overall efficiency. *Journal of Dairy Science, 101*(11), 9670–9679. https://doi.org/10.3168/jds.2018-14430.

Gil, M. I., Aguayo, E., & Kader, A. A. (2006). Quality Changes and Nutrient Retention in Fresh-Cut versus Whole Fruits during Storage. *Journal of Agricultural and Food Chemistry, 54*(12), 4284–4296. https://doi.org/10.1021/jf060303y.

Girard, B., & Fukumoto, L. R. (2000). Membrane Processing of Fruit Juices and Beverages: A Review. *Critical Reviews in Food Science and Nutrition, 40*(2), 91–157. https://doi.org/10.1080/10408690091189293.

Gouws, P. A., Gie, L., Pretorius, A., & Dhansay, N. (2005). Isolation and identification of Alicyclobacillus acidocaldarius by 16S rDNA from mango juice and concentrate. *International Journal of Food Science & Technology, 40*(7), 789–792. https://doi.org/10.1111/j.1365-2621.2005.01006.x.

Groenewald, W. H., Gouws, P. A., & Witthuhn, R. C. (2009). Isolation, identification and typification of Alicyclobacillus acidoterrestris and Alicyclobacillus acidocaldarius strains from orchard soil and the fruit processing environment in South Africa. *Food Microbiology, 26*(1), 71–76. https://doi.org/10.1016/j.fm.2008.07.008.

Han, J. H., Ho, C. H. L., & Rodrigues, E. T. (2005). 9- Intelligent packaging. In J. H. Han (Ed.), *Innovations in Food Packaging* (pp. 138–155). Academic Press https://doi.org/10.1016/B978-012311632-1/50041-3.

Hu, Y., & Wu, W. (2023). Application of Membrane Filtration to Cold Sterilization of Drinks and Establishment of Aseptic Workshop. *Food and Environmental Virology, 15*(2), 89–106. https://doi.org/10.1007/s12560-023-09551-6.

ICMSF, 1998. Soft drinks, fruit juices, concentrates and fruit preserves. Micro-Organisms in Foods: Microbial Ecology of Food Commodities. Springer US (pp. 440–460).

IFU method MM4: Moulds Count Procedure, (1996).

Ilame, S. A., & Singh, S. V. (2015). Application of Membrane Separation in Fruit and Vegetable Juice Processing: A Review. *Critical Reviews in Food Science and Nutrition, 55*(7), 964–987. https://doi.org/10.1080/10408398.2012.679979.

Ishara, S., & Gunasena, D. (2021). Heat Resistant Moulds in Pasteurized Fruit Syrups. *European Journal of Agriculture and Food Sciences, 3*. https://doi.org/10.24018/ejfood.2021.3.1.199.

Ismail, A. F., & Goh, P. S. (2015). Microfiltration Membrane. In S. Kobayashi, & K. Müllen (Eds.), *Encyclopedia of Polymeric Nanomaterials* (pp. 1250–1255). Berlin Heidelberg: Springer https://doi.org/10.1007/978-3-642-29648-2_159.

Jawaheer, B., Goburdhun, D., & Ruggoo, A. (2003). Effect of processing and storage of guava into jam and juice on the Ascorbic acid content. *Materiae Vegetabiles, 58*, 1–12. https://doi.org/10.1023/B:QUAL.0000041161.05123.66.

Jørgensen, C. E., Abrahamsen, R. K., Rukke, E.-O., Johansen, A.-G., Schüller, R. B., & Skeie, S. B. (2016). Optimization of protein fractionation by skim milk microfiltration: Choice of ceramic membrane pore size and filtration temperature. *Journal of Dairy Science, 99*(8), 6164–6179. https://doi.org/10.3168/jds.2016-11090.

Kammerer, J., Carle, R., & Kammerer, D. R. (2011). Adsorption and Ion Exchange: Basic Principles and Their Application in Food Processing. *Journal of Agricultural and Food Chemistry, 59*(1), 22–42. https://doi.org/10.1021/jf1032203.

Kandemir, K., Piskin, E., Xiao, J., Tomas, M., & Capanoglu, E. (2022). Fruit Juice Industry Wastes as a Source of Bioactives. *Journal of Agricultural and Food Chemistry, 70*(23), 6805–6832. https://doi.org/10.1021/acs.jafc.2c00756.

Kechinski, C. P., Guimarães, P. V. R., Noreña, C. P. Z., Tessaro, I. C., & Marczak, L. D. F. (2010). Degradation Kinetics of Anthocyanin in Blueberry Juice during Thermal Treatment. *Journal of Food Science, 75*(2), C173–C176. https://doi.org/10.1111/j.1750-3841.2009.01479.x.

Kelly, N. P., Kelly, A. L., & O'Mahony, J. A. (2019). Strategies for enrichment and purification of polyphenols from fruit-based materials. *Trends in Food Science & Technology, 83*, 248–258. https://doi.org/10.1016/j.tifs.2018.11.010.

Klopotek, Y., Otto, K., & Böhm, V. (2005). Processing Strawberries to Different Products Alters Contents of Vitamin C, Total Phenolics, Total Anthocyanins, and Antioxidant Capacity. *Journal of Agricultural and Food Chemistry, 53*(14), 5640–5646. https://doi.org/10.1021/jf047947v.

Kotzekidou, P. (2014). Byssochlamys. In C. A. Batt, & M. L. Tortorello (Eds.), *Encyclopedia of Food Microbiology* (Second Edition) (pp. 344–350). Academic Press https://doi.org/10.1016/B978-0-12-384730-0.00051-3.

Koupaie, E. H., Leiva, M. B., Eskicioglu, C., & Dutil, C. (2014). Mesophilic batch anaerobic co-digestion of fruit-juice industrial waste and municipal waste sludge: Process and cost-benefit analysis. *Bioresource Technology, 152*, 66–73. https://doi.org/10.1016/j.biortech.2013.10.072.

Kruijf, N. D., Beest, M. V., Rijk, R., Sipiläinen-Malm, T., Losada, P. P., & Meulenaer, B. D. (2002). Active and intelligent packaging: applications and regulatory aspects. *Food Additives & Contaminants, 19*(sup1), 144–162. https://doi.org/10.1080/02652030110072722.

Laorko, A., Li, Z., Tongchitpakdee, S., & Youravong, W. (2011). Effect of gas sparging on flux enhancement and phytochemical properties of clarified pineapple juice by microfiltration. *Separation and Purification Technology, 80*(3), 445–451. https://doi.org/10.1016/j.seppur.2011.05.024.

Lawlor, K. A., Schuman, J. D., Simpson, P. G., & Taormina, P. J. (2009). Microbiological Spoilage of Beverages. In W. H. Sperber, & M. P. Doyle (Eds.), *Compendium of the Microbiological Spoilage of Foods and Beverages* (pp. 245–284). New York: Springer https://doi.org/10.1007/978-1-4419-0826-1_9.

Le, T. T., Vu, L. T., & Le, N. L. (2021). Effects of membrane pore size and transmembrane pressure on ultrafiltration of red-fleshed dragon fruit (Hylocereus polyrhizus) juice. *Journal of Chemical Technology & Biotechnology, 96*(6), 1561–1572. https://doi.org/10.1002/jctb.6672.

Lee, S. Y., Chang, S. S., Shin, J. H., & Kang, D. H. (2007). Membrane filtration method for enumeration and isolation of Alicyclobacillus spp. from apple juice. *Letters in Applied Microbiology, 45*(5), 540–546. https://doi.org/10.1111/j.1472-765X.2007.02229.x.

Li, N., Wei, Y., Li, X., Wang, J., Zhou, J., & Wang, J. (2019). Optimization of deacidification for concentrated grape juice. *Food Science & Nutrition, 7*(6), 2050–2058. https://doi.org/10.1002/fsn3.1037.

Lima Tribst, A. A., de Souza Sant'Ana, A., & de Massaguer, P. R. (2009). Review: Microbiological quality and safety of fruit juices—past, present and future perspectives. *Critical Reviews in Microbiology, 35*(4), 310–339. https://doi.org/10.3109/10408410903241428.

Lipnizki, F. (2010). Cross-Flow Membrane Applications in the Food Industry. *Membrane Technology* (pp. 1–24) https://doi.org/10.1002/9783527631384.ch1.

Lu, C., Bao, Y., & Huang, J.-Y. (2021). Fouling in membrane filtration for juice processing. *Current Opinion in Food Science, 42*, 76–85. https://doi.org/10.1016/j.cofs.2021.05.004.

Lue, S. J., & Chiang, B. H. (1989). Deacidification of passion fruit juice by ultrafiltration and ion-exchange processes. *International Journal of Food Science & Technology, 24*(4), 395–401. https://doi.org/10.1111/j.1365-2621.1989.tb00659.x.

Mai, H. C. (2017). Application of Cross-Flow Filtration Technique in Purification and Concentration of Juice from Vietnamese Fruits. *Beverages, 3*(3), 44. https://www.mdpi.com/2306-5710/3/3/44.

Maksimovic, M., Omanovic-Miklicanin, E., & Badnjevic, A. (2019). *Nanofood and Internet of Nano Things*.

Mandha, J., Shumoy, H., Matemu, A. O., & Raes, K. (2023). Characterization of fruit juices and effect of pasteurization and storage conditions on their microbial, physicochemical, and nutritional quality. *Food Bioscience, 51*, 102335. https://doi.org/10.1016/j.fbio.2022.102335.

Marsellés-Fontanet, À. R., Puig, A., Olmos, P., Mínguez-Sanz, S., & Martín-Belloso, O. (2009). Optimising the inactivation of grape juice spoilage organisms by pulse electric fields. *International Journal of Food Microbiology, 130*(3), 159–165. https://doi.org/10.1016/j.ijfoodmicro.2008.12.034.

McCarron, A., Donnelley, M., McIntyre, C., & Parsons, D. (2016). Challenges of up-scaling lentivirus production and processing. *Journal of Biotechnology, 240*, 23–30. https://doi.org/10.1016/j.jbiotec.2016.10.016.

McKnight, I. C., Eiroa, M. N. U., Sant'Ana, A. S., & Massaguer, P. R. (2010). Alicyclobacillus acidoterrestris in pasteurized exotic Brazilian fruit juices: Isolation, genotypic characterization and heat resistance. *Food Microbiology, 27*(8), 1016–1022. https://doi.org/10.1016/j.fm.2010.06.010.

Mehyar, G., & Han, J. (2011). Active Packaging for Fresh-Cut Fruits and Vegetables. *Modified Atmosphere Packaging for Fresh-Cut Fruits and Vegetables*, 267–283. https://doi.org/10.1002/9780470959145.ch14.

Mejia, J. A. A., & Yáñez-Fernandez, J. (2021). Clarification Processes of Orange Prickly Pear Juice (Opuntia spp.) by Microfiltration. *Membranes, 11*(5), 354. https://www.mdpi.com/2077-0375/11/5/354.

Men, Y., Li, Z., Zhu, L., Wang, X., Cheng, S., & Lyu, Y. (2023). New insights into membrane fouling during direct membrane filtration of municipal wastewater and fouling control with mechanical strategies. *Science of The Total Environment, 869*, 161775. https://doi.org/10.1016/j.scitotenv.2023.161775.

Meyer, A. S., Köser, C., & Adler-Nissen, J. (2001). Efficiency of Enzymatic and Other Alternative Clarification and Fining Treatments on Turbidity and Haze in Cherry Juice. *Journal of Agricultural and Food Chemistry, 49*(8), 3644–3650. https://doi.org/10.1021/jf001297n.

Mirsaeedghazi, H., Mousavi, S. M., Emam-Djomeh, Z., Rezaei, K., Aroujalian, A., & Navidbakhsh, M. (2012). Comparison between ultrafiltration and microfiltration in the clarification of pomgranate juice. *Journal of Food Process Engineering, 35*(3), 424–436. https://doi.org/10.1111/j.1745-4530.2010.00598.x.

Mirsaeedghazi, H., & Emam-Djomeh, Z. (2017). Clarification of Bitter Orange (Citrus Aurantium) Juice Using Microfiltration with Mixed Cellulose Esters Membrane. *Journal of Food Processing and Preservation, 41*(1), e12738. https://doi.org/10.1111/jfpp.12738.

Morittu, V. M., Mastellone, V., Tundis, R., Loizzo, M. R., Tudisco, R., Figoli, A., Cassano, A., Musco, N., Britti, D., Infascelli, F., & Lombardi, P. (2020). Antioxidant, Biochemical, and In-Life Effects of Punica granatum L. Natural Juice vs. *Clarified Juice by Polyvinylidene Fluoride Membrane. Foods, 9*(2), 242. https://www.mdpi.com/2304-8158/9/2/242.

Nath, K., Dave, H. K., & Patel, T. M. (2018). Revisiting the recent applications of nanofiltration in food processing industries: Progress and prognosis. *Trends in Food Science & Technology, 73*, 12–24. https://doi.org/10.1016/j.tifs.2018.01.001.

Norman, S. I. (1995). Juice enhancement by ion exchange and adsorbent technologies. In P. R. Ashurst (Ed.), *Production and Packaging of Non-Carbonated Fruit Juices and Fruit Beverages* (pp. 253–273). Springer US https://doi.org/10.1007/978-1-4615-0949-3_9.

Norrahim, M. N. F., Mohd Kasim, N. A., Knight, V. F., Ong, K. K., Mohd Noor, S. A., Abdul Halim, N., Ahmad Shah, N. A., Jamal, S. H., Janudin, N., Misenan, M. S. M., Ahmad, M. Z., Yaacob, M. H., & Wan Yunus, W. M. Z. (2021). Emerging Developments Regarding Nanocellulose-Based Membrane Filtration Material against Microbes. *Polymers, 13*(19), 3249. https://www.mdpi.com/2073-4360/13/19/3249.

Oke, M., Jacob, J. K., & Paliyath, G. (2012). Biochemistry of Fruit Processing. *Food Biochemistry and Food Processing* (pp. 554–568) https://doi.org/10.1002/9781118308035.ch28.

Omar, J. M., Nor, Z., Basri, M. S. M., & Pa, N. F. (2020). Clarification of guava juice by an ultrafiltration process: analysis on the operating pressure, membrane fouling and juice qualities. *Food Research, 4*, 85–92. https://doi.org/10.26656/fr.2017.4(S1).S30.

Onsekizoglu, P., Bahceci, K. S., & Acar, M. J. (2010). Clarification and the concentration of apple juice using membrane processes: A comparative quality assessment. *Journal of Membrane Science, 352*(1), 160–165. https://doi.org/10.1016/j.memsci.2010.02.004.

Panigrahi, C., Mondal, M., Karmakar, S., Mishra, H. N., & De, S. (2020). Shelf life extension of sugarcane juice by cross flow hollow fibre ultrafiltration. *Journal of Food Engineering, 274*, 109880. https://doi.org/10.1016/j.jfoodeng.2019.109880.

Pap, S., Šolević Knudsen, T., Radonić, J., Maletić, S., Igić, S. M., & Turk Sekulić, M. (2017). Utilization of fruit processing industry waste as green activated carbon for the treatment of heavy metals and chlorophenols contaminated water. *Journal of Cleaner Production, 162*, 958–972. https://doi.org/10.1016/j.jclepro.2017.06.083.

Pereira, A. L. F., Almeida, F. D. L., de Jesus, A. L. T., da Costa, J. M. C., & Rodrigues, S. (2013). Storage Stability and Acceptance of Probiotic Beverage from Cashew Apple Juice. *Food and Bioprocess Technology, 6*(11), 3155–3165. https://doi.org/10.1007/s11947-012-1032-1.

Pereira, A. P. M., & Sant'Ana, A. S. (2018). Diversity and fate of spore forming bacteria in cocoa powder, milk powder, starch and sugar during processing: A review. *Trends in Food Science & Technology, 76*, 101–118. https://doi.org/10.1016/j.tifs.2018.04.005.

Pettipher, G. L., Osmundson, M. E., & Murphy, J. M. (1997). Methods for the detection and enumeration of Alicyclobacillus acidoterrestris and investigation of growth and production of taint in fruit juice and fruit juice-containing drinks. *Letters in Applied Microbiology, 24*(3), 185–189. https://doi.org/10.1046/j.1472-765X.1997.00373.x.

Provesi, J. G., Dias, C. O., & Amante, E. R. (2011). Changes in carotenoids during processing and storage of pumpkin puree. *Food Chemistry, 128*(1), 195–202. https://doi.org/10.1016/j.foodchem.2011.03.027.

Pruksasri, S., Lanner, B., & Novalin, S. (2020). Nanofiltration as a potential process for the reduction of sugar in apple juices on an industrial scale. *LWT, 133*, 110118. https://doi.org/10.1016/j.lwt.2020.110118.

Rastogi, N. K. (2018). Chapter 13- Reverse Osmosis and Forward Osmosis for the Concentration of Fruit Juices. In G. Rajauria, & B. K. Tiwari (Eds.), *Fruit Juices* (pp. 241–259). Academic Press https://doi.org/10.1016/B978-0-12-802230-6.00013-8.

Reis, M. H. M., Madrona, G. S., Ferreira, F. B., de Santana Magalhães, F., Bindes, M. M. M., & Cardoso, V. L. (2019). 6- Membrane Filtration Processes for the Treatment of Nonalcoholic Beverages. In A. M. Grumezescu, & A. M. Holban (Eds.), *Engineering Tools in the Beverage Industry* (pp. 175–207). Woodhead Publishing https://doi.org/10.1016/B978-0-12-815258-4.00006-8.

Rico-Munoz, E. (2017). Heat resistant molds in foods and beverages: recent advances on assessment and prevention. *Current Opinion in Food Science, 17*, 75–83. https://doi.org/10.1016/j.cofs.2017.10.011.

Rodriguez-Saona, L. E., Fry, F. S., McLaughlin, M. A., & Calvey, E. M. (2001). Rapid analysis of sugars in fruit juices by FT-NIR spectroscopy. *Carbohydrate Research, 336*(1), 63–74. https://doi.org/10.1016/S0008-6215(01)00244-0.

Sadeghi, I., Kaner, P., & Asatekin, A. (2018). Controlling and Expanding the Selectivity of Filtration Membranes. *Chemistry of Materials, 30*(21), 7328–7354. https://doi.org/10.1021/acs.chemmater.8b03334.

Sagu, S. T., Karmakar, S., Nso, E. J., Kapseu, C., & De, S. (2014). Ultrafiltration of Banana (Musa acuminata) Juice Using Hollow Fibers for Enhanced Shelf Life. *Food and Bioprocess Technology, 7*(9), 2711–2722. https://doi.org/10.1007/s11947-014-1309-7.

Salehi, F. (2014). Current and future applications for nanofiltration technology in the food processing. *Food and Bioproducts Processing, 92*(2), 161–177. https://doi.org/10.1016/j.fbp.2013.09.005.

Salehinia, S., Mirsaeedghazi, H., & Khashehchi, M. (2021). The effect of laser on the efficiency of membrane clarification of pomegranate juice. *Journal of food science and technology, 58*(5), 1682–1692. https://doi.org/10.1007/s13197-020-04678-x.

Sant'Ana, A. S., Rosenthal, A., & Massaguer, P. R. (2009). Heat resistance and the effects of continuous pasteurization on the inactivation of Byssochlamys fulva ascospores in clarified apple juice. *Journal of Applied Microbiology, 107*(1), 197–209. https://doi.org/10.1111/j.1365-2672.2009.04195.x.

Sarbatly, R., Sariau, J., & Krishnaiah, D. (2023). Recent Developments of Membrane Technology in the Clarification and Concentration of Fruit Juices. *Food Engineering Reviews, 15*(3), 420–437. https://doi.org/10.1007/s12393-023-09346-2.

Satyannarayana, K. V. V., Sandhya Rani, S. L., & Vinoth Kumar, R. (2023). Clarification of citrus fruit juices using microfiltration technique equipped with indigenously developed novel ceramic membrane. *Journal of food science and technology, 60*(7), 2001–2011. https://doi.org/10.1007/s13197-023-05734-y.

Selvaraju, G., & Bakar, N. K. A. (2017). Production of a new industrially viable green-activated carbon from Artocarpus integer fruit processing waste and evaluation of its chemical, morphological and adsorption properties. *Journal of Cleaner Production, 141*, 989–999. https://doi.org/10.1016/j.jclepro.2016.09.056.

Severcan, S. S., Uzal, N., & Kahraman, K. (2020a). Clarification of Apple Juice Using New Generation Nanocomposite Membranes Fabricated with TiO2 and Al2O3 Nanoparticles. *Food and Bioprocess Technology, 13*(3), 391–403. https://doi.org/10.1007/s11947-019-02373-0.

Severcan, S. S., Uzal, N., & Kahraman, K. (2020b). Clarification of pomegranate juice using PSF microfiltration membranes fabricated with nano TiO2 and Al2O3. *Journal of Food Processing and Preservation, 44*(8), e14559. https://doi.org/10.1111/jfpp.14559.

Sharma, S., Biswal, B. K., kumari, D., Bindra, P., Kumar, S., Stobdan, T., & Shanmugam, V. (2018). Ecofriendly Fruit Switches: Graphene Oxide-Based Wrapper for Programmed Fruit Preservative Delivery To Extend Shelf Life. *ACS Applied Materials & Interfaces, 10*(22), 18478–18488. https://doi.org/10.1021/acsami.8b02048.

Sharoba, A. M., & Ramadan, M. F. (2011). Rheological behavior and physiochemical characteristics of goldenberry (Physalis peruviana) juice as affected by enzymatic treatment. *Journal of Food Processing and Preservation, 35*(2), 201–219. https://doi.org/10.1111/j.1745-4549.2009.00471.x.

Sidlagatta, V., Chilukuri, S. V. V., Dasi, S. D., Devana, B., & Rangaswamy, L. (2023). Understanding clarification of sweet orange juice by microfiltration and fouling behaviour. *CIGR, 25*(2).

Singh, R., & Hankins, N. (2016). Introduction to Membrane Processes for Water Treatment. In (pp. 15-52). https://doi.org/10.1016/B978-0-444-63312-5.00002-4

Singh, V., & Das, C. (2021). Clarification of Citrus fruit (Mosambi) juice by hybrid (Pretreatment and Membrane) process. *Materials Today: Proceedings, 47*, 1384–1388. https://doi.org/10.1016/j.matpr.2021.02.435.

Solís-Fuentes, J. A., Galán-Méndez, F., Hernández-Medel, M. d. R., García-Gómez, R. S., Bernal-González, M., Mendoza-Pérez, S., & Durán-Domínguez-de-Bazúa, M. d. C (2019). Effectiveness of bagasse activated carbon in raw cane juice clarification. *Food Bioscience, 32*, 100437. https://doi.org/10.1016/j.fbio.2019.100437.

Soto-Silva, W. E., Nadal-Roig, E., González-Araya, M. C., & Pla-Aragones, L. M. (2016). Operational research models applied to the fresh fruit supply chain. *European Journal of Operational Research, 251*(2), 345–355. https://doi.org/10.1016/j.ejor.2015.08.046.

Spayd, S. E., & Morris, J. R. (1981). Physical and Chemical Characteristics of Puree from Once-over Harvested Strawberries1. *Journal of the American Society for Horticultural Science, 106*(1), 105–109. https://doi.org/10.21273/jashs.106.1.105.

Splittstoesser, D. F., Churey, J. J., & Lee, C. Y. (1994). Growth Characteristics of Aciduric Sporeforming Bacilli Isolated from Fruit Juices. *Journal of Food Protection, 57*(12), 1080–1083. https://doi.org/10.4315/0362-028X-57.12.1080.

Sreedevi, P., Jayachandran, L. E., & Srinivasa Rao, P. (2020). Application of high-pressure processing for extending the shelf life of sugarcane juice under refrigerated conditions. *Journal of Food Processing and Preservation, 44*(12), e14957. https://doi.org/10.1111/jfpp.14957.

Steyn, C. E., Cameron, M., & Witthuhn, R. C. (2011). Occurrence of Alicyclobacillus in the fruit processing environment — A review. *International Journal of Food Microbiology, 147*(1), 1–11. https://doi.org/10.1016/j.ijfoodmicro.2011.03.004.

Su, Y. (2018). Current State-of-the-art Membrane Based Filtration and Separation Technologies. In H. Zhu, & P. Sun (Eds.), *Graphene-based Membranes for Mass Transport Applications* p. 0. The Royal Society of Chemistry https://doi.org/10.1039/9781788013017-00001.

Tajchakavit, S., Boye, J. I., & Couture, R. (2001). Effect of processing on post-bottling haze formation in apple juice. *Food Research International, 34*(5), 415–424. https://doi.org/10.1016/S0963-9969(00)00186-1.

Tayefe, M., Zade, A. N., Hashemi, S. A., & Hasannia, F. (2014). Isolation of Alicyclobacillus acidoterrestris from Commercial Spoiled Apple Juice and Study on Some Influence Parameters on its Growth in Apple Juice. *Egyptian Academic Journal of Biological Sciences, H. Botany, 5*(1), 73–78. https://doi.org/10.21608/eajbsh.2014.16830.

Thompson, S. S., Miller, K. B., Lopez, A. S., & Camu, N. (2012). Cocoa and Coffee. *Food Microbiology* (pp. 881–899) https://doi.org/10.1128/9781555818463.ch35.

Toydemir, G., Capanoglu, E., Gomez Roldan, M. V., de Vos, R. C. H., Boyacioglu, D., Hall, R. D., & Beekwilder, J. (2013). Industrial processing effects on phenolic compounds in sour cherry (Prunus cerasus L.) fruit. *Food Research International, 53*(1), 218–225. https://doi.org/10.1016/j.foodres.2013.04.009.

Tranquillini, R., Scaramuzza, N., & Berni, E. (2017). Occurrence and ecological distribution of Heat Resistant Moulds Spores (HRMS) in raw materials used by food industry and thermal characterization of two Talaromyces isolates. *International Journal of Food Microbiology, 242*, 116–123. https://doi.org/10.1016/j.ijfoodmicro.2016.11.023.

Tylkowski, B., Nowak, M., Tsibranska, I., Trojanowska, A., Marciniak, L., Valls, R. G., Gumi, T., Giamberini, M., & Jastrząb, R. (2017). Concentration and Fractionation of Polyphenols by Membrane Operations. *Curr Pharm Des, 23*(2), 231–241. https://doi.org/10.2174/1381612822666161021124358.

Urošević, T., Povrenović, D., Vukosavljević, P., Urošević, I., & Stevanović, S. (2017). Recent developments in microfiltration and ultrafiltration of fruit juices. *Food and Bioproducts Processing, 106*, 147–161. https://doi.org/10.1016/j.fbp.2017.09.009.

Vaillant, F., Millan, A., Dornier, M., Decloux, M., & Reynes, M. (2001). Strategy for economical optimisation of the clarification of pulpy fruit juices using crossflow microfiltration. *Journal of Food Engineering, 48*(1), 83–90. https://doi.org/10.1016/S0260-8774(00)00152-7.

Vardin, H., & Fenercioğlu, H. (2003). Study on the development of pomegranate juice processing technology: Clarification of pomegranate juice. *Food /Nahrung, 47*(5), 300–303. https://doi.org/10.1002/food.200390070.

Vera, E., Dornier, M., Ruales, J., Vaillant, F., & Reynes, M. (2003). Comparison between different ion exchange resins for the deacidification of passion fruit juice. *Journal of Food Engineering, 57*(2), 199–207. https://doi.org/10.1016/S0260-8774(02)00298-4.

Vieira, A. H., Balthazar, C. F., Guimaraes, J. T., Rocha, R. S., Pagani, M. M., Esmerino, E. A., Silva, M. C., Raices, R. S. L., Tonon, R. V., Cabral, L. M. C., Walter, E. H. M., Freitas, M. Q., & Cruz, A. G. (2020). Advantages of microfiltration processing of goat whey orange juice beverage. *Food Research International, 132*, 109060. https://doi.org/10.1016/j.foodres.2020.109060.

Wakeman, R. (2007). The influence of particle properties on filtration. *Separation and Purification Technology, 58*(2), 234–241. https://doi.org/10.1016/j.seppur.2007.03.018.

Wang, B.-J., Wei, T.-C., & Yu, Z.-R. (2005). Effect of operating temperature on component distribution of West Indian cherry juice in a microfiltration system. *LWT - Food Science and Technology, 38*(6), 683–689. https://doi.org/10.1016/j.lwt.2004.09.002.

Wang, J., Cahyadi, A., Wu, B., Pee, W., Fane, A. G., & Chew, J. W. (2020). The roles of particles in enhancing membrane filtration: A review. *Journal of Membrane Science, 595*, 117570. https://doi.org/10.1016/j.memsci.2019.117570.

Warczok, J., Ferrando, M., López, F., & Güell, C. (2004). Concentration of apple and pear juices by nanofiltration at low pressures. *Journal of Food Engineering, 63*(1), 63–70. https://doi.org/10.1016/S0260-8774(03)00283-8.

Yee, W. P., Nor, Z., Basri, M. S. M., & Roslan, J. (2021). Membrane-based clarification of banana juice: pre-treatment effect on the flux behaviour, fouling mechanism and juice quality attributes. *Food Research, 5*, 57–64. https://doi.org/10.26656/fr.2017.5(S1).046.

Youn, K.-S., Hong, J.-H., Bae, D.-H., Kim, S.-J., & Kim, S.-D. (2004). Effective clarifying process of reconstituted apple juice using membrane filtration with filter-aid pretreatment. *Journal of Membrane Science, 228*(2), 179–186. https://doi.org/10.1016/j.memsci.2003.10.006.

Zarouk, S., Mirsaeedghazi, H., & Massah, J. (2022). The Effect of Magnetic Field on Efficiency and Fouling Mechanisms During Membrane Clarification of Pomegranate Juice. *Iranian Journal of Chemistry and Chemical Engineering, 41*(2), 595–606. https://doi.org/10.30492/ijcce.2020.123797.4057.

Zhong, W., Li, X., Yang, H., & Li, E. (2018). A novel, effective, and feasible method for deacidifying kiwifruit wine by weakly basic ion exchange resins. *Journal of Food Process Engineering, 42*, e12969. https://doi.org/10.1111/jfpe.12969.

Zhu, D., Shen, Y., Wei, L., Xu, L., Cao, X., Liu, H., & Li, J. (2020). Effect of particle size on the stability and flavor of cloudy apple juice. *Food Chemistry, 328*, 126967. https://doi.org/10.1016/j.foodchem.2020.126967.

Zhu, Z., Mhemdi, H., Ding, L., Bals, O., Jaffrin, M. Y., Grimi, N., & Vorobiev, E. (2015). Dead-End Dynamic Ultrafiltration of Juice Expressed from Electroporated Sugar Beets. *Food and Bioprocess Technology, 8*(3), 615–622. https://doi.org/10.1007/s11947-014-1427-2.

Application of filtration in beverage production

12

Ulaş Baysan[a,b], *Özgül Altay*[a], *Özgün Köprüalan Aydın*[a], *Işıl İlter Baysan*[a,c] *and Figen Kaymak Ertekin*[a]
[a] Department of Food Engineering, Faculty of Engineering, Ege University, İzmir, Türkiye, [b] Science and Technology Application and Research Center, Dokuz Eylül University, İzmir, Türkiye, [c] Department of Food Engineering, Faculty of Engineering, Manisa Celal Bayar University, Manisa, Türkiye

12.1 Introduction

Filtration practices are very important in beverage production because this process affects many factors such as sensory properties, microbial stability, and ensuring final product quality, which determines the quality, clarity, hygienic status, and shelf-life of the final product. The filtration process ensures that the beverage is free of unwanted particles and cloudiness. This makes the drink clearer, more visually appealing, and better accepted by the consumer. In beverage production, filtration is used at different stages to control the composition and properties of the product. For example, in fermented beverages, control can be achieved over the final flavor and alcohol content of the product by removing yeast cells after fermentation. Filtration with activated carbon and similar materials removes or reduces unwanted tastes and odors of the beverage. It is especially effective for removing chemicals and organic compounds such as chlorine found in some ingredients such as tap water or raw beverage ingredients (Sutherland, 2008a).

Ensuring microbial safety plays a vital role when evaluating the quality of a food sample. Contamination of microorganisms into the beverage can result in off-flavors, bad odors, and even health risks. Filtration eliminates microorganisms from the beverage, ensuring the microbial stability of the product and prolonging its shelf-life. Thus, it plays a critical role in product quality, safety, and stability besides sterilization and microbial control, which are critical to maintaining consumer health.

Selection and application of the correct filtration methods provide the desired properties of the final product and increase customer satisfaction (Fischer et al., 2006). Filtration applications in the food industry can be categorized into three categories. The initial category involves separating slurries containing significant amounts of insoluble solids. During this process, filtration cake forms on the flow surface. The second category is clarification; when clarification aims to remove small amounts of insoluble solids from valuable liquids, the production of a clear liquid with the removal of unwanted solids is also observed. Here, filtration may result in a cake buildup on the medium or the entrapment of solids within its structure, particularly when the solid content is minimal. The third category, microfiltration, targets the removal of very fine

Mechanical Separation Processes in the Food Industry: Unit Operations and Processing Equipment in the Food Industry.
DOI: https://doi.org/10.1016/B978-0-12-819523-9.00003-2

particles, around 1 μm or less, usually focusing on eliminating microorganisms from liquid foods (Maria Leena et al., 2019).

One unit procedure that is used, is called "solid–liquid filtration," which involves passing a liquid through a septum or porous membrane to remove insoluble solids from a solution. Solid particles are trapped by the membrane either inside its structure or on its upstream surface. The liquid component that passes the membrane is called "filtrate," the suspension is called "feed slurry," and the membrane is called "filter medium." The "filter cake" is the term used to describe the residual solids. Gravity alone can promote the flow of a filtered solution, as can pressure filtration, vacuum filtration, and centrifugal filtration, which uses centrifugal force to create a vacuum downstream of the medium (Cheremisinoff, 2000; Sutherland, 2008a; Svarovsky, 2001a). A different method to change the distribution of particle sizes of a fluid is to incorporate solid material with a certain size distribution, known as filter aids. These compounds are also used as filter media in precoat filtering processes. Filter aids must meet particular requirements, such as proper particle size and shape characteristics to achieve the necessary cake permeability, chemical impermeability during the procedure, and affordability. In this chapter, the importance of the filtration process in the beverage industry is discussed. Additionally, various types of filtrations used in the industry, influencing factors, and equipment used in the industry have been examined.

12.2 Types of mechanical filtration

As seen in the processing of wine, beer, juices, oils, and syrups, filtration is a commonly used technique for clarifying liquids by removing tiny solid particles (Fellows, 2009). The initial particles in a suspension get trapped in the filter media during its passage through the filter, reducing the amount of space available for liquid flow. As a result, fluid flow resistance increases, and a larger pressure difference is required to maintain the filtrate flowrate. Choosing a filter medium that allows liquids to pass through while holding particles in place is essential to any type of filtration, whether it is done under pressure or under vacuum. One can use pressure, vacuum, or gravity as the driving force. Filtration assumes a pivotal role in reducing substantial liquid volumes to more manageable proportions in beverage production. The driving force prompting liquid separation from solids can be categorized into four groups. Filtration operations can be conducted either in batches (discontinuous) or continuously, with equipment applying pressure or vacuum to the filter bed.

A. **Gravity filtration** is relatively slow, and its application in the food industry is restricted. Filtration aids, such as perlite, cellulose, activated carbon, bone char, pumice, bentonites, bleaching clay, and diatomaceous earth are used to enhance the formation of filter cake and improve filtration efficiency. These aids are chemically inert powders, each serving specific purposes like flocculation, adsorption, and stabilization to prevent hazes in various beverages like beer, wine, and juices. Overall, filtration plays a vital role in processing various beverages, such as beer, cider, fruit juices, and wine, among others, ensuring their quality, stability, and clarity (Perlmutter, 2015).

B. **Vacuum filtration** stands out as one of the simplest mechanical driving forces used in beverage production. Typically operating continuously, these filters offer precise control over cake thickness and are versatile in terms of material construction. Vacuum filters excel in solid washing capabilities, with rotary drum and rotary disc filters being two common types. Rotary drum filters consist of a horizontal cylinder that is partitioned into compartments and enveloped in filter cloth. These compartments are linked to a central vacuum pump. As the drum rotates through a liquid bath, the filtrate permeates through the filter cloth and exits through channels within the drum. The filter cake is then removed, washed, and released from the vacuum, with compressed air aiding in cake removal before the cycle restarts. Similarly, rotary vacuum disc filters consist of vertical discs that rotate within a liquid bath. Each disc is segmented into sections with outlets leading to a central shaft. Scrapers are employed to continually remove the cake formed on the discs. These filters boast large filter areas and high capacities, albeit with moderately high moisture content in the produced cake. Although they require significant initial investment, vacuum filters offer compactness and low labor costs. With or without compression, the exertion of this force necessitates mechanical limitations. Pressure filters typically operate in batches, with the exception of the rotary pressure filter. Cake-washing capability can be exceptional, and the resulting cakes are usually as dry as possible without the introduction of heat (Svarovsky, 2001c, b).

C. **Pressure filtration:** Beverage production commonly employs three types of pressure filters: the plate-and-frame filter press, the shell-and-leaf pressure filter, and the rotary filter press. In the plate-and-frame design, filters made from materials like woven nylon, polypropylene, cloth, or paper are supported on vertical plates. These filters form chambers when pressed together, with each chamber separated by a filter cloth acting as a gasket. Liquor is pumped into the press, and pressure gradually increases until it reaches a predetermined value. Filtration occurs due to the pressure difference between the feed pressure and gravity discharge. Filtration improves as a layer of solid particles builds up on the filter cloths. Following the separation of the chambers, gravity releases the filter cakes. The mesh leaves of the shell-and-leaf pressure filter are piled inside a pressure vessel and coated in filter media. The pressure at which feed liquor is injected into the shell is roughly 400 kPa. The greatest use case for this sort of filter is regular filtration of comparable liquors. The filtrate passes through the screens in the rotary filter press because the suspension is supplied between two parallel rotating stainless-steel screens. The outlet restriction results in the extrusion of a dry cake. This type of press offers continuous operation, low-pressure feed, and uniform cake dryness, with advantages in startup/shutdown procedures, wash cycle duration, maintenance, and operating costs. Overall, these pressure filters play a crucial role in beverage production, offering efficient filtration with various features suited to different applications and operational needs (Sutherland, 2008a; Svarovsky, 2001b).

D. **Centrifugal filtration** presents a middle ground between vacuum and pressure filtration, offering both continuous operation and automated batch mode capabilities. The effectiveness of centrifugal filters hinges on the characteristics and behavior of the solids being filtered, with the efficiency of cake washing varying accordingly. Centrifugal forces acting on the fluid promote filtration in centrifugal filters. These filters have a revolving basket with a filter medium inside of it. Centrifugal filtration does not require a density difference between the particulates and the suspending liquid, compared to other filtration techniques. Sedimentation happens in the liquid head above the cake at the beginning of the filtration cycle if such a difference is present. As a result, the cake may experience particle size stratification, with coarser particles positioned closer to the filter media and serving as a precoat for finer particles. A notable aspect of centrifugal filters is the additional pressure from the centrifugal force, caused by the liquid layer atop the cake. This, combined with

Table 12.1 Discontinuous and continuous system filtration types.

Discontinuous filters	Continuous filters
Plate and Frame Filter Press	Rotary Pressure Filter
Horizontal Pressure Leaf Filter	Disc Filter
Membrane Press	Rotary Drum Vacuum Filter
Nutsche Filter/Agitated Nutsche Filter	Pressurized Drum Filter (P-DF)
Tubular Filter	Vacuum Belt Filters
Cartridge Filter	Pan Filter
Bag Filter	Screw Press
Candle Filter	Belt Press
Pressure Plate Filter	
Combination Filter PressMembrane Press	
Tower Press	

centrifugal forces, contributes to the filtration. Overall, centrifugal filters offer unique advantages and considerations compared to other filtration methods, depending on the specific characteristics of the solids and liquids being processed (Svarovsky, 2001a).

In discontinuous (batch) filtration systems, various methods are utilized for cake discharge, each offering unique benefits. These systems offer diverse solutions for cake discharge in batch filtration, each with its advantages and applications. Continuous filtration systems offer ongoing operation for pressure filtration, cake washing, and drying of slurries. These continuous filtration systems offer various options for ongoing operation, providing efficient cake washing, drying, and discharge for different applications and scale requirements (Sutherland, 2008a; Svarovsky, 2001b).

Discontinuous and continuous system filtration types are summarized in Table 12.1.

12.3 Factors affecting mechanical filtration

Problems related to the separation of solids from liquids can be addressed through various processes depending on the properties of the solids and the solid–liquid ratio in the mixture. If the quantity of solids is significantly lower than that of the liquid, the filtration process is applied.

Filtration involves the separation of insoluble substances from a liquid by passing them through a porous medium. This porous medium, known as the filter chamber, serves to separate the solid and liquid components of the mixture. As the liquid passes through, solid material accumulates on the filter medium, forming what is known as the filter cake. The resulting liquid, now free of solid particles, is called the filtrate. Filtration methods include constant pressure filtration, constant-rate filtration, and variable rate-variable pressure filtration. The filtration rate can be influenced by various factors such as particle size, shape of particles, amount of solid material, temperature, pressure, and filtration area. Filtration processes applied in the food industry can vary from simple straining processes to advanced separation processes. One phase may be liquid or gas, while the other may be solid or semi-solid and the solid materials may be coarse or very fine, hard or soft, the mixture may be hot or cold, and the filtration process may be performed under vacuum or pressure. The substance to be filtered may be in granular form, easily filterable or exhibit adhesive properties, may block the filter medium, or may be present in compressible or non-compressible colloidal forms. In filtration processes applied in the food industry—such as in the production of olive oil, sunflower oil, and fish oil following pressing, in the removal of fine particles causing turbidity in wine and fruit juice production, or in the separation of cream from milk in dairy production—partial or complete separation may be desired, depending on the nature of the filtrate and the purpose of filtration (Perry & Green, 1984).

Various factors affect filtration positively or negatively. These factors include clarity of the filtrate, filtration rate, cake thickness, and compressibility of the cake. The size of the particles to be separated plays a significant role in filtration. Coarse particles that are resistant to compression make filtration easier. Separating fine particles, unlike coarse ones, can be very challenging. Fine particles create a dense cake on the filter medium with no pores between the particles, resulting in a very slow filtration process. The crystalline or amorphous structure of the particles is also crucial. Crystalline particles, if hard and angular, form a coarse, porous filter cake on the filter medium that easily allows the filtrate to pass through. However, flat- or leaf-shaped crystals make filtration difficult. Viscous substances quickly clog the pores of the filter medium and form a thin layer on the filter cake that does not allow the filtrate to pass through easily. In such cases, processes like heating or adding hydrophilic gels or electrolytes to facilitate coagulation are carried out to transform fine particles into coarse particles that can be filtered more easily. If it is not possible to retain or coagulate very fine colloid particles with the filter medium, it is necessary to add filtration aids to the feed to facilitate filtration. The amount of solid particles in the feed is important in terms of filtration capacity. It is difficult to filter feeds containing excessive amounts of solid material. Therefore, it is necessary to subject some of the solid materials in the feed to sedimentation or perform coarse pre-filtration. The fluidity of the feed increases with increasing temperature. In aqueous suspensions, increasing the temperature from 10°C to 40°C doubles the filtration rate, and increasing it from 10°C to 70°C triples the filtration rate.

Pressure is applied by increasing the thickness of the feed on the filter medium, working under pressure, or applying a vacuum from the reverse side. Increasing pressure increases the filtration rate. However, maintaining pressure too high can have the opposite effect in some cases. This is more important for cakes that lose porosity due to compression. In such cases, it is not advisable to work at a certain pressure. The same result can be achieved by working under a vacuum instead of pressure (Sutherland, 2007).

12.3.1 Filtration media

To achieve effective filtration, the pores in the filter medium used should be smaller than the size of the solid particles present in the liquid. The type of filter medium used varies depending on the intended purpose, as the filter medium should not have a detrimental effect on the structure of the liquid to be filtered. Additionally, it should be economically feasible. Filter media such as perforated metal sheets, wire mesh, ceramic screens, natural and synthetic fiber weaves, paper, felt, rubber, and leather-like felts, loose fillers such as coke, sand, wood shavings, diatomaceous earth, asbestos, and coal, porous materials made of ceramics, glass, synthetic materials, metal, and rubber, leather, or membrane are used. Filters made of perforated metal (stainless steel) can be removed and cleaned, and they can capture solid and semi-solid phases in the liquid phase that passes through them. In the preliminary filtration of raw milk in the milk intake line and the retention of large-sized foreign materials such as stones, straw, insects, and similar coarse particles, filters made of wire, cloth, and perforated metal sheets are used (Bennett, 2015).

12.3.2 Type of filter medium

A layer of solid particles builds up on top of the filter medium during cake filtration, and the cake serves as the main support for the filter media. These filters, often known as "surface filters," work especially well with suspensions that have a high solid component (Svarovsky, 2001a). There are four primary types of filtrations when it comes to the techniques used to create a pressure drop across the filter media. While gravity static head is not commonly used, gravity filtration, which depends on Earth's fundamental gravitational pull, is usually used for initial dewatering. To speed up the filtration process, vacuum filtration employs suction on the filtrate side of the medium. Pressure filtration drives filtering by using liquid pressure, which is produced by pumping or gas pressure in the feed vessel. Centrifugal filtration uses a spinning perforated basket with a filter material or equivalent material. The centrifugal force creates a pressure decrease through the filter medium, removing liquid from both the cake and the medium itself (Svarovsky, 2001a). Cake filtration occurs when solids accumulate on top of the filter material. Filtration commences after the primary cake layer forms, with the filter mostly providing support. These filters, called "surface filters," are appropriate for suspensions with a high solid component (Svarovsky, 2001a).

Four types of filtrations achieve pressure drop over the filter medium. Gravity filtering uses the Earth's natural gravitational force and is primarily utilized for early dewatering, while gravity static head is less widely used. Vacuum filtration uses suction on the filtrate side of the medium to propel the filtering process. Pressure filtration uses liquid pressure from pumping or gas pressure in the feed vessel to drive filtering. Centrifugal filtration uses a rotating perforated basket with a filter material or equivalent material. The centrifugal field's movement creates a pressure drop through the filter medium, removing liquid from both the cake and the medium (Svarovsky, 2001a).

12.3.3 Structure of filter cake

During filtration, a complex filter cake forms with an inhomogeneous, multilayered structure characterized by a coarse lower part and a finer upper region. During the cake-washing process, the height of the cake varies, especially as fine particles infiltrate. This infiltration leads to a denser structure and a decrease in height, consequently reducing the flux. Raking machines are used to maintain a constant flux by breaking the filter bed and decreasing resistance, albeit at the cost of allowing small particles to pass through, leading to a turbid filtrate. The standard lautering process usually spans a duration of 90 to 180 minutes without encountering any plugging issues (Tippmann et al., 2010). These changes, often sudden decreases or increases, are attributed to alterations in the structure of the filter cake (Chellappah et al., 2010). Additional explanations comprise channeling, which arises from the formation of channels during batch sedimentation, causing an escalation in settling rate over time; wall effects, resulting from solid-wall friction that diminishes effective pressure near the wall, thereby creating a more open cake structure; and compression/binding interaction, where flow resistance escalates due to the deposition of small-scale solids within the cake, leading to reduced porosity and heightened flow resistance (Sorensen et al., 1995). Factors affecting cake structure include suspension pH and filter area. A decrease in suspension pH generally increases the filtration rate, with lower pH values leading to larger particles and a higher tendency for sudden changes in cake structure. Filter area correlates with cake structure, with smaller filter cells promoting a more robust cake construction that resists abrupt changes, while larger cell walls offer less support to cake formation (Tarleton & Morgan, 2001).

12.3.4 *Filter aid materials*

One of the most significant operational challenges during filtration is the clogging of filter medium pores, leading to a slowdown in filtration rate. Another issue encountered during filtration is the failure to achieve the desired clarity in the filtrate. To address these problems, filter aid materials are employed, which are finely powdered, non-compressible substances. The primary function of filter aid materials is to create a porous layer on the filter medium. These pores are so small that they trap colloidal particles in the feed, preventing them from clogging the pores of the filter medium and prolonging the filtration time. Filter aid materials should easily mix with the liquid to be filtered and remain in suspension throughout the filtration process. They should not contain soluble salts, organic matter, or substances that can alter the properties of the liquid being filtered. The layer formed by the filter aid material on the filter medium should not compress under the pressure applied during filtration and should maintain its permeability for an extended period. The particles of the filter aid material should be observable under a microscope, vary in size and shape, and be porous. Only then can the filter cake formed by the filter aid material have high permeability, low resistance to filtrate flow, and maximum solid retention. The most commonly used filter aid material is diatomaceous earth (kieselguhr), which consists of the siliceous skeletons of diatoms, single-celled algae that lived in stagnant waters during the Miocene epoch about 5 million years ago. Diatomaceous earth particles are nearly pure silicon dioxide and are so small that they can only be easily seen under a microscope. They provide extensive surfaces for the adsorption of colloids and vary in size and shape, ranging from three to 100 microns. Another filter aid material is perlite, an amorphous aluminum silicate of volcanic origin, composed of small, spherical, glassy particles. Perlite can be used in liquids with a pH ranging from four to nine, but it is less effective than kieselguhr in terms of permeability and solid retention. Cellulose, derived from beech or pine wood pulp, is also used as a filter aid material. It is recommended to use cellulose fibers in combination with kieselguhr. Other materials such as fibers, asbestos, and activated carbon, among others, are also used as filter aid materials. Filter aid materials can be incorporated into the mixture to be filtered in three ways. Firstly, the filter surface can be coated with these materials to prevent colloidal particles in the feed from clogging the pores of the filter medium and to facilitate the removal of the filter cake after filtration. Secondly, the filter aid materials can be thoroughly mixed with the substance to be filtered in a specific percentage, improving the permeability of the sediment settling on the filter surface and reducing its compressibility, thereby preventing an increase in filter cake resistance during filtration. Thirdly, precoated filters with these aid materials can be used, particularly in vacuum-operated rotary cylindrical filters, when filter cake removal is necessary. Separating the sediment from the filter aid material can be achieved through chemical methods without affecting the chemical properties of the feed, such as its pH and color (Girard & Fukumoto, 2000; Johnston, 1988).

12.3.5 Pressure drop in filtration

During the filtration process, the liquid to be purified encounters three different resistance forces:

a) Resistance is caused by the channels and pores of the filtration device.
b) Resistance is offered by the filtration medium to the flow.
c) Resistance of the filter cake.

The layer formed by the solid particles removed from the liquids creates an increasing resistance to liquid flow for the filtration process. The resistance, different from that of the filtration medium, is termed "filter cake resistance." The diagram of pressure application in the filtration process is shown in Fig. 12.1. Initially zero, the filter cake resistance increases during filtration and reaches its maximum at the end of the process. At the beginning of filtration, solid particles in the mixture fill the pores of the filter, increasing the resistance of the filter medium to the flow of filtrate. The filtration medium refers to the layer of initial sediment particles accumulated on the filter's surface. In theoretical calculations of filtration, the resistances created by the filtration device are generally neglected. Thus, the total resistance (total pressure drop) at any point in filtration is the sum of the resistance of the filter cake and the resistance of the filtration medium.

The primary variable controlled in the filtration of a mixture is the total pressure drop. If the pressure drop is kept constant, the flow rate, initially maximum at the beginning of filtration, gradually decreases. This type of filtration is called constant pressure filtration. On the other hand, if the pressure drop varies, increasing as filtration progresses and reaching a maximum at the end of filtration, and the inlet pressure is continuously increased, it is referred to as constant-rate filtration. There are limitations to constant pressure filtration. High pressure can lead to clogging of the pores by

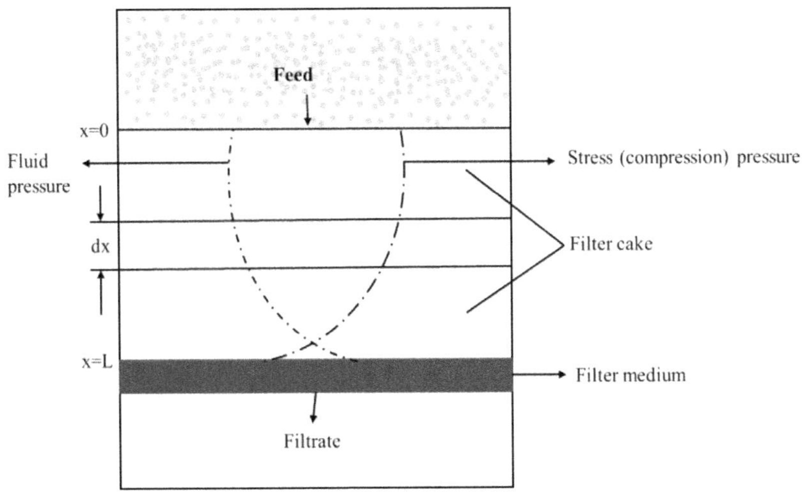

Figure 12.1 Schematic representation of pressure application in the filtration process.

sediment particles, resulting in a low filtrate flow rate for the remainder of the filtration. Additionally, if sediment particles do not have a homogeneous size distribution (a mix of crystal and colloidal particles), high pressure forces the colloidal portion of the sediment to fill the gaps between the initial crystal layers, significantly reducing the filtrate flow rate. Conversely, if the initial pressure is kept low, the first layer of sediment formed on the filtration medium is loose and porous, allowing for a high filtrate flow rate. In this case, the sediment layer cannot fill the holes in the filtration medium, allowing for easy removal of the formed cake from the filtration medium. A low initial pressure results in the initial filtrate being turbid, but this turbidity quickly dissipates, resulting in a high filtration rate and thus a large filtration capacity (Fillaudeau et al., 2007; Fu et al., 2021a).

12.4 Filtration in the beverage industry

The beverage industry is a highly competitive market that demands the highest quality products to satisfy consumer preferences. One essential process in this industry is filtration, which plays a crucial role in ensuring the purity and clarity of beverages. Mechanical filtration is a widely used method in the beverage industry to remove unwanted particles and impurities from liquids. It involves the use of physical barriers, such as screens or membranes, to separate solid particles from liquids based on their size and shape (Sutherland, 2007). One of the main advantages of using mechanical filtration in the beverage industry is the ability to achieve a high level of product clarity. By implementing robust mechanical filtration processes, beverage manufacturers can confidently deliver safe and hygienic products to their consumers. Moreover, mechanical filtration offers versatility in terms of the range of applications it can be used for in the beverage industry. From large-scale production facilities to small craft breweries, mechanical filtration systems can be tailored to meet specific production needs. Whether it is clarifying juices, filtering beer, or removing sediment in wine production, there are various types of mechanical filters available to suit different requirements. This flexibility allows beverage manufacturers to optimize their production processes and achieve desired outcomes efficiently (Sutherland, 2007, 2008a; Werner et al., 2020). Mechanical filtration is an essential process in the beverage industry that helps ensure product clarity, taste consistency, and safety. By effectively removing particles, impurities, and microorganisms, filtration systems play a crucial role in delivering high-quality beverages to consumers. With its versatility and ability to meet specific production needs, mechanical filtration has become an indispensable tool for beverage manufacturers worldwide (Cheremisinoff, 2000).

12.4.1 Alcoholic beverages

Beer production is a complex process that involves several stages, including fermentation, filtration, and packaging. One important step in this process is the use of mechanical filtration pressure or/and vacuum filtration (cartridge and vacuum disc filters) methods are the most commonly used (Santos et al., 2017; Slabý et al., 2018). The first stage in beer production is the fermentation process, where yeast converts sugars into alcohol and carbon dioxide. After fermentation, the beer is often cloudy due to the presence of yeast cells, proteins, and other solids. Mechanical filtration is used to remove these particles, resulting in a clear and visually appealing product. Mechanical filtration has several benefits in beer production (Fillaudeau et al., 2007). Firstly, it improves the appearance of the beer by removing any visible particles and sediment. This is especially important for certain beer styles, such as lagers, where clarity is highly valued. Second, it enhances the stability of the beer by removing substances that can cause haze or spoilage over time. This ensures that the beer maintains its quality and flavor for a longer period. However, it is important to note that while mechanical filtration can be effective in achieving clarity and stability, it can also remove some desirable components from the beer like certain proteins and yeast compounds that can contribute to the flavor and mouthfeel of the beer. Therefore, brewers must carefully consider the balance between clarity and flavor when deciding on the extent of filtration to be used (Hermia & Brocheton, 1994). In addition to their role in improving beer clarity and stability, mechanical filtration methods also contribute to the overall efficiency of the brewing process. By removing impurities that could potentially clog equipment or cause off-flavors, these methods help to maintain consistent production rates and reduce downtime. Furthermore, by ensuring that the final product meets quality standards, mechanical filtration methods help to minimize waste and improve the overall yield (Aadil et al., 2019).

The **wine production process** is multifaceted and comprises several steps, including the growing of grapes, the fermentation of grape juice, and the aging and bottling of the finished product. The production of wine dates back thousands of years and has evolved significantly over time, with different regions and cultures developing their unique methods and traditions. Today, wine production is a multi-billion-dollar industry that spans the globe, with countless vineyards and wineries producing a wide variety of wines to cater to diverse tastes and preferences (Grainger & Tattersall, 2005). One important step in wine production is the filtration process, which helps remove any impurities and unwanted particles from the wine. Mechanical filtration is a commonly used method in the wine industry to achieve this. In addition to pressure or/and vacuum filtration (cartridge filters, vacuum disc filters, filter presses, and rotary vacuum drum filters) used in beer production, the centrifugal filtration method is preferred in wine production. These mechanical filtration methods are highly effective in removing larger particles, such as grape skins, seeds, and stems, as well as any residual yeast or bacteria that may be present in the wine (Rayess et al., 2016). By removing these impurities, mechanical filtration helps improve the clarity and stability of the wine. Among several advantages of using mechanical filtration in wine production is a relatively quick and efficient method, in which wine can be passed through the filter at a controlled rate, allowing for thorough removal of impurities without affecting the overall quality of the wine. Additionally, mechanical filtration is a non-chemical process, meaning it does not introduce any foreign substances into the wine. This is important to maintain the natural flavors and aromas of the wine. However, in some cases, excessive filtration can even strip away desirable components from the wine, resulting in a loss of complexity and character of the wine. To overcome these limitations, winemakers often use a combination of filtration methods (Buzrul, 2012).

To achieve the right taste and quality, the production of spirits is a multi-step, intricate process. Mechanical filtration methods have gained popularity in recent years due to their efficiency and effectiveness in producing high-quality spirits. Vacuum filtration (vacuum disc filters), the mechanical filtration method commonly used to produce this beverage, utilizes physical barriers to separate impurities from the liquid, resulting in a cleaner and more refined product (Corbion et al., 2023). One of the advantages of using mechanical filtration methods in spirits production is their ability to remove both visible and invisible impurities. These impurities can include solids, such as fruit pulp or sediment, as well as dissolved compounds that affect the taste and aroma of the spirit. By employing mechanical filters, distillers can ensure that their final product is free from any unwanted substances, resulting in a more consistent and enjoyable drinking experience for consumers (Corbion et al., 2023; Pielech-Przybylska & Balcerek, 2019). Another benefit of mechanical filtration is its efficiency in large-scale production. With the use of automated filtration systems, distillers can process a significant amount of spirit within a short period (Palmer, 2015). This helps improve productivity and reduces the time required for the production process. Additionally, mechanical filtration methods can be easily customized to meet specific requirements, allowing distillers to achieve their desired level of clarity and purity in their spirits. Furthermore, mechanical filtration methods offer a cost-effective solution compared to other filtration techniques (Fellows, 2009). Traditional methods such as chill filtration or charcoal filtering can be time-consuming and require additional equipment. In contrast, mechanical filters are relatively simple to operate and maintain, making them a more economical choice for distilleries. However, it is important to note that while mechanical filtration can improve the quality of spirits, it should be used judiciously. Over-filtering can lead to a loss of flavor and character in the spirit, resulting in a bland and uninspiring final product. Distillers must strike a balance between removing impurities and preserving the unique characteristics that make their spirits stand out (Palmer, 2015).

12.4.2 Non-alcoholic beverages

Fruit juice production involves the extraction of juice from fruits, which can be used as a refreshing and healthy beverage. To ensure the highest quality of fruit juice, various methods are employed, including mechanical filtration, which is a process that removes solid particles and impurities from the juice with filters or sieves. It is effective in improving the clarity, taste, and shelf life of the fruit juice. In fruit juice production, the selection and preparation of the fruits are the first steps (Lu et al., 2021; Sellick, 1993). After the fruits are carefully chosen based on their ripeness and quality, they are washed and sorted to remove any dirt or damaged parts and, they are crushed or pressed to extract the juice. The extracted juice contains solid particles, such as pulp and seeds, as well as other impurities. To remove these solid particles and impurities, mechanical filtration methods are employed. This involves passing the juice through filters or sieves that have different mesh sizes. The size of the mesh determines the size of particles that can be removed. For example, a fine mesh filter can remove smaller particles, while a coarser mesh filter can remove larger particles (Echavarría et al., 2011). In fruit juice production, vacuum filtration (drum filters, vacuum belt filters) and centrifugal filtration are commonly used as mechanical filtration methods. In addition to removing solid particles, filtration also helps improve the clarity and taste of the fruit juice. By removing impurities, the juice becomes clearer and more visually appealing. It also enhances the flavor by eliminating any unwanted tastes or odors caused by impurities (Girard & Fukumoto, 2000; Ilame & Singh, 2015).

With the increasing demand for clean and safe drinking water, the production of bottled water, consisting of purification and packaging steps has become a thriving industry (Wang et al., 2008). One of the key aspects of bottled water production is the use of mechanical filtration methods to remove impurities and contaminants from the water, ensuring that it meets the required quality standards. Mechanical filtration methods play a crucial role in ensuring the purity and safety of bottled water which involves the use of filters and membranes to remove solid particles, sediments, and other impurities from the water. The filters used in mechanical filtration can be made of different materials such as sand, activated carbon, or ceramic, depending on the specific requirements of the water treatment process. These filters effectively trap contaminants and prevent them from passing through, thereby improving the quality and clarity of the water. Pressure filtration (cartridge filters, filter presses), vacuum filtration (bag filters), and centrifugal filtration are commonly used mechanical filtration methods in bottled water production (Brandt et al., 2017; Berk, 2018; Wang et al., 2008).

Dairy-based beverages have become increasingly popular in recent years, with a wide range of options available on the market. These beverages are made by combining dairy products, such as milk or yogurt, with various flavorings, sweeteners, and additives. The production process for dairy-based beverages involves several steps, including pasteurization, homogenization, and filtration (Richert, 1975). Mechanical filtration methods play a crucial role in ensuring the quality and safety of these beverages. The first step in the production of dairy-based beverages is pasteurization, which involves heating the dairy product to a specific temperature for a predetermined amount of time to kill any harmful bacteria that may be present. Pasteurization helps to extend the shelf life of the beverage and reduce the risk of foodborne illnesses. Mechanical filtration methods are often used during the step to remove any impurities or foreign particles that may be present in the dairy product. After pasteurization, the dairy product is homogenized which involves breaking down the fat particles in the milk or yogurt to create a smooth and uniform texture. Homogenization helps to prevent the separation of fat from the rest of the beverage and enhances its overall consistency (Tetra Pak, 1995). Mechanical filtration may also be employed during this step to further refine the texture of the dairy-based beverage. Once the dairy product has been pasteurized and homogenized, it is ready for flavorings, sweeteners, and additives that can be added. These ingredients are carefully selected to enhance the taste, aroma, and nutritional profile of the beverage. Mechanical filtration is often used during this stage to remove any solid particles or sediments that may be present in the flavorings or additives (Svarovsky, 2001a). Centrifugal filtration is the most important method applied for dairy-based beverage production to produce both delicious and visually appealing product (Lin & Wu, 2022; Parisher & Rhea, 2021).

12.5 Filtration equipment used in the beverage industry

The beverage industry requires dependable and effective filtration equipment to maintain the quality and consistency of its products. In the beverage industry, centrifugal, vacuum, and pressure filters are frequently utilized for this purpose. Manufacturers differentiate their machines based on slight design changes or unique characteristics within each filter type, which makes them ideal for a certain beverage industry application (Sutherland, 2008b, 2010). Through commercial selection, these filtering equipment alternatives have developed throughout time, providing a wide range of manufacturers and filter types to select from (Sparks, 2012). This makes it possible for beverage producers to identify the ideal filtering apparatus to fulfill their unique needs, whether it is the elimination of solid contaminants, the clarification of liquids, or the enhancement of flavor and appearance. Filter components used in mechanical filters are often composed of polymers, metals, ceramics, or other unique materials. By removing different-sized particles from input liquids, they enhance the quality of the final product throughout the beverage production process. For instance, drinking water becomes drinkable when dirt, sand, and other particles are removed from it via filters used in water purification systems (Sutherland, 2008c; Werner et al., 2020).

To fulfill regulatory requirements and preserve product quality, the beverage industry depends heavily on filtration. The kind of beverage being processed, the required level of purity or clarity, flow rate, and capacity requirements, and the entire production process all influence the choice of filtering equipment. The particular needs of the product being processed must be taken into account when selecting filtering equipment for the beverage industry. For instance, many steps of filtering are used in the brewing industry to remove particles and pollutants that might alter the flavor and appearance of beer (Svarovsky, 1990).

In the beverage industry, pressure filters are frequently used to clear liquids and eliminate solid contaminants. These filters use pressure to push liquid through a filter medium, where they collect particles and impurities (Brandt et al., 2017). Vacuum filters, on the other hand, are ideal for delicate items like wine and juice that need little filtration to retain their original flavor and fragrance. Filters accomplish this by producing a pressure differential that draws liquid through the filter medium (Svarovsky, 2001c). Centrifugal filters are perfect for high-flow circumstances in which quick and efficient filtering is needed because they employ centrifugal force to separate liquids from solids. With the development of commercial alternatives throughout time, these filtering equipment options have become more diverse, offering beverage producers the ability to provide customized solutions with unique designs and functionality (Svarovsky, 2001a). In the beverage industry, filtering equipment plays a crucial role in upholding product quality and adhering to legal requirements. Various filter types are chosen based on the properties of the beverage being processed, including its degree of clarity, flow rate, and overall manufacturing process. Because of this variety, beverage producers may select the filtering apparatus that best suits their needs. At this stage, enhancing the filtration processes' customization and enhancement is crucial to improving the beverage industry's competitive advantage.

12.5.1 Pressure filters

To filter the liquid and get rid of undesirable particles, a typical mechanical filter used in the beverage industry is the pressure filters that utilize a certain amount of pressure. Pressure filters provide quick and efficient filtration, which makes them particularly popular in large-scale manufacturing lines. In pressure filters, the driving force for filtration is usually the liquid pressure developed by pumping or by the force of gas pressure in the suspension feed vessel (Svarovsky, 1990). To enhance the purity of the liquid used to produce juice, for instance, a pressure filter in a beverage manufacturing plant can separate fruit particles. Additionally, it eliminates solid particles from sugar syrups and other beverage components, enhancing the consistency and quality of the final product (Perlmutter, 2015; Sutherland, 2008c).

The level of pressure at which the liquid is permitted to flow is controlled by a pressure regulator, and pressure filters frequently include customized filter components. During the filtering process, the required separation level can be attained in this manner. Because of their great effectiveness and little maintenance needs, pressure filters are preferred in areas where they are utilized in the industrial manufacture of beverages (Perlmutter, 2015). Fig. 12.2 shows the industrial and schematic view of pressure filter equipment.

Cartridge filters: The filter material in cartridge filters allows for the mechanical separation and cleaning of solid particles from liquids. A cylindrical cartridge with one or more layers of filter material is what makes up a cartridge filter. Filter media is often composed of polypropylene, polyester, glass fiber, or other materials and has a fine structure at the micron scale. The liquid approaches the filter at the start of

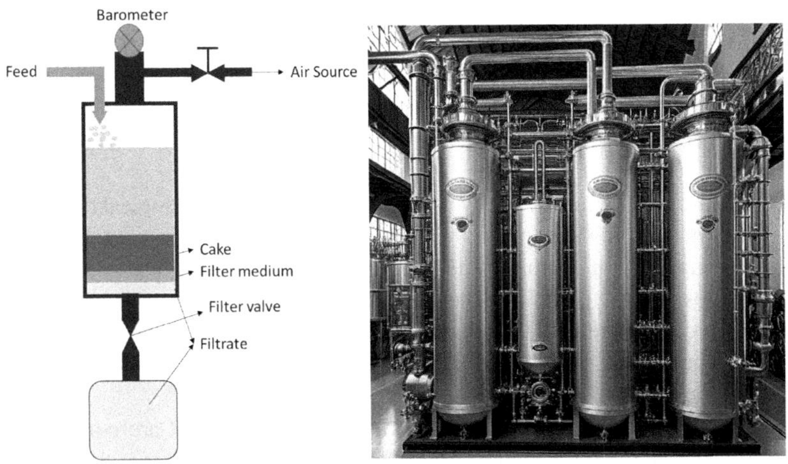

Figure 12.2 Pressure filter equipment.

the procedure. The filter media's holes or pores allow particles to be present in the liquid to be blocked. The cleansed liquid passes through the filter while these particles are trapped in the filter medium. The selection of filter media material is contingent upon the specific needs of the application as well as the characteristics of the liquid. In general, these filters are simple to maintain and may be easily replaced. After a while, dirty cartridges are swapped out for new ones to keep the filter functioning as intended. Pleated cartridge filters offer more surface area and last longer and more effectively than bulk cartridge filters, which are commonly used to separate solid particles from liquid. In the food and beverage business, they can be utilized in the manufacturing of drinking water, beverages, beer, and wine (Palacios et al., 2002).

Drum filters: Large filtering devices like drum filters are frequently employed on an industrial scale. Typically, the drum is shaped like a revolving cylinder, and when the liquid flows across its inner surface, particles are filtered out. While the filtering process continues, the drum is cleaned thanks to the cleaning mechanisms, and continuous filtration is provided. A drum filter is used in the fruit juice manufacturing process, for instance, to remove solid particles and fruit particles from raw fruit juice that contains fruit pulps. Juice quality and clarity are enhanced as a result (Cheremisinoff, 2017).

Plate filters: A sequence of plates and frames make up the construction of plate filters. The particles are held within the plates while the liquid moves through the spaces between them. The pressure between the plates ensures that the particles adhere to the filter material. These filters are typically employed in beverage manufacturing and industrial operations (Perlmutter, 2014). Pressure plate filters are used to remove undissolved solid particles from sugar syrups before they are employed in the manufacturing of alcoholic beverages. Alcoholic beverage uniformity and quality are guaranteed in this technique (Hermia & Brocheton, 1994).

Filter presses: Pressurized presses are used by filter presses to carry out the filtering process. As the liquid moves through the press, the filter material separates it. These filters can operate under high pressure and separate the liquid from the filter material. Chemical operations and industrial wastewater treatment facilities frequently employ these kinds of filters (Berk, 2018; Kinnarinen et al., 2013). Pressurized filter presses are used in wine production to separate the liquid from the grape pomace. As a result, the wines become more transparent, improving their visual appeal and quality (Buzrul, 2012).

***Disc filters*:** Disc filters work by use of revolving disc technological devices. As the liquid moves through the gaps between the discs, particles are trapped. Disc filters are typically employed in chemical and industrial operations. The pressure between the discs ensures that the particles adhere to the disc surface. Pressurized disc filters can be used to remove the carbon dioxide gas from the liquid in carbonated drinks used in their manufacturing. By keeping the beverage's sparkling character, this raises the quality of the final product (Mäkinen, 2002).

Bag Filters: The unique bag-shaped filter components that are contained in bag filters allow them to filter the liquid. Particles are caught in the bags as the liquid moves through the gaps between the bag filters. The filtering of industrial liquids frequently uses these filters. A bag of filter material attached to the end of a pipe can serve as a basic bag filter. The bag can be positioned inside an enclosure, such as a mesh basket, for applications requiring higher pressure and more volume. By capturing different particles and contaminants in the water, pressurized bag filters, which are used in water treatment facilities to generate drinking water, aid in the production of clean drinking water (Perlmutter, 2015).

12.5.2 Vacuum filters

Vacuum filters are an efficient filtering technology for separating particles in liquids. These filters are essential to the beverage industry's efforts to improve clarity and maximize product quality. Vacuum filters, as opposed to pressure filters, draw the liquid's particles toward the filter material, guaranteeing a clean liquid. To effectively separate viscous liquids or tiny particles from the liquid, vacuum filters are typically utilized (Huttunen et al., 2017; Osborne, 1977). The application of suction to the filtered part of the medium drives the filtering process in vacuum filters. While 100 kPa is the potential pressure drop that may be achieved with vacuum filtering, in actuality, it is typically restricted to 70 or 80 kPa or below (Svarovsky, 2001c). When utilized in crucial steps like the separation of yeast cells and sediment materials in the wine production process or the removal of particles in the fruit juice production process, vacuum filters play a significant role in enhancing product quality and guaranteeing clarity in the beverage industry. Vacuum filters are able to provide an effective and dependable filtration process in the beverage industry because of their flexible designs, continuous process capabilities, and adaptability to a range of applications.

Rotary vacuum drum filter: Applying vacuum to the inner surface of a rotary drum allows vacuum drum filters to separate particles in the liquid. Clean liquid can be produced while the filtering process is ongoing because of the particles sticking to the drum surface. The vacuum forces particles stuck to the revolving drum surface apart. Because of its continuous processing capabilities, it is widely employed in industrial

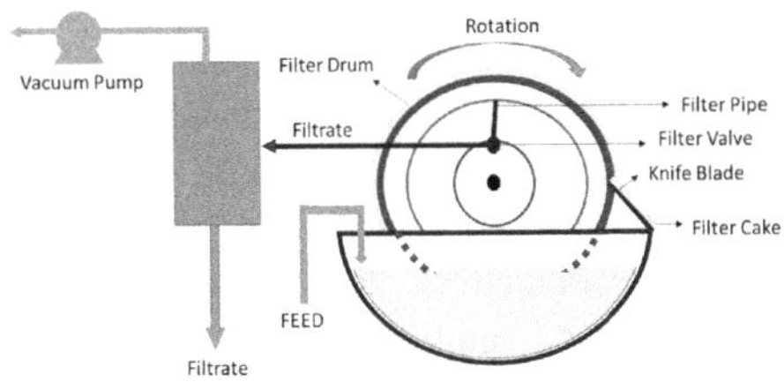

Figure 12.3 Rotary drum vacuum filter equipment.

applications. Vacuum drum filters can be used in the wine-making process to separate yeast cells and other sedimentary material from fermented wine. The wine becomes more transparent as a result (Cheremisinoff, 2017; Sutherland, 2008c). Fig. 12.3 shows the industrial and schematic view of rotary drum vacuum filter equipment.

Vacuum bag filters: Vacuum bag filters capture particles in the liquid in a bag before allowing clean liquid to flow through by applying a vacuum. The bag filters' design is adaptable to a wide range of uses. One or more vacuum bags are used in the filtration process. Its adaptable design allows it to be used in a range of scenarios. Filters using vacuum bags can be used to remove solid particles from both alcoholic and non-alcoholic beverages (Hutten, 2016).

Vacuum filter presses: A vacuum, created by the vacuum filter press is used to draw liquid particles into the filter material. The clean liquid is separated from the particles by forcing them to flow toward the filter material due to the vacuum (Kurita et al., 2010). When compared to a pressure filter press, the filtration process can frequently be slower but still yield more efficient separation. These presses are used in the beverage industry, such as in the winemaking process, to efficiently remove sediments, particles, and contaminants, producing wines that are clearer and more refined (Grainger & Tattersall, 2005).

Vacuum disc filters: To separate solid particles from liquids, the beverage industry uses vacuum disc filters as one sort of filtering apparatus. These filters are made of revolving discs, and the liquid particles can flow through the spaces between the discs and onto the filter material. The clean liquid is produced when the suction generated during the filtering process pulls the particles in between the discs into the filter material. Vacuum disc filters are used in the food and beverage industry to produce wine and spirits, beer, and beverage additives. They have the benefits of eliminating turbidity, enhancing product quality, and supplying clarity. These filters' versatile design enables them to be used with a range of disc combinations and filter materials to suit various beverage applications. By speeding up filtration procedures, vacuum disc filters may often function with great efficiency and contribute significantly to the manufacturing of beverage types (Cheremisinoff, 2000; Sutherland, 2008c).

Vacuum belt filters: The vacuum band filters are one kind of filtering apparatus used for removing solid particles from liquids. Its operation is predicated on the movement of solid particles under vacuum as they are transported through a liquid on a belt conveyor. The filtering process is sped up by the vacuum, which attracts particles towards the belt. After the operation is finished, the compression zone makes sure that the particles on the belt are crushed and cleaned (Berk, 2018). Vacuum belt filters have several applications in the beverage industry, ranging from the manufacturing of syrups and sugar syrups to the processing of fruit juices, alcohol, and liquor to the processing of coffee and tea. By successfully eliminating solid particles from the liquids they contain, these filters may improve the quality of the final product (Bennett, 2015). Vacuum band filters can separate materials effectively, operate quickly and continuously, and offer a variety of design possibilities. These characteristics enable the production processes for beverages to be more industrially efficient while maintaining optimal product quality (Sutherland, 2009).

12.5.3 Centrifugal filters

Centrifugal force is often used for industrial solid–liquid separation because it creates a strong filtration driving force that rapidly separates solid and liquid phases (Fu, Zhou, Sun, Yuan, & Zhu, 2021a, 2021b; Zhou et al., 2018). Originating from its initial application in separating milk on farms (Tetra Pak, 1995), centrifugation has maintained its significance in both industrial and laboratory settings for substance separation.

In the conventional process of centrifugal filtration, the liquid and particles within the working fluid align in the same direction propelled by centrifugal force. During this process, solid particles are captured by the filter while the liquid permeates through, achieving effective filtration. Notably, centrifugal filtration equipment is recognized for its compact size and high processing capacity (Loginov et al., 2017). In comparison to alternative separators, centrifuges not only yield solid components with low water content and a highly pure supernatant, but also demonstrate exceptional sealing capabilities, enabling continuous operation across diverse industries (Svarovsky, 2001a).

The core of a centrifuge comprises a rotating drum, where the liquid undergoes rotation induced by the centrifuge, generating centrifugal force. This force drives the outward movement of solid particles within the liquid, compelling them toward the drum's wall. Consequently, the solid particles separate from the liquid, adhering to the inner surface of the drum, while the purified liquid flows toward the center of the drum. The clarified liquid is then discharged from the drum's center through an outlet point. Following this stage, the liquid becomes more suitable for various applications or undergoes further processing, given its reduced solid particle content. Fig. 12.4 shows the industrial and schematic view of centrifugal filter equipment.

All centrifuge filtration units consist of a cylindrical or conical rotating basket from which the separated solids are discharged at the open end. The basket is supported from its other end by a drive shaft coming from a motor, which can have a fixed or variable

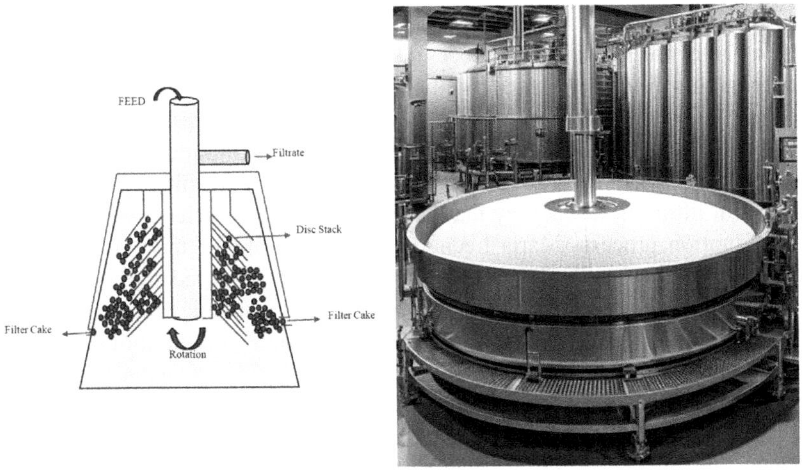

Figure 12.4 Centrifugal filter equipment.

speed. The filtrate passes through the walls of the basket which are typically made of a porous filter medium such as woven wire mesh, perforated plates, or welded wedge wire screen, from the inside to the outside, surrounding a housing, leaving behind solids as a cake on the filter medium (Sutherland, 2008c). Centrifuge filters can be used in alcoholic beverage production, especially in whisky and wine production, to separate solid particles and improve the clarity of liquid products. They can also be effective in the production of carbonated beverages, for the separation of carbonates and sugars, as well as are also used in the clarification of fruit juices and other beverages (Sutherland, 2007).

12.6 Future trends in mechanical filtration technology and conclusion

Mechanical filtration is the process of separating particles from a liquid or gas through a filter material. This process is based on the principle that the particles are too large to pass through the pores of the filter. It is widely used in many different beverage sectors such as milk, oil, beer, and wine and fruit juice in the food industry. In the beverage industry, mechanical filtration provides the opportunity to obtain high quality end products and ensure the purity and clarity of beverages. The effectiveness of mechanical filtration depends on some factors such as the type of filter material, filter media and its properties, flow characteristics, filter types, density, viscosity, and particle size of the material to be filtered, pressure difference of the system, accumulation of filter cake and the resistance force created by the filter. Mechanical filtration should be designed and implemented by taking into account the characteristics of the beverages to be filtered as well as the independent variables that will affect them. In addition, the suitability and features of the filtering equipment are as important as the independent variables affecting the filtration process. Filtration equipment should be designed by determining the filtration mechanism that is suitable for the product to be filtered. As a result, mechanical filtration has a vital role in the beverage industry, and it is predicted that this importance will increase with the development of filtration processes with new equipment and systems. In recent years, there have been significant advancements in mechanical filtration technology that have improved the efficiency and effectiveness of the filtration process (Maria Leena et al., 2020). One of these advancements is the development of new filter media materials with enhanced filtration properties. These new materials, such as ceramic membranes and synthetic fibers, offer improved pore size distribution, higher filtration rates, and increased durability. This allows for more efficient removal of particles and impurities from beverages, resulting in a higher quality end product (Akosile et al., 2020). Another advancement in mechanical filtration technology is the use of automated filtration systems. These systems are equipped with advanced sensors and control mechanisms that continuously monitor and adjust the filtration process to optimize efficiency and ensure consistent product quality. By automating the filtration process, beverage manufacturers can reduce human error and increase production efficiency (Blomenhofer et al., 2013; Ibrahim et al., 2016). Furthermore, there is a growing trend towards the use of modular filtration systems in the beverage industry. These systems are designed to be easily scalable and customizable, allowing beverage manufacturers to adapt their filtration processes to changing production demands. Modular systems also offer improved flexibility in terms of maintenance and replacement of individual filtration components, resulting in reduced downtime and lower operating costs (Campana et al., 2017). Another future trend in filtration technology is the integration of artificial intelligence (AI) and machine learning algorithms into filtration systems. AI can analyze vast amounts of data collected during the filtration process to optimize filtration parameters and predict potential issues or failures. By leveraging AI capabilities, beverage manufacturers can improve process efficiency, reduce waste, and ensure consistent product quality (Li et al., 2021). Mechanical filtration technology continues to advance in the beverage industry, driven by the need for higher product quality, increased efficiency, and improved process control. The development of new filter media materials, automated filtration systems, modular filtration systems, nanofiltration membranes, and the integration of AI are key trends that will shape the future of mechanical filtration technology in the beverage industry (Lowe et al., 2022). These advancements will not only enhance the quality and safety of beverages but also enable beverage manufacturers to meet evolving consumer demands for innovative and tailored products.

References

Aadil, R. M., Madni, G. M., Roobab, U., Ur Rahman, U., & Zeng, X. A. (2019). Quality control in beverage production: An overview. *Quality control in the beverage industry: Volume 17: The science of beverages*. Elsevier Inc https://doi.org/10.1016/B978-0-12-816681-9.00001-1.

Akosile, S. I., Ajibade, F. O., Lasisi, K. H., Ajibade, T. F., Adewumi, J. R., Babatola, J. O., & Oguntuase, A. M. (2020). Performance evaluation of locally produced ceramic filters for household water treatment in Nigeria. *Scientific African, 7*, e00218. https://doi.org/10.1016/j.sciaf.2019.e00218.

Bennett, A. (2015). Innovation trends in food and beverage filtration applications. *Filtration and Separation, 52*(2), 28–33. https://doi.org/10.1016/S0015-1882(15)30091-4.

Berk, Z. (2018). Filtration and expression. *Food process engineering and technology* (pp. 219–242) https://doi.org/10.1016/b978-0-12-812018-7.00008-7.

Blomenhofer, V., Groß, F., Procelewska, J., Delgado, A., & Becher, T. (2013). Water quality management in the food and beverage industry by hybrid automation using the example of breweries. *Water Science and Technology: Water Supply, 13*(2), 427–434. https://doi.org/10.2166/ws.2013.038.

Brandt, M. J., Johnson, K. M., Elphinston, A. J., & Ratnayaka, D. D. (2017). Water filtration. *Twort's water supply* https://doi.org/10.1016/b978-0-08-100025-0.00009-0.

Buzrul, S. (2012). High hydrostatic pressure treatment of beer and wine: A review. *Innovative Food Science and Emerging Technologies, 13*, 1–12. https://doi.org/10.1016/j.ifset.2011.10.001.

Campana, G., Howlett, R., Setchi, R., & Cimati, B. (2017). *Sustainable design and manufacturing 2017: selected papers on sustainable design and manufacturing. Smart innovation, systems and technologies*: 68. Springer.

Chellappah, K., Tarleton, S., & Wakeman, R. (2010). The porosity, permeability and restructuring of heterogeneous filter cakes. *Chem. Eng. Technol., 33*(8), 1283–1289. https://doi.org/10.1002/ceat.201000093.

Cheremisinoff, N. P. (2000). Mechanical separation equipment. *Handbook of chemical processing equipment* (pp. 334–434). Elsevier https://doi.org/10.1016/b978-0-7506-7126-2.x5000-2.

Cheremisinoff, N. P. (2017). Industrial liquid filtration equipment. *Fibrous filter media.* Elsevier Ltd https://doi.org/10.1016/B978-0-08-100573-6.00003-4.

Corbion, C., Smith-Ravin, J., Marcelin, O., & Bouajila, J. (2023). An overview of spirits made from sugarcane juice. *Molecules, 28*(19). https://doi.org/10.3390/molecules28196810.

Echavarría, A. P., Torras, C., Pagán, J., & Ibarz, A. (2011). Fruit juice processing and membrane technology application. *Food Eng. Rev., 3*(3–4), 136–158. https://doi.org/10.1007/s12393-011-9042-8.

Fellows, P. J. (2009). Separation and concentration of food components. *Food Processing Technology, 16*, 188–228. https://doi.org/10.1533/9781845696344.2.188.

Fillaudeau, L., Boissier, B., Moreau, A., Blanpain-avet, P., Ermolaev, S., Jitariouk, N., & Gourdon, A. (2007). Investigation of rotating and vibrating filtration for clarification of rough beer. *J. Food Eng., 80*(1), 206–217. https://doi.org/10.1016/j.jfoodeng.2006.05.022.

Fischer, S., Ruß, W., Buckow, R., Heinz, V., Ulmer, H., Behr, J., Meyer-Pittroff, R., Knorr, D., & Vogel, R. F. (2006). Effects of hydrostatic high pressure on microbiological and technological characteristics of beer. *Monatsschr. Brauwiss., 59*(5–6), 90–99.

Fu, S., Zhou, F., Sun, G., Yuan, H., & Zhu, J. (2021a). Performance evaluation of industrial large-scale cyclone separator with novel vortex finder. *Adv. Powder Technol., 32*(3), 931–939. https://doi.org/10.1016/j.apt.2021.01.033.

Fu, S., Zhou, F., Sun, G., Yuan, H., & Zhu, J. (2021b). Performance evaluation of industrial large-scale cyclone separator with novel vortex finder. *Adv. Powder Technol., 32*(3), 931–939. https://doi.org/10.1016/j.apt.2021.01.033.

Girard, B., & Fukumoto, L. R. (2000). Membrane processing of fruit juices and beverages: A review. *Crit. Rev. Food Sci. Nutr., 40*(2). https://doi.org/10.1080/10408690091189293.

Grainger, K., & Tattersall, H. (2005). Detailed processes of red and white. *Wine production: Vine to bottle* (pp. 73–86) https://doi.org/10.1002/9780470995600.ch12.

Hermia, J., & Brocheton, S. (1994). Comparison of modern beer filters. *Filtration and Separation, 31*(7), 721–725. https://doi.org/10.1016/0015-1882(94)80154-1.

Hutten, I. M. (2016). Standards for nonwoven filter media. *Handbook of nonwoven filter media* (pp. 563–588). Elsevier https://doi.org/10.1016/B978-0-12-811037-9.00011-4.

Huttunen, M., Nygren, L., Kinnarinen, T., Häkkinen, A., Lindh, T., Ahola, J., & Karvonen, V. (2017). Specific energy consumption of cake dewatering with vacuum filters. *Miner. Eng., 100*, 144–154. https://doi.org/10.1016/j.mineng.2016.10.025.

Ibrahim, S. N., Hakim, M. S. L., Asnawi, A. L., & Malik, N. A. (2016). Automated water tank filtration system using LDR sensor. In *Proceedings: 6th International Conference on Computer and Communication Engineering: Innovative Technologies to Serve Humanity* (pp. 195–199). *ICCCE*. https://doi.org/10.1109/ICCCE.2016.51.

Ilame, S. A., & Singh, V. (2015). Application of membrane separation in fruit and vegetable juice processing: A review. *Crit. Rev. Food Sci. Nutr., 55*(7), 964–987. https://doi.org/10.1080/10408398.2012.679979.

Johnston, P. R. (1988). Liquid filtration. *Chem. Eng. Prog., 84*(11), 18–26. https://doi.org/10.1016/b978-1-85617-464-0.00004-3.

Kinnarinen, T., Häkkinen, A., & Ekberg, B. (2013). Steam dewatering of filter cakes in a vertical filter press. *Drying Technol., 31*(10), 1160–1169. https://doi.org/10.1080/07373937.2013.780246.

Kurita, Y., Suwa, S., & Murata, S. (2010). Filter presses: A review of developments in automatic filter presses. *Filtration and Separation, 47*(3), 32–35. https://doi.org/10.1016/S0015-1882(10)70128-2.

Li, L., Rong, S., Wang, R., & Yu, S. (2021). Recent advances in artificial intelligence and machine learning for nonlinear relationship analysis and process control in drinking water treatment: A review. *Chem. Eng. J., 405*, 126673. https://doi.org/10.1016/j.cej.2020.126673.

Lin, C. C., & Wu, J. M. (2022). A novel centrifugal filtration device. *Separations, 9*(5). https://doi.org/10.3390/separations9050129.

Loginov, M., Zierau, A., Kavianpour, D., Lerche, D., Vorobiev, E., Gésan-Guiziou, G., Mahnic-Kalamiza, S., & Sobisch, T. (2017). Multistep centrifugal consolidation method for characterization of filterability of aggregated concentrated suspensions. *Sep. Purif. Technol., 183*, 304–317. https://doi.org/10.1016/j.seppur.2017.03.067.

Lowe, M., Qin, R., & Mao, X. (2022). A review on machine learning, artificial intelligence, and smart technology in water treatment and monitoring. *Water (Switzerland), 14*(9), 1–28. https://doi.org/10.3390/w14091384.

Lu, C., Bao, Y., & Huang, J. Y. (2021). Fouling in membrane filtration for juice processing. *Current Opinion in Food Science, 42*, 76–85. https://doi.org/10.1016/j.cofs.2021.05.004.

Mäkinen, M. T. M. (2002). *Disc filter external models in an integrated simulation environment.* Lappeenranta University of Technology.

Maria Leena, M., Gover Antoniraj, M., Moses, J. A., & Anandharamakrishnan, C. (2020). Three fluid nozzle spray drying for co-encapsulation and controlled release of curcumin and resveratrol. *J. Drug Delivery Sci. Technol., 57*, 101678. https://doi.org/10.1016/j.jddst.2020.101678.

Maria Leena, M., Vimala Bharathi, S. K., Moses, J. A., & Anandharamakrishnan, C. (2019). Potential applications of nanofibers in beverage industry. Nanoengineering in the beverage industry*: Volume 20: The* science of beverages. Elsevier Inc https://doi.org/10.1016/B978-0-12-816677-2.00011-9.

Osborne, D. G. (1977). Vacuum filtration. *Solid–liquid separation*. Butterworth-Heinemann (Vols. 1, Solids) https://doi.org/10.1016/b978-0-408-70795-4.50018-8 .

Palacios, V. M., Caro, I., & Pérez, L. (2002). Comparative study of crossflow microfiltration with conventional filtration of sherry wines. *J. Food Eng., 54*(2), 95–102. https://doi.org/10.1016/S0260-8774(01)00189-3.

Palmer, G. H. (2015). Beverages: Distilled. *Encyclopedia of food grains*: 3–4 (2nd *ed*). Elsevier Ltd https://doi.org/10.1016/B978-0-12-394437-5.00233-3.

Parisher, R. A., & Rhea, R. A. (2021). Chapter 6: Mechanical equipment. *Pipe drafting and design* (4th ed., pp. 119–153). Elsevier.

Perlmutter, B. (2014). *Framework for selecting solid–liquid filtration technologies for particle fines removal* (pp. 56–61). *Chemical Industry Digest*.

Perlmutter, B. A. (2015). Types of filtration systems. *Solid–liquid filtration* (pp. 35–45) https://doi.org/10.1016/b978-0-12-803053-0.00003-1.

Perry, R, & Green, D. (1984). Perry's chemical engineers' handbook *(6th ed.)*. McGraw-Hill Book Company. https://scholar.google.com.tr/scholar?hl=tr&as_sdt=0%2C5&q=PERRY%2C+R.H.+and+GREEN%2C+D.+1984.+Perry%27s+Chemical+Engineers%27+Handbook%2C+Sixth+Edition.+McGraw-Hill+Book+Company%2C+Singapore&btnG=.

Pielech-Przybylska, K., & Balcerek, M. (2019). New trends in spirit beverages production. *Alcoholic beverages: Volume 7: The science of beverages*. Elsevier Inc https://doi.org/10.1016/B978-0-12-815269-0.00003-9.

Rayess, Y. El, Manon, Y., Jitariouk, N., Albasi, C., Peuchot, M. M., Devatine, A., & Fillaudeau, L (2016). Wine clarification with Rotating and Vibrating Filtration (RVF): Investigation of the impact of membrane material, wine composition and operating conditions. *J. Membr. Sci., 513*, 47–57. https://doi.org/10.1016/j.memsci.2016.03.058.

Richert, S. H. (1975). Current milk protein manufacturing processes. *J. Dairy Sci., 58*(7), 985–993. https://doi.org/10.3168/jds.S0022-0302(75)84670-4.

Santos, L. M. R., Oliveira, F. A., Ferreira, E. H. R., & Rosenthal, A. (2017). Application and possible benefits of high hydrostatic pressure or high-pressure homogenization on beer processing: A review. *Food Sci. Technol. Int., 23*(7), 561–581. https://doi.org/10.1177/1082013217714670.

Sellick, I. (1993). Filtration technologies in food and beverage processing. *Filtration and Separation, 30*(6). https://doi.org/10.1016/0015-1882(93)80395-D.

Slaby, M., Šterba, K., & Olsovska, J. (2018). Fitration of beer: A review. *Kvasny Prumysl, 64*(4), 173–184. https://doi.org/10.18832/kp201823.

Sorensen, P. B., Christensen, J. R., & Bruus, J. H. (1995). Effect of small scale solids migration in filter cakes during filtration of wastewater solids suspensions. *Water Environ. Res., 67*(1), 25–32. https://doi.org/10.2175/106143095x131169.

Sparks, T. (2012). Understanding filter presses and belt filters. *Filtration and Separation, 49*(4), 20–24. https://doi.org/10.1016/S0015-1882(12)70193-3.

Sutherland, K. (2007). Separation equipment: The food and beverage industries. *Filtration and Separation, 44*(2), 13–15. https://doi.org/10.1016/S0015-1882(07)70049-6.

Sutherland, K. (2008a). Filter selection. *Filters and Filtration Handbook*, 489–506. https://doi.org/10.1016/b978-1-85617-464-0.00008-0.

Sutherland, K. (2008b). Filters and filtration handbook. *Filters and Filtration Handbook*, 1–523. https://doi.org/10.1016/B978-1-85617-464-0.X0001-6.

Sutherland, K. (2008c). Types of filter. *Filters and Filtration Handbook*, 97–207. https://doi.org/10.1016/b978-1-85617-464-0.00003-1.

Sutherland, K. (2009). Process filters: What's new in process filtration? *Filtration and Separation, 46*(6), 15–17. https://doi.org/10.1016/S0015-1882(09)70224-1.

Sutherland, K. (2010). Food and beverage: Filtration in the food and beverage industries. *Filtration and Separation, 47*(2), 28–31. https://doi.org/10.1016/S0015-1882(10)70078-1.

Svarovsky, L. (1990). *Part I—Batch pressure filtration Solid-Liquid Separation*, 379–401 https://doi.org/10.1016/b978-0-408-03765-5.50017-8.

Svarovsky, L. (2001a). Centrifugal filtration. *Solid–liquid separation* (4th ed, pp. 432–441). Butterworth-Heinemann https://doi.org/10.1108/eb053336.

Svarovsky, L. (2001b). Pressure filtration: Part II—Continuous filtration. *Solid-Liquid Separation (4th Edition)* (pp. 393–408). Butterworth-Heinemann.

Svarovsky, L. (2001c). Vacuum filtration. *Solid–liquid separation* (4th ed, pp. 409–431). Butterworth-Heinemann https://doi.org/10.1007/978-1-59259-996-7_16.

Tarleton, S., & Morgan, S. (2001). An experimental study of abrupt changes in cake structure during dead-end pressure filtration. *Filtration Society, 1*(4), 93–100. https://www.researchgate.net/profile/Es-Tarleton/publication/48353170_An_experimental_study_of_abrupt_changes_in_cake_structure_during_deadend_pressure_filtration/links/55cccbe108aeeaab209b4a72/An-experimental-study-of-abrupt-changes-in-cake-structure-dur.

Tetra Pak (1995). In *Dairy processing handbook: G3* (pp. 331–352). Tetra Pak Processing Systems.

Tippmann, J., Scheuren, H., Voigt, J., & Sommer, K. (2010). Procedural investigations of the lautering process. *Chem. Eng. Technol., 33*(8), 1297–1302. https://doi.org/10.1002/ceat.201000109.

Wang, Y., Hammes, F., & Egli, T. (2008). The impact of industrial-scale cartridge filtration on the native microbial communities from groundwater. *Water Res., 42*(16), 4319–4326. https://doi.org/10.1016/j.watres.2008.07.027.

Werner, R. A., Geier, D. U., & Becker, T. (2020). The challenge of cleaning woven filter cloth in the beverage industry—Wash jets as an appropriate solution. *Food Eng. Rev., 12*(4), 520–545. https://doi.org/10.1007/s12393-020-09228-x.

Zhou, F., Sun, G., Han, X., Zhang, Y., & Bi, W. (2018). Experimental and CFD study on effects of spiral guide vanes on cyclone performance. *Adv. Powder Technol., 29*(12), 3394–3403. https://doi.org/10.1016/j.apt.2018.09.022.

Application of filtration in oil refining

13

Ranko Romanić and Tanja Lužaić
Faculty of Technology Novi Sad, University of Novi Sad, Blvd. cara Lazara 1, Novi Sad, Serbia

13.1 Introduction

Vegetable oils and fats or "oils" are a very important group of lipids. "Lipids" are traditionally classified into acyl- and non-acyl lipids. Vegetable oils are neutral acyl-lipid class, mainly fatty acids esters, and glycerol, triacylglycerols (98%–99%), which also contain 1%–2% of minor compounds. The quality of edible oil depends on the quality of the raw material, which, in turn, is influenced by the maturity of the oilseed, harvesting conditions, seed/fruit handling, storage conditions, and seed infection by insects (Bockish, 1998; Gupta, 2017; Hernandez & Kamal-Eldin, 2013).

According to (FAO, 2024; USDA, 2024) data, global oilseed production over the past 5 years has been around 600 million metric tons (mmt). The most cultivated oilseeds worldwide include soybeans, followed by rapeseed (canola), sunflower, peanuts, and cottonseed. The highest production yields palm oil, followed by soybean oil, rapeseed oil, and sunflower oil (FAO, 2024; USDA, 2024). Various raw materials are used to produce edible vegetable oils and fats. However, economically profitable, sustainable oil and fat production is based on raw materials containing sufficient oil content, oilseeds (Gupta, 2017; Krist, 2020). Worldwide, several plant species are used for obtaining vegetable oils and fats, with only 10–12 raw materials (oilseeds) having significant technological and economic importance. Crude oil is refined in order to produce edible refined oil. The technological process of crude oil production involves two fundamental phases: preparing the raw material for oil extraction and extracting the oil using an appropriate extraction method or combination (Bockish, 1998).

Vegetable oils are carefully refined to produce high-quality edible refined oils. The success in obtaining high-quality refined oil greatly depends on the quality of the crude oil being refined. The quality of crude oil can vary and is influenced by various factors beyond the direct control of operators. Poor-quality crude oil creates specific challenges during the refining process. The objective of refining is to reduce and eliminate specific undesirable components from crude oil. The refining process includes the following steps: degumming (prerefining), neutralization (chemical and/or physical), bleaching, winterization (dewaxing), and deodorization (or deodorization/deacidification) (Gupta, 2017). Certain oil refining stages, such as bleaching and winterization, include filtration as an essential part. Filtration is a crucial step in the bleaching of edible oils. The method of filtration depends on the type of oil being processed, the bleaching temperature, the amount of impurities in the oil, the type of "bleaching earth" used, and the

Mechanical Separation Processes in the Food Industry: Unit Operations and Processing Equipment in the Food Industry.
DOI: https://doi.org/10.1016/B978-0-12-819523-9.00004-4

filtration aids employed. Filtration during the bleaching and winterization of edible oils requires constant monitoring by operators. The duration of the filtration cycle during the bleaching step depends on the oil being processed, the type of "bleaching earth" used, the bleaching temperature, and the content of soap, phospholipids, and moisture in the oil. During winterization, filtration is performed at lower temperatures, so the length of the filtration cycle also depends on the content of waxes and triacylglycerols with lower melting points that remain in the filter cake. The frequency of filter replacement affects the cost of the refining process, with longer filter cycles reducing costs. Precoating aids extend filtration cycles by minimizing their chemical cleaning. Filtration aids also allow for less frequent filter replacements and can facilitate the separation of filter cake. The choice of filter type and size is determined by their purpose in the refining process, the type of particles removed by filtration, as well as capital and operational costs. Oil before filtration, as mentioned earlier, contains various undesirable components, including moisture. Moisture evaporates during vacuum bleaching, while other undesirable components, such as phospholipids, are removed through adsorption using bleaching clay. In physically refined oil, the maximum level of phospholipids should not exceed 25 ppm. Chemically refined oil should be filtered with no more than 0.1% moisture, 25 ppm soap, and 5 ppm phospholipids. The content of these undesirable components significantly affects the filtration process itself and the frequency of filter leaf cleaning. During the selection of the type of filter and supplier, it is advisable to avoid purpose-specific designs that rely solely on a specific filtration aid or filters that prevent the use of other products.

Filters can be either continuous (including centrifuges and decanters, vacuum belt filters, vacuum drum filters, continuous settlers, hydro cyclones, and belt presses) or discontinuous (including sand and media filters, filter presses, cartridge, bag, and other disposable element filters, pressure leaf/tube type filters, and single plate Nutsch filters). Pressure leaf or tube filters can be used for applications ranging from 0.001% to 5% solid content, up to a maximum of 15%. With 0.3% or more solid content, oil can be filtered directly without filter aid. Below 0.3% solid material, precoating or body feed is necessary. The flow rate typically ranges from 0.02 to 2.5 $m^3/m^2/h$. Polishing filters serve as safety filters that inform the processor that the main filtration is functioning correctly. They may have disposable elements such as bags or cartridges for applications with approximately 1–100 ppm solid content, and they work most economically at <25 ppm solid content. There are good reasons to minimize the use of disposable elements because replacing bags, cartridges, or paper sheets requires opening the filter, which takes time and can lead to quality issues. Bags and cartridges will also become saturated with oil, resulting in oil loss. These drawbacks can be overcome by using pulse-type filters that utilize filter cloths as the filtration medium, which can be used for 12–24 months (Veldkamp, 2022).

Considering the previously stated facts, this chapter aims to present, in addition to the significance of oils and fats in human nutrition, the basic components of oils and fats, as well as the principles and methods of production and refining of raw oils. Primarily, it focuses on the application of filtration and filters in the crude oil refining and production of edible refined vegetable oils.

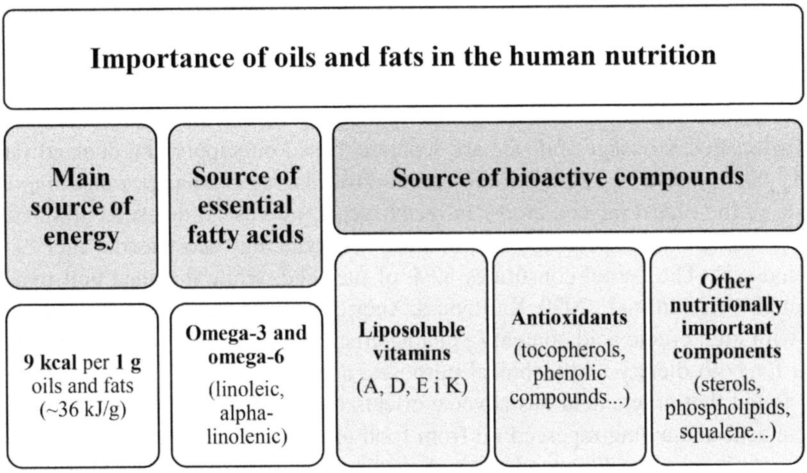

Figure 13.1 Importance of oils and fats in the human nutrition.

13.2 Vegetable oils

Vegetable oils are the most important source of energy in human nutrition. Compared to other macronutrients (carbohydrates and proteins), the energy value of fats is the highest. Additionally, from a nutritional perspective, they are important due to other components, such as essential fatty acids and fat-soluble vitamins, which are vital and found in oils and fats, (Fig. 13.1) (Bockish, 1998; Hernandez & Kamal-Eldin, 2013).

Worldwide, several plant species are used for obtaining vegetable oils and fats, with only 10–12 raw materials (oilseeds) having significant technological and economic importance. Approximately 95% of the total global production comes from these specific raw materials (Karleskind, 1996; Krist, 2020; Ramadan, 2019).

Soybean (*Glycine max.* L., *Merril, Fabaceae*) is an annual plant. Soybeans dominate the oilseed market due to their high protein content, which is significantly higher than other oilseeds (Gerde et al., 2020). The seed kernel typically constitutes 90%–93% of the seed, while the seed hull varies from 7% to 10%. Oil content in the seed varies depending on the variety and growing conditions, ranging from 12% to 24%. In commercial domestic (Republic of Serbia) varieties, it usually falls between 19% and 22% (Hrustić et al., 1998). Soybean oil is highly susceptible to oxidation, leading to a fishy odor after rancidity sets in. The occurrence of reversibility is mainly attributed to the content of alpha-linolenic acid, which is why mild selective hydrogenation is recommended to reduce or eliminate this negative effect (Gerde et al., 2020; Oštrić-Matijašević & Turkulov, 1980). The mesocarp of the palm fruit (*Elaeis guineensis* Jacq.) is used for palm oil production. Palm fruit grows on trees from the palm family (*Arecaceae*) and is cultivated for oil extraction. Oil palm is grown in West and Central Africa, Southeast Asia (Malaysia and Indonesia), as well as in Central and Latin America. It is essential to differentiate palm oil and palm kernel oil. The palm kernel oil is obtained from the kernel of the oil palm fruit, while palm oil is obtained from the mesocarp of the fruit, as previously mentioned. These two oils have different chemical compositions, primarily the fatty acids composition, and physical characteristics, so usage and sale are separate, based on supply and demand (Hamm et al., 2013; Lai et al., 2020). Rapeseed, known also as canola (*Brassica napus* L.), is one of the oldest oilseed crops. In recent years, interest in rapeseed production in Europe has grown, especially due to its use in producing "eco-friendly fuel," known as biodiesel. The kernel constitutes 82% of the seed, while the seed hull makes up about 18% (Eskin et al., 2020; Karlović & Andrić, 1996). Traditional rapeseed (canola) oil, with high erucic acid content (a characteristic fatty acid for this type of oil), was used for both dietary and technical purposes, accounting for over 40%. However, it was found that erucic acid has adverse effects on health, leading to the exclusion of erucic acid-containing rapeseed oil from food use (Bockisch, 1998). Edible rapeseed oil must adhere to quality standards and contain less than 2% erucic acid (Eskin et al., 2020). Sunflower (*Helianthus annuus* L.) is a one-year plant. The kernel represents the most valuable part of sunflower seeds, and before processing, sunflower seeds are dehulled, seed hull is partially removed. Hybridization plays a crucial role in improving sunflower seed characteristics, such as high oil content in the seeds (known as "oily" hybrids), disease and pest tolerance, altered oil quality, and resistance to specific herbicide and pesticide groups. Continuous development of new sunflower seed varieties and hybrids is ongoing (Jocić et al., 2015; Lužaić et al., 2022). Peanut (*Arachis hypogaea*) is a one-year plant that originated from South America. Only about 40% of total peanut production is used for oil production. Peanut oil is primarily used for cooking. In addition to its sensory properties, peanut oil has excellent oxidative stability. Peanut oil is obtained through pressing or a combination of pressing and extraction (Pattee, 2005).

Oils and fats consist mostly of triacylglycerols (98%–99%), which are commonly referred to as triglycerides, and different minor compounds (1%–2%), (Fig. 13.2).

Triacylglycerols consist of three fatty acids, substituting the hydroxyl (alcoholic) positions of the trihydroxy alcohol glycerol. Depending on the number of the hydroxyl groups of glycerol replaced by fatty acids, the resulting compounds can be mono-, di-, or triacylglycerols. These compounds are also known as neutral oils (Gupta, 2017). Fatty acids are an integral part of triacylglycerols. They typically contain 4–22 carbon atoms and have linear structures. Fatty acids are sometimes categorized as short-chain (4–8 carbon atoms), medium-chain (10–12 carbon atoms), and long-chain (14 or more carbon atoms). Among saturated fatty acids in vegetable oils, the most prevalent are: lauric (C12), palmitic (C16), stearic (C18), arachidic (C20), and behenic (C22). Unsaturated fatty acids, such as oleic (C18:1), linoleic (C18:2), and alpha-linolenic (C18:3), are also abundant. Carbon atoms with all four reaction sites react with other elements, and these bonds are called saturated. Unsaturated fatty acids have lower hydrogen atoms than needed to fully satisfy the valency of each carbon atom. Therefore, certain carbon atoms are connected by a double bond. Most hydrogen atoms in the double bonds of natural fatty acids are on the same side of the double bond, indicating a *cis* position (or *cis* isomer). However, some hydrogen atoms can shift to

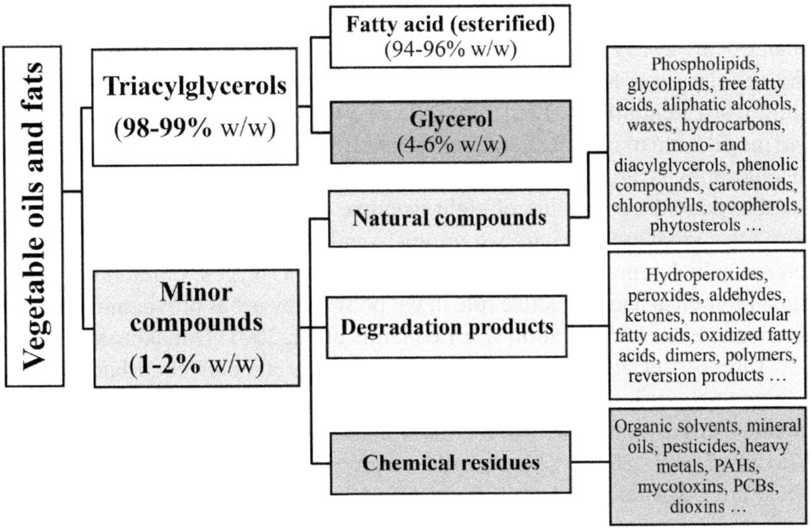

Figure 13.2 Vegetable oils and fats constituents.

the other side of the bond during the process of hydrogenation (chemical saturation of double bonds) to produce *trans* isomers. Unsaturated fatty acids are unstable and highly susceptible to oxidation even at room temperature. They tend to oxidize easily when exposed to air or oxygen, forming aldehydes, ketones, oxidative polymers, cyclic compounds, etc. On the other hand, saturated fatty acids are not subject to oxidation to such an extent, but will decompose under high temperatures. They can lead to the formation of thermal polymers and other toxins (Gunstone, 2005; Gupta, 2017).

Besides triacylglycerols, other elements soluble or dispersed in fats include phospholipids (phosphatides), glycolipids, free fatty acids, aliphatic alcohols, waxes, hydrocarbons, mono- and diacylglycerols, phenolic compounds, carotenoids, chlorophylls, trace metals, tocopherols, and phytosterols, (Fig. 13.2) (Cuesta & Sánchez-Muniz, 2003; Grompone, 2005; Grompone, 2020). Pesticides can also be found in fats (Gunstone, 2005). Some minor constituents, such as tocopherols and phytosterols, have a positive effect on the nutritional and technological properties of oils (Fernández-Martínez et al., 2009). Other minor components negatively affect the mentioned properties of oils. The presence of phospholipids, free fatty acids, and waxes is not desirable in the oil (Kamal-Eldin, 2005). However, these minor elements (eg, phospholipids) serve as by-products with significant applications in various industries (Salas et al., 2006). Tocopherols are a group of fat-soluble components composed of a polar part, a chromanol ring, and a hydrophobic side, which is a derivative of the phytyl chain. Tocopherols are naturally present in four isomeric forms: α-, β-, γ-, and δ-tocopherol, differing in the number and position of methyl substituents on the chromanol ring (DellaPenna & Mène-Saffrané, 2010; Romanić et al., 2009). They are the main constituents with antioxidant activity in oilseeds and oils extracted, significantly influencing their characteristics. Both, tocopherols and tocotrienols (collectively known as tocochromanols), exhibit vitamin E activity (Brigelius-Flohé & Traber, 1999; Romanić et al., 2021). Tocopherols avert oxidative, metabolic, and inflammatory harm at the cellular level and the subsequent onset of various chronic illnesses (Azzi & Galli, 2010; Eitenmiller & Lee, 2004). From a technological standpoint, tocopherols shield the oil from oxidative alterations (Shahidi & Zhong, 2010). Carotenoids together with chlorophylls are the most abundant pigments in vegetable oils. Carotenoids are tetraterpenoids, consisting of eight isoprene molecules and containing 40 carbon atoms. They are classified into two groups: xanthophylls, which contain oxygen, and carotenes, which represent pure hydrocarbons without oxygen. Carotenoids play a significant biological and protective role in the body. They act as provitamins and provide protection against UV radiation (De Leonardis et al., 2001; Dimakou & Oreopoulou, 2012; Tuberoso et al., 2007). In unrefined oils, carotenoids contribute not only to the color of the oil, an important sensory quality parameter but also exhibit strong antioxidant effects similar to tocopherols (Choe & Min, 2006; Dimakou & Oreopoulou, 2012; Lužaić et al., 2023). Chlorophylls are green pigments that influence the color of oils and are essential for sensory evaluation (Brühl & Matthäus, 2004). Chlorophyll and its derivatives are the most potent agents in promoting photooxidation in oils exposed to light, significantly contributing to their susceptibility to oxidative processes. Recent research also indicates the antioxidant properties of chlorophylls present in oil (Tynek et al., 2012). Phenols are a highly valued group of compounds. Due to their significant antioxidant properties, phenolic compounds contribute significantly to the favorable nutritional and chemical characteristics of oils. The antioxidant activity of polyphenols is attributed to the presence of the *o*-dihydroxy phenolic structure, which has a strong capability to form chelates with metal ions, thereby inhibiting the formation of oxygen radicals. Polyphenols exhibit antioxidant activity against alkoxyl and peroxyl radicals and regenerate α-tocopherol by reducing tocopherol radicals. Their potential role in preventing cardiovascular diseases and other therapeutic effects can be attributed not only to their antioxidant characteristics but also to other metabolic processes (Choe, 2008; Kroon & Williamson, 2005; Virgili et al., 2001). In addition to their antioxidant properties, phenolic compounds also exhibit other important characteristics, such as hormonal effects (estrogenic or antiestrogenic) (Fruhwirt et al., 2003). Furthermore, among them, there are compounds with potential anticancer and cardioprotective indications (protective effects on the cardiovascular system) (Coni et al., 2000; Fruhwirt et al., 2003), while some polyphenols possess antimicrobial, antiviral, and anti-inflammatory properties. Phytosterols, also known as plant sterols, are part of a broad category of steroids that hold significant physiological importance in both flora and fauna. In plants, this category includes essential compounds such as triterpenoids, highly important in defense against pathogens - (Flores-Sánchez et al., 2002); brassinosteroids, campesterol derivatives, plant hormones necessary for regular growth and development (Bishop & Yokota, 2001); and various secondary plant metabolites, like glycoalkaloids and saponins (Hartmann, 1998). Plant sterols have a tetracyclic cyclopentafenantrene structure and an alkyl side chain with 8–10 carbon atoms (Hartmann, 1998). Phospholipids are the main compounds of biological membranes. Structurally, phospholipids are esters of glycerol, fatty acids (in positions 1 and 2 of the glycerol molecule), and phosphate groups (esterified in position 3) to

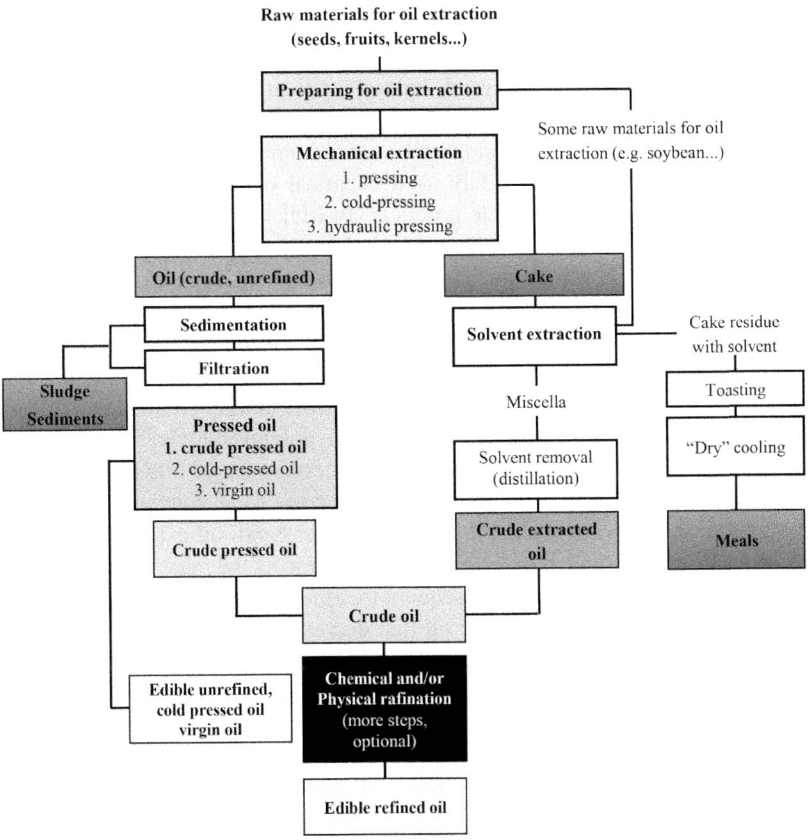

Figure 13.3 Technological scheme of crude and unrefined oil production.

which various phosphatidic groups are further attached. Phosphoacylglycerols (phospholipids) have a synergistic effect with tocopherols, exhibiting antioxidant activity. They also have metal-binding capabilities and catalytic activity during the degradation of hydroperoxides (Carelli et al., 1997; Smouse, 1995). However, phospholipids are more sensitive to oxidation than triacylglycerols because they contain a higher degree of unsaturated fatty acids necessary for maintaining membrane permeability (Shahidi & Zhong, 2010). These fatty acids in oil can develop an unpleasant flavor and taste during storage, reducing the oil's shelf life (Desai et al., 2002). Waxes are esters of fatty acids and long-chain alcohols. They are removed during the refining process, specifically in the winterization step. Waxes with a higher number of carbon atoms cause greater cloudiness (turbidity) in crude (unrefined) and refined oils, due to their higher melting points (Añón & Martini, 2005).

Edible vegetable oils are obtained through mechanical pressing, chemical (solvent) extraction, or a combination of both processes. The technological scheme for chemical and physical oil refining is shown in (Fig. 13.3) (Bockish, 1998; Popović et al., 2020).

The method of oil extraction affects the content and composition of minor compounds that contribute to the oil's functional properties and oxidative stability. These methods produce crude pressed, crude extract, or their combination when it is called crude oil. Unrefined, cold-pressed, and virgin oils are considered to have high nutritional value. Recently, cold-pressing and hydraulic pressing technology has become a desirable alternative to the widely used chemical oil extraction due to consumer demand for safe and natural edible products (El Makawy et al., 2019; Kiralan et al., 2018; Ramadan, 2013). The advantages of industrial-scale cold-pressing technology include lower energy consumption and reduced investment costs. Cold-pressed oils are favored over refined oils due to their higher concentrations of antioxidants and bioactive substances like sterols, carotenoids, and phenols. (Bhatnagar & Gopala Krishna, 2014; Prescha et al., 2014). The main shortcoming of mechanical oil extraction using cold-pressing technology is its low efficiency and variable quality (Ramadan, 2020). Most cold-pressed oils contain significant amounts of polyunsaturated fatty acids, which adversely affect oxidative stability. Additionally, cold-pressed oils have higher levels of prooxidative compounds, resulting in a shorter shelf life compared to refined oils (Brühl, 1996; Rotkiewicz et al., 1999). On the other hand, oil extraction by chemical methods is more economical in terms of yield and meets the mass requirements of the population. However, it requires subsequent treatment and oil refining. Refining seeks to eliminate components from crude oil that decrease the sensory properties, quality, and shelf life of the refined oil. These undesirable components have various origins and include: characteristic components, specific to each oil type and may include protein residues, phospholipids, pigments, sulfur compounds, etc; degradation products, form during seed cultivation or storage, including free fatty acids, peroxides, aldehydes, ketones, and non-hydratable phospholipids; chemical residues, result from chemicals used during plant growth, such as pesticides, herbicides, desiccants, and phosphoric acid; contaminants, coming from equipment or other oils, including solvents, metals, and soaps; environmental pollutants, soil, water, and air pollution during seed drying can introduce compounds like polycyclic aromatic hydrocarbons (PAHs), dioxins, and 3-MCPD, (Fig. 13.3). During refining, it is essential to achieve the best sensory and chemical quality of oil, high nutritional value, and good oil sustainability while minimizing losses. Additionally, environmental protection and energy efficiency are crucial considerations.

13.3 Vegetable oil refining

Crude oil serves as the raw material for the production of refined oil. The technological process of crude oil production (Fig. 13.3), in the previous chapter) involves two fundamental phases: preparing the raw material for oil extraction and extracting the oil using an appropriate extraction method or their combination (Bockish, 1998; Popović et al., 2020).

The oil extraction process needs to be adjusted primarily to the characteristics of the raw material. The raw materials should be prepared so that the oil can be easily separated while ensuring the highest oil quality. Depending on the type of raw material, preparation may involve cleaning, fractionation, dehulling, milling, and hydrothermal treatment. After raw material preparation, crude oil is extracted either by pressing or by combining pressing with solvent extraction (the purpose of solvent extraction is to extract as much oil as possible from the raw material after pressing) (Bockish, 1998).

The quality of crude oil varies and depends on a few factors, not directly under the control of raw material processors and oil producers. Poor-quality crude oil poses challenges during the refining process in terms of achieving high-quality end products (Gupta, 2017). Parameters for identifying and assessing the quality of crude oils obtained from typically used raw materials are presented in Table 13.1 and Table 13.2.

Regardless of whether it is obtained by pressing or solvent extraction, crude oil contains different impurities that need to be eliminated to ensure a satisfactory shelf life of the oil (Dijkstra, 2015). As previously mentioned, the quality of refined oil is conditioned by the quality of seeds or fruits containing oil. The crude oil's quality is influenced by several factors, including maturity, harvest conditions handling, storage conditions, and raw material insect control. Crude oil has varying fatty acid content and other component variations within the raw material. These factors can impact the refining process. Immature seeds and fruits (raw material) have lower activity of enzymes like lipoxygenase, trypsin inhibitors, and urease compared to mature raw material. Immature raw material may also have higher levels of free fatty acids and chlorophyll compared to mature raw material. The oil and total protein content do not significantly differ between mature and immature raw materials. Raw soybean, sunflower, cottonseed, and rapeseed oils may exhibit a more pronounced green color if harvested seeds are not at their technological maturity or if the harvest season is excessively wet. Crude oils will require additional efforts to remove chlorophyll during degumming and bleaching. In cases of dry harvest conditions due to drought, physical seed damage can result in higher free fatty acid content and oxidation of crude oil. The oil will then have reduced stability. If seeds are damaged during harvest, transportation, or storage, they can lead to higher free fatty acid content in the crude oil and increased oxidation. Such oil will demand additional stages in the refining process and generally is less oxidative stabile (Gupta, 2017).

The objective of refining is to reduce and eliminate specific undesirable components from crude oil (Gupta, 2017). Since these undesirable components differ in chemical characteristics, various methods have been established in order to remove them. These methods include degumming (pre-refining), neutralization (chemical and/or physical), bleaching, winterization (dewaxing), and deodorization. Degumming aims to remove phospholipids from crude oil. It also eliminates components soluble in water such as sugars and glycolipids, as well as most crudities like meals, seeds, and fruit residues. The phosphatide sludge also contains oil, the amount of sludge depends on the specific degumming process used. In the industry, several different processes are employed. In the chemical neutralization process, phosphatides and, primarily, free fatty acids (FFA) from crude oil are removed by treating the oil with a water solution of sodium or potassium hydroxide, which reacts with FFA to form soaps that separate from the oil using a centrifuge (separators), known as soapstock. Neutral oil also remains in the soapstock (Nedić Grujin et al., 2023; Wen et al., 2023). The aim of the bleaching process is to remove colored compounds, such as carotenoids and chlorophylls, from the oil by mixing it with an adsorbent ("bleaching earth"). After bleaching, the adsorbent is removed from the filters. This refining step also eliminates residual impurities. Spent

Table 13.1 Identification parameters for some crude oils.

Parameter	Soybean oil (degummed)	Palm oil	Rapeseed oil (with low erucic acid)	Sunflower oil			Coconut oil	Cottonseed oil
				Linoleic	High oleic	Mid oleic		
Iodine value [gI_2/100g]	124–139	50.0–55.0	105–126	118–141	78–90	94–122	6.3–10.6	100-123
Refractive index, n_D 25°C	1.466–1.470	1.454–1.456 at 50°C	1.465–1.467	1.461–1.468	1.467–1.471	1.461–1.471	1.448–1.450 at 40°C	1.458-1.466 at 40°C
Relative density, 20°C/water 20°C	0.919–0.925	0.891–0.899 at 50°C	0.914–0.920	0.918–0.923	0.909–0.915	0.914–0.916	0.908–0.921 at 40°C	0.918-0.926
Saponification value [mg KOH/g]	189–195	190–209	182–193	188–194	182–194	190–191	248–265	189-198
Unsaponifiable matter [% w/w]	≤ 1.5	≤ 1.2	≤ 2.0	≤ 1.5	≤ 1.5	≤ 1.5	≤ 1.5	≤ 1.5
Fatty acid composition [% w/w]								
C 6:0	nd	nd	nd	nd	nd	nd	nd-0.7	nd
C 8:0	nd	nd	nd	nd	nd	nd	4.6–10.0	nd
C 10:0	nd	nd	nd	nd	nd	nd	5.0–8.0	nd

(continued on next page)

C 12:0	nd-0.1	nd-0.5	nd	nd-0.1	nd	nd	45.1–53.2	nd-0.2
C 14:0	nd-0.2	0.5–2.0	nd-0.2	nd-0.2	nd-0.1	nd-1.0	16.8–21.0	0.6-1.0
C 16:0	8.0–13.5	39.3–47.5	2.5–7.0	5.0–7.6	2.6–5.0	4.0–5.5	7.5–10.2	21.4–26.4
C 16:1	nd–0.2	nd–0.6	nd–0.6	nd–0.3	nd–0.1	nd–0.05	nd	nd–1.2
C 17:0	nd–0.1	nd–0.2	nd–0.3	nd–0.2	nd–0.1	nd–0.05	nd	nd–0.1
C 17:1	nd–0.1	nd	nd–0.3	nd–0.1	nd–0.1	nd–0.06	nd	nd–0.1
C 18:0	2.0–5.4	3.5–6.0	0.8–3.0	2.7–6.5	2.9–6.2	2.1–5.0	2.0–4.0	2.1–3.3
C 18:1	17.0–30.0	36.0–44.0	51.0–70.0	14.–39.4	75.0–90.7	43.1–71.8	5.0–10.0	14.7–21.7
C 18:2	48.0–59.0	9.0–12.0	15.0–30.0	48.3–74.0	2.1–17.0	18.7–45.3	1.0–2.5	46.7–58.2
C 18:3	4.5–11.0	nd–0.5	5.0–14.0	nd–0.3	nd–0.3	nd–0.5	nd–0.2	nd–0.4
C 20:0	0.1–0.6	nd–1.0	0.2–1.2	0.1–0.5	0.2–0.5	0.2–0.4	nd–0.2	0.2–0.5
C 20:1	nd–0.5	nd–0.4	0.1–4.3	nd–0.3	0.1–0.5	0.2–0.3	nd–0.2	nd–0.1
C 20:2	nd–0.1	nd	nd–0.1	nd	nd	nd	nd	nd–0.1
C 22:0	nd–0.7	nd–0.2	nd–0.6	0.3–1.5	0.5–1.6	0.6–1.1	nd	nd–0.6
C 22:1	nd–0.3	nd	nd–2.0	nd–0.3	nd–0.3	nd	nd	nd–0.3
C 22:2	nd	nd	nd–0.1	nd–0.3	nd	nd–0.09	nd	nd–0.1
C 24:0	nd–0.5	nd	nd–0.3	nd–0.5	nd–0.5	0.3–0.4	nd	nd–0.1
C 24:1	nd	nd	nd–0.4	nd	nd	nd	nd	nd

nd: not detected.
From CodexAlimentarius. (1999); Przybylski and Eskin (2011); Wang (2008); Warner et al. (2003); Zou et al. (2012).

Table 13.2 Quality parameters for some crude oils.

Parameter	Soybean oil (degummed)	Palm oil	Rapeseed oil (with low erucic acid)	Sunflower oil			Coconut oil	Cottonseed oil
				Linoleic	High oleic	Mid oleic		
Lovibond color (red, R yellow, Y and blue, B Lovibond units, cell 25,4 mm)	maximum 3 R and maximum 50 Y	-	maximum 3 R and maximum 40 Y	maximum 3 R and maximum 30 Y	maximum 3 R and maximum 30 Y	maximum 3 R and maximum 30 Y	maximum 3 R and maximum 50 Y	60.5 R and 4.7 B
Free fatty acid content (as dominant fatty acid) [% w/w]	≤ 1.5	1–5	≤ 2.0	≤ 3.00	≤ 3.00	≤ 3.00	≤ 4.0	5
Moisture content [% w/w]	≤ 0.30	0.23	≤ 0.30	≤ 0.30	≤ 0.30	≤ 0.30	0.20	0.20
Total phosphorus content, P [mg/kg]	≤ 150	200–500	≤ 250	≤ 250	≤ 250	≤ 250	22.2	700-900
Total phospholipids content [% w/w]	≤ 0.45	0.50–1.25	≤ 0.625	≤ 0.625	≤ 0.625	≤ 0.625	0.055	1.75-2.25
(calculated based on the P content)	as P × 30	as P × 25	as P × 25	as P × 25	as P × 25	as P × 25	as P × 25	as P × 25
Total wax content [mg/kg]	-	-	-	≤ 1000	≤ 1000	≤ 1000	-	700
Total chlorophyll content [mg/kg]	1.0–1.5	0.3–1.8	5–50	0–1	0–1	0–1	0.098	-
Total phytosterols content [%]	0.33	0.04–0.08	0.7–1.8	0.25–0.50	0.25–0.50	0.25–0.50	0.08	0.27-0.64
Total tocopherols content [mg/kg]	600–3370	150–870	700–1200	440–1520	440–1520	440–1520	55	1000
Total tocotrienols content [mg/kg]	-	530–740	-	-	-	-	-	140-800
Flash point [°C]	324	240	275	316	316	316	168	293

From Akhmedov et al. (2022); CodexAlimentarius (1999); Eskin et al. (2020); Gerde et al. (2020); Gordon and Rahman (1991); Grompone (2020); Kochhar et al. (2020); Lai et al. (2020); Przybylski and Eskin (2011); Rukmini and Raharjo (2010); Salimath et al. (2021); Soomro and Sherazi (2013); Usman et al. (2023); Wang (2008); Warner et al. (2003); Withana-Gamage et al. (2020); Zou et al. (2012).

"bleaching earth" contains at least 35% organic matter (w/w). Some raw vegetable oils (such as sunflower oil) contain significant amounts of waxes, which need to be removed during the winterization stage (Broughton et al., 2018; Lužaić et al., 2024). Waxes with higher melting points crystallize when the oil is stored at low temperatures, forming sediment (Nedić Grujin et al., 2023). Waxes are not harmful, but cloudiness and sediment caused by waxes are not desirable when selling oil in bottles, especially to consumers. Waxes with higher melting points crystallize when the oil is cooled, and these crystals are then removed from the oil through filtration or centrifugation. In the deodorization process, compounds with unpleasant odors are removed from the oil under vacuum conditions using steam at high temperatures. During deodorization, free fatty acids (FFA) are also eliminated from the oil. Therefore, it is possible to avoid chemical neutralization, provided that the previous degumming step was efficient enough to reduce the residual phosphorus level to less than 10 ppm (Dijkstra, 2015).

Depending on the neutralization method used, refining can be either chemical or physical. The technological scheme for chemical and physical oil refining is shown in Fig. 13.4.

The chemical refining process is widely used, especially for oils obtained from seeds. Crude oil is first treated with phosphoric acid and then neutralized using a sodium or potassium hydroxide solution. FFA and hydroxide form soap. The soap is later separated using a primary centrifuge. The remaining soap is removed by washing with water and separating the oil and water in a water-washing centrifuge (separators). Chemical refining is convenient for oils with a higher phospholipid content (such as crude soybean oil, degummed soybean oil, degummed rapeseed oil, crude sunflower oil, crude and degummed corn oil, crude and degummed cottonseed oil, and crude safflower oil). However, chemical refining is not convenient for palm or coconut oil due to their low phospholipid content. The chemical refining process is debatable from an environmental point of view. As a byproduct of the chemical refining process, soap is obtained, which is treated with sulfuric acid to recover fatty acids. This creates wastewater with high biological and chemical oxygen demand (COD and BOD). To mitigate this situation, the oil processing industry has developed the process of physical refining. It has been found that this process is suitable for crude oils with low levels of nonhydratable phospholipids and those with high levels of free fatty acids. The physical refining is particularly convenient for crude palm and coconut oils, but also for the refining of crude sunflower oil, due to their low phospholipid content and high free fatty acid content. Chemical refining of any of these oils results in an oil loss in the produced soapstock. The physical refining is challenging for oils with high levels of nonhydratable phospholipids. However, recent advancements in enzymatic degumming using more efficient enzymes have improved its effectiveness (Gupta, 2017; Lamas, 2014; Lamas, 2016).

13.4 Filtration

Filtration is a mechanical technological operation used for the separation of solid particles from a fluid by passing through a filtration medium that retains the solid

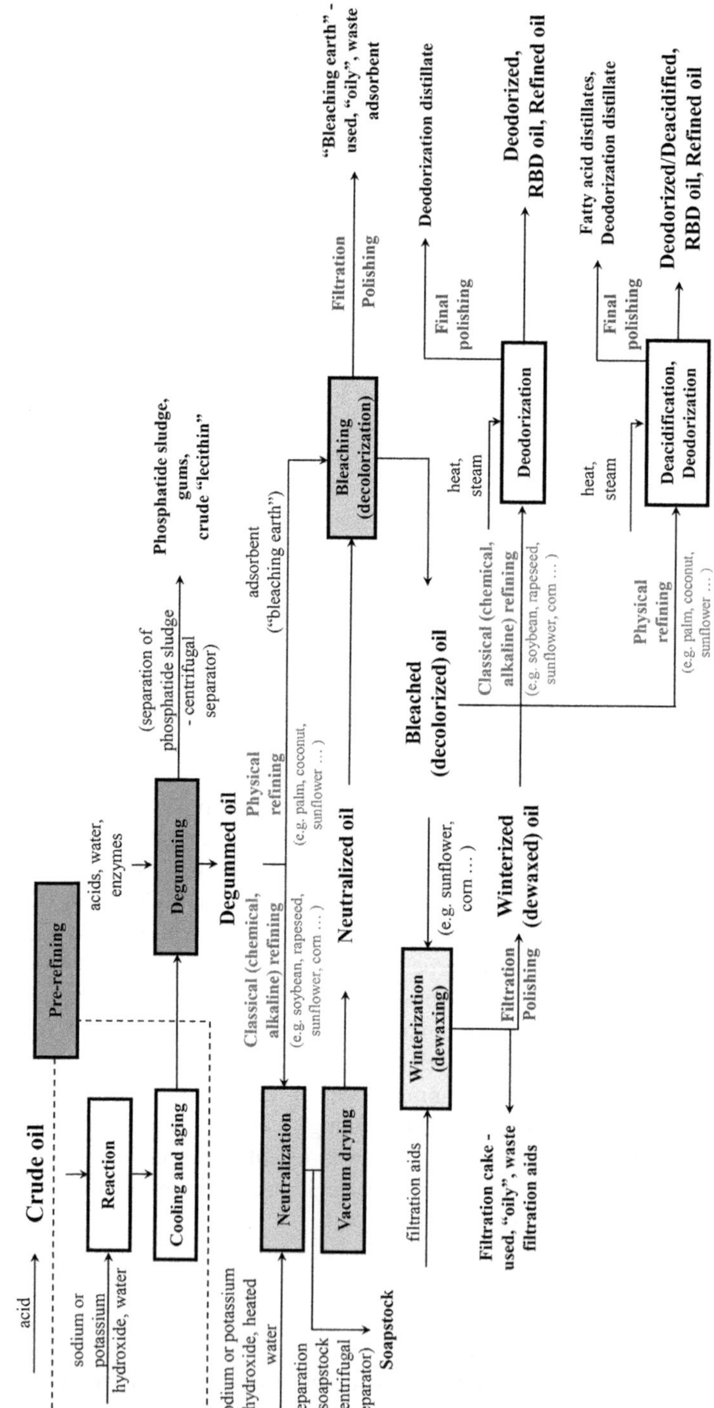

Figure 13.4 Technological scheme of chemical and physical oil refining.

particles (Ripperger et al., 2011; Van Der Bruggen et al., 2003). The process of filtration is far more complex under operating conditions and is influenced by a large number of physicochemical parameters (Hwang & Lin, 2002; Ripperger et al., 2011). Two types of filtration are distinguished. Filtration through filter cake is the first type. A formed filtration cake serves as the filtration medium (porous barrier made of paper, metal, ceramics, etc.). It is applied to relatively concentrated suspensions (1% volume or more of solid substance in the suspension). The second filtration type uses a filtration medium (fabric, filter paper, membrane). In this case, the pores of the filtration medium clog much earlier, before conditions for the formation of a filtration cake are met. It is suitable for suspensions with very low concentrations (below 0.1% volume of solid substance in the suspension). For suspensions with average concentrations (between 0.1% and 1% volume of solid substance), both types of filtration can be applied.

The filtration rate is defined as the ratio of driving force to resistance. Since the filtration cake typically represents a porous layer of fine particles, the flow through such a layer is expected to be laminar. In this case, the Carman-Kozeny equation (Richardson et al., 2002) can be applied (Equation 13.1):

$$\frac{dV}{dt} = \frac{\Delta p\ A}{\alpha \frac{G\ V\ \mu}{A}} \tag{13.1}$$

Where: dV/dt is the filtration rate (volume of filtrate passing through the filter per unit time, m^3/s), Δp is the pressure drop across the filtration cake (N/m^2), A is the filtration area (m^2), α is the specific resistance of the filtration cake (m/kg), μ is the fluid viscosity (Ns/m^2), G is the mass of solid phase in the suspension per unit volume of liquid in the suspension (kg/m^3).

The driving force of filtration is the pressure difference (Δp), and in addition to the resistance of the filtration cake (α), there is also resistance from the filtration medium and the filtration device itself. Considering all these resistances (connected in series), the basic filtration equation takes the form presented at (Equation 13.2):

$$\frac{dV}{dt} = \frac{\Delta p_t A}{\mu\left(\frac{\alpha G V}{A} + R_m\right)} \tag{13.2}$$

Where: Δpt is the total pressure difference across the filtration device during filtration (N/m^2), R_m is the resistance of the filtration medium and the filtration device (m^{-1}).

The specific resistance of the filtration cake is given by the following (Equation 13.3):

$$\alpha = \frac{5(1-\varepsilon)A_0^2}{\rho_1 \varepsilon^3} \tag{13.3}$$

Where: (1-ε) represents the fraction of the solid phase in the cake, A_0 is the specific surface area of the particle, ρ_1 is the density of the solid phase, and ε denotes the porosity.

The specific cake resistance depends on the particle surface area (A_0) and the porosity of the layer (ε), and it may not remain constant during filtration. It primarily depends on the type and pressure resistance of the particles, as well as their tendency to flocculate (Hwang & Lin, 2002).

In most cases, during filtration, α remains constant, especially in the case of incompressible filtration cakes. Under pressure, particles can respond in two ways: if the resistance of the particles to pressure is greater than the applied pressure, they will be compacted into interstitial spaces without changing their shape. In the second case, particles with lower resistance than the pressure will deform. In both cases, the porosity of the filtration layer decreases, leading to an increase in specific resistance. In practice, this results in shorter filtration cycles (Outmans, 1963; Sorrentino et al., 2024).

Practical cake filtration is carried out in two ways: either at a constant pressure difference (where the filtration rate decreases with increasing cake thickness) or at a constant filtration rate (with the condition that pressure can increase with cake thickness, i.e., proportional to increased resistance).

Generally, the particles removed from oil are small and compressible, so after a short time of filtration, the pores of the filtration medium become clogged, affecting the filter capacity and filtrate quality. This scenario corresponds to compressible filtration cakes. In order to extend the filtration cycle time, a filtration aid is used. This aid must consist of solid, finely porous, incompressible particles and must not affect the chemical composition, odor, or taste of the oil. Before filtration begins, the filtration aid is applied as a precoat layer or is continuously dosed during filtration before the oil enters the filter. This significantly reduces the resistance of the filtration layer through which the oil flows, ensuring both high flow rates and a high degree of clarification. During filtration, the thickness of the filtration cake increases, while porosity decreases (Buyel et al., 2015; Nedić Grujin et al., 2023; Smith, 1998). This largely depends on the quantity and fineness of the filtration aid used in relation to the particles being removed from the oil, which causes oil turbidity. Filtration is facilitated by pressure differences created by vacuum, gravity, or centrifugal force. In recent times, various models have been developed to describe possible filtration methods, but four basic models are highlighted, as shown in Fig. 13.5 (Ripperger et al., 2011).

Cake filtration is the most common filtration model. Solid particles that need to be removed are distributed on the inlet side of the filtering medium, forming a homogeneous porous layer with constant permeability. If the filtration rate (dV/dt) is constant, the pressure change (Δp) is linear and proportional to the amount of separated solid particles (Dahlstrom et al, 1997).

In the case of surface (blocking) filtration, the pressure drop (Δp) is caused by particles blocking the pores. Soft, gelatinous particles retained by the filter lead to pore clogging. If the flow rate (dV/dt) is constant, the pressure drop increases exponentially with the amount of filtrate, and the number of open pores decreases asymptotically toward zero. Pores can belong to the filtration medium (such as a sieve or filtration layer) or can be within the filtration cake. In-depth (deep-bed) filtration, solid particles are retained in a deep filtration layer through the adhesion of the particles to the structure of the filtration layer itself. Only relatively large particles are retained. As

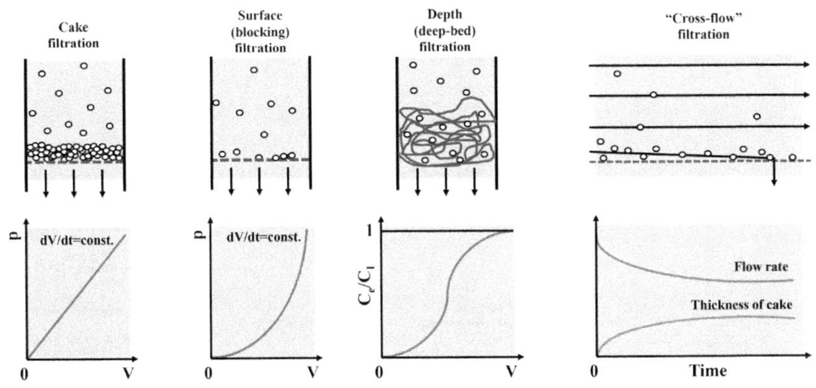

Figure 13.5 Basic filtration models.

the filter layer becomes saturated with solid particles, their concentration in the filtrate leaving the layer approaches the concentration of the input suspension (Anlauf, 2019; Altmann & Ripperger, 2002).

In "cross-flow" filtration, the suspension flows tangentially at high speed across the surface of the filtration medium, preventing the formation of a filtration cake. Only a small quantity of liquid passes through the filtration medium. The layer of solid material accumulates in the boundary layer on the filter surface, reducing the flow of filtrate (Ripperger et al., 2011).

In the production of oils and fats, mechanical technological operations have a significant application. Filtration as a mechanical separation operation is used in the production of crude oils, oil refining, modification of oils and fats, production of inedible products, oleochemicals, biodiesel, etc. The application of filtration within the application of the mechanical separation process in the production of oils and fats is shown in (Fig. 13.6).

13.5 Bleaching

In addition to phospholipids, free fatty acids, waxes, oxidation products, traces of metals, and other compounds, crude vegetable oils also contain pigments (Seçilmiş et al., 2021). Pigments (mainly carotenoids and chlorophylls) are not harmful, however, due to consumer habits, they are removed in the refining process, especially in the bleaching step. Bleaching (or decolorization) is the most important and critical step in edible oil refining. During this complex physical and chemical process, adsorbent (activated "bleaching earth" or bleaching clays, activated carbon, special silica and silicates, etc., or a combination of these) is used to remove and reduce levels of carotenoids and chlorophylls, as well as other undesirable components from neutralized oil under specific conditions. In addition to pigments, bleaching also removes residues of phospholipids, soaps, heavy metals, waxes, oxidation products, and other impurities from oil. Finally, oil deodorization/deacidification and deodorization, known as "thermally

Production	Separation	Technique	Application
• Crude oil	• Crude oil from insoluble residues, particles, impurities, sediment etc.	• Settling • Clarification • Filtration	• Tanks • Decanters • Vertical leaf filters • Belt filters
• **Oil refining**	• Degummed oil from gums, sludge, lecithin etc.	• Centrifugation • Separations	• Centrifuges • Separators
	• Neutralized oil from sopstock, water etc.	• Centrifugations • Separations	• Centrifuges • Separators
	• **Bleached oil from bleaching earth**	• **Filtration**	• **Vertical leaf filter** • **Horizontal pressure leaf filter** • **Plate filter press** • **Candle filter** • **Disk filter**
	• **Winterized oil from filtration aid with waxes, solid TAGs**	• **Filtration** • Centrifugations • Separations	• **Horizontal pressure leaf filter** • **Membrane filter press** • **Pulse tube filter**
	• Deodorization or deodorization/ deacidification	• No filtration	• No filter
	• **Refined oil from cloudiness**	• **Polishing filtration**	• **Paper filter** • **Single/multi bag filter** • **Cartridge filter**
• Modification of oils and fats, specialty oils and fats production	• Fractionation • Interesterification • Olein and stearin separation • Solvent separation • Hydrogenation	• Filtration	• Rotation vacuum filter • Plate filter press • Membrane filter press • Belt filter • Vertical leaf filter • Pulse tube filter • Cricket filter
• Non-edible products, oleochemicals, biodiesel etc.	• Fatty acid separation, catalyst filtration, glycerin filtration, biodiesel pre-tretment, glycerin filtration, winterization and methyl ester/water separation	• Filtration • Centrifugation	• Vertical leaf filter • Pulse tube filter • Plate filter press • Chamber filter press • Centrifuge (high-speed)

Figure 13.6 Application of filtration as a mechanical separation process in the production of oils and fats.

or heat bleaching", is also used for the reduction of total pigments content (Gharby, 2022; Gupta, 2017; Seçilmiş et al., 2021; Taylor, 2005; Taylor et al., 2020).

Natural, non-activated bleaching clays are obtained from clay mineral deposits, dried, ground, and sieved to achieve the desired particle size range. Bentonite, a type of clay, is found in various forms worldwide, usually near the surface. At least 85% of this material consists of a form of aluminum silicate known as montmorillonite. Montmorillonite is characterized by its three-layered structure, with alternating layers of silicon dioxide above and below a central aluminum layer (List, 2009). Activated clay with acid is used for vegetable and animal fats. It also serves as a catalyst for polymerization and isomerization reactions. Activated clays can take the form of fixed layers or powdered additives (Fahn et al., 2004; List, 2009).

Mineral acids are used to increase the active surface area of bleaching clay. As the clay's surface area increases, the contact area between the clay and oil also increases, facilitating the removal of pigments from the oil (Seçilmiş et al., 2021). Before activation with mineral acids, the clay needs preparation. At the mining site, the clay is separated from foreign materials, then dispersed in a water and acid solution, mixed in the appropriate ratio. Typically, HCl or H_2SO_4 is used. Nevertheless, activated clays are more efficient compared to neutral ones, making these steps economically justified (List, 2009; Taylor et al., 2020).

Due to the significant consumption of bleaching earth in the bleaching process, and losses of oil with the bleaching earth, a large proportion of oil refining costs are incurred in the bleaching step. (Alhamed & Al-zahrani, 1999). On average, bleaching clay absorbs approximately 0.36 kg of oil per kg of clay (Gupta, 2017). The consumption of bleaching clay leads to environmental concerns (Low et al., 2022). Additionally, acid-activated clay causes the formation of oxidation products. Apart from clay, oil bleaching can also be accomplished using activated carbon, as mentioned earlier. Powdered activated carbon (PAC) is particularly advantageous in the fats and oils industry. Essential characteristics of an adsorbent include surface area, capillary volume per unit mass, cross-sectional surface area of different capillaries, particle size, chemical surface properties, and the nature of the adsorbent itself (List, 2009; Santoso et al., 2020). They are well-suited for adsorbing larger molecules (e.g., pigments), making them particularly useful for removing pigments or other undesirable minor components from oil.

The main factors affecting the bleaching stage include oil quality, the quantity and type of bleaching clay, and the process conditions. Oil of good quality is characterized by low levels of phospholipids, free fatty acids, and soaps. Other important factors include moisture content, the absence of air, temperature, and decolorization time (Seçilmiş et al., 2021; Taylor, 2005; Taylor et al., 2020). The effect of moisture on the decolorization process is difficult to understand since bleaching clay and oil have different moisture contents. (Erickson, 1995). For instance, Zschau (2001) reported that adding a specific level of moisture has a favorable effect on canola oil bleaching. Conversely, Shaw et al. (1985) concluded that adding water was not beneficial in the case of soybean oil bleaching. Reduced bleaching efficiency was observed with varying moisture concentrations. Since oils coming from vacuum dryers typically have nearly optimal moisture content, the best advice is to use moisture addition judiciously, only when specific circumstances and experiences demand it. Moisture content in the range of 0.1%–0.2% is close to optimal.

During the bleaching stage, it is important to exclude air, which is achieved by conducting bleaching under reduced pressure conditions. The presence of oxygen during bleaching increases the possibility to form reddish-yellow compounds (chroman-5,6-quinones). If oxidative changes occur during bleaching, new pigments resistant to adsorption develop, and the adsorbent strongly catalyzes the oxidation process. During bleaching, the pressure (<50 mmHg), and recommended vacuum levels are in the range of 20–28 mmHg or even lower (Taylor et al., 2020). The equipment used in the bleaching step is shown in the process flow diagram in (Fig. 13.7).

Generally, the bleaching process is carried out at temperatures between 90 and 120°C for 20–30 min, using 0.5%–2% clay (Zschau, 2001). Pigment content rapidly decreases during the initial 15–20 min of bleaching, reaching equilibrium conditions during that time and then gradually stabilizing (Henderson, 1993; Taylor et al., 1989; Taylor et al., 2020). Peroxide content, as the primary oxidation product, slowly decreases, while free fatty acid content gradually increases during the 30-minute bleaching time. Although some oils, especially palm oil, may require more time for

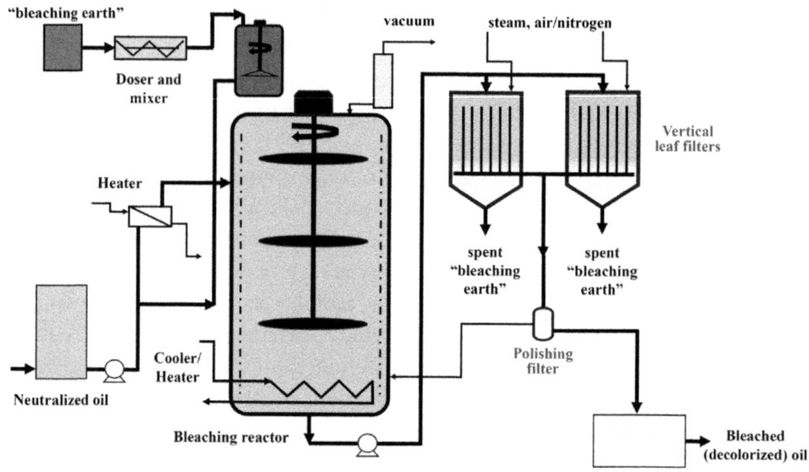

Figure 13.7 Process flow diagram of bleaching process.

bleaching, the above pattern will hold. Specifically, adsorption and catalytic processes are initially strong but then begin to equalize until moisture is removed from the clay and oil through heat and vacuum. After a certain time, most of these processes will be completed, and longer retention times will yield diminishing returns. Higher temperatures enhance pigment adsorption and various catalytic reactions. Consequently, pigment and peroxide levels decrease with increasing temperature, while free fatty acid content dramatically increases due to intensified triacylglycerol hydrolysis reactions. Higher temperatures during bleaching are desirable because they allow for increased pigment removal and peroxide decomposition with a smaller amount of bleaching clay. However, a challenge arises from elevated levels of free fatty acids generated at higher bleaching temperatures. Although these can be removed during deodorization, they have a relatively low value compared to triacylglycerols and represent a significant economic loss. Summarizing the presented data, general guidelines for optimizing the bleaching process include maintaining low oil moisture content (0.1%–0.2%), though not necessarily zero, maintaining adequate pressure (<50 mmHg), minimizing bleaching time and ensuring that the bleaching temperature does not exceed approximately 120°C (Taylor et al., 2020).

The efficiency of bleaching improves as the average particle size of the bleaching clay decreases, suggesting that using finely ground clay would reduce the required quantity for achieving a specific bleaching effect (Henderson, 1993; Richardson, 1978). However, the flow rate of oil through the filter cake decreases as the average particle size of the bleaching clay decreases. Additionally, oil retention increases with reduced particle size. Nevertheless, the oil loss can be partially offset by using less clay. The loss of filtration speed is, in any case, a more serious problem as it impacts bleaching capacity.

The essential part of the bleaching process is filtration. The method of filtration and the choice of filter type depend on the type and quality of oil being refined, the bleaching temperature, the amount of impurities removed from the oil, the type of "bleaching earth" used, and the filtering aids employed. The duration of the filtration cycle during the bleaching step varies based on the oil being processed, the type of "bleaching earth" used, the bleaching temperature, and the content of soap, phospholipids, and moisture in the oil (Veldkamp, 2022). Adequate filtration equipment and operation are essential for an efficient and cost-effective filtration process. The clay provides a barrier or initial coating layer sufficient to retain all suspended solid matter from the filtered liquid. Additionally, ensuring maximum porosity is crucial so that the maximum quantity of suspended solids can be retained before clogging occurs. When the septum is clogged, cleaning becomes necessary, which, in turn, disrupts the flow. Clearly, when suspended solid matter is lower in size than the cake pores, it will pass through them. Hence, commercially available adsorbents come in various classes, allowing retention of different particle sizes. Precoating is achieved by recirculating the clay suspension in oil through the filter. A minimal amount of cake is needed to form a stable layer used for further filtration. The minimum required quantity pertains to the filter surface area – at least 0.5 kg of clay/m^2 of filter surface. When the clay content falls within the range of 0.5%–2%, oil clarity should be achieved after 10–15 min of recirculation (Taylor et al., 2020). Various filter designs are available today, including filter presses, candle filters, leaf filters (horizontal pressure leaf filters and vertical pressure leaf filters), and centrifugal discharge filters (disk-filters), as shown in (Fig. 13.6). Old types of filter presses used in the bleaching step had plates made of cast iron, and a suspension was introduced into the frame. This suspension needed to be filtered through a canvas that covered the profiled surface of the frames. This design allowed the filtrate to flow out through an outlet tap located on each frame. Once the space for the filter cake was filled, the presses had to be opened, and the used "bleaching earth" fell into special catchers and was further transported by a screw conveyor. For the proper functioning of filter presses, the quality of the filter cloth is crucial. It has to have sufficient mechanical strength to withstand pressure differences that occur during filtration and a large number of repeated washings. The pore size of the filter cloth needs to be appropriate. There is a compromise between achieving high filtration efficiency (which requires larger pores) and the need for a shorter flow time for turbid filtrate (which demands smaller pores). During filtration, a filter cake layer forms, which later takes on the role of the filter cloth, and further filtration occurs through this layer. In the initial phase of filtration, solid particles pass through with the filtrate, resulting in a turbid filtrate. Therefore, there has to be a way to return this cloudy filtrate back for further filtration until clear filtrate is obtained. Additionally, the filter cloth should have minimal pore-clogging tendencies and a long service life. Continuous operation filter presses are closed-type filters that are fully automated. Even with these filters, interruptions occur during the emptying period, so there is no completely continuous operation. Due to the emptying period, two filters have to be installed alternately with each bleaching apparatus. Candle filters contain vertical tubes hanging in the filtration space. These tubes have a large number of circular openings, and filter cloth is pulled over them. The upper part of all tubular filtration elements is connected to the filtrate outlet pipes. The filter is filled with a suspension of bleaching clay/oil, and an initial filtration layer is formed through which further filtration occurs. When a thick filter

Figure 13.8 Vertical pressure leaf filter used in the bleaching step.

cake layer forms and filtration slows down due to pressure drop, the filter is emptied by blowing nitrogen or steam in the opposite direction to the filtrate flow. Leaf filters have filtration elements in the form of hanging leaves in the filtration space filled with suspension. In principle, these filters work similarly to candle filters, with the only difference being the method of removing the filter cake, which is done by shaking in this case. The construction of these filters can be either vertical or horizontal (Bokish, 1998; Gupta, 2017). The industry has embraced vertical pressure leaf filters (Fig. 13.8) due to the best performance in terms of costs, size, and handling. Filter sizes range up to 85 m^2 of filter surface. However, horizontal pressure leaf filters (Fig. 13.9) are also prevalent. The photographs in these figures show vertical and horizontal pressure leaf filters in industrial conditions. The great advantage of these filters is that units with a size of 200 m2 of filtration surface are available (Taylor et al., 2020).

When the filter cake reaches the designed thickness (between 18 and 22 mm), the filter tank is emptied, and the residual oil is removed by blowing with dry steam or air (or air/inert gas). In conventional filter systems, clay particles must be separated from the clay suspension in the oil by passing through a clay precoat. This precoat provides a barrier or initial coating layer sufficient to retain all suspended solid matter from the filtered liquid while still allowing economically justified oil quantities to pass through. If the pores are too small, the filtration speed decreases to the point

Figure 13.9 Horizontal pressure leaf filter used in the bleaching step.

where the filtration process is no longer economical. To overcome this issue, a process called electro-filtration was developed, allowing the use of fine-grained (<10 mm) bleaching clays. Electro filtration employs an electric field to agglomerate super-fine clay particles before their filtration in conventional filters. With this process, clay usage has been reduced by up to 40% (Transfeld, 1998). Disk-filters use horizontal plates as filtration elements, which are placed on a large rotating shaft. The filtration elements are housed in a vessel designed to withstand operating pressures of five bars (up to 30 bars) and operating temperatures of 100°C. The filtration element has a stainless-steel mesh with pore sizes ranging from 50 to 130 μm. Filtration capacity ranges from 300 to 600 l $m^{-2}h$. The filter cake is removed by rotating the central shaft and the action of centrifugal force. Cake removal is very fast, taking only 1–2 m.

13.6 Winterization (dewaxing)

Winterization or dewaxing refers to the process where waxes and/or triacylglycerols with higher melting points are removed from the oil. Essentially, winterization simplifies the process of fractionation. In some cases, winterization is necessary only to improve the oil's appearance. For many oils, cooling them to refrigerator temperatures causes waxes and triacylglycerols with higher melting points to separate, resulting in oil cloudiness and compromising its sensory properties (Adhvaryu et al., 2002; Ghosh et al., 2015). Residual waxes in the oil can lead to cloudiness in the refined edible oil. In addition to waxes, oil clouding (cloudiness, turbidity) can also be caused by the presence of high-melting triacylglycerols, as well as various products of oil polymerization reactions (Fig. 13.10). If the oil undergoes further processing, such as hydrogenation, these waxes can adversely affect the catalyst and create filtration problems for hydrogenated fats. Winterization is not essential if oils are used as

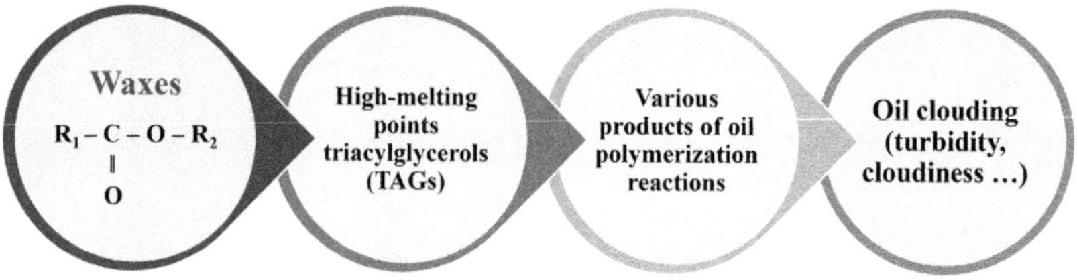

Figure 13.10 Oil cloudiness causes.

components in food or margarine, where the need for winterization is typically absent (Bockisch, 1998; Dijkstra, 2013).

Since waxes are considered undesirable components in oil due to their tendency to crystallize at room temperature and cause cloudiness in refined oils (Martini & Añón, 2003; Nedić Grujin et al., 2023), the classic winterization method involves gradual crystallization and separation using filters. The first phase involves cooling the oil with gentle mechanical mixing, leading to wax crystallization. Subsequently, the wax crystals are separated from the oil through filtration using appropriate filters (Fig. 13.6) (Chalapud et al., 2017).

Numerous factors influence the winterization process. Notably, accompanying substances present in the oil play a significant role in wax removal during winterization. These components include phospholipids and free fatty acids, which, due to their emulsifying properties, contribute to more effective wax removal. Therefore, the most rational approach would be to perform winterization after all refining stages, including deodorization or deodorization/deacidification. Temperature and crystallization period play a crucial role in the effectiveness of winterization. Based on oil producers' experience and data from various authors, it is concluded that the crystallization temperature has to be below 15°C, with a minimum retention time of 4 h (Gharby, 2022; Gupta, 2017). Optimal results are achieved when crystallization is carried out at 10°C, with a crystallization period of 10 h.

The most common method for wax removal from oil is classical winterization (Fig. 13.11). During this process, the oil undergoes crystallization followed by filtration to separate waxes after crystallization. The crystallization process is carried out in two phases. In the first phase, the oil is rapidly cooled to the desired temperature with intense mixing, leading to the formation of crystalline nuclei. In the second phase, slow crystal growth occurs. The oil is gently stirred and maintained at a constant temperature. Due to its low thermal conductivity, a large cooling surface area is required. When the entire oil quantity cools rapidly, small crystal nuclei form, which then continue to grow rapidly, yielding large crystals suitable for separation through filtration. If the cooling rate is inadequate, a mass of small crystals forms, which is unsuitable for filtration. Some authors suggest that the required crystallization time ranges from 18 to 36 h, while others believe that complete wax separation occurs by holding the oil at 10°C for 4 h (Bockisch, 1998). The final stage involves separating the wax crystals using filtration. Despite the seemingly straightforward procedure,

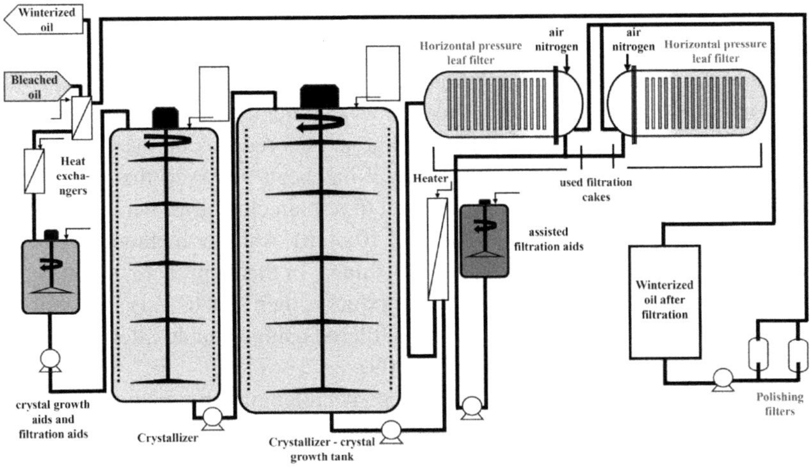

Figure 13.11 Process flow diagram of classical winterization process.

classical winterization presents several challenges. It is labor-intensive and time-consuming, and waxes are often not fully separated, resulting in relatively high oil losses (significant oil residue remains in the filter cake after filtration) (Bockisch, 1998). Additionally, oil cloudiness due to residual waxes may reappear after a certain length of time.

Solvent winterization is applied to more effectively separate waxes, resulting in oil losses up to three times lower. Hexane is commonly used as a solvent. Filtration is carried out at lower temperatures, typically ranging from −5 to −7°C. The solvent is used to reduce the oil's viscosity, facilitating wax crystallization. Although the entire process is somewhat more complex, the lower oil losses justify the use of this method. For sunflower oil, classical winterization results in losses of approximately 18 kg per ton, while winterization using a solvent reduces losses to 9 kg per ton (Chalapud et al., 2017). The viscosity and the amount of removed wax are lower, resulting in overall reduced oil losses during refining (Bockisch, 1998).

An alternative winterization process has been developed using centrifugal separators for crude oils with higher wax content. In this process, neutral oil is cooled, and a specific amount of low-concentration sodium hydroxide solution is added to increase the soap content. The oil is then sent to crystallizers equipped with mixers, where waxes agglomerate around soap particles. After aging, the mixture is transferred to another crystallizer, where a small amount of chilled water is added (Landucci et al., 2013). Following the aging period, the oil is heated using a heat exchanger before separation in a centrifugal separator. The centrifugal separator removes a significant portion of the wax. Finally, the oil is washed with heated water. This process is known as "wet" winterization. Another variation involves combined cold neutralization and wax removal using centrifugal separators. In this process, the oil is cooled before neutralization. The reaction occurs at low temperatures with extended retention times, but a small excess of sodium hydroxide is used. After the process, high-quality oil is obtained, but the losses are relatively high, and the separator capacity is significantly reduced. Additionally, soap or soapstock is challenging to handle. Although centrifugation is excellent for oils with high wax content, this method does not guarantee oil without the possibility of subsequent cloudiness. To further reduce wax content, one variation of the process involves cold water washing after the separator. This reduces the wax content in the oil by an additional 10–20 ppm. However, it has been observed that even after this process, the oil often remains cloudy. For oil to remain clear after winterization, it has to have extremely low wax content (around 10 ppm). Another method used to ensure clear oil is post-winterization filtration (polishing). In this approach, the oil is cooled, and an assisted filtration aid is added. The mixture is then sent to a crystallization tank. After an appropriate aging period, the oil is filtered using suitable filters, like classical winterization (Manjula & Subramanian, 2009).

Oil filtration during the winterization step is challenging due to the presence of oil turbidity-causing particles. During filtration, solid substances separate from the liquid, remaining on the filter medium (Ripperger et al., 2011). Typically, these particles causing oil cloudiness are small and compressible. These impurities very quickly lead to clogging of the pores of the filtration medium, thereby reducing the capacity of the filter and the quality of the filtered oil. To prevent filter medium clogging and reduce its specific resistance, thereby extending the filtration cycle duration, a filtering aid has to be used. This aid should not affect the oil's chemical composition, aroma, or taste and must consist of solid, finely porous, non-compressible particles. An assisted filtration aid should be of suitable material, chemically resistant, and of suitable purity (such as perlite, diatomaceous earth, cellulose, starch, etc.). Additionally, the choice of the filtration aid's granulation is essential: the particles should be coarse enough to provide suitable porosity for the filtration cake but fine enough to prevent the passage of smaller particles. The amount of assisted filtration aid required for filtration should match the volume of impurities being separated. The filterability of this mixture is then determined, and if the flow rate is low, a larger quantity of the assisted filtration aid should be used (Ripperger et al., 2011). Before filtration, the assisted filtration aid is applied to the filter media as a precoat layer. During filtration, it is continuously dosed into the oil before the filter. This significantly reduces the resistance of the layer and ensures both high flow rates and effective separation (Buyel et al., 2015; Smith, 1998). Filtration without assisted aids would be impossible because using the filter alone would result in surface filtration (Anlauf, 2019). Nowadays, filtration aids primarily consist of natural cellulose fibers (Nechyporchuk et al., 2016). Cellulose-based filter aids have low specific weight and reduce the consumption of assisted filtration aids, minimizing oil losses. Due to the low fiber density, the filtration aid remains easily suspended without settling. Precoat filtration, which utilizes a thin layer of $\approx$0.5 1.0 kg m^{-2} deposited on the filter medium before initiating the feed into the filter, is typically employed to protect the filter medium from contamination by capturing solid materials. This process also ensures a finer matrix for capturing fine solid particles and provides clarity to the filtrate (Dahlstrom et al., 1997). The filtration cake resulting from this process can be used in livestock feed production because it does not contain toxic substances or silicate crystals. Cellulose fibers do not cause wear and tear on filters, pumps, or filtration equipment. Additionally, the fibrous mesh structure

Figure 13.12 Horizontal pressure leaf filter used in the winterization step.

of the filtration cake makes it easier to remove from the filter (Di Pretoro & Manenti, 2020; Du et al., 2011; Nedić Grujin et al., 2023).

Membrane filtration can also be applied for wax removal. Waxes are removed through microfiltration using membranes with pore sizes ranging from 0.5 to 1 μm. The oil temperature has to be below 20°C. After microfiltration, very few waxes remain in the filtrate (Ghosh, 2007). Membranes used for this purpose must have sufficiently small pores to retain wax crystals, be resistant to oil exposure, be physically neutral toward the oil, prevent swelling, withstand pressure from both the external and internal sides, and allow for both forward and reverse filtration and backflushing. The fundamental element of membrane winterization is a cylinder made of alloyed steel, with parallel membrane tubes positioned inside. These membrane tubes are open at both ends and are secured and sealed using epoxy resin, completely separating the inner part of the membrane tubes from the space containing the wax suspension in the oil. The suspension is filtered through the membranes, and the clear oil fills the inner part of the membrane tubes, while the wax collects on the outer surface of the membrane tubes (Bockisch, 1998). In addition to the above-mentioned filters, the horizontal pressure leaf filter (Fig. 13.12), membrane filter press, and pulse tube filter can also be used in the waxes removal process (Fig. 13.6).

Although high-quality oil is obtained at the end of the refining process, it is advisable to perform additional polishing filtration on the oil at low temperatures, just before packaging edible refined oil. After oil deodorization or deodorization/deacidification using steam, it often happens that corrosion particles from devices and equipment, e.g. tubing and boilers end up in the oil. For this reason, the oil undergoes final filtration, i.e., polishing (Fig. 13.13). Polishing can be described as giving a final polish to edible oil. For this type of filtration, paper filters, cartridge filters, and bag filters (single/multi) are commonly used, Fig. 13.6. Filter cartridges made of polypropylene could retain particles in the range of 15–25 μm, providing excellent results and relatively long

Figure 13.13 Polishing filters.

filtration cycles. However, if the cartridges are impregnated with solid material, they need to be regenerated, which increases costs. Bag filters have a small surface area and behave like surface filters, requiring frequent bag replacements. They can also be washed several times, but after a certain number of washes, they must be replaced with new filters.

Recent advancements in the development of filters used in edible oil production have aimed at enhancing process efficiency from both economic and oil quality perspectives. Significant progress has been made with the introduction of nano- and ultrafiltration technologies, which enable lower energy consumption and the ability to operate at lower temperatures, thereby helping to preserve the nutritional quality of the oil. Despite these advancements, several challenges remain in the application of filtration in edible oil refining. Foremost among these are environmental concerns related to the management of waste and by-products generated during the refining process. The treatment of oily wastewater, which contains a complex mixture of fats, oils, and other contaminants, continues to pose a significant environmental issue. Traditional treatment methods, such as coagulation and flocculation, are often insufficient, prompting the exploration of more advanced techniques like electrocoagulation and membrane filtration (Kaya & Hung, 2021; Sonawane & Waghmode, 2023).

13.7 Conclusion

Crude oil's refining and the production of edible vegetable oils would not be possible without the inclusion of filtration. Filtration is a mechanical technological operation used to separate solid particles from liquids by passing them through a filter medium that retains the solid particles. Filtration is an integral part of oil bleaching, winterization, and final oil filtration. The choice of filtration method and filter type depends on several factors: the refining steps where filtration is used, process conditions (temperature, time), oil type and quality, impurity content, the type of "bleaching earth" used, and the filtration aids employed. The manufacturer's goal is to select filters with maximum efficiency while keeping the process economical. From this perspective, it has been determined which filters are most commonly used. In the bleaching step, vertical leaf filters, horizontal pressure leaf filters, plate filter presses, candle filters, and disk filters are commonly used. For winterization, horizontal pressure leaf filters, membrane filter presses, and pulse tube filters are preferred. Finally, for polishing filtration, paper filters, single/multi-bag filters, and cartridge filters are most frequently employed. Further research on the application of various filter constructions should focus on optimizing production processes. Producers of edible refined oil must tailor optimization efforts to their specific requirements, including capacity, crude oil quality, refining methods, and economic constraints. This optimization can be achieved through the selection of appropriate filter constructions and the adjustment of process parameters. The application of machine learning techniques, which are becoming increasingly prevalent, can significantly enhance this optimization process.

Acknowledgement

This research was supported by the Science Fund of the Republic of Serbia, #Grant No 7752847, Value-Added Products from Maize, Wheat, and Sunflower Waste as Raw Materials for the Pharmaceutical and Food Industry – PhAgroWaste.

References

Adhvaryu, A., Erhan, S. Z., & Perez, J. M. (2002). Wax appearance temperatures of vegetable oils determined by differential scanning calorimetry: Effect of triacylglycerol structure and its modification. *Thermochim. Acta, 395*(1–2), 191–200. https://doi.org/10.1016/S0040-6031(02)00180-6.

Akhmedov, A. N., Atakulova, D. T., Kholmurodova, Z. D., & Shaymardonova, G. X. (2022). Analysis of chlorophylls in lowquality cottonseed oil. *The Austrian Journal of Technical and Natural Sciences, 9-10*, 18–21. https://doi.org/10.29013/ajt-22-9.10-18-21.

Alhamed, Y., & Al-Zahrani, A. (1999). Techno-economical evaluation of oil recovery and regeneration of spent bleaching clay. *Journal of King Abdulaziz University-Engineering Sciences, 11*(2), 115–126. https://doi.org/10.4197/eng.11-2.10.

Anlauf, H. (2019). *Wet cake filtration: Fundamentals, equipment, and strategies* (pp. 1–349). Wiley https://doi.org/10.1002/9783527820665.

Bishop, G. J., & Yokota, T. (2001). Plant steroid hormones, brassinosteroids: Current highlights of molecular aspects on their synthesis/metabolism, transport, perception and response. *Plant Cell Physiol., 42*, 114–120. https://doi.org/10.1093/pcp/pce018.

Bhatnagar, A. S., & Gopala Krishna, A. G. (2014). Lipid classes and subclasses of cold-pressed and solvent-extracted oils from commercial Indian Niger (*Guizotia abyssinica* (L.f.) Cass.) seed. *Journal of the American Oil Chemists' Society, 91*(7), 1205–1216. https://doi.org/10.1007/s11746-014-2461-y.

Bokish, M. (1998). The importance of fats. In M. Bokish (Ed.), *Fats and oils handbook* (pp. 1–52). AOCS Press https://doi.org/10.1016/B978-0-9818936-0-0.50006-8.

Brigelius-Flohé, R., & Traber, M. G. (1999). Vitamin E: Function and metabolism. *FASEB J., 13*, 1145–1155. https://doi.org/10.1096/fasebj.13.10.1145.

Broughton, R., Ruíz-Lopez, N., Hassall, K. L., Martínez-Force, E., Garcés, R., Salas, J. J., & Beaudoin, F. (2018). New insights in the composition of wax and sterol esters in common and mutant sunflower oils revealed by ESI-MS/MS. *Food Chem., 269*, 70–79. https://doi.org/10.1016/j.foodchem.2018.06.135.

Brühl, L. (1996). Determination of trans fatty acids in cold pressed oils and in dried seeds. *Fett/Lipid, 98*(11), 380–383.

Buyel, J. F., Opdensteinen, P., & Fischer, R. (2015). Cellulose-based filter aids increase the capacity of depth filters during the downstream processing of plant-derived biopharmaceutical proteins. *Biotechnol. J., 10*(4), 584–591. https://doi.org/10.1002/biot.201400611.

Carelli, A. A., Brevedan, M. I. V., & Crapiste, G. H. (1997). Quantitative determination of phospholipids in sunflower oil. *Journal of the American Oil Chemists' Society, 74*(5), 511–514. https://doi.org/10.1007/s11746-997-0173-2.

Chalapud, M. C., Baümler, E. R., & Carelli, A. A. (2017). Characterization of waxes and residual oil recovered from sunflower oil winterization waste. *Eur. J. Lipid Sci. Technol., 119*(2). https://doi.org/10.1002/ejlt.201500608.

Choe, E. (2008). Effects and mechanisms of minor compounds in oil on lipid oxidation. In C. C. Akoh, & D. B. Min (Eds.), *Food lipids: Chemistry, nutrition and biotechnology* (3rd ed.) (pp. 449–511). CRC Press/Taylor & Francis Group.

Choe, E., & Min, D. B. (2006). Mechanisms and factors for edible oil oxidation. *Compr. Rev. Food Sci. Food Saf., 5*(4), 169–186. https://doi.org/10.1111/j.1541-4337.2006.00009.x.

CodexAlimentarius. (1999). *Standard for named vegetable oils codex Stan 210-1999*. Codex Alimentarius.

Coni, E., Di Benedetto, R., Di Pasquale, M., Masella, R., Modesti, D., Mattei, R., & Carlini, E. A. (2000). Protective effect of oleuropein, an olive oil biophenol, on low density lipoprotein oxidizability in rabbits. *Lipids, 35*(1), 45–54. https://doi.org/10.1007/s11745-000-0493-2.

Dahlstrom, D. A, Bennett, R. C., Emmett, R. C., Harriott, P., Jr., Laros, T., Leung, W., McCleary, C. S., Morey, A. M. B., Oldshue, J. Y., & Priday, G. (1997). Liquid-Solid operations and equipment. In R. H. Perry, & D. W. Green (Eds.), *Perry's chemical engineers' handbook* (7th ed.) (pp. 1–133). McGraw-Hill Companies Inc.

De Leonardis, A., Mcciola, V., & De Felice, M. (2001). Chemical and commercial characteristics of cold pressed sunflower oils. *Italian Food and Beverage Technology, 25*, 46–52.

Desai, N. C., Mehta, M. H., Dave, A. M., & Mehta, J. N. (2002). Degumming of vegetable oil by membrane technology. *Indian J. Chem. Technol., 9*(6), 529–534.

Di Pretoro, A., & Manenti, F. (2020). Filtration. In A. Di Pretoro, & F. Manenti (Eds.), *Non-conventional unit operations: Solving practical issues* (1st ed.) (pp. 47–57). Springer.

Dijkstra, A. J. (2013). *Edible oil processing from a patent perspective* (pp. 1–275). Springer US https://doi.org/10.1007/978-1-4614-3351-4-1.

Dijkstra, A. J. (2015). Interesterification, chemical or enzymatic catalysis. *Lipid Technology, 27*(6), 134–136. https://doi.org/10.1002/lite.201500029.

Dimakou, C., & Oreopoulou, V. (2012). Antioxidant activity of carotenoids against the oxidative destabilization of sunflower oil-in-water emulsions. *LWT - Food Sci. Technol., 46*(2), 393–400. https://doi.org/10.1016/j.lwt.2011.12.013.

Du, L., Chen, X., Li, W., & Zhu, Q. (2011). A study on enhancement of filtration process with filter aids diatomaceous earth and wood pulp cellulose. *Chin. J. Chem. Eng., 19*, 792–798. https://doi.org/10.1016/S1004-9541(11)60058-X.

Eitenmiller, R., & Lee, J. (2004). *Vitamin E. Food chemistry, composition, and analysis.* Marcel Dekker.

El Makawy, A. I., Ibrahim, F. M., Mabrouk, D. M., Ahmed, K. A., & Ramadan, M. F. (2019). Effect of antiepileptic drug (*Topiramate*) and cold pressed ginger oil on testicular genes expression, sexual hormones and histopathological alterations in mice. *Biomed. Pharmacother., 110*, 409–419. https://doi.org/10.1016/j.biopha.2018.11.146.

Erickson, D. R. (1995). Bleaching/adsorption treatment. In D. R. Erickson (Ed.), *Practical handbook of soybean processing and utilization* (pp. 203–217). Elsevier https://doi.org/10.1016/b978-0-935315-63-9.50016-4.

Eskin, M., Aladedunye, F., Unger, E., Shah, S., Chen, G., & Jones, P. (2020). Canola oil. In F. Shahidi (Ed.), *Bailey's industrial oil and fat products* (7th ed.) (pp. 1–63). John Wiley & Sons.

Fahn, Y. Y., Shen, P., & Su, A. C. (2004). Biomimetic co-assembly of gibbsite and mesophases: A novel synthesis route by devitrification of gel derived from hydrothermal aluminosilicate/CTAB solution. *Clays Clay Miner., 52*(4), 451–461. https://doi.org/10.1346/CCMN.2004.0520406.

FAO (2024). Food outlook: Biannual report on global food markets. Retrieved March 20, 2024, from https://www.fao.org/3/cc9701en/cc9701en.pdf.

Fernández-Martínez, J. M., Pérez-Vich, B., & Velasco, L. (2009). Sunflower. In J. Vollmann, & I. Rajcan (Eds.), *Oil crops* (pp. 155–232). Springer.

Flores-Sánchez, I. J., Ortega-López, J., Montes-Horcasitas, M. D. C., & Ramos-Valdivia, A. C. (2002). Biosynthesis of sterols and triterpenes in cell suspension cultures of *Uncaria tomentosa. Plant Cell Physiol., 43*(12), 1502–1509. https://doi.org/10.1093/pcp/pcf181.

Fruhwirth, G. O., Wenzl, T., El-Toukhy, R., Wagner, F. S., & Hermetter, A. (2003). Fluorescence screening of antioxidant capacity in pumpkin seed oils and other natural oils. *Eur. J. Lipid Sci. Technol., 105*(6), 266–274. https://doi.org/10.1002/ejlt.200390055.

Galli, F., & Azzi, A. (2010). Present trends in vitamin E research. *Biofactors, 36*, 33–42. https://doi.org/10.1002/biof.75.

Gerde, J. A., Hammond, E. G., Johnson, L. A., Su, C., Wang, T., & White, P. J. (2020). Soybean oil. In F. Shahidi (Ed.), *Bailey's industrial oil and fat products* (7th ed.). John Wiley & Sons.

Gharby, S. (2022). Refining vegetable oils: Chemical and physical refining. *Hindawi, The Scientific World Journal, 6627013*, 1–10. https://doi.org/10.1155/2022/6627013.

Ghosh, M. (2007). Review on recent trends in rice bran oil processing. *Journal of the American Oil Chemists' Society, 84*, 315–324. https://doi.org/10.1007/s11746-007-1047-3.

Ghosh, S., Pradhan, M., Patel, T., Haj-shafiei, S., & Rousseau, D. (2015). Long-term stability of crystal-stabilized water-in-oil emulsions. *J. Colloid Interface Sci., 460*, 247–257. https://doi.org/10.1016/j.jcis.2015.08.074.

Gordon, M. H., & Rahman, I. A. (1991). Effect of processing on the composition and oxidative stability of coconut oil. *J. Am. Oil Chem. Soc., 68*(8), 574–576. https://doi.org/10.1007/bf02660153.

Grompone, M. A. (2005). Sunflower oil. In F. Shahidi (Ed.), *Edible oil and fat products: Chemistry, properties, and health effects* (1, pp. 655–730). John Wiley & Sons.

Grompone, M. A. (2020). Sunflower and high-oleic sunflower oils. In F. Shahidi (Ed.), *Bailey's industrial oil and fat products* (pp. 1–54). John Wiley & Sons https://doi.org/10.1002/047167849x.bio017.pub2.

Gunstone, F. D. (2005). Vegetable oils. In F. Shahidi (Ed.), *Edible oil and fat products: Chemistry, properties, and health effects* (1, pp. 213–267). John Wiley & Sons.

Gupta, M. K. (2017). *Practical guide to vegetable oil processing* (2nd ed., pp. 1–488). AOCS Press https://doi.org/10.1016/B978-1-63067-050-4.00018-0.

Hartmann, M. A. (1998). Plant sterols and the membrane environment. *Trends Plant Sci., 3*, 170–175. https://doi.org/10.1016/S1360-1385(98)01233-3.

Henderson, J. H. (1993). A laboratory study of the press effect in adsorptive bleaching. *Journal of the American Oil Chemists' Society, 70*(9), 831–835. https://doi.org/10.1007/BF02545339.

Hernandez, E. M., & Kamal-Eldin, A. (2013). *Processing and nutrition of fats and oils*. Wiley https://doi.org/10.1002/9781118528761.

Hamm, W., Hamilton, R., & Calliauw, G. (2013). *Edible oil processing*. Wiley https://doi.org/10.1002/9781118535202.

Hrustić, M., Vidić, M., & Jocković, Đ. (1998). *Soja, STP astrodesign*. Belgrade.

Hwang, K. J., & Lin, T. T. (2002). Effect of morphology of polymeric membrane on the performance of cross-flow microfiltration. *J. Membr. Sci., 199*(1), 41–52. https://doi.org/10.1016/S0376-7388(01)00675-5.

Jocić, S., Miladinović, D., & Kaya, Y. (2015). Breeding and genetics of sunflower. In E. Martínez-Force, N. T. Dunford, & J. J. Salas (Eds.), *Sunflower chemistry, production, processing, and utilization* (pp. 1–25). AOCS Press https://doi.org/10.1016/B978-1-893997-94-3.50007-6.

Kamal-Eldin, A. (2005). Minor components of fats and oils. In F. Shahidi (Ed.), *Edible oil and fat products: Chemistry, properties, and health effects* (3, pp. 319–359). John Wiley & Sons.

Karleskind, A. (1996). *Oils and fats manual*. Intercept Ltd.

Karlović, Đ., & Andrić, N. (1996). *Kontrola kvaliteta semena uljarica/quality control of oilseeds*. University of Novi Sad, Faculty of Technology Novi Sad, Federal Institute for Standardization.

Kaya, D., & Hung, Y.-T. (2021). *Advances in treatment of vegetable oil refining wastes* (pp. 325–375) https://doi.org/10.1007/978-3-030-54626-7_8.

Kiralan, M., Çalik, G., Kiralan, S., & Ramadan, M. F. (2018). Monitoring stability and volatile oxidation compounds of cold-pressed flax seed, grape seed and black cumin seed oils upon photo-oxidation. *Journal of Food Measurement and Characterization, 12*(1), 616–621. https://doi.org/10.1007/s11694-017-9674-3.

Kochhar, S., Azadmard-Damirchi, S., Nasirpour-Tabrizi, P., & Dutta, P. (2020). Sterols, stanols, and waxes. In F. Shahidi (Ed.), *Bailey's industrial oil and fat products* (pp. 1–65). John Wiley & Sons.

Krist, S. (2020). *Vegetable fats and oils*. Springer International Publishing https://doi.org/10.1007/978-3-030-30314-3.

Kroon, P., & Williamson, G. (2005). Polyphenols: Dietary components with established benefits to health. *J. Sci. Food Agric., 85*, 1239–1240. https://doi.org/10.1002/jsfa.2204.

Lai, O., Phuah, E., Lee, Y., & Basiron, Y. (2020). Palm oil. In F. Shahidi (Ed.), *Bailey's industrial oil and fat products* (7th ed.) (pp. 1–54). John Wiley & Sons.

Lamas, D. L., Constenla, D. T., & Raab, D. (2016). Effect of degumming process on physicochemical properties of sunflower oil. *Biocatal. Agric. Biotechnol., 6*, 138–143. https://doi.org/10.1016/j.bcab.2016.03.007.

Lamas, D. L., Crapiste, G. H., & Constenla, D. T. (2014). Changes in quality and composition of sunflower oil during enzymatic degumming process. *LWT—Food Science and Technology Journal, 58*(1), 71–76. https://doi.org/10.1016/j.lwt.2014.02.024.

Landucci, G., Pannocchia, G., Pelagagge, L., & Nicolella, C. (2013). Analysis and simulation of an industrial vegetable oil refining process. *J. Food Eng., 116*(4), 840–851. https://doi.org/10.1016/j.jfoodeng.2013.01.034.

List, G. R. (2009). *Bleaching and purifying fats and oils: Theory and practice* (pp. 1–276). Elsevier Inc https://doi.org/10.1016/C2015-0-02456-X.

Low, A., Shamsuddin, R., & Siyal, A. A. (2022). Economic analysis of waste minimisation and energy recovery from spent bleaching earth. *Cleaner Engineering and Technology, 7*, 100418. https://doi.org/10.1016/j.clet.2022.100418.

Lužaić, T., Grahovac, N., Hladni, N., & Romanić, R. (2022). Evaluation of oxidative stability of new cold-pressed sunflower oils during accelerated thermal stability tests. *Food Sci. Technol., 42*, e67320. https://doi.org/10.1590/fst.67320.

Lužaić, T., Kravić, S., Stojanović, Z., Grahovac, N., Jocić, S., Cvejić, S., Pezo, L., & Romanić, R. (2023). Investigation of oxidative characteristics, fatty acid composition and bioactive compounds content in cold pressed oils of sunflower grown in Serbia and Argentina. *Heliyon, 9*, e18201. https://doi.org/10.1016/j.heliyon.2023.e18201.

Lužaić, T., Nedić Grujin, K., Pezo, L., Nikolovski, B., Maksimović, Z., & Romanić, R. (2024). Implementation of cellulose-based filtration aids in industrial sunflower oil dewaxing (Winterization): Process monitoring, prediction, and optimization. *Foods, 13*(18), 2960. https://doi.org/10.3390/foods13182960.

Manjula, S., & Subramanian, R. (2009). Simultaneous degumming, dewaxing and decolorizing crude rice bran oil using nonporous membranes. *Sep. Purif. Technol., 66*(2), 223–228. https://doi.org/10.1016/j.seppur.2009.01.004.

Martini, S., & Añón, M. C. (2003). Crystallization of sunflower oil waxes. *Journal of the American Oil Chemists' Society, 80*(6), 525–532. https://doi.org/10.1007/s11746-003-0732-y.

Martini, S., & Añón, M. C. (2005). Storage of sunflower seeds: Variation on the wax content of the oil. *Eur. J. Lipid Sci. Technol., 107*(2), 74–79. https://doi.org/10.1002/ejlt.200401044.

Matthäus, B., & Brühl, L. (2004). Cold-pressed edible rapeseed oil production in Germany. *INFORM—International News on Fats, Oils and Related Materials, 15*(4), 266–268.

Mène-Saffrané, L., & DellaPenna, D. (2010). Biosynthesis, regulation and functions of tocochromanols in plants. *Plant Physiol. Biochem., 48*(5), 301–309. https://doi.org/10.1016/j.plaphy.2009.11.004.

Nechyporchuk, O., Belgacem, M. N., & Bras, J. (2016). Production of cellulose nanofibrils: A review of recent advances. *Ind. Crops Prod.; 93*, 2–25. https://doi.org/10.1016/j.indcrop.2016.02.016.

Nedić Grujin, K., Lužaić, T., Pezo, L., Nikolovski, B., Maksimović, Z., & Romanić, R. (2023). Sunflower oil winterization using the cellulose-based filtration aid—Investigation of oil quality during industrial filtration probe. *Foods, 12*(12), 2291. https://doi.org/10.3390/foods12122291.

Outmans, H. D. (1963). Mechanics of static and dynamic filtration in the borehole. *Soc. Pet. Eng. J., 3*(03), 236–244. https://doi.org/10.2118/491-pa.

Oštrić-Matijašević, B., & Turkulov, J. (1980). *Tehnologija ulja i masti/oil and fat technology, I part*. University of Novi Sad, Faculty of Technology Novi Sad.

Pattee, H. E. (2005). Peanut oil. In F. Shahidi (Ed.), *Edible oil and fat products: Chemistry, properties, and health effects* (2, pp. 319–359). John Wiley & Sons.

Popović, S., Hromiš, N., Šuput, D., Bulut, S., Romanić, R., & Lazic, V. (2020). Valorization of by-products from the production of pressed edible oils to produce biopolymer films. In M. F. Ramadan (Ed.), *Cold pressed oils: Green technology, bioactive compounds, functionality, and applications* (pp. 15–30). Elsevier https://doi.org/10.1016/B978-0-12-818188-1.00003-7.

Prescha, A., Grajzer, M., Dedyk, M., & Grajeta, H. (2014). The antioxidant activity and oxidative stability of cold-pressed oils. *Journal of the American Oil Chemists' Society, 91*(8), 1291–1301. https://doi.org/10.1007/s11746-014-2479-1.

Przybylski, R., & Eskin, N. A. M (2011). Oil composition and properties. In J. K. Daun, N. A. M. Eskin, & D. Hickling (Eds.), *Canola, chemistry, production, processing, and utilization* (pp. 189–227). AOCS Press.

Ramadan, M. F. (2013). Healthy blends of high linoleic sunflower oil with selected cold pressed oils: Functionality, stability and antioxidative characteristics. *Ind. Crops Prod., 43*(1), 65–72. https://doi.org/10.1016/j.indcrop.2012.07.013.

Ramadan, M. F. (2019). *Fruit oils: Chemistry and functionality*. Springer Nature.

Ramadan, M. F. (2020). Introduction to cold pressed oils: Green technology, bioactive compounds, functionality, and applications. In M. F. Ramadan (Ed.), *Cold pressed oils: Green technology, bioactive compounds, functionality, and applications* (pp. 1–5). Academic Press https://doi.org/10.1016/b978-0-12-818188-1.00001-3.

Richardson, L. L. (1978). Use of bleaching, clays, in processing edible oils. *Journal of the American Oil Chemists' Society, 55*(11), 777–780. https://doi.org/10.1007/BF02682647.

Richardson, J. F., Harker, J. H., & Backhurst, J. R. (2002). Flow of fluids through granular beds and packed columns. In J. F. Richardson, J. H. Harker, & J. R. Backhurst (Eds.), *Chemical engineering series, chemical engineering* (5th ed.) (pp. 191–236). Butterworth-Heinemann https://doi.org/10.1016/B978-0-08-049064-9.50015-1.

Ripperger, S., & Altmann, J. (2002). Crossflow microfiltration—State of the art. *Sep. Purif. Technol., 26*(1), 19–31. https://doi.org/10.1016/S1383-5866(01)00113-7.

Ripperger, S., Gösele, W., & Alt, C. (2011). Filtration, 1. Fundamentals. In C. Goldmann (Ed.), *Ullmann's encyclopedia of industrial chemistry* (pp. 678–708). Wiley-VCH Verlag GmbH & Co: KGaA, Weinheim, Germany.

Romanić, R., Dimić, E., Lazić, V., & Vujasinović, V. (2009). Oxidative stability and tocopherol content of refined sunflower oil during long-term storage in different commercial packagings. *Acta Alimentaria, 38*(3), 319–327. https://doi.org/10.1556/aalim.2009.0001.

Romanić, R., Lužaić, T., & Radić, B. (2021). Enriched sunflower oil with omega 3 fatty acids from flaxseed oil: Prediction of the nutritive characteristics. *LWT—Food Science and Technology Journal, 150*, 112064. https://doi.org/10.1016/j.lwt.2021.112064.

Rotkiewicz, D., Konopka, I., & Zylik, S. (1999). State of works on the rapeseed oil processing optimization. I. Oil obtaining. *Rośliny Oleiste/Oilseed Crops, 20*, 151–168.

Rukmini, A., & Raharjo, S. (2010). Pattern of peroxide value changes in virgin coconut oil (VCO) due to photo-oxidation sensitized by chlorophyll. *Journal of the American Oil Chemists' Society, 87*(12), 1407–1412. https://doi.org/10.1007/s11746-010-1641-7.

Santoso, E., Ediati, R., Kusumawati, Y., Bahruji, H., Sulistiono, D. O., & Prasetyoko, D. (2020). Review on recent advances of carbon based adsorbent for methylene blue removal from waste water. *Materials Today Chemistry, 16*, 100233. https://doi.org/10.1016/j.mtchem.2019.100233.

Salas, J. J., Martínez-Force, E., & Garcés, R. (2006). Accumulation of phospholipids and glycolipids in seed kernels of different sunflower mutants (*Helianthus annuus*). *Journal of the American Oil Chemists' Society, 83*(6), 539–545. https://doi.org/10.1007/s11746-006-1237-4.

Salimath, S. S., Romsdahl, T. B., Konda, A. R., Zhang, W., Cahoon, E. B., Dowd, M. K., & Chapman, K. D. (2021). Production of tocotrienols in seeds of cotton (*Gossypium hirsutum*

L.) enhances oxidative stability and offers nutraceutical potential. *Plant Biotechnol. J., 19*(6), 1268–1282. https://doi.org/10.1111/pbi.1355.

Seçilmiş, Ş. S., Koçak Yanık, D., Fadiloğlu, S., & Göğüş, F. (2021). A comparative study on performance of industrial and microwave techniques for sunflower oil bleaching process. *Food Chem., 365*, 130488. https://doi.org/10.1016/j.foodchem.2021.130488.

Shahidi, F., & Zhong, Y. (2010). Lipid oxidation and improving the oxidative stability. *Chem. Soc. Rev., 39*(11), 4067–4079. https://doi.org/10.1039/b922183m.

Shaw, D. B., Taylor, R. S., Sidebottom, M. J., & Ball, M. G. (1985). The function of bleaching earths in the processing of soybean oil. *Journal of the American Oil Chemists' Society, 32*(4), 619.

Smith, G. (1998). Filter aid filtration. In T. H. Meltzer, & M. W. Jornitz (Eds.), *Filtration in the biopharmaceutical industry* (1st ed.) (1, pp. 1–69). Marcel Dekker Inc.

Smouse, T. H. (1995). Factors affecting oil quality and stability. In K. Warner, & N. A. M. Eskin (Eds.), *Methods to assess quality and stability of oils and fat-containing foods* (pp. 17–36). AOCS Press.

Sonawane, A., & Waghmode, S. (2023). A review on vegetable oil refining: Process, advances and value addition to refining by-products. https://doi.org/10.5772/intechopen.108752

Soomro, R. K., & Sherazi, S. T. (2013). Extraction and characterization of seed oil waxes by using chromatographic techniques. *International Journal of Industrial Chemistry, 4*(1), 9. https://doi.org/10.1186/2228-5547-4-9.

Sorrentino, G., Gajjar, P., Withers, P. J., Chellappah, K., & Biscontin, G. (2024). Experimental investigation of filter cake deposition under cross-flow and static filtration conditions. *Chem. Eng. Sci., 283*, 119417. https://doi.org/10.1016/j.ces.2023.119417.

Sánchez-Muniz, F. J., & Cuesta, C. (2003). Sunflower oil. In B. Caballero (Ed.), *Encyclopedia of food sciences and nutrition* (2nd ed.) (pp. 5672–5680). Academic Press.

Taylor, D. R. (2005). Bleaching. In F. Shahidi (Ed.), *Edible oil and fat products: Chemistry, properties, and health effects* (1, pp. 285–339). John Wiley & Sons.

Taylor, D. R., Jenkins, D. B., & Ungermann, C. B. (1989). Bleaching with alternative layered minerals: A comparison with acid-activated montmorillonite for bleaching soybean oil. *J. Am. Oil Chem. Soc., 66*(3), 334–341. https://doi.org/10.1007/BF02653285.

Taylor, D. R., Toro-Vazquez, J. F., & Charo-Alonso, M. A. (2020). Bleaching. In F. Shahidi (Ed.), *Bailey's industrial oil and fat products* (pp. 1–51). John Wiley & Sons.

Transfeld, P. (1998). Electrofiltration cuts bleaching earth use. *INFORM—International News on Fats, Oils and Related Materials, 9*(5), 502–506.

Tuberoso, C. I. G., Kowalczyk, A., Sarritzu, E., & Cabras, P. (2007). Determination of antioxidant compounds and antioxidant activity in commercial oilseeds for food use. *Food Chem., 103*(4), 1494–1501. https://doi.org/10.1016/j.foodchem.2006.08.014.

Tynek, M., Pawłowicz, R., Gromadzka, J., Tylingo, R., Wardencki, W., & Karlovits, G. (2012). Virgin rapeseed oils obtained from different rape varieties by cold pressed method—Their characteristics, properties, and differences. *Eur. J. Lipid Sci. Technol., 114*(3), 357–366. https://doi.org/10.1002/ejlt.201100296.

USDA. (2024). United States Department of Agriculture—Foreign Agricultural Service. *Oilseeds: World Markets and Trade*. Retrieved March 20, 2024, from https://apps.fas.usda.gov/psdonline/circulars/oilseeds.pdf .

Usman, I, U. , Kafle, S. , R., Abubakar, A. M., Shamsuddin, R., Itamah, E. I., & Abubakar, M. A (2023). Assessing coconut oil properties as a potential replacement of the transformer oil. *SPIN-Jurnal Kimia & Pendidikan Kimia, 5*(2), 200–208.

Van Der Bruggen, B., Vandecasteele, C., Van Gestel, T., Doyen, W., & Leysen, R. (2003). A review of pressure-driven membrane processes in wastewater treatment and drinking water production. *Environ. Prog., 22*(1), 46–56. https://doi.org/10.1002/ep.670220116.

Veldkamp, R. (2022). Optimising filtration. *Ofimagazine*, 20–24. Retrieved March 18, 2024, from https://www.ofimagazine.com/content-images/news/Filtration.pdf .

Virgili, F., Scaccini, C., Packer, L., & Rimbach, G. (2001). Cardiovascular disease and nutritional phenolics. In J. Pokorny, N. Yanishlieva, & M. Gordon (Eds.), Antioxidants in *food practical applications* (pp. 87–99). CRC Press.

Wang, T. (2008). Minor constituents and phytochemicals of soybeans. In L. A. Johnson, P. J. White, & R. Galloway (Eds.), *Soybeans, chemistry, production, processing and utilization* (pp. 1297–1329). AOCS Press.

Warner, K., Vick, B. A, Kleingartner, L., Isaac, I., & Doroff, K. (2003). Composition of sunflower, NuSun (mid-oleic sunflower), and high oleic sunflower oils, In: Proc. Sunflower Res. Workshop, Fargo, ND, 16–17 January 2003. National Sunflower Assoc., Mandan, ND, Available at http://www.sunflowernsa.com/research/research-workshop/documents/107.PDF.

Wen, C., Shen, M., Liu, G., Liu, X., Liang, L., Li, Y., Zhang, Y., & Xu, X. (2023). Edible vegetable oils from oil crops: Preparation, refining, authenticity identification and application. *Process Biochem., 124*, 168–179. https://doi.org/10.1016/j.procbio.2022.11.017.

Withana-Gamage, T., Perera, S., & Wanasundara, U. (2020). Coconut oil. In F. Shahidi (Ed.), *Bailey's industrial oil and fat products* (pp. 1–37). John Wiley & Sons.

Zou, Y., Jiang, Y., Yang, T., Hu, P., & Xu, X. (2012). Minor constituents of palm oil: Characterization, processing and application. In O. M. Lai, C. P. Tan, & C. C. Akoh (Eds.), *Palm oil: Production, processing, characterization, and uses* (pp. 471–526). AOCS Press.

Zschau, W. (2001). Bleaching of edible fats and oils. *Eur. J. Lipid Sci. Technol., 103*, 505–551. https://doi.org/10.1002/1438-9312(200108)103:8<505::AID-EJLT505>3.0.CO;2-7.

Application of filtration in sugar refining

14

Samiye Adal[a], *Özge Süfer*[b], *Gülşah Çalışkan Koç*[c] *and Yasemin Çelebi*[d]

[a] Department of Food Engineering, Faculty of Engineering, Afyon Kocatepe University, Afyonkarahisar, Türkiye, [b] Department of Food Engineering, Osmaniye Korkut Ata University, Osmaniye, Türkiye, [c] Department of Food Processing, Esme Vocational School, Usak University, Usak, Türkiye, [d] Food Technology Program, Department of Food Processing, Esme Vocational School, Usak University, Usak, Türkiye

14.1 Introduction

Sugar (commonly referred to as sucrose) has been a fundamental component of human diets for a considerable duration, and serves as a crucial ingredient for the food, beverage, and pharmaceutical sectors (Qi, 2022). Presently, sugarcane products consist predominantly of white sugar and noncentrifugal sugar, with the industrialization of by-products such as bagasse, filter mud (sludge), and molasses also being partially implemented (Wang et al., 2022). Noncentrifugal sugar, a substance derived from the evaporation of water in sugarcane juice, is recognized by different terms such as whole cane sugar, unrefined muscovado, jaggery, and kokuto. While scientific studies have validated the numerous health benefits of noncentrifugal sugar, it remains somewhat marginalized in the realm of functional foods and nutraceuticals (Jaffé, 2012). Sugarcane, which employs C4 photosynthesis, holds the capacity to produce a diverse array of bio-based products including ethanol, electricity, paper, vinasse, and environmentally sustainable goods. Additionally, it can also provide food, animal feed, bio-fertilizers, bio-plastics, and bio-molecules. The expansion of the sugar industry to include these diverse products could enhance its competitiveness in the market, reduce waste production, and have positive environmental impacts. Despite these benefits, many sugarcane-producing countries have been slow to adopt diversification strategies (Aguilar-Rivera et al., 2022), as raw sugar production is predominantly carried out in developing nations and subsequently shipped in large quantities to developed nations. The sugar is then refined and exported in diverse formats, such as railcars and bags. The advent of liquid sugar has introduced a novel dynamic, prompting brokers to represent buyers and sellers in the tangible market (Abbott, 2022). Embracing sugarcane diversification is crucial for optimizing the value chain of sugarcane, promoting sustainable agriculture practices, and working towards achieving sustainable development goals by the year 2030 (Aguilar-Rivera et al., 2022).

In this section of the book, the sugar-refining process was briefly summarized, and the filtration step was comprehensively reviewed. Therefore, the objective of this chapter is to thoroughly evaluate the effectiveness and potential of different filtration

Mechanical Separation Processes in the Food Industry: Unit Operations and Processing Equipment in the Food Industry.
DOI: https://doi.org/10.1016/B978-0-12-819523-9.00012-3

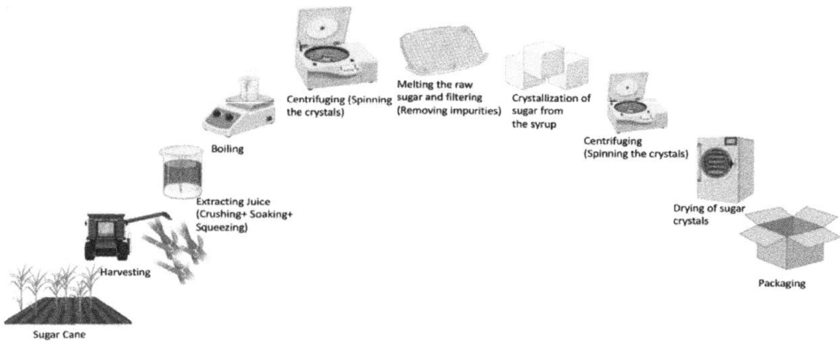

Figure 14.1 General processing steps of sugar refining process.

methods, specifically membrane technology, in improving operations in the sugar manufacturing industry. It aims to investigate the wide range of uses of filtration techniques, improve the quality of sugar, minimize environmental impact, and potentially identify opportunities for advancement and cost-effectiveness.

14.2 General view of sugar-refining process

The global sugar demand is primarily met by two sources: sugarcane and sugar beet. Sugarcane accounts for almost 60% of global sugar production, while the remaining portion is derived from sugar beets (Cheremisinoff et al., 2013). Sugarcane has been recognized for at least 2,000 years, with documented production of crystal sugar from it for a comparable duration. This is evidenced by the Sanskrit term “sharkara” which refers to a substance in a granular state, and serves as the etymology of the term “sugar” in contemporary languages. Production of sugar derived from the beet sugar plant emerged relatively recently, with its development dating back to the early 19th century (Jenkins, 2013). As the 21st century dawned, the sugar industry faced a myriad of opportunities and challenges: environmental pressures, diminishing resources (such as food and water), restricted energy supply, concerns regarding health and safety, adherence to regulatory standards, scrutiny of the nutritional value of refined sugar, global competition, and a constantly changing global economy (Chou & Engel, 2009).

Fig. 14.1 outlines the general processing steps of the sugar-refining process. The initial step of the sugar-refining process is harvesting the sugarcanes. The sugar-producing industry is characterized by its strong seasonality, with beets having a season length of $\approx$6–18 weeks and cane having a season length of 20–32 weeks. Approximately 10% of the entire sugarcane yield holds the potential for conversion into commercially viable sugar, requiring an estimated water usage of around 20 m^3 ton^{-1} of processed cane. Sugarcane comprises 70%, 14%, 13.3%, and 2.7.% of water, fiber, sucrose ($\approx$10% to $-$15% sucrose content), and soluble impurities, respectively (Cheremisinoff et al., 2013).

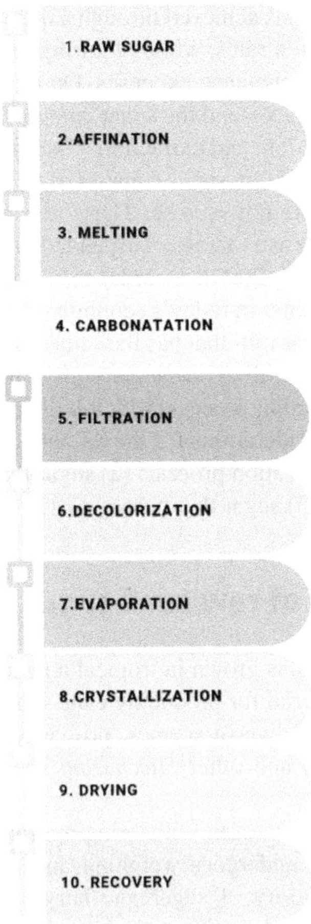

Figure 14.2 Flow diagram of a typical sugar refining process.

The main objective of the cane sugar-refining process is to remove colored compounds that can negatively impact the final product's quality. The traditional refining method removes 40–50% of the pigmented impurities through carbonation or phosphatation and eliminates an additional 40–60% of the remaining color through decolorization using resins, bone char, or granular carbon (Cartier et al., 1997). As depicted in Fig. 14.2, the sugar-refining process involves the following stages: affinitive, melting, carbonatation, filtration, decolorization, evaporation, crystallization, drying, and recovery. Sugarcanes regularly go through a washing process, followed by the extraction of juice from them. The juice undergoes various stages of processing, starting with clarification to remove mud, followed by evaporation to concentrate the syrup, crystallization to separate the liquor, and centrifugation to separate molasses from the crystals. Sugar crystals undergo a drying process and may undergo additional refining prior to being packaged for transportation. In certain regions, such as

South Africa, juice extraction is achieved through a diffusion method, which offers the advantage of higher extraction rates, while requiring less energy, thereby resulting in decreased operating and maintenance expenses. On the other hand, the processing of sugar beets (water content is 75% and the sugar content is 17%) involves three distinct processes: washing, preparation, and extraction (Cheremisinoff et al., 2013).

The typical composition of raw sugar consists of ≈ 97.52% sucrose, 0.48% water, and 2% contaminants that are not sucrose. These nonsucrose elements consist of reducing sugars (23%), soluble ash (38.5%), suspended impurities (3%), and unspecified substances (35.3%) (El-Syiad, 2000). In order to produce refined sugar in the most efficient and sustainable manner in today's economy, it is recommended to (a) connect a simplified refinery to a sugar mill that has the same production capacity, and (b) if an independent refinery is used, utilize raw sugar with a maximum color rating of 1,000 ICU as the input for the refining process, which will streamline the process (Chou & Engel, 2009). The main considerations for the refiner are: (1) sugar that exhibits good solubility during the crystallization process; (2) sugar that performs effectively during the clarification stage, and (3) sugar that possesses a light color.

14.3 Manufacture of raw cane sugar

Sugarcane, a variety of grass, is grown in tropical and subtropical areas and serves as the main source of raw material for producing cane sugar. The hybrid in question is a multifaceted combination of diverse species, primarily originating from *Saccharum officinarum* (*S. officinarum*) and other *Saccharum* species (Hassan & Musa, 2017; Thaval & Kent, 2012).

The manufacturing process starts with bringing cane to the mill, following the harvesting period, where it undergoes weighing and processing at automated cane receiving stations. The majority of sugarcane harvested globally is still manually reaped with a cane knife or machete. Next, the sugarcane is conveyed from the field to the factory. Cane billets are carefully placed into a cane transporter and subsequently conveyed to a shredder. The shredder is responsible for the reduction and grinding of the cane into fibrous material, as well as the bursting of the juice cells. A sequence of mills is utilized to feed the cane through pairs of rollers. The mill is comprised of three sizable rollers that are organized in a triangular configuration, typically connected to pressure feeds (Payne, 2013).

Juice from sugarcane is extracted through a process called milling. Cane preparation and extraction are the two main stages of this process. The sugar industry employs two primary extraction methods, namely diffusion and grinding (Thaval & Kent, 2012). Diffusion, the method of separating juice from fiber in sugarcane, entails moving juice from disintegrated cane through the countercurrent flow of water, unlike the expulsion of juice under pressure, which occurs in milling (Payne, 2013). The milling process primarily involves extracting juice from sugarcane by compressing the cane between two sets of sizable cylindrical rolls within a series of milling machines known as a milling train. After passing through a pair of rolls and extracting juice, the remaining sugarcane residue is commonly referred to as bagasse. Following the milling process, bagasse generally comprises ≈30%–50% fiber, 45%–60% water, and a decreasing amount of brix as consecutive milling units handle the bagasse. During imbibition, water or diluted juice is introduced to the bagasse before its entry into the milling unit, aiming to facilitate the extraction of juice from the significantly drier bagasse. The substance introduced into the bagasse is commonly referred to as imbibition water or imbibition juice. Compound imbibition refers to a commonly used variation of the imbibition process (Thaval & Kent, 2012). Achieving effective juice separation in a milling tandem requires proper cane preparation either by using preparatory machinery like shredders or by preparing it directly in the mills. Effective preparation entails extracting a significant amount of juice from the cellular structure of the cane while maintaining the integrity of the fiber size to ensure it remains suitable for milling. Ideal preparation typically occurs at the start of a milling tandem, with a robust shredder being a preferred equipment. This shredder efficiently removes fiber from cane, releasing juice, while preserving a somewhat elongated fiber structure. Such preparation also guarantees optimal feedability (Payne, 2013).

The predominant components of sugarcane juice are water and organic molecules, primarily sucrose, dextrose, and levulose, with the latter two assumed to be present in equal proportions. The mill tandem produces an opaque liquid that ranges in color from greenish-grey to dark green. It contains suspended fine bagasse (bagacillo), gums, albumin, wax, coloring matter, and particles of soil, sand, clay, and muck. The pH of raw cane juice typically ranges from 5.2 to 5.4 (Baikow, 2013). The milling process generates a cloudy liquid with a color spectrum from greenish-grey to dark green. This liquid consists of suspended fine bagasse (bagacillo), gums, albumin, wax, pigments, and particles of soil, sand, clay, and muck. The pH level of raw cane juice usually falls within the range of 5.2–5.4 (Baikow, 2013).

The elimination of insoluble non-sugar contaminants from sugarcane juice is a crucial aspect of the manufacturing of raw sugar, and this is often accomplished through the process of juice clearing. Lime saccharate and polymeric flocculants are utilized to eliminate suspended matter that undergoes precipitation and then exits the clarifier through the underflow (Rein & Attard, 2007). Lime has two commercial forms; quick lime (CaO) and hydrated lime ($Ca(OH)_2^-$) (Delden, 2015). In most cases, lime is introduced into raw cane juice in the form of milk of lime to enhance dispersion and accelerate the process. Additionally, there are other reducing chemicals of lesser significance present. Approximately 0.5% of the cane juice is composed of inorganic compounds or ashes. After the process of crushing cane juice, a specific quantity of fiber, primarily cellulose, remains. This fiber, known as "bagacillo" goes through the cush-cush screen. In order to create clarified juice with a pH of roughly 6.8–7.2, the raw cane juice is typically treated with lime to raise its pH to a level over 8.0 (Baikow, 2013).

The target of the clarifying process is to eliminate extraneous substances from the juice at the earliest stage of the procedure. The juice consists of a significant amount of colloidal and fine suspended matter, which is primarily eliminated during the clearing process. Additionally, certain soluble elements are also removed. The procedure, concerning suspended matter, bears a strong resemblance to the treatment of water supplies. The fine suspended particles are coagulated, resulting in particles that will settle at an acceptable pace. Subsequently, the liquid is transferred to settling vessels to facilitate the settling of the coagulated matter (Jenkins, 2013). Pressure filtration and chemical treatment are the two most often utilized ways of clarification; chemical clarification is the recommended approach. There are three primary categories of clarification processes that can be identified:

(a) *Defecation*, also known as lime defecation, in which lime is the sole substance introduced (with the exception of minor additives), phosphoric acid, lime (as lime sucrate to boost solubility), and polyacrylamide flocculent are used in the phosphatization process to create a calcium phosphate floc. The floc skimmed from the liquor surface and the floc separated from the liquor is typically separated by air flotation.
(b) *Sulphitation*, includes introducing sulfur dioxide gas into the juice along with lime, leading to the creation of a calcium sulphite precipitate.
(c) *Carbonatation*, also known as carbonation, is a process that involves the addition of carbon dioxide gas, resulting in the formation of a precipitate composed of calcium carbonate.

Sulphation and carbonatation are exclusively employed, with only a few insignificant deviations, for the manufacturing of white sugar intended for direct consumption (Jenkins, 2013).

The resulting clarified juice is then directed to the evaporators for the purpose of concentration. The clarified juice is condensed to achieve a syrupy texture before being transferred to vacuum pans for the crystallization process to produce raw sugar. This is done using a multiple effect, which involves connecting evaporators in series and gradually increasing the amount of vacuum (Baikow, 2013). Evaporating the water concentrates the sugar content. The technique involves two steps: first, water is evaporated to a concentrated liquid condition, and then the concentration is increased until sugar crystallizes in the super-saturated solution. The process involves two steps: first, evaporation and then boiling. Both steps are performed under a vacuum to minimize boiling temperature and prevent sucrose from inverting at high temperatures for extended periods of time (Delden, 2015). Sugarcane honey is a syrup that is obtained through the process of evaporating sugarcane juice at a temperature range of 105–110 °C, resulting in a total solids content ranging from 65% to 75%. Carbohydrate-rich sugarcane syrup exhibits a viscous, transparent, and light or lustrous brown hue or amber-yellow shade. It is a nourishing food that provides an abundance of minerals, including chlorine, iron, calcium, potassium, sodium, phosphorus, and magnesium. Additionally, it is comprised of naturally occurring antioxidants, including flavonoids and vitamins (Abdel-Aleem, 2020). Following the process of evaporation in a multiple-effect evaporator until consistency of syrupy is achieved, it is necessary to further evaporate the clarified juice in order to facilitate the crystallization of sugar. The process is carried out within a vacuum pan, which is a receptacle used to boil syrup under vacuum conditions, resulting in the formation of a dense amalgamation of crystals and mother liquor, commonly referred to as massecuite. Vacuum pans are evaporators that have a single effect and can have a diameter of up to 14 feet (4.270 m) and occasionally even greater. Various vacuum pans exist, such as the coil, standard calandria, and ribbon calandria (Baikow, 2013).

The mud that flows downstream from the clarifier often consists of a precipitate suspension ranging from 5% to 10%, and a sugar juice ranging from 10% to 15%. Filtration equipment is used to remove the dirt gathered with sugarcane from the underflow juice, which is then returned to the cane fields. In the past, manufacturers mostly utilized conventional rotary vacuum filters (RVFs) to separate more than 85% of the solid particles in the mud juice, resulting in a filter mud product with a moisture content of 75%–80% (Rein & Attard, 2007).

Commercially acceptable raw sugar necessitates crystals with a generally consistent size ranging from 0.8 to 1.0 mm in length and from 98.8 to 99.3 pol. Among the many quality variables, namely color, ash, and filterability, color is the sole element that holds significant value for the refiner. Efficient production of marketable sugar that meets these standards can be achieved using a two-step process of crystallization from syrup. It is essential to incorporate these processes because of the inherent limitations posed by the centrifugal separation of sugar crystals from the mother liquor or molasses. In the initial phase, A-massecuites are formed using syrup with a purity level between 83 and 88. When more than 60% of the sucrose has crystallized, the massecuite becomes a solid mass unsuitable for centrifugation. Thus, crystallization is controlled below this point, and the molasses is returned for the subsequent process known as B-massification. The highest allowable percentage of sucrose in crystal form is limited in this crystallization process, due to the increased mother liquor viscosity, resulting in a maximum limit of 50% (Payne, 2013). Both batch and continuous centrifugal are employed to segregate the solid and liquid components of unprocessed sugar mixtures. In the sugar industry, batch centrifugals of various types are essentially the same, differing only in size, capacity, and accessories (Baikow, 2013).

The commercial sugar that exits the centrifugal, intended for sale or export, typically has a moisture percentage ranging from 0.5% to 2%. When the moisture level of sugar is above a specific threshold, especially when it exceeds 1%, it greatly affects the preservation of its properties. The upper limit is contingent upon the relative abundance of contaminants inside the sugar. The presence of a higher-purity sugar has a greater negative impact when a specific percentage of water is used (Rein & Attard, 2007). The initial stage in the process of drying refined sugar involves the use of batch centrifugal forces to separate the dry and liquid phases of the refinery massecuites. After the liquid phase, also known as 'run-off,' is diluted and brought back for reboiling. The dry phase of the massecuite, after being released from the centrifugal, is then dried, screened, granulated, conditioned, or compressed into tablets or cubes and dried (Baikow, 2013). Finally, the inferior liquors and materials obtained during the refining process are gathered and subjected to further processing in order to extract the highest feasible quantity of sugar. In order to recover the sucrose that has been lost in both the affinition syrup and granulated syrup, a series of crystallizations are performed, resulting in the production of a by-product known as molasses or blackstrap. The sucrose content in this molasses remains below 50%. However, conducting a fresh crystallization process to extract additional sugar is not economically feasible due to the substantial energy expenses involved (Costa, 2018).

14.4 Outlines in filtration

14.4.1 Filtration of Mud (Carbonate cake)

Filtration is mostly used in the raw sugar industry for treating clarifier mud. The cane sugar industry has employed rotary vacuum filters since the 1930s, resulting in comparatively minimal sucrose losses. The sugar loss in cake, namely the sucrose that enters the cake production process in cane, has exhibited a decline from 0.5% to 0.8%, during the 1960s to slightly above 0.2% in 1995. Similarly, the percentage of cane in filter cake has fallen from ≈ 4.6% in 1980/81 to ≈ 2.6% in 1995/96 (Lionnet, 1996). The process of mud filtration represents the final chance to retrieve the residual sugar present in mill muck. The quantity of mud can range from less than 2% of the weight of sugarcane for uncontaminated sugarcane to over 5% for sugarcane that has been forcefully loaded. The moisture content of the mud cake typically ranges from 60% to 85% on average. The polarization of the cake is influenced by the quantity of bagacillo employed for filtration. The level of polarization will increase as the quantity of bagacillo employed increases. The utilization of mechanical loaders has led to a higher concentration of sand, clay, muck, and other insoluble constituents of soil in the mud cake, which is transported alongside sugarcane from the fields. This is the reason why the cake exhibits a greater volume and weight when the cane is subjected to mechanical loading rather than being washed prior to grinding (Baikow, 2013).

The goals of mud filtration encompass two main aspects: (1) the retrieval of sugar and its subsequent inclusion in the filtrate, and (2) the optimization of mud solids retention, with the aim of reducing the quantity of mud that is recycled back into the process. These two objectives are accomplished through effective water washing (Rackemann et al., 2015). A typical clarification process is illustrated in Fig. 14.3. During the clarification process, the suspended matter and heavy particles, which have coagulated due to the presence of lime and heat, settle on the surface of clarifier trays. The mud is extracted from the clarifier trays and transferred to the mud thickening compartment using elongated arms that are equipped with squeegees. The mud is elevated from the thickening compartment of the clarifier using suction diaphragm pumps and directed into the mud box. From there, it is transported by gravity to the mud filter station. The viscosity of the mud exhibits a range from a highly viscous state to a notably fluid state. To enhance filtration, a tiny quantity of lime milk can be incorporated into the mud, resulting in a more granular texture. However, it is important to note that the pH level of the limed mud should not surpass 8.5 (Baikow, 2013). Decantation was likely the oldest technique used to extract sugar from subsider mud. After removing the clarified juice, the mud remaining in the open defecators was mixed with water, heated with the addition of lime to aid in coagulation, and permitted to settle. After the water layer was transformed into a diluted solution of sugar, it was reintroduced into the processing system, while the secondary mud layer was disposed of. The aforementioned procedure is occasionally referred to as double decantation, wherein the initial decantation involves the separation of mud from the juice. Plate and frame filter presses have been widely employed and continue to be utilized in certain nations. The filter presses are of the conventional plate and frame design, including

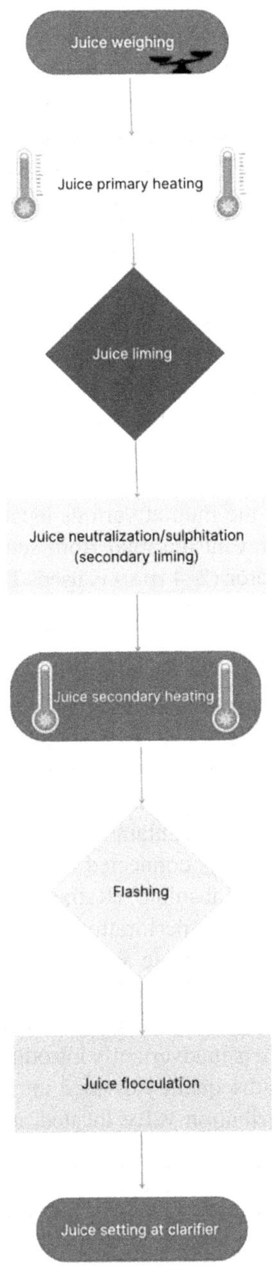

Figure 14.3 Typical steps of clarification process (Keerthi, 2002).

frames that are approximately two inches thick and can reach a square area of up to three feet. These presses are designed to facilitate systematic cleaning by rotating plates through the cake on both sides (Rackemann et al., 2015).

The primary characteristic of filters designed for cane mud involves the utilization of a perforated copper or stainless-steel sheet as a screen, replacing the conventional cotton cloth. A conventional fabric would rapidly become obstructed by minute mud particles, resulting in a significant decrease in the filtering efficiency to nearly negligible levels within a few rotations of the drum. In contrast, the metal screen facilitates the smooth removal of the cake, so ensuring the screen's optimal performance. In cane sugar factories, rotary vacuum filters are widely employed for the purpose of filtration of mud from clarifiers. The fundamental principle underlying all rotary vacuum filters is identical to that of the Oliver-Campbell rotary vacuum filter. The Oliver-Campbell filter, which was already widely used in the chemical sector, was initially employed for the filtration of cane mud ≈1930. The filter consists of a hollow drum that rotates along a horizontal axis and is immersed in the mud feed, also known as the filter boot, in order to facilitate the filtration process. The capacity and performance of the filter are enhanced by introducing the mud at various locations along the whole length of the filter boot. In order to prevent the mud from settling and becoming stagnant in the filter boot, a rotating agitator (2–4 rpm) is used. This agitator is positioned on the filter's axis and is powered by a separate motor. Its purpose is to disrupt the liquid surface. The filter drum is propelled by a compact electric motor to achieve a drum speed ranging from 4 to 10 rpm. The outer edge of the drum functions as the filtration surface, which is partitioned into 24 distinct parts and spans the entire length of the drum. Every individual sector is linked to a vacuum system through a short metallic pipe that terminates in a distributing valve located at one end of the drum. This valve serves three distinct tasks: one that is linked to the atmosphere; the second connection is established with a chamber that maintains a low vacuum, often ranging from 20 to 40 kPa (abs). The third chamber is connected to a chamber that maintains a greater vacuum, typically above 60 kPa (absolute) (Rainey et al., 2014).

Due to the excessive size of the perforations in the screen (≈0.5 mm), bagacillo is employed as a filter assist. A suitable amount of fine bagacillo is extracted from the final bagasse and combined with the mud prior to being fed into the filter. The bagacillo particles serve as the primary filter medium on the screen. However, a certain quantity of contaminated juice is inadvertently introduced prior to the formation of the filtering mat. The filtrate is subsequently divided into two pieces, known as the hazy and clear filtrates, at the distribution valve located at the end of the filter. The hazy filtrate denotes the initial phase of the filtration process, occurring when the filtering surface encounters the mud in the filter tank and is initially connected to the vacuum. The transparent filtrate port or route is responsible for gathering the filtrate from the residual components of the cycle, which includes the washing process (Jenkins, 2013).

A vacuum pressure of 18 cm (7 in.) of mercury, referred to as pickup, is applied to the filter sections immersed in the tank containing clarifier settlings mixed with bagasse fines. In this setup, the liquid passes through the screen, leading to the formation of a cake on the screen. Once the cake is formed, the bagasse fibers begin to trap the insoluble particles, causing the passing liquid to become relatively clear compared to the initial unfiltered influx. In this phase, the cake experiences a rotational process within the settlings, gradually increasing in thickness until it eventually emerges from the settlings. In this context, the valving is connected to a high vacuum system with a maximum pressure of 50 cm (20 in.) of mercury. Subsequently, water is sprayed onto the surface, permeating the cake and extracting the juice. Water is introduced into the drum via drop pipes at the top to maintain moisture in the cake and sustain the vacuum until it is released, allowing for the removal of the cake from the drum. The liquid and residue are combined, while the solid waste is disposed of (Payne, 2013). The rotating vacuum filter is not a true filter, but rather a crude separator of dry and liquid phases of mud, unless covered with filtering cloth or precoated with filter aid. Because rotary vacuum filters have perforated copper plates on their drums, their effluent juices are typically not clear enough to be fed directly to evaporators alongside clarified juices without further processing. Mud is prepared for filtration on rotary vacuum filters by mixing it with bagacillo to a thick consistency in a mud-bagacillo mixer. The amount of bagacillo in clarifier mud varies inversely with its thickness. The recommended amount of bagacillo is 15%–30% of the cake's dry content, or 6–12 pounds per ton of clean cane. Using mechanically loaded cane requires more bagacillo. In order to dilute heavy mud, a little water has to be added (Baikow, 2013).

The sugarcane industry is also familiar with Eimco and BMA drum rotating vacuum filters. The Oliver-Campbell rotary vacuum filter drum is comprised of 24 distinct sections that are subjected to vacuum. The drum undergoes three distinct phases within a single spin. A low vacuum of $\approx$10 in is applied to the filtering chambers when they are immersed in the constant-level mud feed tank. The vacuum is capable of effectively drawing muddy juice into the filtering chamber and, by suction, retaining the mud on the perforated plates of the drum, resulting in the formation of a mud cake material. The hazy juice is directed into the receiver designed for receiving cloudy juice. Upon exiting the constant level muck tank, the chambers experience an increase in vacuum to $\approx$20 in. Simultaneously, the filtering chamber is directed into the washing zone, where the cake undergoes a water wash. Currently, the juice, which has been combined with wash water, is passed through the filtering chamber. The diluted juice exhibits a higher level of cleanliness compared to the initial contaminated juice. However, it lacks the necessary level of cleanliness to be combined with clarified juice and subsequently sent to evaporators. Typically, the turbid juices and juices derived from the subsequent phase are combined and subsequently reintroduced into the liming tank. After the completion of the washing and drying zones, the vacuum within the filtering chamber is disrupted, allowing the desugared cake to either fall freely or be scraped off the surface of the drum (Rackemann et al., 2015).

The mud filter cake has been identified to contain a substantial quantity of sugar that is lost throughout the manufacturing process. The principal approach for mitigating sugar loss involves the utilization of an adequate quantity of wash water during the filtration process. Nevertheless, the efficacy of water application is contingent upon the porosity of the cake and the method employed for its application. If the addition of wash water is done efficiently, increasing the rate of wash water will result in a reduction in the quantity of sugar that is lost. The majority of mud filtration methods utilize displacement washing, a process in which wash water propels the juice via capillaries inside the mud cake with minimal mixing. This approach is based on the Kozeny-Carman relation, which suggests that the juice is not diffused and diluted within the cake. Excessive compression of the cake leads to a decrease in its porosity, resulting in increased resistance to the addition of wash water. Consequently, the cake becomes less permeable, preventing the displacement of sugar. Conversely, a cake with high porosity promotes the channeling of water, so compromising the effectiveness of the washing process. Therefore, in both scenarios, the recovery of sugar would be constrained. In the case of RVF operation, when the cake resistance is elevated, any surplus water will be carried away from the cake's surface and deposited into the mud boot, where the process of dilute washing is less efficient (Rackemann et al., 2015).

The filter area required should be proportional to the amount of mud solids that enter the factory. In sugar plants equipped with milling trains, it is typically seen that around two-thirds of the mud solids that enter the factory are retained inside the juice, while the remaining portion is transported out of the factory along with bagasse. Milling factories typically used RVF filter area ratings ranging from 0.5 to 0.8 m^2 per metric ton of cane. However, diffuser factories had lower rates (Rein & Attard, 2007). Enhanced filter station capacity ratings offer a filter area ranging from 85 to 100 m^2 per 1 ton h^{-1} of mud solids input. The anticipated pol loss in a cake for this particular number is around 0.3% for a cake wash water percentage of 200. The loading of filter mud solids (t/h/100 m^2) is quantified by dividing the dry mud solids output by the filter area (Rainey et al., 2014).

Vacuum belt filters (VBFs) have been employed in several sectors for the purpose of manufacturing highly desiccated filter cakes, and have also been utilized in sugar mills for the clarification of ash. During the 1980s, Crees and Willersdorf (1983) and Kruger (1984) evaluated VBFs for mud filtration and showed that adequate feed conditioning might lead to satisfactory filtering performance. The inclusion of a flocculent was crucial in mitigating the loss of pol (Rainey et al., 2014). Kruger (1984) determined that the most efficient operation of the VBFs was achieved by using thin cakes measuring 6 mm. This allowed for more successful cake washing at lower vacuum levels ranging from 20 to 25 kPa, resulting in cakes with a pol. content of 0.5% and moisture levels of 70%–75%. The performance results exhibit a small improvement compared to the standard RVF operation. The investigation of VBFs was terminated in the mid-1980s as a result of the comparatively elevated expenses associated with capital and maintenance in comparison to RVFs during that period (Kruger, 1984).

The main disadvantage of pressure filters is the unclean filter station caused by mud cake and high-pressure water hoses used to clean the filter cloth. The usage of rotary vacuum filters significantly improved results. However, a vacuum filter is actually a strainer rather than a filter (Delden, 2015). The vacuum applied at the filter pickup location needs to be adequate to support the cake without inducing compression, which could reduce permeability. It is advised to maintain an optimal vacuum level of 18 cm (7 in.). Using a higher vacuum level in this scenario can lead to cake compression, which would hinder the penetration of the subsequent wash water required to displace the juice. During the wash cycle, it is necessary for the vacuum to be sufficiently elevated in order to facilitate the movement of juice through the cake. The normal minimum vacuum level is 38 cm (15 in.), while the highest limit is 50 cm (20 in.) to prevent excessive flashing (Payne, 2013).

There are different approaches for mitigating the recycling of mud particles, one of which involves enhancing filter retention. Rotary vacuum filters are the primary choice for processing the clarifier underflow in sugar mills, however, there are various other types of filter equipment available. Improving filter retention can be achieved by slowing down the speed of the filter drum to produce denser cakes, recycling a portion of the filtrate for mud dilution, and incorporating lime. These methods show some effectiveness in the short term and depend on the capacity of the filter stage. Nevertheless, the inclusion of lime in the process results in an increased presence of scale-forming components, specifically calcium, and also leads to the blinding of filter screens. This phenomenon has the potential to prolong the duration of maintenance and diminish the efficiency and recuperation of the filter station. Enhancing the retention of filters does not yield a substantial effect in mitigating the presence of soluble impurities, which are mostly accountable for the creation of scale and other issues related to processing and sugar quality (Rackemann et al., 2015).

Weighing the vacuum filter cake in a sugar mill can improve control and accurately determine sugar losses. To ensure accuracy, the cake should be discharged without water onto a belt or drag conveyer before being put into a mud cake bin for vehicles. Before leaving the factory to take mud to the fields, the truckloads might be weighed and recorded. To dispose of the mud, dilute it with water upon discharge from the drum and pump it to a pond. To determine cake weight, an attendant can remove a sample from the filter drum surface using a tared wooden spatula measuring exactly 6" × 6". This spatula with cake can be weighed using a tiny scale nearby. To estimate the weight of the discharged cake, multiply the average net weight of multiple mud cakes weighing by four, factor in the drum's filtering area in square feet, and count the number of drum revolutions. All may be estimated for 24 h of manufacturing. This method accurately estimates the weight of a wet cake (Baikow, 2013).

14.4.2 Filtration in refining raw cane sugar

The results of the separation process include a fluid stream (filtrate) that is either particle-free or virtually particle-free, a solid phase that contains some entrained liquid, and potentially a solid-fluid mixture that has a higher concentration of solids (as observed in crossflow cake filtration) (Tien, 2012). Particles that cannot flow through the openings in the fluid are trapped and accumulate to create a progressively thicker layer known as the filter cake. Fabric filter cloths, meshes, screens composed of polymers or metals, and layers of solid particles generate the required small openings for filtration (Varzakas & Tzia, 2014). When the aim is to reclaim the residue, the process is termed cake filtration, whereas when the goal is to obtain a clear filtrate, it is referred to as clarification (Farmanullah & Hussain, 2022). There are primarily two kinds of filtration; surface filtration and depth filtration. Surface filtering involves the collection of particles on a surface that allows the passage of fluids. As the particles accumulate, they form a layer on the filter, known as a cake, which hinders the flow by increasing resistance. Depth filtration effectively eliminates particles from both the surface and the entire volume of the media (Dunnett, 2013).

The driving force comprises two key elements: filtration area and pressure differential. The filtration rate is directly impacted by the filtration area available and the pressure difference across the filter medium (Farmanullah & Hussain, 2022). One area with the potential for enhancing the sugar manufacturing process is the clarification stage. Traditional clarifiers are bulky equipment, leading to increased operational costs and associated environmental concerns. Membrane filtration can be used as an alternate form of clarification to improve these concerns. It can also decrease or eliminate the need for chemicals, especially lime, and result in higher-quality clarified juice (Jegatheesan et al., 2012).

Industrial filters come in several forms, ranging from versatile units suitable for varied filtration tasks to specialized ones designed for specific fluids and process conditions. They are available in a wide range of sizes, suitable for both small and large-scale operations, and can be used for continuous or batch processing. Continuous filters can perform filter cake creation, deliquoring, washing, and discharge in sequence without disrupting the flow of the process feed. On the other hand, batch filters can carry out the same activities but require stopping to remove the cake. Separation occurs on surfaces called filter media, which can be oriented horizontally or vertically and can be flat or cylindrical. Woven media are commonly utilized in cake filters; however, there are other choices available with a substantial range of variations (Tarleton & Wakeman, 2016).

Membrane filters typically possess a clearly defined pore structure and a consistent range of porosity. While depth filters are manufactured in controlled environments, the inherent unpredictability of the fibrous material leads to an indistinct porous structure, unlike the well-defined structure observed in membrane filters. These filters, commonly known as microfiltration (MF) microporous membrane filters, provide a higher level of control over porosity compared to traditional depth filters. The filters are graded on a scale ranging from 0.04 to 8 Jlm, with the most often used filter having a rating of 0.2 Jlm for sterilization purposes (Levy & Jornitz, 2006). Prefiltration of raw sugarcane juice before membrane separation increases yield and reduces costs for future ultrafiltration (UF)/MF (Hinkova et al., 2002).

The conventional purification process entails the addition of lime and sulfur dioxide to the juice, followed by boiling. Sedimentation and vacuum filtration are then used to remove sludge. However, this traditional refining method suffers from drawbacks such as inferior product quality, high operational and chemical expenses, and significant environmental challenges related to solid waste. These issues arise from chemical usage and manual, batch-based operations. As a result, alternative techniques including activated carbon adsorption, electrodialysis, ion exchange, and membrane filtration have been explored to address these challenges. Among these alternatives, membrane filtration, particularly UF, shows the most promise, as it eliminates the need for chemicals, ensures continuous and automated production, and yields high-quality juice (Luo et al., 2016). On the other hand, the sulphitation technique is employed at numerous sugar factories to generate plantation white sugar. The sugar generated using the standard sulphitation method contains SO_2 at levels ranging from 20 to 100 ppm, which exceeds the FDA's guideline of fewer than 10 ppm. As a result, it is not appropriate for use in the beverage sector (Chou et al., 2000). Consequently, there is a growing need to enhance the methods used in cane sugar production in order to obtain a highly purified product, specifically refined sugar, through a single crystallization process. Ideally, this process would take place in a single facility rather than being divided between sugar mills and refineries. This can be achieved by incorporating membrane filtration, as well as implementing new separation techniques like chromatographic separation and ion exchange (Gil & Wright, 1994).

The handling and disposal of filter cakes now cost more due to new environmental standards that place more restrictions on the traditional filtration process. Furthermore, the ongoing decline in membrane costs and the membrane's longer lifespan encourage sugar technologists to think about substituting conventional filtration with membrane separation and to create new uses for membrane separation. However, when introducing a novel membrane design, certain requirements must be considered: achieving a high permeate flux, maintaining a temperature above 85 °C to inhibit bacterial proliferation, accounting for the presence of abrasive constituents in the raw juice streams, and aiming for high concentration factors to minimize sugar losses (Abbara et al., 2003).

14.4.3 Membrane operations in sugar-refining

A membrane is an interface that segregates components based on their structure and size. In a separation process, a membrane acts as a selective barrier, permitting certain components to pass through while obstructing others in a mixture under the influence of an applied force. The solution that traverses the membrane and contains solvent and solutes not retained by the membrane is termed the permeate. Conversely, the solution containing the species retained (rejected) by the membrane is known as the retentate stream or concentrate. Permeation through the membrane can occur due to different driving forces, and typically, the rate of permeation (selected transport) is directly linked to the driving force. Several types of membranes, such as polymeric, stainless steel, ceramic, and carbon, are under development for use in the sugar industry. The prevalent configurations for sugar industry applications include tubular designs for ceramic, carbon, and stainless-steel membranes, as well as spiral wound and hollow fiber configurations for polymeric membranes (Abbara et al., 2003). Table 14.1 presents the main membrane process along with the characteristics of the feed and permeate, the driving force behind the separation, and an approximate classification of pressure-driven membranes (Abbara et al., 2003; Saska, 2000; Steindl & Doyle, 1999).

Another efficient approach to improve the quality of diffused juice involves employing membrane filtration to clarify press water. Pressed water is a significantly diluted stream containing about 3%–4% of the total sugar entering the mill. Following the pulp pressing process, it undergoes diffusion and is subsequently recycled. Membrane filtration is a versatile technique suitable for both sterilizing press water and eliminating suspended particles (Abbara et al., 2003; Hinkova et al., 2002). Reverse osmosis (RO) can produce a permeate that is nearly pure water. This water has the potential for utilization in the extraction unit. The concentrate, encompassing all components, can be directly routed to the low-grade crystallization phase, bypassing the high-grade crystallization stage, and subsequently be efficiently removed in molasses. As depicted

Table 14.1 Process characteristics of membranes and features of various pressure-driven membrane types.

Membrane processes	Feed	Permeate	Driving force	Pore size	MWCO	Operating pressure (bar)	Reject
Microfiltration	Liquid	Liquid	Pressure	0.05–10 μm	Larger than 400,000	Up to 3	Particle substance, suspension
Ultrafiltration	Liquid	Liquid	Pressure	2000 °A	20,000 400,000	7	Macromolecules or colloidal matter
Nanofiltration	Liquid	Liquid	Pressure	10–80 °A	200 20,000	20	Divalent ions-organic solutes (sugars)
Reverse osmosis	Liquid	Liquid	Pressure	< 10 °A	Smaller than 200	20 100	Permit solely the flow of water

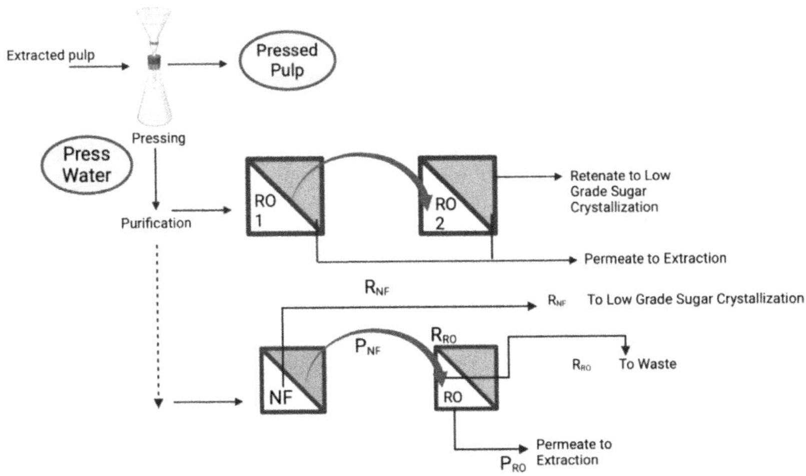

Figure 14.4 Flow diagram of press water treatment with RO or NF, (R: retenate, P: Permeate).

in the upper section of Fig. 14.4, trials were conducted on the two-stage RO system incorporating a prefiltration stage. The spiral wound module 4" × 40" utilized the high retention RO membrane OSMO 411T-MS10. This membrane achieved a NaCl retention rate of 99%, under operating conditions of 3 MPa and 60 °C (Hinkova et al., 2002).

Membrane technology is widely used in several industries such as the food and dairy sector, water purification, liquid effluent stream treatment, and corn refining. Recent research has demonstrated that, membrane technology has significant potential in reducing energy consumption, minimizing or eliminating chemical clarifying, and enhance the quality of the end product (Rafik et al., 2015). Membrane techniques, comprising MF, UF, nanofiltration (NF), RO, osmotic distillation, and membrane distillation, have proven to be valuable alternative methods in fruit juice manufacturing for purposes like clarification, stabilization, depectinization, and concentration (Cassano & Drioli, 2014). Moreover, electrodialysis (ED) is a process used to separate or concentrate ions in solution by selectively moving them through semipermeable membranes under the influence of an electrical gradient. The primary use of ED currently lies in desalinating brackish water to produce drinking water and removing impurities from milk whey to extract valuable ingredients for infant nutrition products (Fidaleo & Moresi, 2006). Despite the research conducted by Danico a/s in the late 1960s on the potential uses of membranes in the sugar industry, only a few industrial installations were operational in beet and cane factories until the 1990s (Madsen, 1990). Steindl (2001) obtained a comprehensive overview of the data collected from several research projects conducted at multiple locations worldwide. The investigations determined that, membranes effectively eliminate polysaccharides, turbidity, and colloidal contaminants, leading to syrups and molasses with reduced viscosity. Additionally, membranes enhance growth rates and improve fatigue (Steindl, 2001). Despite its significant industrial utility, many research findings rely on laboratory or pilot plant-scale equipment. Experimentation with small-scale equipment serves as the preliminary step in developing a large-scale approach (Urošević et al., 2017).

The sugar industry has had an ongoing interest in using membrane filtration to enhance product quality and as a preliminary step in producing higher-value goods (Steindl & Rackemann, 2010). Sugar processing is a highly energy-consuming activity in the food industry, posing a difficulty for membrane separation techniques such as MF, UF, NF, and RO. However, the use of membrane separation methods in sugar production is limited by factors such as the large volumes of sugar juices, their high viscosity, and the high osmotic pressure (Hinkova et al., 2001). MF and UF offer several advantages over traditional methods. These include the ability to separate substances without altering the temperature or pH of the solution, as well as the elimination of chemical additives. This results in reduced production costs, improved waste treatment, enhanced product quality, and decreased labor expenses (Urošević et al., 2017). UF functions as a selective process where substances with large molecular weights, like proteins and suspended particles, are blocked from passing through the membrane. Meanwhile, substances with lower molecular weights can traverse the membrane unhindered. Consequently, there is a minimal rejection of mono- and di-saccharides, salts, amino acids, organic compounds, inorganic acids, or sodium hydroxide. MF is a selective technique that primarily removes suspended particles, allowing proteins to pass through the membrane without obstruction. Nevertheless, there exists a significant disparity between the implementation of a task and the desired ideal scenario (Wagner, 2001).

MF membranes and ultrafiltration membranes with high molecular weight cut-off (MWCO) are frequently utilized in the clarification of sugarcane juice, mainly to eliminate turbidity. Nevertheless, UF with low MWCO and NF can also be employed for decolorization purposes, resulting in a quantifiable enhancement in purity. This would also facilitate the separation of salts from sugar by the utilization of NF membranes. RO membranes have the potential to be employed for additional concentration, hence facilitating potential savings in energy usage during the crystallization process. Membranes tolerant to high temperatures are ideal for filtering sugarcane juice because of their elevated operating temperatures. Various materials such as polysulfone, polyethersulfone, polyvinylidene fluoride, and polytetrafluoroethylene can effectively accomplish the filtration of sugar juice. Inorganic membranes are well-suited for sugar juice clearing, because they possess excellent resistance to chemical degradation and mechanical abrasion, making them highly favorable for this application. Nevertheless, it is imperative to handle them with caution owing to their fragility. There are some well-known uses of membrane filtration for the direct generation of white sugar from raw juice, and they are illustrated in Table 14.2.

14.5 Conclusion

The sugar industry faces numerous challenges and opportunities in its quest to produce high-quality sugar, while minimizing environmental impact and operational costs. Traditional methods of refining sugar have served well for many years, but are now

Table 14.2 Membrane filtration examples of sugar industry applications.

Membrane detail and reference	Type/module	Remark	Results
The facility comprises 7.38 m^2 of spiral-wound polymeric membranes with MWCO of ≈200,000, along with two tubular carbon membranes featuring pore sizes of 0.1 and 0.5 μm, totaling 1.0 m^2 in surface area. Steindl and Doyle (1999)	Spiral wound Tubular carbon	Polymeric membranes feature wide feed spacing and are capable of continuous operation at temperatures as high as 95 °C. These membranes have been successfully operating at temperatures of 93–94 °C for more than 200 h.	Results from further studies show that polysaccharides, including dextrans, are completely removed, heat and mass transfer coefficients in evaporators and pans are increased, pan boiling time is reduced by 20%–30%, and the color of the resulting sugar is reduced by up to 60%. The degree of separation between suspended solids and soluble contaminants is similar.
*50 kDa, polyvinyledene fluoride (Koch, USA) Saska et al. (1999)	Spiral wound	Experiments were conducted with clarified juice at a temperature of 95^{0}C.	Spiral polymeric membrane systems are cheaper than other membrane solutions and may be a good choice for sugar companies seeking high-quality raw cane sugar. Membrane filtration and resin absorption methods like chromatography could make premium white sugar production in a raw mill cost-effective.

(continued on next page)

Table 14.2 Membrane filtration examples of sugar industry applications—cont'd

Membrane detail and reference	Type/module	Remark	Results
*UF, 5 and 30 kDa YM series (Amicon, USA) Balakrishnan et al. (2000)	Stirred	Experimental tests were conducted in a laboratory setting at a temperature of 85^{0}C both raw and limed juice.	UF of cane juice results in a clear and sparkling filtrate that is notably better than the clear juice obtained through liming-sulphitation. In the absence of sulphitation, permeate clarity improved threefold, and juice color decreased by up to fivefold compared to conventionally clarified juice. Moreover, the calcium oxide content decreased by 25%–30%.
*Filtration area of 2 × 0.2 m^2 and a porosity range of 20–100-UF, *0.2–5 μm- MF (Membralox) Hinkova et al. (2002)	Ceramic	The pilot plant experimental parameters include temperature conditions up to 85 °C, pressure conditions up to 0.6 MPa, and a wide pH range from 0.5 to 13.5.	MF and UF can produce juice with high quality, allowing for direct crystallization. Filtered juice yields better crystal quality, especially in terms of color and colloid content, compared to nontreated juice.
*50 kDa–0.45 μm TiO2/a-Al_2O_3 *ZiO_2 coated ceramic CarbosepTM 40 (TechSep, France) Nene et al. (2000)	Tubular-ceramic	Laboratory experiments at 90 °C with limed mixed juice.	NA

(*continued on next page*)

* The tubular membrane system utilized eight modules, each with a membrane area of 2 m^2. These modules contained PVDF tubular UF membranes with a pore size ranging from 30 to 50 nm, supplied by X-Flow in the Netherlands. *One pressure vessel with a UF or NF module (effective membrane area of 5 m^2) spiral wound membrane (Sepro Membranes, USA, provided polyethersulphone (PES) *UF membranes, 5 nm pore size, and polyamide NF membranes, 0.5 nm pore size Luo et al. (2016)	Tubular Spiral wound	For tight UF and NF, the crossflow velocity was at approximately 65 L min^{-1} and 35 L min^{-1}, and the operating temperature was maintained at 60°C using circulating cooling water, except during cleaning.	A combined membrane system, comprising two stages of UF and NF, was demonstrated to effectively clarify, decolorize, and pre-concentrate raw sugarcane juice at a pilot-plant scale. The process successfully eliminated 96.55% of color and 99.99% of turbidity, resulting in a final product with a color value below 800 IU, making it suitable for further processing through multi-effect evaporation and crystallization to produce high-quality white sugar.

(continued on next page)

Table 14.2 Membrane filtration examples of sugar industry applications—cont'd

Membrane detail and reference	Type/module	Remark	Results
*Polyacrylonitrile-based MF membrane with 0.027 m^2 filtration area and 0.1 μm pore size. Panigrahi et al. (2018)	Hollow fiber	A research experiment examined how operating parameters trans membrane pressure (35, 69, 104, and 138 kPa) and crossflow velocity (CFV) (0.123, 0.246, and 0.369 $m\ s^{-1}$) affect membrane production and juice quality. The steady-state permeate flux was 5.41 to 6.23 $l\ m^{-2}$ h under the operating parameters examined.	The microfiltered permeate retained the total soluble solids (TSS), sucrose, and vitamin C present in the centrifuged juice. MF achieved a five-log reduction in microbial load and a four-log reduction in yeast and mold count. Operating conditions of 104 kPa and Re 249 yielded juice with favorable physical characteristics, such as low solid content, high clarity, low turbidity, and high TSS, while maintaining productivity. The research recommends implementing a pre-clarification step to enhance sugarcane juice filtration for extended shelf life.

(*continued on next page*)

A ceramic membrane with a pore size of 20 nm Shi et al. (2019)	Tubular Ceramic	The study compared two membranes, A and B, with the same 20 nm pore size but distinct production procedures. Juice pretreatment strategies evaluated included heating (60, 75, and 90 °C) with or without sedimentation at pHs (7.2, 7.5, and 7.8) and utilizing evaporator supply juice.	After partially liming the juice at 75 °C, excellent permeate flow was achieved at a volumetric concentration factor of 20. The study highlights the necessity for developing solid/liquid separation systems that function under softer circumstances for food processing, enhancing performance and quality. A concentration-mode mixed cake filtration-complete blocking fouling model accurately predicts ceramic membrane filtration performance for partially clarified juice processing.

(continued on next page)

Table 14.2 Membrane filtration examples of sugar industry applications—cont'd

Membrane detail and reference	Type/module	Remark	Results
A 0.02 μm pore size SCEPTER cross-flow membrane was employed to purify sugarcane juice. Membrane tubes, manufactured of 316 L stainless steel, have a diameter of 25 mm and a length of 6 m, offering 800 m^2 of surface area. Vu et al. (2020)	*316 L stainless steel *Crossflow	To the carbonated juice, 100 ppm of sodium bicarbonate was added after pre-treatment. Next, the 2-gpm UF membrane unit treated the carbonated juice. Before crystallization, UF membrane retentate is recycled to a secondary clarifier. At least pH 8 is maintained during the UF process. A laboratory evaporator was used to evaporate the membrane filtering permeate.	UF treatment resulted in decreased viscosity and turbidity, as well as enhanced crystal development and larger crystal size in the juice. Moreover, UF contributed to improved color reduction. The UF process enabled the production of refined sugar with reduced energy consumption, operating costs, and capital investment compared to conventional methods. Furthermore, this approach offers environmental advantages and maintains product quality without compromising on commercial processes. Implementing a UF system provides an alternative to sulfitation, enhancing sugar product quality while minimizing sulfur dioxide emissions and improving working conditions for mill employees.

(*continued on next page*)

Ceramic membrane; lanthanum phosphate nanofibrils of 6 μm thickness are deposited on multichannel porous alumina supports using sol-gel deposition. A 50 cm long, 3.4 cm diameter α-alumina support with 19–3 mm diameter channels was employed. Mumbai-based HR Johnson supplied the tubes. Akhtar et al. (2020)	Nanofibril-coated ceramic UF membrane	The experiment maintained consistent pressure at 135 psi (9.3 bar), CFV at 0.02 m s^{-1}, and temperature at 25 °C throughout. Permeate volume, conductivity, pH, and flow rate were monitored every 5 min in the tank. Different feed flow rates and pressure conditions were tested in the trials.	During juice clarification, physical cleaning resulted in a significant 50-fold increase in membrane flow. This method effectively tackled the primary fouling problem caused by the formation of a cake layer on the UF membrane surface. The permeate qualities achieved through this technique met acceptable standards. The hybrid cleaning approach for membrane regeneration presents a feasible alternative for clarifying sugarcane juice.
Stainless steel membranes with 100 and 20 nm hole diameters were used to clear limed sugarcane juice, 18 mm channel diameter, 0.35 m^2 membrane area. Du et al. (2022)	Tubular/ stainless steel	The research evaluates the pure water flux of 100 nm and 20 nm membranes across different operational parameters, along with the flux of filtered pretreatment mixed juice (transmembrane pressure of 3 bar, crossflow velocity of 3 m s^{-1}, and temperature of 70 °C).	The study demonstrated the production of high-quality filtered juice, showing an average increase in purity of 1.26 units over time. Turbidity was reduced by 99.99%, and there was an average decrease in color of 29.3%. Analysis using four filtering mathematical models and SEM indicated that, flow reduction was primarily due to cake formation. Using FTIR and EDX spectroscopy, organic fouling was identified as the main cause of fouling. The research confirms the efficacy of stainless steel membranes in filtering raw sugarcane juice.

being reconsidered in light of evolving environmental regulations, concerns about chemical usage, and the desire for improved product quality. While traditional methods of sugar production have been effective, increasing environmental regulations, quality demands, and cost pressures necessitate the adoption of innovative techniques. Membrane filtration, particularly MF, UF, and NF, emerges as a promising solution to address these challenges. The adoption of innovative techniques, such as membrane filtration presents a promising solution to many of these challenges. Membrane filtration offers the potential for improved clarity, reduced color, and lower viscosity in sugar juice, while also reducing the need for chemicals and streamlining production processes. Furthermore, membrane filtration can aid in the clarification of press water, enhancing the overall efficiency of the sugar-refining process. When choosing a membrane technology for sugar filtration, several factors need consideration, including pore size, operating pressure, compatibility with sugar solutions, throughput, initial investment, maintenance costs, and operational expenses. Pore size is significant for impurity removal while operating pressure impacts both energy consumption and filtration efficiency. The ability of the membrane material to work effectively with sugar solutions is essential for ensuring long-term performance. Besides that, it is advisable to consider the initial investment, maintenance costs, and operational expenses in order to make a well-informed decision. By considering these factors in relation to the specific sugar filtration process, it is possible to achieve the desired level of filtration and purity. Additionally, there is a need for continued research and development to optimize membrane technologies for sugar industry applications and address specific challenges such as high temperatures, abrasive materials, and high concentration factors. Overall, by embracing membrane filtration and other innovative technologies, the sugar industry can enhance sustainability, improve product quality, and remain competitive in the global market.

List of abbreviations

CFV	Crossflow velocity
ED	Electrodialysis
EDX	Energy-dispersive X-ray
FDA	Food and Drug Administration
FTIR	Fourier transform infrared spectroscopy
MF	Microfiltration
MWCO	High molecular weight cut-off
NA	Not applicable
NF	Nanofiltration
P	Permeate
PES	Polyethersulphone
PVDF	Polyvinylide fluoride
R	Retenate
Re	Reynolds number
RO	Reverse osmosis
RVF	Rotary vacuum filter

SEM	Scanning electron microscope
TSS	Total soluble solids
UF	Ultrafiltration
VBF	Vacuum belt filter

References

Abbara, A., Abdel-Rahman, A. K., & Bayoumi, M. R. (2003). Utilizing of membrane filtration in the sugar Industry: A review. In *The 33rd Annual Conference of the Egyptian Society of Sugar Technologists*. Egypt: At El-Hawamdia.

Abbott, G. C. (2022). *Sugar* (1st ed., p. 414). Routledge, Taylor and Francis Group.

Abdel-Aleem, W. (2020). Effect of sugarcane juice pre-treatment on the quality and crystallization of sugarcane syrup (treacle). *Journal of Food Processing & Technology, 11*(7), 834.

Aguilar-Rivera, N., Khan, M. T., Khan, I. A., Iqbal, R., & Aslam, M. M. (2022). Sustainable conversion of wastes into green bioproducts to introduce diversification and green economy in the sugar industry: A review. *Sugar Technology, 24*(4), 1198–1211.

Akhtar, A., Subbiah, S., Mohanty, K., Sundar, R., Unnikrishnan, R., & Hareesh, U. S. (2020). Sugarcane juice clarification by lanthanum phosphate nanofibril coated ceramic ultrafiltration membrane: PPO removal in absence of lime pre-treatment, fouling and cleaning studies. *Separation and Purification Technology, 249*, 117157.

Baikow, V. E. (2013). Manufacture and refining of raw cane sugar: (2nd ed., 2). Elsevier.

Balakrishnan, M., Dua, M., & Bhagat, J. J. (2000). Ultrafiltration for juice purification in plantation white sugar manufacture. *International Sugar Journal, 102*(1213), 823–830.

Cartier, S., Tatoud, L., Theoleyre, M., & Decloux, M. (1997). Sugar refining process by coupling flocculation and crossflow filtration. *Journal of Food Engineering, 32*(2), 155–166.

Cassano, A., & Drioli, E. (2014). *Integrated membrane operations in the food production* p. 376. Walter de Gruyter GmbH & Company, KG De Gruyter.

Cheremisinoff, N., Rosenfield, P., & Davletshin, A. (2013). *Responsible care: A new strategy for pollution prevention and waste reduction through environment management*. Elsevier.

Chou, C., Iqbal, K., Min, Y., Gao, D., & Duffaut, E. (2000). An innovative process for white sugar production||. In *Proceddings Conference on Sugar Processing Research*.

Chou, C. C., & Engel, L. (2009). Simplified "green" sugar refining process for the 21st century. In *Sugar Industry Technologists Annual Technical Conference*. Louisiana.

Costa, P. D. C. R. (2018). Chemical sugar loss reduction in a sugar refinery. *Dissertation for the degree of master in chemical and biochemical engineering* p. 105. Universidade Nova De Lisboa Faculdade de Ciências e Tecnologia Departamento de Química.

Crees, O., & Willersdorf, A. (1983). An evaluation of a horizontal vacuum filter. In Proceedings of the Australian Society of Sugarcane Technologists: 5, (pp. 215–218). Australian Society of Sugar Cane Technologists (ASSCT).

Delden, E. (2015). Standard fabrication practices for cane sugar mills: (1st ed., 1). Elsevier.

Du, N., Pan, L., Liu, J., Wang, L., Li, H., Li, K., & Li, W. (2022). Clarification of limed sugarcane juice by stainless steel membranes and membrane fouling analysis. *Membranes, 12*(10), 910.

Dunnett, S. (2013). Filtration mechanisms. *Aerosol Science: Technology and Applications*, 89–117.

El-Syiad, S. I. (2000). Egyptian raw cane sugar quality in relation to refining requirements. *Food Chemistry, 68*(3), 253–257.

Farmanullah, & Hussain, S (2022). Filtration. In S. A. Khan (Ed.). In *Essentials of Industrial Pharmacy. AAPS Advances in the Pharmaceutical Sciences Series. vol 46: 46*. Cham.: Springer https://doi.org/10.1007/978-3-030-84977-1_4.

Fidaleo, M., & Moresi, M. (2006). Electrodialysis applications in the food industry. *Advances in Food and Nutrition Research, 51*, 265–360.

Gil, E., & Wright, H. (1994). Process for producing refined sugar from raw juices. United States Patent, No. US 5281279, 14 pp.

Hassan, K. S., & Musa, A. (2017). Investigation into the filterability of raw sugars from different geographical regions of the world. *International Journal of Computational and Theoretical Chemistry, 5*(6), 53–58.

Hinkova, A., Bubnik, Z., Kadlec, P., Pour, V., & Starhova, H. (2001). Membrane filtration in the sugar industry. *Chemical Papers-Slovak Academy of Sciences, 54*(6A), 375–382.

Hinkova, A., Bubnik, Z., Kad–lec, P., & Pridal, J. (2002). Potentials of separation membranes in the sugar industry. *Seperation and Purification Technology, 26*(1), 101–110.

Jaffé, W. R. (2012). Health effects of non-centrifugal sugar (NCS): A review. *Sugar Technology, 14*(2), 87–94.

Jegatheesan, V., Shu, L., Keir, G., & Phong, D. (2012). Evaluating membrane technology for clarification of sugarcane juice. *Reviews in Environmental Science and Bio/Technology, 11*, 109–124.

Jenkins, G. H. (2013). *Introduction to cane sugar technology* p. 478. Elsevier.

Keerthi, S. S. (2002). *Study on Sugar Process Manufacturing Plant Pelwatta Sugar Industries Limited. Bachelor Thesis, Physical Sciences of the Faculty of Applied Sciences*. Sabaragamuwa University of Sri Lanka, Buttala, Sri Lanka.

Kruger, G. (1984). The application of a horizontal vacuum belt filter to smuts dewatering and cane mud filtration. In *Proceedings of the South African Sugar Technologists Association* (pp. 34–38).

Levy, R. V., & Jornitz, M. W. (2006). Types of filtration. *Advances in Biochemical Engineering/Biotechnology, 98*, 1–26.

Lionnet, G. (1996). Mud filtration. *Proceedings of the Annual Congress—South African Sugar Technologists' Association, 70*, 280–282.

Luo, J., Hang, X., Zhai, W., Qi, B., Song, W., Chen, X., & Wan, Y. (2016). Refining sugarcane juice by an integrated membrane process: Filtration behavior of polymeric membrane at high temperature. *Journal of Membrane Science, 509*, 105–115.

Madsen, R. (1990). New developments in the purification of beet sugar. *International Sugar Journal, 92*(1103), 221–223.

Nene, S., Karode, S., Courtois, T., Mietton-Peuchot, M., Gupta, B., & Ben-Aim, R. (2000). Filtration of raw cane sugar solutions with ceramic membranes. In *The 8th World Filtration Conference* 3–7 April 2000, Brighton, UK.

Panigrahi, C., Karmakar, S., Mondal, M., Mishra, H. N., & De, S. (2018). Modeling of permeate flux decline and permeation of sucrose during microfiltration of sugarcane juice using a hollow-fiber membrane module. *Innovative Food Science and Emerging Technologies, 49*, 92–105.

Payne, J. H. (2013). Unit operations in cane sugar production (1st ed., 4 p. 218). Elsevier.

Qi, A. (2022). Sugar beet production under changing climate: Opportunities and challenges. In V. Misra, S. Srivastava, & A. K. Mall (Eds.), *Sugar beet cultivation, management and processing* (pp. 407–427). Springer.

Rackemann, D., Bakir, H., Edwards, J., Cripps, L., & Doherty, W. (2015). Evaluation of mud filtrate clarification to improve factory performance. In *Proceedings of the 37th Annual Conference of the Australian Society of Sugarcane Technologists—ASSCT, Australia* (pp. 416–424).

Rafik, M., Qabli, H., Belhamidi, S., Elhannouni, F., Elkhedmaoui, A., & Elmidaoui, A. (2015). Membrane separation in the sugar industry. *Journal of Chemical and Pharmaceutical Research, 7*(9), 653–658.

Rainey, T., Thaval, O., & Rackemann, D. (2014). *Developments in mud filtration technology in the sugarcane industry. Sugarcane: Production, consumption and agricultural management systems* (pp. 263–292). Nova Science Publishers Inc.

Rein, P., & Attard, R. (2007). *Cane sugar engineering* (2nd ed., p. 943). Berlin: Bartens.

Saska, M. (2000). Method for producing white sugar. U.S. Patent No. 6,096,136. Washington, DC: U.S. Patent and Trademark Office.

Saska, M., McArdle, J., & Eringis, A. (1999). Filtration of clarified cane juice using spiral polymeric membrane configuration. In *Proceedings of International Society of Sugarcane Technologists.*

Shi, C. R., Rackemann, D. W., Moghaddam, L., Wei, B. Y., Li, K., Lu, H. Q., & Doherty, W. O. S (2019). Ceramic membrane filtration of factory sugarcane juice: Effect of pretreatment on permeate flux, juice quality and fouling. *Journal of Food Engineering, 243*, 101–113.

Steindl, R. (2001). Membrane filtration technology in the cane sugar industry. In *Proceedings of International Society of Sugarcane Technologists: 24* (pp. 3–14).

Steindl, R., & Doyle, C. (1999). Applications and benefits of membrane filtration for the Australian sugar industry. In *Proceedings of the Australian Society of Sugarcane Technologists: 21* (pp. 406–411).

Steindl, R. J., & Rackemann, D. W. (2010). Membrane filtration of clarified juice. In *Proceedings of the 27th International Society of SugarCane Technologists Congress,* 7–11 March 2010, Veracruz, Mexico (PDF) Membrane filtration of clarified juice. Available from: https://www.researchgate.net/publication/277826324_Membrane_filtration_of_clarified_juice [accessed Feb 13 2025].

Tarleton, S., & Wakeman, R. (2016). *Particle separations by filtration and sedimentation* (pp. 327–387). Springer.

Thaval, O. P., & Kent, G. A. (2012). Modelling the flow of cane constituents through the milling process of a raw sugar factory. *Proceedings of South African Sugar Technologists' Association* (85, pp. 435–453).

Tien, C. (2012). *Principles of filtration* (1st ed.). Elsevier.

Urošević, T., Povrenović, D., Vukosavljević, P., Urošević, I., & Stevanović, S. (2017). Recent developments in microfiltration and ultrafiltration of fruit juices. *Food and Bioproducts Processing, 106*, 147–161.

Varzakas, T., & Tzia, C. (2014). *Food engineering handbook: Food process engineering* p. 672. CRC Press.

Vu, T., LeBlanc, J., & Chou, C. C. (2020). Clarification of sugarcane juice by ultrafiltration membrane: Toward the direct production of refined cane sugar. *Journal of Food Engineering, 264*, 109682.

Wagner, J. (2001). *Membrane filtration handbook: Practical tips and hints*: 129. Osmonics Cambrige.

Wang, Q., Xie, C. F., Shi, C. R., Li, Y., Shang, Y. H., Hou, C. X., Hang, F.-X., Li, J.-B., & Li, K. (2022). Sugar product diversification and its opportunities in China. *Sugar Technology, 24*(4), 1093–1106.

Index

Page numbers followed by "*f*" and "*t*" indicate, figures and tables respectively.

G

H

I

9780128195239